Bioinorganic Chemistry: Inorganic Elements in the Chemistry of Life

An Introduction and Guide

Wolfgang Kaim
Universität Stuttgart, Stuttgart, Germany
and
Brigitte Schwederski
Universität Stuttgart, Stuttgart, Germany

John Wiley & Sons
Chichester · New York · Brisbane · Toronto · Singapore

© B.G. Teubner, Stuttgart, 1991.
First published under the title *Bioanorganische Chemie*.
Translation arranged with approval of the publisher B.G. Teubner, Stuttgart.

Copyright © 1994   John Wiley & Sons Ltd, The Atrium, Southern Gate, Chichester,
West Sussex PO19 8SQ, England

Telephone   (+44) 1243 779777

Email (for orders and customer service enquiries): cs-books@wiley.co.uk
Visit our Home Page on www.wileyeurope.com or www.wiley.com

Reprinted with corrections April 1995, January 1996.
Reprinted June 1997, November 2001, April 2004, August 2005

Other Wiley Editorial Offices

John Wiley & Sons Inc., 111 River Street, Hoboken, NJ 07030, USA

Jossey-Bass, 989 Market Street, San Francisco, CA 94103-1741, USA

Wiley-VCH Verlag GmbH, Boschstr. 12, D-69469 Weinheim, Germany

John Wiley & Sons Australia Ltd, 33 Park Road, Milton, Queensland 4064, Australia

John Wiley & Sons (Asia) Pte Ltd, 2 Clementi Loop #02-01, Jin Xing Distripark, Singapore 129809

John Wiley & Sons Canada Ltd, 22 Worcester Road, Etobicoke, Ontario, Canada M9W 1L1

Library of Congress Cataloging-in-Publication Data

Kaim, Wolfgang, 1951-
     [Bioanorganische Chemie. English]
     Bioinorganic chemistry : inorganic elements in the chemistry of
life : an introduction and guide / Wolfgang Kaim, Brigitte
Schwederski.
          p.   cm.— (Inorganic chemistry)
     Includes bibliographical references and index.
     ISBN 0-471-94368-1 : ISBN 0-471-94369-X (pbk.)
     1. Bioinorganic chemistry. I. Schwederski, Brigitte, 1959-
     II. Title. III. Series: Inorganic chemistry (John Wiley & Sons)
     QP531.K3513   1994

     574.19'214 — dc20   94-6718
                                                                          CIP

British Library Cataloguing in Publication Data

A catalogue record for this book is available from the British Library

ISBN 10: 0-471-94369-X (P/B)
ISBN 13: 978-0-471-94369-3 (P/B)
ISBN 10: 0-471-94368-1 (H/B)
ISBN 13: 978-0-471-94368-6 (H/B)

Bioinorganic Chemistry

Inorganic Chemistry
A Textbook Series

Wolfgang Kaim
Curriculum Vitae

Wolfgang Kaim was born in 1951 near Frankfurt am Main, Germany, and studied chemistry at the universities of Frankfurt and Konstanz. After obtaining his PhD with H. Bock in 1978, he spent a postdoctoral year with F. A. Cotton at Texas A & M University. In 1987 he moved from the University of Frankfurt to a Full Professorship at the University of Stuttgart. Main research interests focus on the charge and electron transfer reactivity of molecular compounds and various aspects of coordination chemistry.

Brigitte Schwederski
Curriculum Vitae

Brigitte Elisabeth Schwederski was born in 1959 in Recklinghausen, Germany. From 1977 to 1983 she studied chemistry and biology at the University of Bochum and in 1988 completed her PhD in the research group of Dale W. Margerum at Purdue University, Indiana. Since 1988 she has been a Research Assistant at the University of Stuttgart. Main research interests include inorganic model complexes of bioinorganic systems, their characteristics and reactivity.

Contents

Preface xi

1 Historical Background, Current Relevance and Perspectives 1
   References 5

2 Some General Principles 6
   2.1 Occurrence and Availability of Inorganic Elements in Organisms 6
   2.2 Biological Functions of Inorganic Elements 15
   2.3 Biological Ligands for Metal Ions 16
       2.3.1 Coordination by Proteins — Comments on Enzymatic Catalysis 17
       2.3.2 Tetrapyrrole Ligands and Other Macrocycles 23
       2.3.3 Nucleobases, Nucleotides and Nucleic Acids (RNA, DNA) as Ligands 33
   2.4 Relevance of Model Compounds 35
   References 36

3 Cobalamins Including Vitamin and Coenzyme B_{12} 39
   3.1 History and Structural Characterization 39
   3.2 Reactions of the Alkylcobalamins 41
       3.2.1 One-Electron Reduction and Oxidation 41
       3.2.2 Co—C Bond Cleavage 42
       3.2.3 Mutase Activity of Coenzyme B_{12} 46
       3.2.4 Alkylation Reactions of Methylcobalamin 52
   3.3 Model Systems and the Role of the Apoenzyme 53
   References 54

4 Metals at the Center of Photosynthesis: Magnesium and Manganese 56
   4.1 Volume and Total Efficiency of Photosynthesis 56
   4.2 Primary Processes in Photosynthesis 57
       4.2.1 Light Absorption (Energy Acquisition) 58
       4.2.2 Exciton Transport (Directed Energy Transfer) 59
       4.2.3 Charge Separation and Electron Transport 63
   4.3 Manganese-Catalyzed Oxidation of Water to O_2 68
   References 79

5 The Dioxygen Molecule, O_2: Uptake, Transport, and Storage of an Inorganic Natural
   Product 82
   5.1 Molecular and Chemical Properties of Dioxygen, O_2 82
   5.2 Oxygen Transport and Storage through Hemoglobin and Myoglobin 88

5.3   Alternative Oxygen Transport in Some Lower Animals: Hemerythrin and
       Hemocyanin   100
       5.3.1   Magnetism   100
       5.3.2   Light Absorption   101
       5.3.3   Vibrational Spectroscopy   101
       5.3.4   Mössbauer Spectroscopy   102
       5.3.5   Structure   103
       References   105

6   **Catalysis Through Hemoproteins: Electron Transfer, Oxygen Activation and Metabolism
    of Inorganic Intermediates** **107**
    6.1   Cytochromes   109
    6.2   Cytochrome P-450: Oxygen Transfer from O_2 to Nonactivated Substrates.   112
    6.3   Peroxidases: Detoxification and Utilization of Doubly Reduced Dioxygen   119
    6.4   Controlling the Reaction Mechanism of the Oxyheme Group — Generation and
           Function of Organic Free Radicals   120
    6.5   Hemoproteins in the Catalytic Transformation of Partially Reduced Nitrogen and
           Sulfur Compounds   122
    References   125

7   **Iron–Sulfur and Other Nonheme Iron Proteins** **128**
    7.1   Biological Relevance of the Element Combination Iron/Sulfur   128
    7.2   Rubredoxins   133
    7.3   [2Fe–2S] Centers   133
    7.4   Polynuclear Fe/S Clusters: Relevance of the Protein Environment and Catalytic
           Activity   134
    7.5   Model Systems for Iron–Sulfur Proteins   139
    7.6   Iron-Containing Enzymes without Porphyrin or Sulfide Ligands   141
           7.6.1   Iron-Containing Ribonucleotide Reductase (RR)   141
           7.6.2   Soluble Methane Monooxygenase (MMO)   143
           7.6.3   Purple Acid Phosphatases (Fe/Fe and Fe/Zn)   144
           7.6.4   Mononuclear Nonheme Iron Enzymes   144
    References   146

8   **Uptake, Transport and Storage of an Essential Element as Exemplified by Iron** **150**
    8.1   The Problem of Iron Mobilization — Oxidation States, Solubility and Medical
           Relevance   150
    8.2   Siderophores: Iron Uptake by Microorganisms   152
    8.3   Phytosiderophores: Iron Uptake by Plants   160
    8.4   Transport and Storage of Iron   161
           8.4.1   Transferrin   162
           8.4.2   Ferritin   165
           8.4.3   Hemosiderin   169
    References   170

9   **Nickel-Containing Enzymes: The Remarkable Career of a Long Overlooked Biometal** .   **172**
    9.1   Overview   172
    9.2   Urease   173

9.3   Hydrogenases.   174
9.4   CO Dehydrogenase = CO Oxidoreductase = Acetyl–CoA Synthase.   176
9.5   Methyl–Coenzyme M Reductase   180
9.6   Model Compounds   183
References   184

10   Copper-Containing Proteins: An Alternative to Biological Iron.   187
10.1   Type 1: 'Blue' Copper Centers.   192
10.2   Type 2 and Type 3 Copper Centers in O_2-Activating Proteins:
        Oxygen Transport and Oxygenation   197
10.3   Copper Proteins as Oxidases/Reductases.   202
10.4   Cytochrome c Oxidase.   206
10.5   Cu,Zn and Other Superoxide Dismutases: Substrate-Specific Antioxidants.   209
References   212

11   Biological Functions of the 'Early' Transition Metals: Molybdenum, Tungsten,
     Vanadium and Chromium.   215
11.1   Oxygen Transfer through Tungsten- and Molybdenum-Containing Enzymes. . .   215
        11.1.1   Overview   215
        11.1.2   Enzymes Containing the Molybdopterin Cofactor   217
11.2   Metalloenzymes in the Biological Nitrogen Cycle: Molybdenum-Dependent
        Nitrogen Fixation   224
11.3   Alternative Nitrogenases   232
11.4   Biological Vanadium Outside Nitrogenases.   235
11.5   Chromium(III) in the Metabolism?   238
References   238

12   Zinc: Structural and Gene-Regulatory Functions and the Enzymatic Catalysis of
     Hydrolysis or Condensation Reactions   242
12.1   Overview   243
12.2   Carboanhydrase (CA).   246
12.3   Carboxypeptidase A (CPA) and Other Hydrolases   251
12.4   Catalysis of Condensation Reactions by Zinc-Containing Enzymes.   256
12.5   Alcohol Dehydrogenase (ADH) and Related Enzymes.   257
12.6   The 'Zinc Finger' and Other Gene Regulatory Zinc Proteins   260
12.7   Insulin, hGH, Metallothionein and DNA Repair Systems as Zinc-Containing
        Proteins   262
References   264

13   Unequally Distributed Electrolytes: Function and Transport of Alkali and Alkaline Earth
     Metal Cations.   267
13.1   Characterization of K^+, Na^+, Ca^{2+} and Mg^{2+}.   267
13.2   Complexes of Alkali and Alkaline Earth Metal Ions with Macrocycles.   273
13.3   Ion Channels   277
13.4   Ion Pumps   281
References   284

14 **Catalysis and Regulation of Bioenergetic Processes by the Alkaline Earth Metal Ions Mg^{2+} and Ca^{2+}** **286**
    14.1 Magnesium: Catalysis of Phosphate Transfer by Divalent Ions 286
    14.2 The Ubiquitous Regulatory Role of Ca^{2+} 293
    References 300

15 **Biomineralization: The Controlled Assembly of 'Advanced Materials' in Biology** **303**
    15.1 Overview 303
    15.2 Nucleation and Crystal Growth 308
    15.3 Examples of Biominerals 309
          15.3.1 Calcium Phosphate in the Bones of Vertebrates 309
          15.3.2 Calcium Carbonate 312
          15.3.3 Amorphous Silica 313
          15.3.4 Iron Biominerals 314
          15.3.5 Strontium and Barium Sulfates 315
    References 316

16 **Biological Functions of the Nonmetallic Inorganic Elements** **318**
    16.1 Overview 318
    16.2 Boron 318
    16.3 Silicon 318
    16.4 Arsenic and PH$_3$ 319
    16.5 Bromine 320
    16.6 Fluorine 320
    16.7 Iodine 321
    16.8 Selenium 323
    References 328

17 **The Bioinorganic Chemistry of the Quintessentially Toxic Metals** **330**
    17.1 Overview 330
    17.2 Lead 332
    17.3 Cadmium 335
    17.4 Thallium 338
    17.5 Mercury 338
    17.6 Aluminum 343
    17.7 Beryllium 346
    17.8 Chromium as Chromate(VI) 346
    References 348

18 **Biochemical Behavior of Inorganic Radionuclides: Radiation Risks and Medical Benefits** **351**
    18.1 Overview 351
    18.2 Natural and Man-Made Radioisotopes Outside Medical Applications 352
    18.3 Bioinorganic Chemistry of Radiopharmaceuticals 356
          18.3.1 Overview 356
          18.3.2 Technetium — a 'Man-Made Bioinorganic Element' 358
    References 362

19 Chemotherapy with Compounds of Some Nonessential Elements **363**
    19.1 Overview 363
    19.2 Platinum Complexes in Cancer Therapy 363
          19.2.1 Discovery, Application and Structure–Effect Relationships 363
          19.2.2 Cisplatin: Mode of Action 367
    19.3 Cytotoxic Compounds of Other Metals 372
    19.4 Gold-Containing Drugs Used in the Therapy of Rheumatoid Arthritis 373
          19.4.1 Historical Development 373
          19.4.2 Gold Compounds as Anti-Rheumatic Agents 374
          19.4.3 Hypotheses on the Mode of Action of Gold-Containing Anti-Rheumatic
                Drugs 375
    19.5 Lithium in Psychopharmacological Drugs 376
    References 376

Bibliography **379**

Glossary **381**

List of Inserts **385**

Index **387**

Preface

This book originated from a two-semester course offered at the Universities of Frankfurt and Stuttgart (W.K.). Its successful use requires a basic knowledge of the modern sciences, especially of chemistry and biochemistry, at a level that might be expected after one year of study at a university or its equivalent. Despite these requirements we have decided to explain some special terms in a glossary and, furthermore, several less conventional physical methods are briefly described and evaluated with regard to their practical relevance at appropriate positions in the text.

A particular problem in the introduction to this highly interdisciplinary and not yet fully mature or definitively circumscribed field lies in the choice of material and the depth of treatment. Although priority has been given to the presentation of metalloproteins and the electrolyte elements, we have extended the scope to therapeutically, toxicologically and environmentally relevant issues because of the emphasis on functionality and because several of these topics have become a matter of public discussion.

With regard to details, we can frequently only offer hypotheses. In view of the explosive growth of this field there is implicit in many of the statements regarding structure and mechanisms the qualification that they are 'likely' or 'probable'. We have tried to incorporate relevant literature citations up to the year 1993.

Another difficult aspect when writing an introductory and, at the same time, fairly inclusive text is that of the organization of the material. For didactic reasons we follow partly an organizational principle focused on the elements of the periodic table. However, living organisms are opportunistic and could not care less about such systematics; to successfully cope with a problem is all that matters. Accordingly, we have had to be 'nonsystematic' in various sections, for example, treating the hemerythrin protein in connection with the similarly O_2-transporting hemoglobin (Chapter 5) and not under 'diiron centers' (Section 7.6). Several sections are similarly devoted to biological-functional problems such as biomineralization or antioxidant activity and may thus include several different elements or even organic compounds. The simplified version of the P-450 monooxygenase catalytic cycle which we chose for the cover picture illustrates the priority given to function and reactivity as opposed to static-structural aspects.

We regret that the increasingly available color-coded structural representations of complex proteins and protein aggregates cannot be reproduced here. General references to the relevant literature are given in the Bibliography at the end of the book while specific references are listed at the end of each chapter in the sequence of appearance.

For helpful comments and encouragement during the writing and correction of manuscripts we thank many of our colleagues. Recent results have become available to us through participation in the special program 'Bioanorganische Chemie' of the Deutsche Forschungsgemeinschaft (DFG). We also thank Teubner-Verlag and John

Wiley & Sons for their patience and support. Very special thanks are due to Mrs Angela Winkelmann for her continued involvement in the processing of the manuscript.

Stuttgart, December 1993   Wolfgang Kaim
                                                                     and
                                                        Brigitte Schwederski

1 Historical Background, Current Relevance and Perspectives

'The progress of an inorganic chemistry of biological systems has had a curious history.'
R. J. P. Williams, *Coord. Chem. Rev.*, **100**, 573 (1990).

The description of a rapidly developing [1] field of chemistry as 'bio-inorganic' seems to involve a contradiction in terms which, however, simply reflects a misconception going back to the beginning of modern science. In the early nineteenth century chemistry was still divided into an 'organic' chemistry, which included only substances isolated from living organisms, and an 'inorganic' chemistry of 'dead matter'. This distinction became meaningless after Wöhler's synthesis of 'organic' urea from 'inorganic' ammonium cyanide in 1828. Nowadays, organic chemistry is defined as the chemistry of carbon compounds, especially of hydrocarbons and their derivatives, with the possible inclusion of certain heteroelements such as N, O or S, regardless of the origin of the material.

The increasing need for a collective, not necessarily substance-oriented designation of the chemistry of living organisms then led to the new term of 'biochemistry'. For a long time, classical biochemistry was concerned mainly with organic compounds; however, the two areas are by no means identical.* Improved trace analytical methods have demonstrated the importance of quite a number of 'inorganic' elements in biochemical processes and have thus revealed a multitude of 'inorganic natural products'. As a matter of fact, some (by today's definition) 'inorganic' elements had been established quite early as essential components of living systems. Some well-known examples include the extraction of potassium carbonate (K_2CO_3, potash) from plants and of iron-containing complex salts $K_{3,4}[Fe(CN)_6]$ from animal blood in the eighteenth century, or the discoveries of elemental phosphorus (as P_4) by dry distillation of urine residues in 1669 and of elemental iodine from the ashes of marine algae in 1812.

In the middle of the nineteenth century, Liebig's studies on the metabolism of inorganic nutrients, especially of nitrogen, phosphorus and potassium salts, significantly improved agriculture so that this particular field of science gained enormous practical importance. However, the theoretical background and the analytical methods of that time were not sufficient to obtain detailed information on the mechanism of action of such essential elements, several of which only occur in trace amounts. Some very conspicuous compounds that include inorganic elements like iron-containing hemoglobin and magnesium-containing chlorophyll, the 'pigments of life', were analyzed and characterized later within a special subfield of organic chemistry,

---

* The term 'bioorganic chemistry' is increasingly being used for studies of organic compounds that are directly relevant for biochemistry.

the chemistry of natural products. It was only after 1960 that bioinorganic chemistry
became an independent and highly interdisciplinary research area.

The following factors have been crucial for this development:

(a) Biochemical isolation and purification procedures, e.g. chromatography, and the
new physical methods of trace element analysis such as atomic absorption or
emission spectroscopy require ever smaller amounts of material and have greatly
improved over the last decades. These advances in methodology have made it
possible to not only detect but also to chemically and functionally characterize
trace elements or otherwise inconspicuous metal ions in biological materials. An
adult human being, for example, contains about two grams of zinc in ionic form
(Zn^{2+}). Although zinc cannot be regarded as a true trace element, the unambiguous proof of its existence in enzymes was established only in the 1930s. Genuine
bioessential trace elements such as nickel (Figure 1.1 and Chapter 9) or selenium
(Section 16.8) have been known to be present as constitutive components in

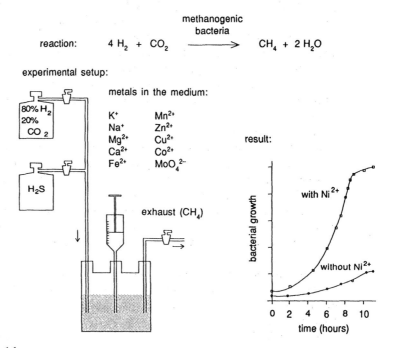

Figure 1.1
Discovery of nickel as an essential trace element in the production of methane by archaea
(archaebacteria). The biological reduction of carbon dioxide by hydrogen to give methane has
been investigated since the 1970s by studying the relevant archaebacteria, these being found,
for example, in sewage plants. Even though the experiments were carried out under strictly
anaerobic conditions and all 'conventional' trace elements were supplied, the results were only
partly reproducible. Eventually it was discovered that during sampling with a syringe
containing a supposedly inert stainless steel (Fe/Ni) tip, minute quantities of nickel had
dissolved. This inadvertent addition of nickel salts led to a distinctive increase in methane
production [2] and, in fact, several nickel-containing proteins and coenzymes have since been
isolated (see Chapter 9). Incidentally, a similar unexpected dissolution effect of an apparently
inert metal led to the discovery of the inorganic anti-tumor agent cis-$PtCl_2(NH_3)_2$ ('cisplatin',
Section 19.2.1)

several important enzymes only since about 1970. There is now a discussion on whether elements such as arsenic may also be bioessential, if only in ultra-trace amounts (Section 16.4).

(b) Efforts to elucidate the mechanisms of organic, inorganic and biochemical reactions have led to an early understanding of the specific biological functions of some inorganic elements. Nowadays, many attempts are being made to mimic biochemical reactivity through studies of the reactivity of model compounds which are prepared outside of living cells (Section 2.4). Theories of the specific function of an inorganic component should ideally be accompanied by experimental confirmation such as the practical application of biochemical reaction sequences in these low molecular weight models.

(c) It is generally acknowledged that interdisciplinary research has become increasingly important for scientific progress. This commonplace statement is particularly true for a field like bioinorganic chemistry (Figure 1.2) which already includes in its name two classical chemical disciplines. In addition, the recent rapid progress in bioinorganic chemistry [1] has been made possible through contributions from:

(i) physics (→techniques for detection and characterization),

(ii) various areas of biology (→supply of material and, more recently, specific modifications based on site-directed mutagenesis),

(iii) agricultural and nutritional sciences (→effects of inorganic elements and their mutual interdependence),

(iv) pharmacology (→interaction between drugs and endogeneous or exogeneous inorganic substances),

(v) medicine (→diagnostic aids, tumor therapy),

(vi) toxicology and the environmental sciences (→potential toxicity of inorganic compounds, the 'concentration problem' [3]).

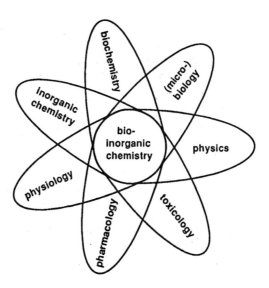

Figure 1.2
Bioinorganic chemistry as a highly interdisciplinary research field

A particularly spectacular example of applied bioinorganic chemistry is the successful use of the simple inorganic complex cis-diamminedichloroplatinum, cis-$Pt(NH_3)_2Cl_2$ ('cisplatin'), in the therapy of certain tumors (Section 19.2). With estimated profits exceeding US$55 million since 1970, this compound has been the subject of one of the most successful patent applications ever granted to an American university.

Generally speaking, even those areas of chemistry that are not primarily biologically oriented can profit from the research in bioinorganic chemistry. Due to the relentless pressures of evolutionary selection, biological processes show a high efficiency under preset conditions. These continuously self-optimizing systems can therefore serve as useful models for problems in modern chemistry. Among the most current topics of this type are:

(a) the efficient collection, conversion and storage of energy,
(b) the catalytic activation of inert substances, especially of small molecules such as H_2O, O_2 or N_2, under mild conditions,
(c) the (stereo-)selective synthesis of high-value substances while minimizing the yield of unwanted by-products and
(d) the degradation and recycling of substances with as few problems as possible, especially the detoxification or recycling of chemical elements from the periodic table (Figure 1.3).

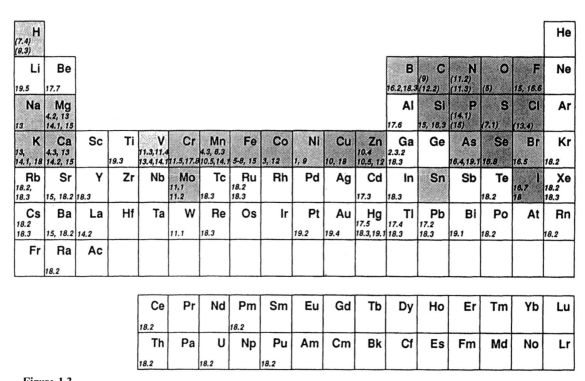

Figure 1.3
Periodic table of the elements. Indicated are the chapters (italicized) in which the corresponding element is discussed in this book: ▨ essential element ▧ presumably essential element for human beings

With such objectives in mind, it is not sufficient to just present and describe bioinorganic systems. A major purpose of this book is the correlation of function, structure and actual reactivity of inorganic elements in organisms. The rather biological than chemical question of 'why?' should eventually stimulate a more purposeful use of chemical compounds also in nonbiological areas.

References

1. S. J. Lippard, Bioinorganic chemistry: a maturing frontier, *Science*, **261**, 699 (1993).
2. P. Schönheit, J. Moll and R. K. Thauer, Nickel, cobalt, and molybdenum requirement for growth of *Methanobacterium thermoautotrophicum*, *Arch. Microbiol.*, **123**, 105 (1979).
3. G. Tölg and R. P. H. Garten, Large anxiety for small quantities—significance of analytical chemistry in the modern industrialized society as exemplified by trace determination of elements, *Angew. Chem. Int. Ed. Engl.*, **24**, 485 (1985).

2 Some General Principles

2.1 Occurrence and Availability of Inorganic Elements in Organisms

'Life' is a process which, for an adult organism, can be characterized as a controlled stationary flow equilibrium maintained by energy consuming chemical reactions ('dissipative system'). *Input* and *output* are essential requirements for such open systems which differ very much from the more familiar and mathematically more easily described 'dead' thermodynamic equilibria (see Figure 2.1).

In addition to the energy flux, life requires a continuous material exchange which, in principle, includes *all* chemical elements (Figure 1.3). The occurrence of these elements in organisms depends on external and endogeneous conditions; elements can be 'bioavailable' to variable extents but can also be enriched ('bioaccumulated') by organisms using active, i.e. energy-consuming processes involving a local reduction of entropy. Some trends are obvious from looking at the most familiar example, the elemental composition of an adult human being (see Table 2.1).

The values for O and H in Table 2.1 reflect the high content of (inorganic) water; the 'organic' element carbon only comes in third. Calcium as the first metallic element ranks fifth, its main quantitative use being the stabilization of the endoskeleton. Table 2.1 further shows relatively large quantities of potassium, chlorine, sodium and magnesium ('mass elements', 'macro nutrients'), followed by iron and zinc, two distinctly less abundant inorganic elements. According to one definition, 'trace elements' with regard to the human body involve a daily requirement of less than 25 mg (see Table 2.3). Fluorine and silicon, two structurally important nonmetallic

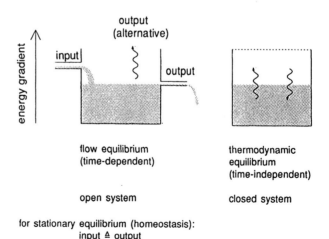

Figure 2.1
Two kinds of 'equilibrium'

Table 2.1 Average elemental composition of a human body (adult, 70 kg [1])

element and symbol		mass (g)	year of discovery as an essential element
oxygen	O	45500	
carbon	C	12600	
hydrogen	H	7000	
nitrogen	N	2100	
calcium	Ca	1050	
phosphorus	P	700	
sulfur	S	175	
potassium	K	140	
chlorine	Cl	105	
sodium	Na	105	
magnesium	Mg	35	
iron	Fe	4.2	17th century
zinc	Zn	2.3	1896
silicon	Si	1.4	1972
rubidium[a]	Rb	1.1	
fluorine	F	0.8	1931
zirconium[a]	Zr	0.3	
bromine[b]	Br	0.2	
strontium[a]	Sr	0.14	
copper	Cu	0.11	1925
aluminum[a]	Al	0.10	
lead[b]	Pb	0.08	
antimony[a]	Sb	0.07	
cadmium[b]	Cd	0.03	(1977)
tin[b]	Sn	0.03	(1970)
iodine	I	0.03	1820
manganese	Mn	0.02	1931
vanadium[b]	V	0.02	(1971)
selenium	Se	0.02	1957
barium[a]	Ba	0.02	
arsenic[b]	As	0.01	1975
boron[b]	B	0.01	
nickel[b]	Ni	0.01	(1971)
chromium	Cr	0.005	1959
cobalt	Co	0.003	1935
molybdenum	Mo	<0.005	1953
lithium[b]	Li	0.002	

[a] Not essential. [b] Essentiality uncertain.

elements, fall within the range of iron and zinc with amounts totalling about 1 g each. The relative amounts of the genuine trace elements is smaller by at least another order of magnitude, and some of them have not yet been unambiguously defined with regard to amount, essential character and function [2]. Strictly speaking, elements should be called *essential* only if their total absence in the organism causes severe, irreversible damage. Frequently, essentiality is now already being invoked if the optimal functioning of organisms is impaired; in such instances the corresponding elements may better be referred to as 'beneficial'. Table 2.1 also illustrates the occurrence of nonnegligible quantities of obviously nonessential elements such as Rb, Zr, Sr, Br, Al or Li in the human body. These elements are probably incorporated owing to a chemical similarity (indicated by $\leftrightarrow$) with important essential elements (Li^+, Rb^+, $Cs^+ \leftrightarrow Na^+$, K^+; Sr^+, $Ba^{2+} \leftrightarrow Ca^{2+}$; $Br^- \leftrightarrow Cl^-$; Al^{3+}, $Zr^{4+} \leftrightarrow Fe^{3+}$). Elements such as As, Pb or Cd which are known to be mainly toxic deserve special attention; a positive effect of traces (detection limit!) has been discussed recently for some of these elements, pointing to the ambivalence of many trace elements and to the problem of threshold values (see Figures 2.3 and 2.4). Possibly a physiological—if not always essential—function has developed for all naturally occurring elements during the evolution of life [3].

Figure 2.2 illustrates the compilation from Table 2.1 through a logarithmic representation of molar concentrations, thus giving a more precise idea of the relative abundance of the various elements. A comparison with elemental compositions outside the biosphere shows a rather good correlation with the composition of sea water [4,5], which is often taken to be an indication for the most probable site of the evolution of life.

It is remarkable that elements such as silicon, aluminum or titanium, which are so prominent as components of minerals in the earth's crust, play only a marginal role in the biosphere. One reason is that, in general, the physiological conditions for life processes include pH values of about 7 in aqueous solutions. Under these conditions, the aforementioned elements in their usual high oxidation states exist as nearly insoluble oxides or hydroxides and are therefore not (bio)available. Molybdenum, on the other hand, is a rare element in the earth's crust but is quite soluble at pH 7 as MoO_4^{2-} and thus rather abundant in sea water; therefore, it has been found as an essential element in many organisms. As a rule, metallic elements are soluble in neutral aqueous media and thus bioavailable either in low oxidation states ($+I$, $+II$), i.e. as hydrated cations, or in very high oxidation states ($+V$, $+VI$, $+VII$), e.g. as hydrated oxo-anions such as MO_4^{n-}. However, despite the good correlation between the elemental composition of organisms and that of sea water one should not underestimate the ability of such organisms to actively transport and accumulate inorganic substances. As pointed out in Chapter 13, living systems have developed complicated mechanisms and use much energy to create and maintain concentration gradients for inorganic ions between membrane-separated compartments inside organisms. Similarly, efficient biological mechanisms exist to accumulate silicate or Fe^{3+} ions, both of which are practically insoluble at pH 7, and thus make them bioavailable for structural or other purposes (see Chapters 8 and 15). Not surprisingly, the elemental compositions are highly variable for different species and even different parts of higher organisms, depending *inter alia* on the kind of metabolism and on the biotope.

The flow equilibrium character of life processes was illustrated in Figure 2.1. Thus, the individual inorganic elements are continuously excreted and replenished even though their overall stationary concentration remains approximately constant ('homeostasis'). The rate of exchange is strongly dependent on the type of compound (chemical speciation) and on the site of action or storage in the individual organism. According to established principles of reaction kinetics, ions of low charge are exchanged relatively quickly (K^+, Mn^{2+}, MoO_4^{2-}) whereas more highly charged species such as Fe^{3+} have longer biological or physiological half-lives. It is not surprising that elements such as Ca^{2+} which find their main quantitative use in the

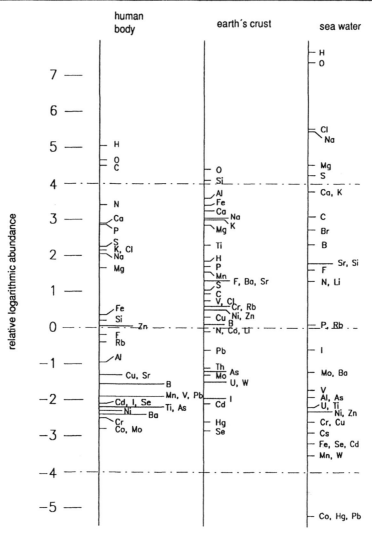

Figure 2.2
Logarithmic diagrams of relative molar concentrations of the elements in different environments (arbitrary units) (data from [1] and [4])

solid-state skeleton are exchanged very slowly. The biological half-life can then amount to several years; nevertheless, a continuous metabolism takes place even for 'biominerals' (see Chapter 15).

Which elements are essential, which are beneficial and which are toxic for a certain organism? Such questions are continuously being discussed under various aspects, particularly for humans, in popular science. Quantitatively, this is a matter of the physiological state, i.e. of the ability to 'function' properly (behavior) or even of the individual disposition of an organism, depending on the presence, the dose or concentration of an element, often related to its share in the food supply. A dose–response diagram of the type in Figure 2.3 can thus be discussed which shows

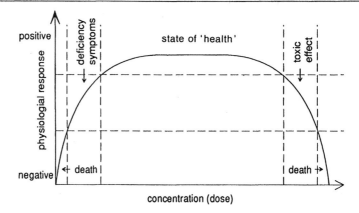

Figure 2.3
Schematic dose–response diagram for an essential element (compare Figure 17.1 for exclusively toxic elements)

the ambivalent effects of many substances and illustrates the principle of Paracelsus: 'The dose makes the poison.' An important term here is that of the *therapeutic width* which characterizes the concentration range causing advantageous physiological effects.

In a more detailed approach, the following aspects have to be considered:

(a) The chemical compound of which the element is a part is often crucial for the response of the organism (chemical speciation). The pathway, the extent and rate of uptake, metabolism, storage and excretion can differ greatly; poor utilization of an otherwise bioavailable essential trace element may thus be responsible for deficiency symptoms. The absorption of inorganic compounds by the organism depends primarily on the solubility and, therefore, on the charge of the system; humans resorb molybdate MoO_4^{2-} to about 70–80% from food whereas slowly reacting Cr^{3+} is resorbed only to a small extent ($<1\%$).

(b) It cannot be expected that higher organisms react uniformly within a population or in the course of their individual development. Therefore, only average statements can be made with regard to a certain situation, e.g. for the adult state of a preferentially homogeneous population (Figure 2.4).

(c) The concentration variation of one particular element generally affects the concentrations and physiological effects of other substances, including other inorganic elements. This multidimensional interdependence has been known qualitatively for a number of elemental nutrients since the experiments of Liebig. Two components can interact by mutually promoting corresponding effects (synergism) or by competing and suppressing each other's effects (antagonism, Figure 2.5).

An antagonistic relationship in a two-component system can be the result of displacement ($Zn^{2+} \leftrightarrow Cd^{2+}$, Pb^{2+}, Cu^{2+} or Ca^{2+}) or mutual deactivation: $Cu^{2+} + S^{2-} \rightarrow CuS$ (insoluble). With three components, e.g. in the system Cu/Mo/S (Chapter 10 and Section 11.1.2), matters become more complicated and, in reality, there is a multidimensional network of synergistic and antagonistic relationships which is further complicated by the spatially unsymmetrical distribution of inorganic

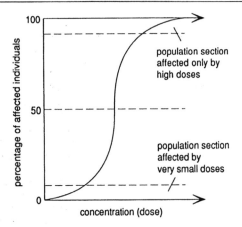

Figure 2.4
Typical variance of the (toxic) effect of a substance within a population

elements in organisms [7]. In Chapter 13 we shall discuss a well-known example, the strongly contrasting distributions of soluble monocations (Na^+ vs. K^+), dications (Ca^{2+} vs. Mg^{2+}) and monoanions (Cl^- vs. $H_2PO_4^-$) in extra- and intracellular regions.

Despite this complexity, some deficiency symptoms of individual inorganic elements are quite familiar, particularly when concerning human beings [2,8] (see the

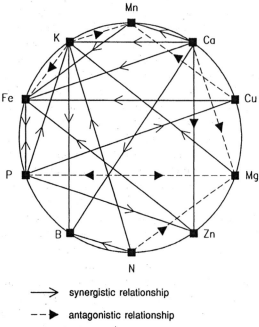

Figure 2.5
Network of interaction between some elemental plant nutrients (from [6])

Table 2.2 Some characteristic symptoms of chemical element deficiency in humans

deficient element	typical deficiency symptoms
Ca	retarded skeletal growth
Mg	muscle cramps
Fe	anemia, disorders of the immune system
Zn	skin damage, stunted growth, retarded sexual maturation
Cu	artery weakness, liver disorders, secondary anemia
Mn	infertility, impaired skeletal growth
Mo	retardation of cellular growth, propensity for caries
Co	pernicious anemia
Ni	growth depression, dermatitis
Cr	diabetes symptoms
Si	disorders of skeletal growth
F	dental caries
I	thyroid disorders, retarded metabolism
Se	muscular weakness, esp. cardiomyopathy
As	impaired growth (in animals)

incomplete list in Table 2.2). As far as causal connections are known for the single elements, these will be discussed in the corresponding chapters within this book. A general syndrome of (trace) element deficiency is growth retardation; the number of truly essential elements seems to be smaller in fully developed organisms than during growth periods. This assumption was confirmed by the pioneering experiments in the 1960s which were designed to guarantee a nutritionally complete diet for astronauts during longer space flights. The inorganic contents of such synthetic food are summarized in Table 2.3 in the form of the RDA (recommended dietary allowances) values of the American Food and Drug Administration [9]. Whether such a composition is really sufficient, how much of it occurs in today's food supply and how far it can be exceeded via increased uptake or separate supplementation without detrimental consequences are still open questions in dietetics, particularly from the popular scientific point of view.

According to Figure 2.3 there are not only deficiency symptoms from the lack of essential elements but also toxic effects resulting from an excess of these, whether caused by insufficient excretion or by excessive uptake [10]. Such poisoning can be treated using 'bioinorganic' measures, viz. through the application of antagonists or a 'chelate therapy' [11,12] which involves the complexation and excretion of acutely toxic metal ions using multidentate chelate ligands (2.1 and Table 2.4). Considering the large number of essential metal ions present in organisms, the problem of selectivity is obvious; rather selectively coordinating ligands have thus been developed for some specific heavy metal ions. The most successful such ligands offer selectivity (i) according to the preferred size of the coordinated ion or (ii) with respect to favored coordinating atoms (S for 'soft' heavy metals, N especially for Cu^{2+}, O for 'hard' metal centers; see Figure 2.6). Furthermore, suitable chelate ligands must (iii) form kinetically and thermodynamically stable complexes and (iv) facilitate rapid renal excretion, e.g. by containing hydrophilic hydroxyl groups.

Table 2.3 Essential elements in food for adults and infants

inorganic constituents	recommended daily allowances (in mg)	
	adult[a]	infant[b]
K	2000 – 5500	530
Na	1100 – 3300	260
Ca	800 – 1200	420
Mg	300 – 400	60
Zn	15	5
Fe	10 – 20	7.0
Mn	2.0 – 5	1.3
Cu	1.5 – 3	1.0
Mo	0.075 – 0.250	0.06
Cr	0.05 – 0.2	0.04
Co	ca. 0.2 (vitamin B_{12})	0.001
Cl	3200	470
PO_4^{3-}	800 – 1200	210
SO_4^{2-}	10	
I	0.15	0.07
Se	0.05 – 0.07	
F	1.5 – 4.0	0.6

[a] Mainly from Recommended Dietary Allowances, RDA; National Academy of Sciences, USA.
[b] Estimated from producers' tables of contents of typical SL(Sine Lacte) food for infants.

Table 2.4 Chelate ligands for detoxification after metal poisoning

ligand (formulae 2.1)	trade or trivial name	preferably coordinated metal ions	detailed description in Chapter
(a) 2,3-dimercapto-1-propanol	dimercaprol, BAL	Hg^{2+}, As^{3+}, Sb^{3+}, Ni^{2+}	17
(b) D-2-amino-3-mercapto-3-methylbutyric acid (D-β,β-dimethylcysteine)[a]	D-penicillamine	Cu^{2+}, Hg^{2+}	10, 17
(c) ethylenediaminetetraacetate	EDTA	Ca^{2+}, Pb^{2+}	
(d) deferrioxamine B	DFO, desferal	Fe^{3+}, Al^{3+}	8.2, 17.6
(e) 3,4,3-LICAMC		Pu^{4+}	18.2

[a] The L-enantiomer is toxic.

(a)

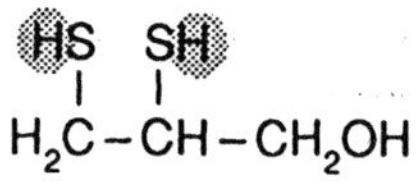

$$\underset{\underset{H_2C - CH - CH_2OH}{}}{\overset{HS \quad SH}{}}$$

(b)

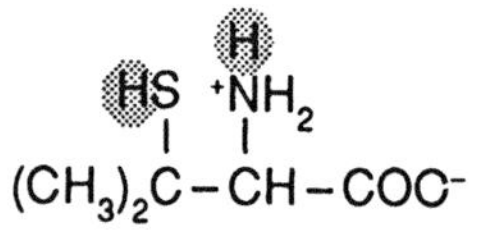

$$\underset{(CH_3)_2C - CH - COO^-}{\overset{HS \quad ^+NH_2}{}}$$

(c)

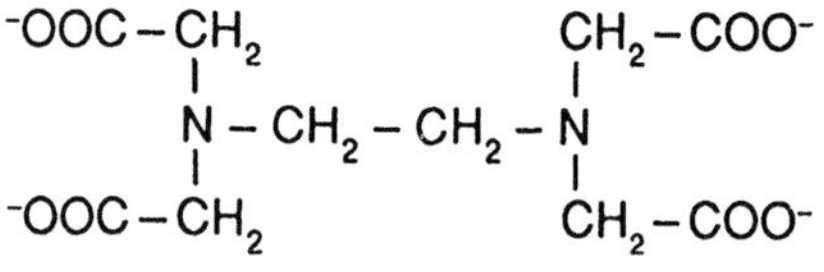

$$\underset{^-OOC - CH_2}{\overset{^-OOC - CH_2}{}} N - CH_2 - CH_2 - N \underset{CH_2 - COO^-}{\overset{CH_2 - COO^-}{}}$$

(d)

$$H_2N - (CH_2)_5 - N - [C - (CH_2)_2 - C - NH - (CH_2)_5 - N]_2 - C - CH_3$$

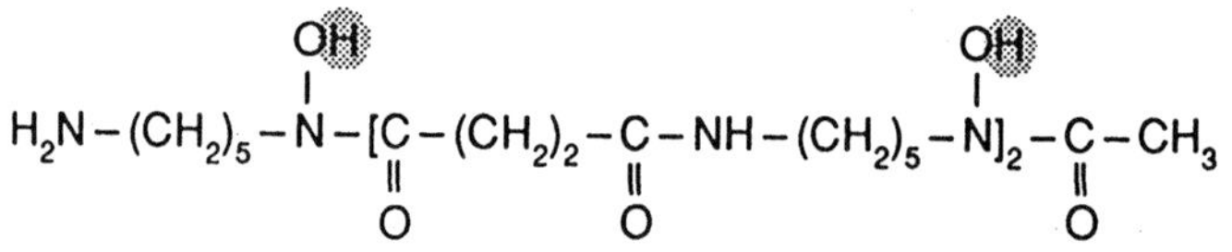

(e)

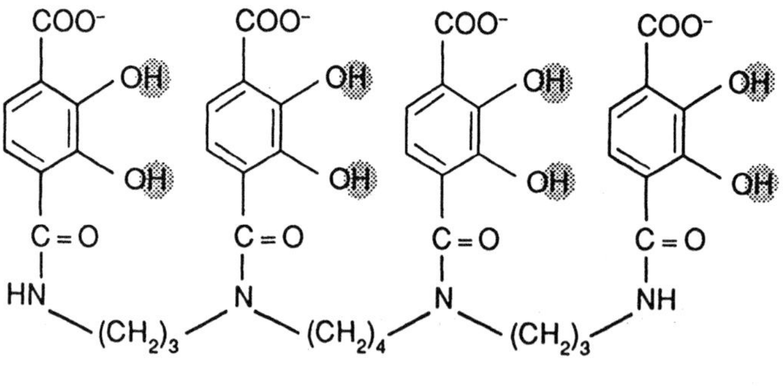

H : acidic protons
which may be sub-
stituted by metal ions

(2.1)

'Hard' and 'Soft' Coordination Centers

The susceptibility of atoms and ions to experience a charge shift in their electron shell through interaction with a coordination partner differs considerably. This has led to an often loosely used distinction between little affected 'hard' and easily polarizable 'soft' coordination centers. Among the soft electron-pair donors are thiolates (RS^-), sulfides (S^{2-}) and selenides; on the other hand, the

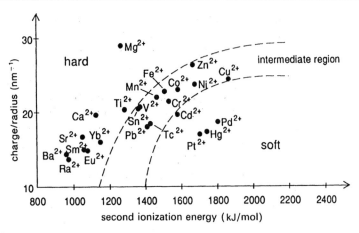

Figure 2.6
Regions of hard and soft metal ions M^{2+} (according to [13])

fluoride anion (F^-) or ligands with negatively charged oxygen donor centers are classified as hard. In many cases, the observed affinities between metal ions and ligand atoms can be interpreted in such a way that interactions between centers of the same type, i.e. hard/hard (highly ionic bond) and soft/soft (partly covalent bond), are preferred. One possible quantitative approach to this rather intuitively used concept is based on a correlation between the ratio charge/ionic radius of a metal dication and the measurable second ionization energy (Figure 2.6).

2.2 Biological Functions of Inorganic Elements

The great efforts made by organisms to take up, accumulate, transport and store inorganic elements is justified only by their important and otherwise not guaranteed function. For living organisms, the arbitrary distinction between 'organic' and 'inorganic' compounds is irrelevant since it is solely based on a historically grown definition. However, there are functions like those listed below for which compounds or ions of the metals are particularly well suited:

(a) The assembly of hard structures in the form of endo- or exoskeletons via biomineralization certainly falls into this category (Chapter 15). Another aspect of this *structural function* is that cell membranes require the presence of metal ions to cross-link the organic 'filling material' and thus maintain the membrane integrity. Even the double helical structure of DNA (a polyanion) is maintained only in the presence of mono- and divalent cations which significantly reduce the otherwise dominating electrostatic repulsion forces between the negatively charged nucleotide phosphate groups [14]. Solid-state/structural functions are represented mainly by the elements Ca, Mg (as dications) and P, O, C, S, Si, F (as parts of anions).

(b) Simple atomic ions are superbly suited as *charge carriers* for very fast *information transfer*. Starting with a transmembrane concentration gradient which has to be actively maintained by integral membrane ion pumps, information units in the form of electrical potential jumps can be created via diffusion, i.e. with maximal speed (biological selection) along ion channels. Electrical impulses in nerves as well as more complex trigger mechanisms, e.g. in the control of muscle contractions, are thus initiated with the fastest possible effect by sudden fluxes (diffusion control) of atomic, i.e. chemically and biologically nondegradable inorganic ions of different size and charge (Na^+, K^+, Ca^{2+}; see Chapter 13 and Section 14.2).

(c) *Formation, metabolism and degradation of organic compounds* in organisms often require acid or base catalysis. Since the physiological pH is generally limited to about 7, except for certain special compartments such as the stomach, the rate enhancement of such reactions cannot be accomplished by simple proton or hydroxide catalysis but requires *Lewis acid/Lewis base catalysis* involving metal ions. Many hydrolytically active enzymes thus contain the relatively small, positively charged metal ions Zn^{2+} and Mg^{2+} (Chapter 12 and Section 14.1).

(d) The *transfer of electrons* which is essential for the short-term *energy conversion* in organisms is mainly, but not exclusively, dependent on redox-active metal centers. A number of corresponding redox pairs has thus been found, some of which involving oxidation states that seem quite unusual under physiological conditions (marked in **bold** in the following). Specific modifications induced by 'bioligands' are largely responsible for the stabilization of such 'unusual' oxidation states. Biologically relevant are in particular the following oxidation states of redox-active metals: $Fe^{II}/Fe^{III}/\mathbf{Fe^{IV}}$, Cu^{I}/Cu^{II}, $Mn^{II}/\mathbf{Mn^{III}}/Mn^{IV}$, $Mo^{IV}/Mo^{V}/Mo^{VI}$, $Co^{I}/Co^{II}/Co^{III}$, $Ni^{I}/Ni^{II}/Ni^{III}$.

(e) The *activation of small, highly symmetrical molecules* with large bond energies places stringent demands on the required catalysts. The ability of transition metal centers to provide unpaired electrons and to simultaneously accept and donate electronic charge (π back-bonding; see Chapters 5 and 11) allows organisms to carry out energetically and mechanistically difficult reactions under physiological conditions, e.g.

  (i) the reversible uptake, transport, storage and conversion (Fe, Cu) and also the generation (Mn) of the paramagnetic dioxygen molecule, $^3O_2$ (Chapters 4 to 6 and 10),
  (ii) the fixation of molecular nitrogen, N_2 (Chapter 11), and its conversion to ammonia (Fe, Mo, V), or
  (iii) the reduction of CO_2 with hydrogen to give methane (Ni, Fe; Chapter 9 and Figure 1.1).

(f) Typical '*organometallic*' *reactivity* such as reductive alkylation or the *facile generation of radicals* for rapid rearrangement of substrate molecules is found for cobalamin coenzymes which contain a σ bond between the transition metal cobalt and primary alkyl groups (Chapter 3).

2.3 Biological Ligands for Metal Ions

The major part of bioinorganic chemistry is concerned with compounds of the metallic elements; non-metallic inorganic bioelements will be discussed in Chapter

16. Within the biological context, metals mainly appear in oxidized form as formally ionized centers which are therefore surrounded by electron-pair donating ligands. Since the chemical elements — 'inorganic' as well as 'organic' — cannot experience a biological evolution by themselves, it is their often highly complex coordination chemistry that is biologically relevant. What kinds of organic–biological material can serve as 'natural' coordination partners for metal centers, in addition to simple and complex phosphates, XPO_3^{2-}, purely inorganic sulfide, S^{2-}, or water, H_2O, together with its deprotonated forms, OH^- and O^{2-}? Relatively little is known about the relevance of metal coordination to *lipids* and *carbohydrates*, although the potentially negatively charged oxygen functions can bind cations electrostatically and even undergo chelate coordination via polyhydroxy groups [15,16]. Likewise, few molecular details are known about the *in vivo* interaction of low molecular weight *coenzymes* (vitamins; see 3.12), *hormones* or other products of the metabolism, e.g. citrate, with metal ions. The complexes formed are frequently labile which makes detection, isolation and structural characterization very difficult. Even so, it has long been known that, for example, the physiological function of ascorbate (vitamin C; see 3.12 [17]) is connected with the Fe^{II}/Fe^{III} redox equilibrium. The *in vitro* coordination chemistry of the redox active isoalloxazine ring of flavins (2.2) was examined more closely; coordinating via the atoms O(4) and N(5), it can form chelate complexes with 'soft' metal ions in its oxidized form and with 'hard' metal centers in its half reduced, i.e. semiquinone, form [18,19].

flavin

flavosemiquinone complex of zinc(II)

R' = CH$_3$, R = CH$_2$(HCOH)$_3$CH$_2$OH:
riboflavin, vitamin B$_2$

(2.2)

In the following, the three most important classes of bioligands will be discussed in greater detail: *peptides* (*proteins*) with amino acid side chains usable for coordination, specially biosynthesized *macrocyclic chelate ligands*, and *nucleobases* as components of nucleic acids.

2.3.1 Coordination by Proteins — Comments on Enzymatic Catalysis

Proteins, including enzymes, consist of α-amino acids which are connected via peptide bonds —C(=O)—N(—H)—. This carboxamide function itself is only a poor metal coordination site (compare, however, Figure 14.4); as in corresponding solvents such as *N*-methyl- or *N,N*-dimethylformamide, the high local concentration of amide functions leads to a relatively high dielectric constant (the protein as a medium) and therefore reduces ionic attraction and repulsion forces within proteins and protein complexes.

Table 2.5 The most important metal-coordinating amino acids

α-amino acid $R-^{\alpha}CH(NH_3^+)CO_2^-$	side chain, R
histidine (His)	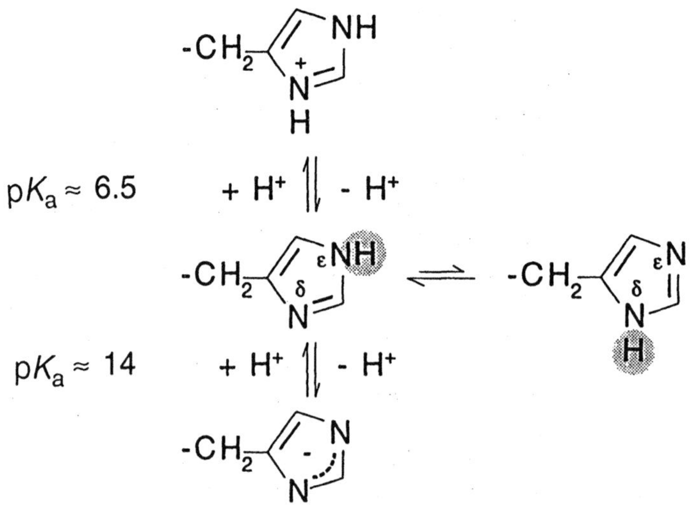
	$pK_a \approx 6.5$ $+ H^+ \Updownarrow - H^+$
	$pK_a \approx 14$ $+ H^+ \Updownarrow - H^+$
methionine (Met)	$-CH_2CH_2SCH_3$
cysteine (Cys)	$-CH_2SH$
selenocysteine (SeCys)	$-CH_2SeH$
tyrosine (Tyr)	 $-CH_2-\!\!\bigcirc\!\!- OH$
aspartic acid (Asp)	$-CH_2COOH$
glutamic acid (Glu)	$-CH_2CH_2COOH$

H : acidic protons which may be substituted by metal cations

The functional groups in the side chains of the following amino acids are particularly well suited for metal coordination (see Table 2.5):

(a) Histidine (His) coordinates mainly through the basic δ imine and sometimes through the ε imine nitrogen center of the imidazole ring (tautomeric forms). *Both nitrogen atoms become available for coordination after metal-induced deprotonation, producing a metal–metal bridging μ-imidazolate* (see Cu,Zn superoxide dismutase, Section 10.5).

(b) Methionine (Met) which often is the limiting essential amino acid in animal feed binds via the neutral δ sulfur atom of the thioether.

(c) Cysteine (Cys) contains a negatively charged γ thiolate center after deprotonation ($pK_a = 8.5$) to 'cysteinate'; it can bridge two metal centers (see the P-clusters of nitrogenase, 7.11).

(d) Selenocysteine (SeCys) is a relatively recently recognized amino acid and features a negatively charged selenolate center after deprotonation ($pK_a \approx 5$; see Section 16.8).

(e) Tyrosine (Tyr) coordinates mainly via the negatively charged phenolate oxygen atom after deprotonation to tyrosinate ($pK_a = 10$; see Figure 8.4); however, metal binding can also occur by the neutral phenolic form or by the simultaneously deprotonated and oxidized species, i.e. the tyrosyl radical.

(f) Glutamate (Glu) and

(g) Aspartate (Asp) bind via the negatively charged carboxylate functions ($pK_a \approx 4.5$). Carboxylates can act as terminal (η^1), as chelating (η^2) or as bridging $\mu\text{-}\eta^1:\eta^1$ ligands (2.3; see also 7.14 [20]); a further distinction concerns the *syn* or *anti* positioning of the binding electron pairs.

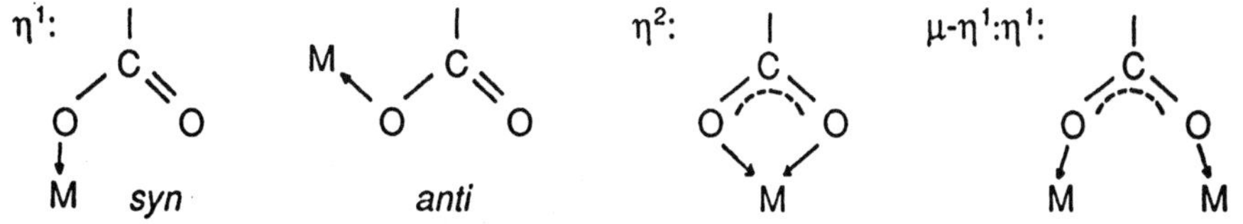

$$(2.3)$$

η^2-Coordination involving four-membered chelate rings is found mainly in complexes with large metal ions, e.g. in Ca^{2+}-containing proteins (Figure 14.4). The sometimes multiple $\mu\text{-}\eta^1:\eta^1$ coordination of glutamate or aspartate has been observed in iron or manganese dimers where the metal centers are often additionally bridged by μ-oxo or μ-hydroxo ligands (4.14, 7.14 or Figure 5.9). In all cases where the coordination is not η^1 (2.3), a strong deviation of the angle (C—O—M) from the ideal sp^2 angle of $120°$ is typical.

(h) Less important for metal-ion coordination are amino acids with simple hydroxy or amino functions like serine (Ser), threonine (Thr), lysine (Lys) or tryptophane (Trp).

The affinities of individual amino acid side chains to certain oxidation states of metals are often characteristic, resulting in a typical selectivity pattern. Metal complex formation constants of free amino acids as well as observations made for proteins led to the following preferred amino acid/metal ion combinations;

His: Zn^{II}, Cu^{II}, Cu^{I} or Fe^{II};
Met: Fe^{II}, Fe^{III}, Cu^{I} or Cu^{II};
Cys^-: Zn^{II}, Cu^{II}, Cu^{I}, Fe^{III}, Fe^{II}, Mo^{IV-VI} or Ni^{I-III};
Tyr^-: Fe^{III};
Glu^-, Asp^-: Fe^{III}, Mn^{III}, Fe^{II}, Zn^{II}, Mg^{II} or Ca^{II}.

Conversely, individual oxidation states of the biometals exhibit characteristic coordination environments. The most important of these as determined from structural studies of metalloproteins are listed in Table 2.6.

Table 2.6 Typical coordination environments of metal centers in proteins (according to [21])

metal oxidation state	bond stability	typical number and type of side chain ligands	typical coordination geometry
Zn(II)	high	3: His, Cys⁻, (Glu⁻)	severely distorted tetrahedron
Cu(I)	high	3,4: His, Cys⁻, Met	severely distorted tetrahedron
Cu(II)	high	3,4: His, (Cys⁻)	distorted square planar arrangement
Fe(II), Ni(II) Co(II), Mg(II)	low	4-6: His, Glu⁻, Asp⁻	distorted octahedron
Fe(III)	high	4-6: Glu⁻, Asp⁻, Tyr⁻, Cys⁻	distorted octahedron

Two features are characteristic of these arrangements and unusual when compared to conventional coordination compounds:

(a) The metal centers are often coordinatively unsaturated with regard to ligation by amino acid side chains, i.e. one such residue is often missing in relation to a 'regular' coordination number of 4 (tetrahedron, square) or 6 (octahedron). For catalytic activity, however, such an open site is essential for coordination of the substrate; in most instances, that potentially open coordination site is temporarily occupied in the 'resting state' by an easily replaced ligand such as H_2O. Proteins that exclusively transfer electrons do not have this requirement of coordinative unsaturation since they do not have to coordinate the substrate directly at the metal center; however, other conditions apply for these proteins in terms of feature (b) (see Sections 6.1 and 10.1).

(b) For the most common coordination numbers 4 and 6, the coordination geometries of many protein-bound metal centers turn out to be not regular and even deviate considerably from ideal tetrahedral, square planar or octahedral symmetries. Of course, a certain degree of deviation must be expected simply because of different amino acid ligands and the generally unsymmetrical environment provided by the protein. Figure 2.7 shows the typical structural representation of a relatively small metalloprotein resulting from X-ray crystallography.

The catalytic metal centers of enzymes are usually found deep within the more or less globular polypeptide which thus acts as a huge chelate ligand [23] employing metal-binding amino acid residues. Access of substrates is often possible via specific channels in the tertiary structure, ensuring the desired substrate selectivity (lock-and-key model of enzymatic catalysis).

In many cases, however, the actual distortion of the metal coordination geometry is so pronounced in the enzyme that it cannot be regarded as coincidental. In 1968 Vallee and Williams [24] rationalized this fact with their 'theory of the entatic state' of a catalytically active enzyme.

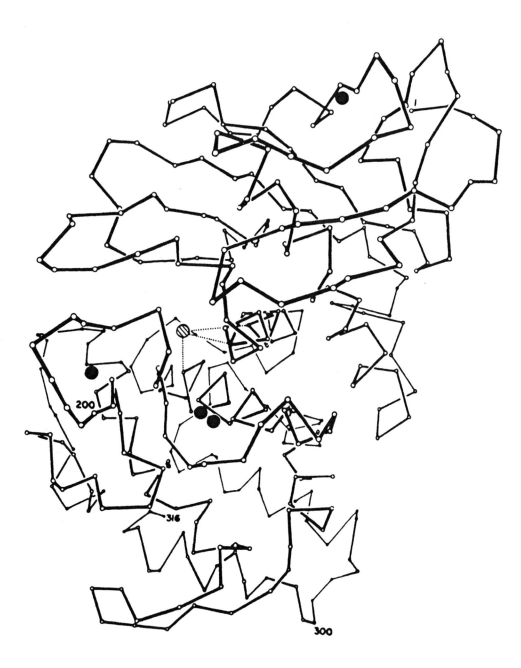

Figure 2.7
Structure of the proteolytic enzyme thermolysin (see Section 12.3) as determined by X-ray
crystallography. The folding of the polypeptide chain of 316 amino acids (molecular mass
34 kDa) is shown in α carbon-backbone representation, i.e. without depicting the side chains.
Represented by small spheres are the positions of four structure-stabilizing Ca^{2+} ions (black)
and of the Zn^{2+} ion (hatched) which is directly involved in the catalysis (from [22])

The 'Entatic State' in Enzymatic Catalysis

Catalysts accelerate chemical reactions, inhibitors retard them. In an energy
potential diagram (Figure 2.8), catalysis corresponds to a reduction in activation
energy; in other words, the transition state to be overcome is changed in such
a way that it is reached from the initial state with less amount of energy. In
metalloenzyme catalysis, a ternary complex is usually formed consisting of
substrate, enzymally modified metal center and the second reactant, e.g. a
coenzyme or an acidic or basic group. The catalytic function of the metal center
is at least twofold, viz. to *electronically activate* one or both reacting species as
intermediately bound ligands and to *position them in space, most often in a
specific, unsymmetrical fashion* (three-point fixing). The latter role of the catalyst
may be regarded simply under statistical aspects; the probability for a successful
encounter of the reactants increases greatly through coordination which involves a restriction of the degrees of freedom for translation and rotation and
thus produces a high 'effective concentration'. The energy difference between
the initial and the final state is not affected at this stage.

According to the hypothesis of the entatic state, the often surprisingly efficient
enzymatic catalysis is explained by a 'preformation of the transition state'; i.e.
the active center of the enzyme already largely features the (complementary)
geometry necessary to reach the critical high-energy transition state of the
substrate. In the 'entatic' (strained) state of the enzyme, much of the energy

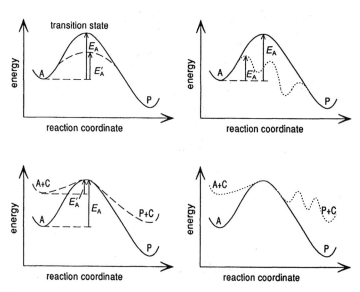

Figure 2.8
Energy profiles for catalyzed reactions at various degrees of sophistication. Upper left:
conventional representation of the reduction of activation energy $E_A \rightarrow E'_A$ upon transition from the initial state, A, to products, P. Lower left: reduction of E_A by introduction
of an 'entatic' (strained, high-energy) enzymatic catalyst, C, which largely provides the
preformed transition-state geometry of the substrate/catalyst complex. Upper right: real,
multistep catalysis involving a new reaction path [29]. Lower right: real enzymatic
catalysis

needed to reach that transition state is already stored and distributed over many chemical bonds. Small remaining geometrical changes between the initial and transition state of the *enzyme/substrate complex* then result in only a small activation energy (energy aspect) which means that productive encounters between reaction partners occur more often (statistical aspect) and that the reaction proceeds more rapidly (kinetic aspect).

For this reason, the active state of a metalloenzyme should not contain a regular (= low-energy, relaxed) coordination environment of the metal that is involved in the catalysis; on the contrary, the main goal should be a destabilization of the initial state [25]. Similar concepts to that of the 'entatic state' had been introduced before by Haldane and Pauling [26] and by Lumry and Eyring [27] in the form of the 'rack mechanism', indicating that the protein serves as a structure-enforcing scaffold. The determination of enzyme structures can thus provide valuable information on otherwise experimentally inaccessible transitionstate geometries of important types of chemical reactions. Since small changes in geometry are usually associated with low activation energies, the efficient enzymatic catalysis of reactions that involve small molecules has to take place stepwise in both biochemical and technical catalysis. Many small di- or triatomic molecules are particularly hard to activate because of their small number of geometrical degrees of freedom (e.g. the problem of N_2 fixation; see Section 11.2). In some cases, enzymatic catalysis involving very reactive intermediates means inhibition of unwanted side reactions ('negative catalysis' [28]).

The concept of the entatic state has been particularly well demonstrated for hydrolyzing metalloenzymes (see Chapter 12). At least 30% of all known enzymes are metal-dependent (especially oxidoreductases and hydrolases).

Summarizing, the protein can act toward metal ions as multidentate chelate ligand via its coordinatively active amino acid side chains. Additional functions of the protein are to provide spatial fixation (with the protein as a scaffold; compare Figure 2.7) and to serve as a medium with defined dielectrical, (a)protic or (non)polar properties [23]. Based on these principles, there are now increasing attempts to design new metal-coordinating proteins as 'minibodies' [30] or parts of antibodies [31].

With certain metal ions such as Mg^{2+}, Fe^{2+}, Ni^{2+} or Co^{2+} the amino acid side chains from Table 2.5 form complexes that are thermodynamically stable but kinetically labile (Table 2.6); i.e. in spite of a favorable equilibrium for complex formation the activation energy for dissociation is so small that exchange reactions are frequent. Naturally, such a situation is not acceptable for efficient catalytic functioning of a metalloenzyme; the kinetic stabilization of such metal ions thus requires special macrocyclic chelate ligands.

2.3.2 Tetrapyrrole Ligands and Other Macrocycles

The tetrapyrroles [32] are at least partially unsaturated tetradentate macrocyclic ligands (2.4) which, in their deprotonated forms, can tightly bind even substitutionally labile divalent metal ions.

The resulting complexes [33–36] are among the most common and best-known bioinorganic compounds (2.5):

(a) Chorophylls contain (otherwise labile) Mg^{2+} as the central ion and partially hydrogenated and substituted porphyrin ligands with an annelated five-membered ring (see Section 4.2).

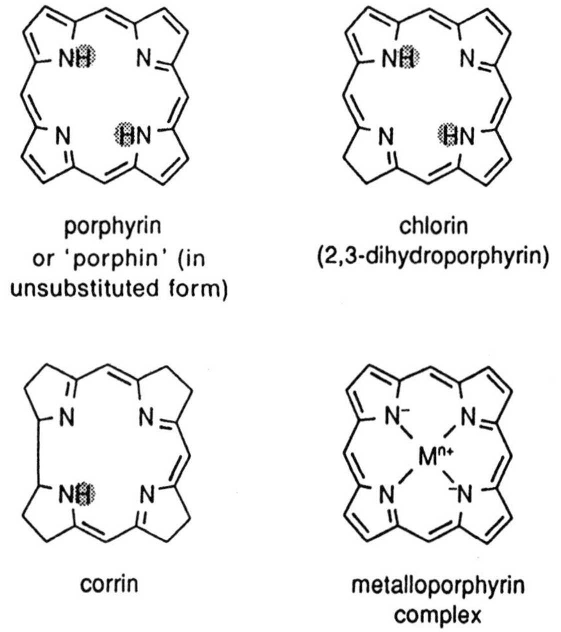

porphyrin
or 'porphin' (in
unsubstituted form)

chlorin
(2,3-dihydroporphyrin)

corrin

metalloporphyrin
complex

⬤: acidic protons which may be substituted by metal ions   (2.4)

(b) Cobalamins (Chapter 3), the coenzymatically active forms of vitamin B_{12}, contain cobalt and a partially conjugated 'corrin' ring π system which contains one ring member less than the porphyrin macrocycles.
(c) The heme group which consists of an iron center and a substituted porphyrin ligand is found, for example, in hemoglobin, myoglobin, cytochromes and peroxidases (compare 6.20). 'Green heme d' contains a chlorin macrocycle [37] and siroheme, another 'hydroporphyrin' complex, features two partially hydrogenated pyrrole rings in the cis position (see 6.20).
(d) In 1980 a porphinoid nickel complex, 'factor 430' (coenzyme F430), was isolated from methane-producing microorganisms (compare Figure 1.1) and characterized. The discovery of such porphinoid complexes in archaebacteria and the relative independence of their function from proteins makes them good candidates for 'first hour catalysts' [38] in biochemistry. A nickel-containing 'tunichlorin' which shares similarities with chlorophyll a has been isolated from certain marine animals (tunicates [39]).

As the complexes (2.5) belong to the most important and, due to their intense color, most conspicuous bioinorganic compounds, their structural and functional

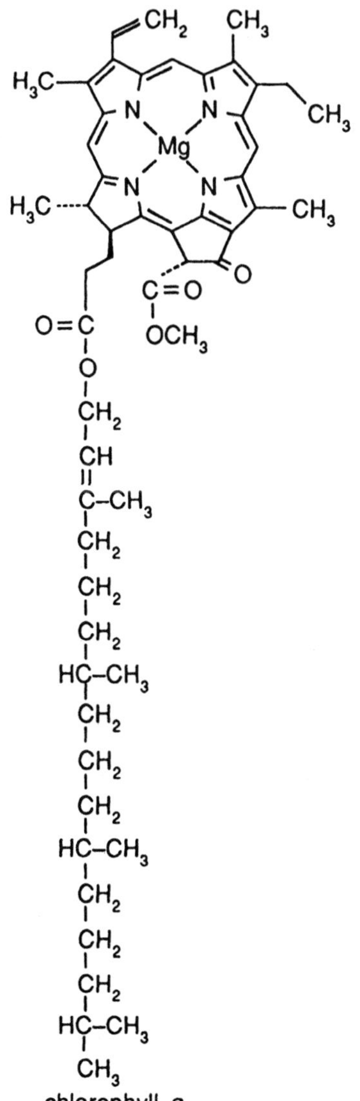

chlorophyll *a*

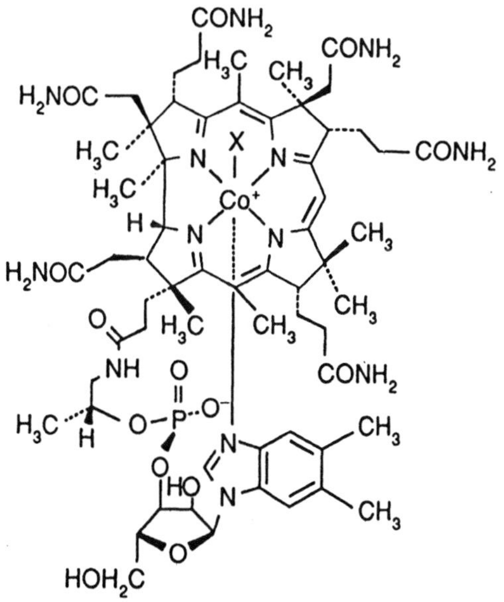

vitamin B$_{12}$ (X = CN)

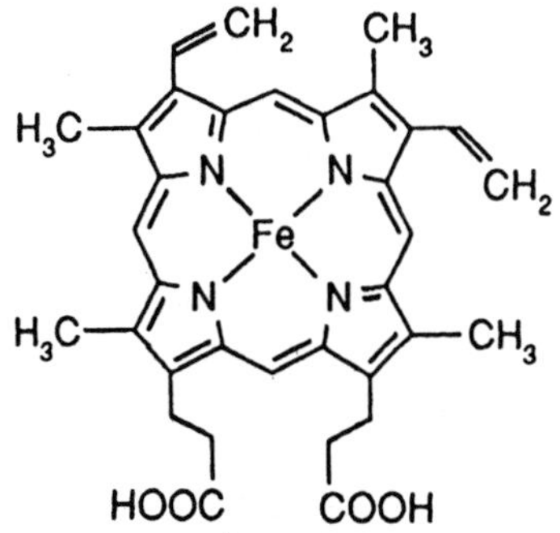

heme
(Fe-protoporphyrin IX)

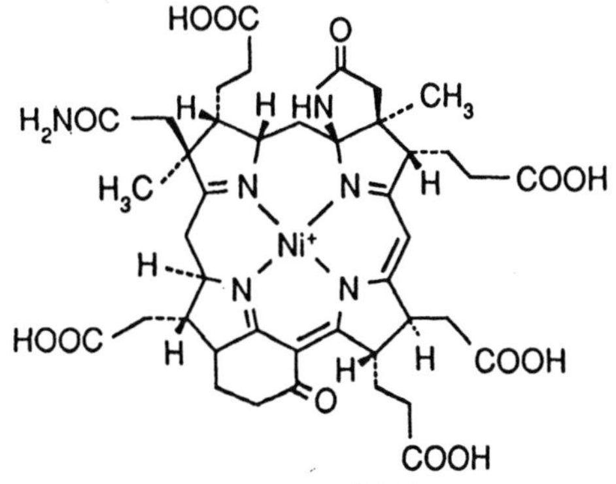

coenzyme F430

(2.5)

investigation and corresponding organic syntheses were rewarded with several Nobel prizes in chemistry:

(a) R. Willstätter (1915): research on the constitution of chlorophyll;
(b) H. Fischer (1930): studies on the constitution of the heme system;
(c) J. C. Kendrew and M. F. Perutz (1962): X-ray structure analysis of myoglobin and hemoglobin;
(d) D. Crowfoot-Hodgkin (1964): X-ray structure analysis of vitamin B_{12} and derivatives;
(e) R. B. Woodward (1965): natural product syntheses including chlorophyll and later vitamin B_{12} (in cooperation with the research group of A. Eschenmoser);
(f) J. Deisenhofer, R. Huber and H. Michel (1988): X-ray structure analysis of a heme- and chlorophyll-containing photosynthetic reaction center from bacteria.

What are the characteristic features of the unique tetrapyrrole bioligands (2.4) which require an elaborate [40] Zn^{2+}-dependent and Pb^{2+}-inhibited biosynthesis (see Sections 12.4 and 17.2)?

(a) The underlying planar or nearly planar (Figure 2.9) ring system is obviously very stable, as illustrated by the presence of porphyrin complexes in sediments [42] and in crude mineral oil. In contrast to R. Willstätter's objections to the first formulation of such macrocyclic structures by W. Küster in 1912, there is no geometric stress; all bond lengths (134–145 pm) and angles (107–126°) as well as torsional angles ($<10°$) are normal for neighboring sp^2-hybridized carbon and nitrogen centers.
(b) As tetradentate chelate ligands which, after deprotonation, carry a single (corrin, F430) or double negative charge, the tetrapyrrol macrocycles can even bind coordinately labile metal ions. The kinetic effect of chelate complex stability can be rationalized by considering that dissociation is only possible if *all* metal-to-ligand bonds are broken simultaneously (which is highly unlikely).
(c) Macrocyclic ligands are usually quite selective with regard to the size of the coordinated ion. This is especially true for the tetrapyrroles because they are rather rigid due to the presence of conjugated double bonds (Figure 2.9). Structural data and model calculations show that spherical ions with a radius of 60–70 pm are optimally suited to fit into the central cavity of tetrapyrrole macrocycles ('*in-plane*' coordination; see Figure 2.9). A survey of different metal ions from the periodic table illustrates this fact (Table 2.7).
(d) Most tetrapyrrole ligands contain an extensively conjugated π system. The Hückel rule for 'aromatic' and thus particularly stabilized cyclic π systems is fulfilled for porphyrins since they feature $18 = 4n + 2$ π electrons in the inner 16-membered ring. This heteroaromaticity presumably contributes to the thermal stability of the ring system [33], which was mentioned earlier. As a consequence of the extensive π conjugation, the ligands and their metal complexes show intense absorption bands in the visible region of the electromagnetic spectrum which led to their designation as tetrapyrrole 'pigments' or 'pigments of life'. Furthermore, the uptake and release of electrons by these heterocycles, i.e. one-electron reduction or oxidation processes, are facilitated because of the narrowing of the π frontier orbital gap; in fact, the resulting anion or cation radicals can be quite stable. Both of these properties, light absorption and redox behavior in terms of

Table 2.7 Ionic radii and (biological) complexation of metal ions by tetrapyrrole ligands

metal ion	ionic radius[a] (pm)	suitability as metal center in complexes with tetrapyrrole macrocycles
Be^{2+}	45	too small
Mg^{2+}	72	proper size, $\rightarrow$ chlorophyll (Chap. 4.2)
Ca^{2+}	100	too big
Al^{3+}	53	rather small
Ga^{3+}	62	gallium(III) porphyrin complexes have been found in crude mineral oil but not in living organisms (very rare element)
In^{3+}	80	rather large, rare element
$O=V^{2+}$ (not spherical)	ca. 60	vanadyl porphyrins are relatively abundant in certain crude oil fractions where they interfere with the catalytic removal of N and S in refineries; they have not been observed in living organisms
Mn^{2+}(h.s.)[b]	83	too large (?)
Mn^{3+}	ca. 60	proper size, use in synthetic oxidation catalysts
Fe^{2+}(h.s.)	78	too large (*out-of-plane* structure, compare Fig. 5.4)
Fe^{2+}(l.s.)[c]	61	proper size
Fe^{3+}(h.s.)	65	proper size
Fe^{3+}(l.s.)	55	rather small
average value for $Fe^{2+/3+}$	65	$\rightarrow$ heme system with Fe^{n+} in various oxidation and spin states (Chapters 5 and 6)
Co^{2+}(l.s.)	65	proper size, $\rightarrow$ cobalamins (Chap. 3)
Ni^{2+}	69	proper size, $\rightarrow$ F430 (Chap. 9.5), tunichlorin
Cu^{2+}	73	relatively large; Cu porphyrins have not been found in organisms, strong bonds are formed mainly with histidine in proteins
Zn^{2+}	74	relatively large; Zn porphyrins have not been found in organisms, strong bonds are formed e.g. with histidine or cysteinate in proteins

[a] For coordination number 6, from [43] [b] h.s.: high-spin. [c] l.s.: low-spin.

electron buffering and storage, render complexes of the tetrapyrrole macrocycles as essential components in the most important biological energy transformations: photosynthesis and respiration (see Chapters 4 to 6).

(e) Tetrapyrrole macrocycles are tetradentate chelate ligands which prefer a planar or nearly planar arrangement (Figure 2.9) around a metal center. Assuming a total coordination number of 6 in an approximately octahedral arrangement, that

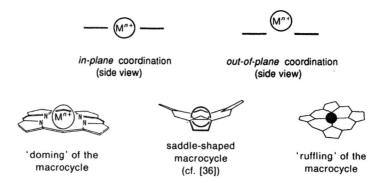

Figure 2.9
Typical geometrical deviations for complexes of tetrapyrrole macrocycles (cf. [41])

situation leaves two axial coordination sites, X and Y, available at the metal center (2.6). For a controlled stoichiometric or catalytic activation of substrates there is indeed a need for two such open coordination sites: one for the actual binding of the substrate and another for the regulation of catalytic activity, e.g. using the 'trans-effect'. Some examples illustrating the useful functions of axial ligands in biological tetrapyrrole complexes follow.

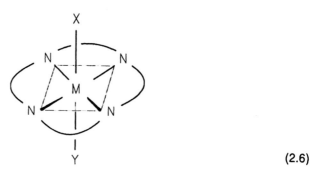

$$(2.6)$$

The substrate of hemoglobin is molecular oxygen, $O_2 = X$, which has to be reversibly coordinated for transportation to the tissues. A functionally useful 'proximal' histidine is coordinated in the sixth position, Y, at the thus tunable iron atom (Section 5.2).

Coenzymatically active cobalamins contain primary alkyl groups, $CH_2R = X$, coordinated directly to the cobalt center. The homolytic cleavage of these metal–alkyl bonds to yield radicals is possibly effected in the enzyme via changes in the coordination of the sixth ligand, Y (a benzimidazole derivative in the original coenzyme; see 3.1).

Chlorophylls exist as highly aggregated and spatially oriented species in 'antenna pigments' (Figure 4.2) or in the 'special pair' of photosynthetic reaction centers (Figure 4.5). This well-controlled aggregation requires multiple coordinative interactions in which the Mg^{2+} centers can act as bifunctional acceptors (Lewis acids) for electron pair-containing ligands X and Y while, for example, the carbonyl groups of the chlorophyll molecules can function as electron-pair donors (Lewis bases).

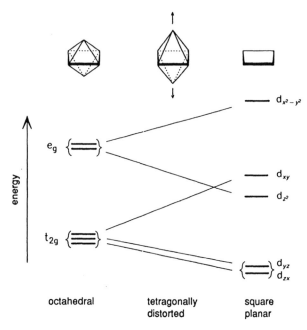

Figure 2.10
Correlation diagram for the splitting of transition metal d orbitals, depending on the extent
of tetragonal distortion (here, axial elongation) of octahedral symmetry

(f) The potentially considerable tetragonal distortion of the octahedral symmetry
through 'strong' dianionic tetrapyrrole ligands causes a characteristic splitting of
the d orbitals of coordinated transition metal centers (Figure 2.10). Some
consequences of this effect for chemical reactivity are discussed in detail for
vitamin B_{12} and cobalamins in Chapter 3. Although the equatorial ligand field
strength of planar tetrapyrrole dianions should stabilize low-spin configurations
versus their high-spin equivalents, there are species such as deoxy-hemoglobin
and deoxy-myoglobin which feature a very critical high-spin iron(II) center. Due
to the rather large size of high-spin Fe^{II} (see Table 2.7), this metal ion does not
fit completely into the cavity of the macrocycle which results in *out-of-plane*
complexation (Figures 2.9 and 5.4) and hence a diminished ligand-field effect from
the porphyrin. A similar relationship between the spin state and geometry of the
tetrapyrrole ligand was found for the d^8 ion nickel(II) in coenzyme F430 (see
Section 9.5).

Electron Spin States in Transition Metal Ions

The terms 'high-spin' and 'low-spin' are derived from *ligand field theory*. In a
ligand field with octahedral symmetry, the five d orbitals (which are energetically equivalent in the free, spherically symmetrical transition metal ion) split
into two energetically different groups of energy levels (2.7).

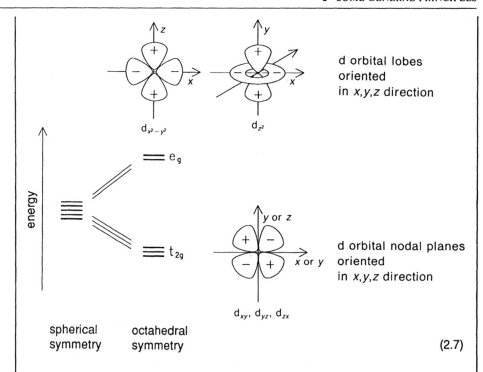

d orbital lobes
oriented
in x,y,z direction

d orbital nodal planes
oriented
in x,y,z direction

spherical
symmetry

octahedral
symmetry

(2.7)

When the d orbitals of such transition metal centers are consecutively filled with electrons, the occupation pattern follows the *Aufbau* principle. At first, each of the three equivalent low-energy t_{2g} orbitals is occupied with one electron of the same spin (Hund's rule of maximal multiplicity). From the fourth electron on, the difference in orbital energies (e_g vs. t_{2g}) and the spin pairing energy have to be compared in order to decide on the most favorable orbital for that next electron (2.8). If the difference in orbital energies is larger than the spin pairing energy, the electron will be placed in an already occupied orbital of the t_{2g} level such that those two spins are paired; two unpaired electrons of the same spin remain in a 'low-spin' situation. If, on the other hand, the energy difference between e_g and t_{2g} orbitals is smaller than the spin pairing energy, the occupation of an e_g level becomes more favorable which, according to Hund's rule, leads to four unpaired electrons in a 'high-spin' configuration.

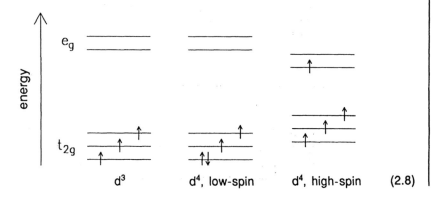

d^3 d^4, low-spin d^4, high-spin   (2.8)

Octahedrally coordinated transition metal ions with an electronic configuration of d^4 to d^7 can thus exist either as low-spin or high-spin species. Which of these alternatives is realized in a complex is determined by the orbital splitting as induced by the 'strength' of the ligand field. Examples:

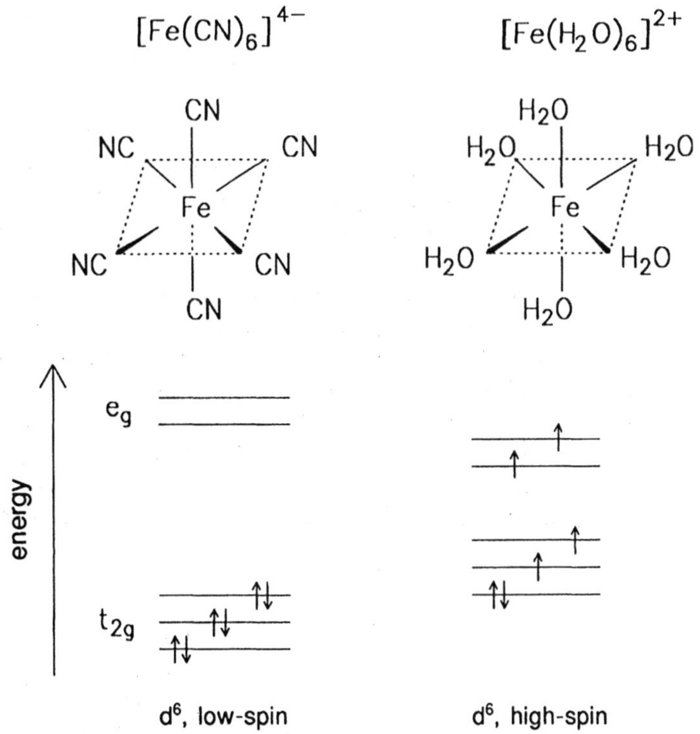

$$[Fe(CN)_6]^{4-} \qquad\qquad [Fe(H_2O)_6]^{2+}$$

d^6, low-spin d^6, high-spin   (2.9)

It is obvious that high-spin vs. low-spin alternatives can also exist for less symmetrical coordination geometries and then also for other d^n configurations. For ions such as Fe^{2+} or Fe^{3+} in low-symmetry environments there is also the alternative of 'intermediate-spin' states.

Despite the slightly domed, ruffled or saddle-shaped structures (Figure 2.9) which permit some fine-tuning of the electronic structure and reactivity, the tetrapyrrole ligands essentially feature a planar two-dimensional ring system. For complexation of the extremely labile metal monocations there are other, multidentate and much more three-dimensionally structured macrocycles available as biological ligands, the ionophores. These ligands, their complexes and their synthetic analoga are discussed in detail in Section 13.2; however, some important features will be pointed out here.

The ionophores are multidentate (≥ 6) chelate ligands which either exist as macrocycles (2.10) or can at least form quasi-macrocycles after coordination-induced ring closure via hydrogen bond interactions.

$$\left[\begin{array}{c} \overset{HC(CH_3)_2}{|} \quad \overset{CH_3}{|} \qquad\qquad \overset{HC(CH_3)_2}{|} \quad \overset{HC(CH_3)_2}{|} \\ \cdot NH-CH-\underset{\underset{O}{\parallel}}{C}-O-CH-\underset{\underset{O}{\parallel}}{C}-NH-CH-\underset{\underset{O}{\parallel}}{C}-O-CH-\underset{\underset{O}{\parallel}}{C}- \end{array}\right]_3$$

valinomycin   (2.10)

The alkali metals which generally form only highly labile complexes and the rather labile Ca^{2+} ion can be bound in the polar inner cavity of such complex ligands (Figure 2.11) by several strategically positioned heteroatoms (N or O) according to their size (size selectivity [45]; chelate/crown ether/cryptate effect [46,47]). Dissociation is possible only when *all* coordinative bonds are broken simultaneously and when a substantial conformational change occurs. The often lipophilic outside of these complexes allows the corresponding ions to be transported through biological membranes. This ability renders such macrochelating natural products useful as antibiotics because of the effect on the ion distribution between both sides of bacterial membranes (see Chapter 13). In 1987 the Nobel prize in chemistry was awarded to C. J. Pedersen, J.-M. Lehn and D. J. Cram for the development of synthetic analogues of such macrocyclic natural products which greatly improved an understanding of the underlying principles of *molecular recognition* mechanisms.

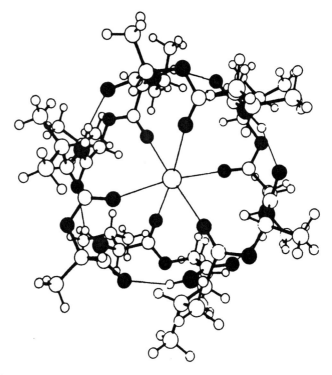

Figure 2.11
Molecular structure of the K^+/valinomycin complex (heteroatoms as filled spheres) (from [44])

2.3.3 Nucleobases, Nucleotides and Nucleic Acids (RNA, DNA) as Ligands

Not only proteins as biochemical function carriers but also the information-carrying oligo- and polynucleotides are suitable ligands for metal ions. As constituents of nucleic acids, the heterocyclic nucleobases (2.11) have long been recognized as potential metal coordination sites [48]. Formation, replication and cleavage of nucleic acid polymers (RNA, DNA [49]) as well as their structural integrity, e.g. the double-helical arrangement of conventional DNA, require the presence of metal ions either in enzymatically bound (e.g. Zn^{2+}) or 'free', charge neutralizing form (e.g. Mg^{2+}). The latter, however, pertains mainly to the anionic phosphate groups of nucleotides.

Nucleobases (2.11) are ambidentate ligands which, even as constituents of nucleosides or nucleotides, offer several different coordination sites for metal ions. Depending on the characteristics of the coordination center (e.g. the type of atom, hybridization, basicity, chelate assistance), on external conditions (e.g. pH) and on the size and nature of the metal center, monodentate or multidentate coordination is possible to imine, amino, amido, oxo or hydroxo functions (compare 17.3). An important aspect is the ability of nucleobases to exist in different tautomeric forms as shown in (2.12) by the example of cytosine.

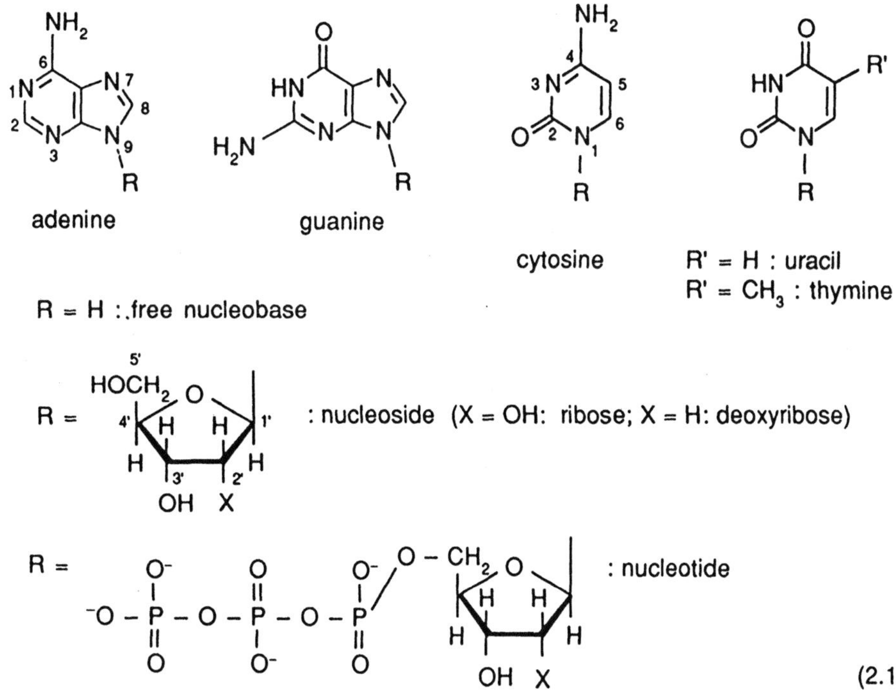

adenine   guanine

cytosine   R' = H : uracil
R' = CH_3 : thymine

R = H : free nucleobase

R = : nucleoside (X = OH: ribose; X = H: deoxyribose)

R = : nucleotide

(2.11)

The presence of positively charged metal ions in the cell nucleus can affect the hydrogen-bond interactions which are essential for 'natural' DNA base pairing to such an extent that the intermediate formation of a 'false' tautomer is favored. This may lead to a 'mispairing' of nucleobases (2.13) and eventually, if not repaired, to altered genetic information transfer (cf. the potentially mutagenic or even carcinogenic effects of metal ions [50]).

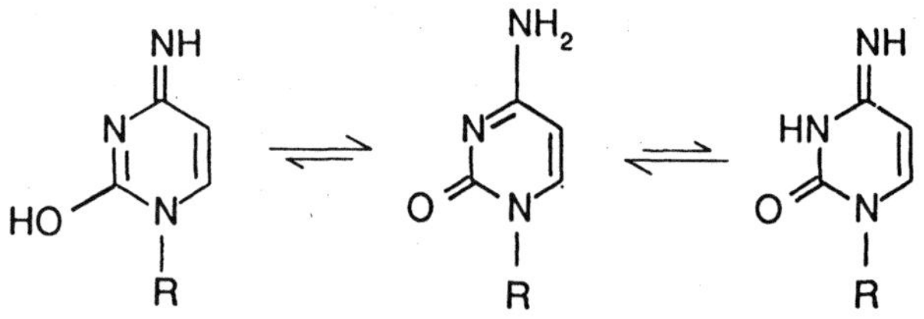

possible tautomers of N(1)-substituted cytosine   (2.12)

The intensive studies aimed at understanding the anti-tumor activity of cisplatin (see Section 19.2) have shown that the coordination of inert metal complexes to DNA can be very specific with regard to certain base-pair sequences in the double helix. These not necessarily covalent metal complex/nucleotide interactions have received particular interest because small, sequence-specific compounds can possibly be developed for selective modification of DNA and thus for potential use in tumor chemotherapy. Sequence specificity, including the recognition of chiral structures, can result from direct metal coordination (see Figures 19.2 and 19.3), from specific coordination geometries in complexes and resulting shape selection (2.14) or from specific interactions, e.g. intercalation, of ligands. The function of the metal can be to oxidatively generate bond-cleaving radicals [52] or to form hydrolytically active centers [53]. A long-range but nevertheless very attractive goal is the design of artificial restriction reagents for selective manipulation of DNA; optically active

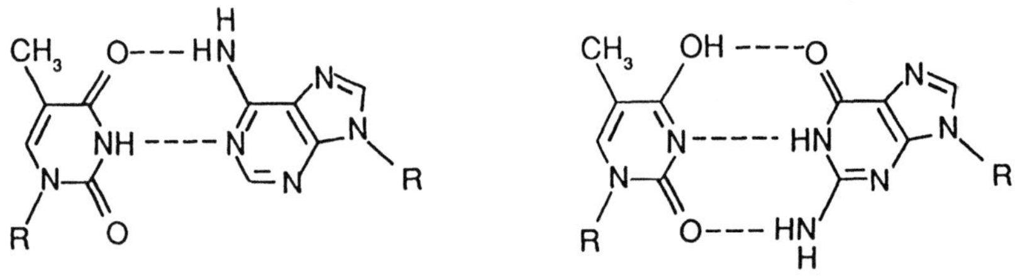

correct base pairing between
thymine and adenine

R: see formula (2.11)

mispairing between 'false' thymine
tautomer and guanine [50]

(2.13)

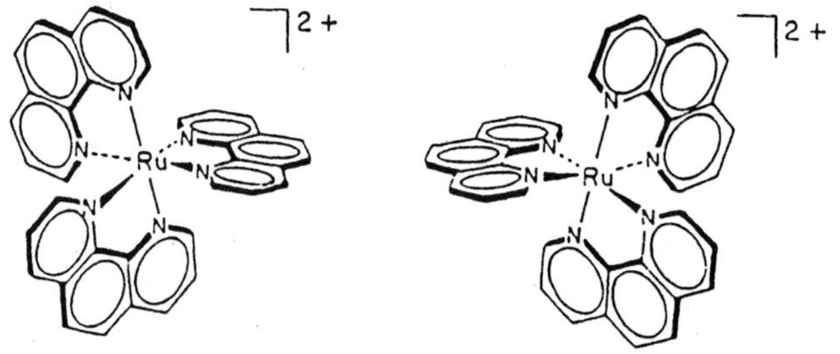

Λ enantiomer   Δ enantiomer
enantiomorphous complexes (compare 8.10)   (2.14)

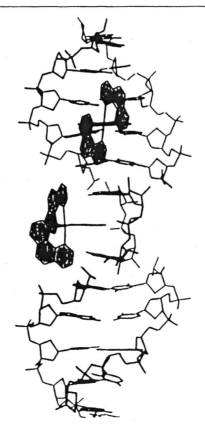

Figure 2.12
Possible DNA intercalation of complexes (2.14) (according to [54]; compare, however, [55])

(chiral) and light-activated ruthenium(II) (2.14) or rhodium(III) complexes with potentially intercalating ligands (Figure 2.12) are being thoroughly investigated [54–56].

2.4 Relevance of Model Compounds

Research on low molecular weight 'model compounds' is a widely used approach by inorganic chemists to simulate the main spectroscopic, structural and reactivity features of large bioinorganic systems. Such studies are particularly appropriate when structural details, e.g. of metalloproteins, are not available or not yet unambiguous. Several states of sophistication can be distinguished for modeling actual biological systems [57,58].

As a minimal requirement, *physical, in particular spectroscopic properties* and *basic structural characteristics* of the natural system should be adequately represented by the model. This approach implies that the spectroscopic behavior is determined mainly by the first coordination sphere around a metal ion. If very little is known about the structure of the system to be modeled, this approach can only help to exclude alternatives.

Even the next step, the *qualitative simulation of the reactivity* of the natural system, has been accomplished only in relatively few model studies. Such a simulation is very desirable because biochemical mechanisms often parallel technical processes and efficient synthetic catalysts are attractive targets. Most often, however, the synthetic enzyme models exhibit only a stoichiometric, non-catalytic reactivity towards the natural substrate.

The last stage of modeling, the approximately *quantitative simulation of reactivity* with respect to reaction rate and substrate specificity, is nearly impossible to attain with low molecular weight systems. As pointed out in Section 2.3.1, the desired high selectivity (lock-and-key analogy) *and* high reactivity (entatic state situation) require the highly complex structure of biochemical compounds. In this introductory text we can mention only a few model compounds which are either historically important or of very recent relevance.

It has now become possible to modify bioinorganic systems, particularly proteins, by specific recombinant DNA techniques, i.e. via site-directed mutagenesis (1993 Nobel prize in chemistry for M. Smith). The effects of altering single amino acids in the peptide sequence can thus be elucidated and new substrate specificities may result [59]. Model studies of this type (see Section 6.1) have a promising potential for application in biotechnology, e.g. with respect to a transfer of N_2-fixation capability or to the microbial degradation of toxic waste.

A more inorganic and long practised modification of metalloproteins is the substitution of the 'natural' metal ion which may be hard to characterize spectroscopically by an isotope or very similar 'ersatz' ion which is more suitable for certain physical methods (isomorphous substitution; see Sections 12.1 and 13.1).

References

1. F. Kieffer, in *Metals and Their Compounds in the Environment — Occurrence, Analysis and Biological Relevance* (ed. E. Merian), VCH, Weinheim, 1991, p. 481.
2. F. H. Nielsen, Nutritional requirements for boron, silicon, vanadium, nickel and arsenic: current knowledge and speculation, *FASEB J.*, **5**, 2661 (1991).
3. C. T. Horovitz, Is the major part of the periodic system really inessential for life?, *J. Trace. Elem. Electrolytes Health Dis.*, **2**, 135 (1988).
4. P. A. Cox, *The Elements: Their Origin, Abundance, and Distribution*, Oxford University Press, Oxford, 1989.
5. E. I. Ochiai, *Bioinorganic Chemistry: An Introduction*, Allyn and Bacon, Boston, 1977.
6. A. B. Fiabane and D. R. Williams, *The Principles of Bio-inorganic Chemistry*, The Chemical Society, London, 1977, p. 68.
7. R. J. P. Williams, Inorganic elements in biological space and time, *Pure Appl. Chem.*, **55**, 1089 (1983).
8. E. J. Underwood and several authors, Metabolic and physiological consequences of trace element deficiency in animals and man, *Phil. Trans. Roy. Soc. London*, **B294**, 1–213 (1981).
9. National Academy of Science USA, National Research Council, Food and Nutrition Board, Recommended Dietary Allowances, 10th edn, National Academy Press, Washington, 1989.
10. H. G. Seiler, H. Sigel and A. Sigel (eds.), *Handbook on Toxicity of Inorganic Compounds*, Marcel Dekker, New York, 1988.

11. R. A. Bulman, The chemistry of chelating agents in medical sciences, *Struct. Bonding (Berlin)*, **67**, 91 (1987).

12. M. M. Jones, Newer chelating agents for *in vivo* toxic metal mobilization, *Comments Inorg. Chem.*, **13**, 91 (1992).

13. R. J. P. Williams, Bio-inorganic chemistry: its conceptual evolution, *Coord. Chem. Rev.*, **100**, 573 (1990).

14. R. J. P. Williams, The chemical elements of life, *J. Chem. Soc. Dalton Trans.*, 539 (1991).

15. W. I. Weis, K. Drickamer and W. A. Hendrickson, Structure of a C-type mannose-binding protein complexed with an oligosaccharide, *Nature (London)*, **360**, 127 (1992).

16. D. M. Whitfield, S. Stojkovski and B. Sarkar, Metal coordination to carbohydrates. Structures and function, *Coord. Chem. Rev.*, **122**, 171 (1993).

17. M. B. Davies, Reactions of L-ascorbic acid with transition metal complexes, *Polyhedron*, **11**, 285 (1992).

18. P. Hemmerich and J. Lauterwein, The structure and reactivity of flavin-metal complexes, in *Inorganic Biochemistry* (ed. G. L. Eichhorn), Elsevier, Amsterdam, 1973, p. 1168.

19. M. J. Clarke, Electrochemical effects of metal ion coordination to noninnocent, biologically important molecules, *Comments Inorg. Chem.*, **3**, 133 (1984).

20. R. L. Rardin, W. B. Tolman and S. J. Lippard, Monodentate carboxylate complexes and the carboxylate shift: implications for polymetalloprotein structure and function, *New J. Chem.*, **15**, 417 (1991).

21. R. J. P. Williams, Symbiotic chemistry of metals and proteins, *Chem. Br.*, **19**, 1009 (1983).

22. B. W. Matthews, J. N. Jansonius, P. M. Colman, B. P. Schoenborn and D. Dupourque, Three-dimensional structure of thermolysin, *Nature (London)*, **238**, 37 (1972).

23. R. Huber, A structural basis for the transmission of light energy and electrons in biology (Nobel address), *Angew. Chem. Int. Ed. Engl.*, **28**, 848 (1989).

24. B. L. Vallee and R. J. P. Williams, Metalloenzymes. Entatic nature of their active sites, *Proc. Natl. Acad. Sci. USA*, **59**, 498 (1968).

25. R. J. P. Williams, Metallo-enzyme catalysis: the entatic state, *J. Mol. Catalysis—Review Issue*, 1 (1986).

26. L. Pauling, Nature of forces between large molecules of biological interest, *Nature (London)*, **161**, 707 (1948).

27. R. Lumry, H. Eyring, Conformation changes of proteins, *J. Phys. Chem.*, **58**, 110 (1954).

28. J. Retey, Reaction selectivity of enzymes through negative catalysis or how enzymes work with highly-reactive intermediates, *Angew. Chem. Int. Ed. Engl.*, **29**, 355 (1990).

29. A. Haim, Catalysis: new reaction pathways, not just a lowering of the activation energy, *J. Chem. Educ.*, **66**, 935 (1989).

30. A. Pessi, E. Bianchi, A. Crameri, S. Venturini, A. Tramontano and M. Sollazzo, A designed metal-binding protein with a novel fold, *Nature*, **362**, 367 (1993).

31. W. S. Wade, J. S. Koh, N. Han, D. M. Hoekstra and R. A. Lerner, Engineering metal coordination sites into the antibody light chain, *J. Am. Chem. Soc.*, **115**, 4449 (1993).

32. G. P. Moss, Nomenclature of tetrapyrroles, *Pure Appl. Chem.*, **59**, 779 (1987).

33. J. H. Fuhrhop, Reactivity of the porphyrin ligands, *Angew. Chem. Int. Ed. Engl.*, **15**, 321 (1974).

34. D. Dolphin (ed.), *The Porphyrins*, Vols. I to VII, Academic Press, New York, from 1978.

35. J. W. Buchler (ed.), Metal complexes with tetrapyrrole ligands I, *Struct. Bonding (Berlin)*, **64**, 1–268 (1987).

36. B. Kräutler, The porphinoids—versatile biological catalyst molecules, *Chimia*, **41**, 277 (1987).

37. R. Timkovich, M. S. Cork, R. B. Gennis and P. Y. Johnson, Proposed structure of heme *d*, a prosthetic group of bacterial terminal oxidases, *J. Am. Chem. Soc.*, **107**, 6069 (1985).

38. A. Eschenmoser, Vitamin B_{12}: experimental work on the question of the origin of its molecular structure, *Angew. Chem. Int. Ed. Engl.*, **27**, 5 (1988).

39. K. C. Bible, M. Buytendorp, P. D. Zierath and K. L. Rinehart, Tunichlorin: a nickel chlorin isolated from the Caribbean tunicate *Trididemnum solidum*, *Proc. Natl. Acad. Sci. USA*, **85**, 4582 (1988).

40. H. A. Dailey, C. S. Jones and S. W. Karr, Interaction of free porphyrins and metalloporphyrins with mouse ferrochelatase. A model for the active site of ferrochelatase, *Biochim. Biophys. Acta*, **999**, 7 (1989).

41. O. Q. Munro, J. C. Bradley, R. D. Hancock, H. M. Marques, F. Marsicano and P. W. Wade, Molecular mechanics study of the ruffling of metalloporphyrins, *J. Am. Chem. Soc.*, **114**, 7218 (1992).

42. P. Schaeffer, R. Ocampo, H. J. Callot and P. Albrecht, Extraction of bound porphyrins from sulfur-rich sediments and their use for reconstruction of palaeoenvironments, *Nature (London)*, **364**, 133 (1993).

43. W. L. Jolly, *Modern Inorganic Chemistry*, McGraw-Hill, New York, 1984.

44. K. Neupert-Laves and M. Dobler, The crystal structure of a K^+ complex of valinomycin, *Helv. Chim. Acta*, **58**, 432 (1975).

45. R. D. Hancock, Chelate ring size and metal ion selection, *J. Chem. Educ.*, **69**, 615 (1992).

46. J. M. Lehn, Supramolecular chemistry–molecules, supermolecules, and molecular functional units (Nobel address), *Angew. Chem. Int. Ed. Engl.*, **27**, 89 (1988).

47. F. Vögtle, *Supramolecular Chemistry*, Wiley, New York, 1993.

48. L. G. Marzilli, Metal complexes of nucleic acid derivatives and nucleotides: binding sites and structures, *Adv. Inorg. Biochem.*, **3**, 47 (1981).

49. T. D. Tullius (ed.), *Metal-DNA Chemistry*, ACS Symposium Series 402, 1989.

50. B. Lippert, H. Schöllhorn and U. Thewalt, Metal-stabilized rare tautomers of nucleobases. 4. On the question of adenine tautomerization by a coordinated platinum(II), *Inorg. Chim. Acta*, **198–200**, 723 (1992).

51. H. Schöllhorn, U. Thewalt and B. Lippert, Metal-stabilized rare tautomers of nucleobases, *J. Am. Chem. Soc.*, **111**, 7213 (1989).

52. D. P. Mack and P. B. Dervan, Nickel-mediated sequence-specific oxidative cleavage of DNA by a designed metalloprotein, *J. Am. Chem. Soc.*, **112**, 4604 (1990).

53. V. Dange, R. B. van Atta and S. M. Hecht, A Mn^{2+}-dependent ribozyme, *Science*, **248**, 585 (1990).

54. A. M. Pyle and J. K. Barton, Probing nucleic acids with transition metal complexes, *Prog. Inorg. Chem.*, **38**, 413 (1990).

55. M. Eriksson, M. Leijon, C. Hiort, B. Norden and A. L. Gräslund, Minor groove binding of $[Ru(phen)_3]^{2+}$ to $[d(CGCGATCGCG)]_2$ evidenced by two-dimensional nuclear magnetic resonance spectroscopy, *J. Am. Chem. Soc.*, **114**, 4933 (1992).

56. A. E. Friedman, J.-C. Chambron, J.-P. Sauvage, N. J. Turro and J. K. Barton, Molecular 'light switch' for DNA: $Ru(bpy)_2(dppz)^{2+}$, *J. Am. Chem. Soc.*, **112**, 4960 (1990).

57. J. A. Ibers and R. H. Holm, Modeling coordination sites in metallobiomolecules, *Science*, **209**, 223 (1980).

58. K. D. Karlin, Metalloenzymes, structural motifs, and inorganic models, *Science*, **261**, 701 (1993).

59. H. Sigel and A. Sigel (eds.), Interrelations among metal ions, enzymes, and gene expressions, *Metal Ions in Biological Systems*, Vol. 25, Marcel Dekker, New York, 1989.

3 Cobalamins Including Vitamin and Coenzyme B₁₂

3.1 History and Structural Characterization

Coenzyme B_{12} and its derivatives (3.1), including vitamin B_{12}, are well suited as introductory examples in bioinorganic chemistry for various reasons. First of all, several milestones in the development of the whole field have been associated with vitamin B_{12} and later on with the coenzyme. This pertains to the immediate therapeutical benefit, the early use of chromatographic purification methods, the structural elucidation by X-ray crystallography and the relationship between enzymatic and coenzymatic reactivity. Furthermore, modern natural product synthesis as well as bioorganic and organometallic chemistry have strongly profited from studies of the B_{12} system.

Coenzyme B_{12} (3.1) is a medium-sized molecule with a molecular mass of about 1580 Da which exhibits its characteristic specificity and high reactivity only in combination with corresponding apoenzymes (3.2).

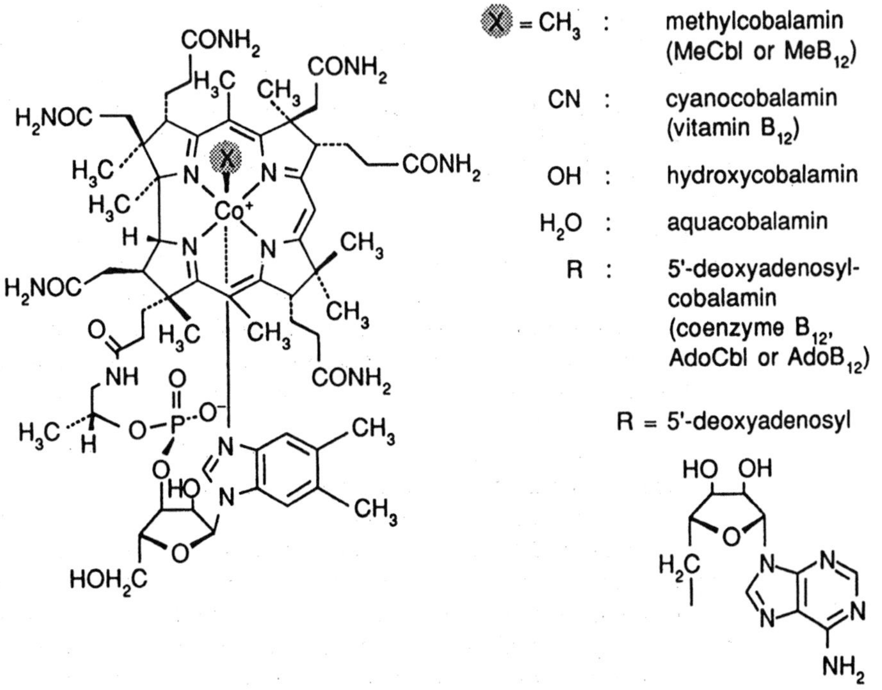

$X = CH_3$: methylcobalamin (MeCbl or MeB₁₂)

CN : cyanocobalamin (vitamin B₁₂)

OH : hydroxycobalamin

H_2O : aquacobalamin

R : 5'-deoxyadenosyl-cobalamin (coenzyme B₁₂, AdoCbl or AdoB₁₂)

R = 5'-deoxyadenosyl

(3.1)

coenzyme	+	apoenzyme	⟶	holoenzyme
low molecular mass, determines the type of reaction		high molecular mass (protein), determines substrate specificity (selectivity) and the reaction rate		complete enzyme, fully functional

$$(3.2)$$

The incorporation of the element cobalt into the coenzyme is quite surprising because cobalt is the least abundant first-row (3d) transition metal in the earth's crust and in sea water (Figure 2.2). Therefore, a very special functionality is to be expected.

The corrin ligand (2.4) is also unique, particularly with regard to its smaller ring size as compared to the porphin systems. Cobalt-containing porphyrin complexes, although stable, are *not* suitable to mimic the actions of coenzyme B$_{12}$.

The sixth, axial metal coordination site in coenzyme B$_{12}$ and methylcobalamin features a primary alkyl group (3.1) which makes these complexes the only fully established examples (see Section 9.5) of 'natural' organometallic compounds in biochemistry (see Sections 3.2.4 and 17.3 on the possible bioalkylation of heavy metals). The configuration Co—CH$_2$R of alkylcobalamins is unusually stable toward hydrolysis in neutral aqueous solution. On the other hand, the cobalt–carbon bond shows a very special reactivity, viz. the *enzymatically controlled* formation of reactive primary alkyl radicals. This unusual reactivity has prompted chemists from various fields beyond biochemistry to thoroughly study these remarkable complexes [1–3].

In retrospect, during the 1920s it was found that injections of extracts from animal liver were able to cure a very malignant ('pernicious') form of anemia which could otherwise be lethal. Improved methods of trace analysis soon showed that the essential component of these extracts contained cobalt. Since the substance was synthesized only by microorganisms and the trace element cobalt had to be supplied in any case, the factor was called 'vitamin B$_{12}$'. Enrichment and isolation turned out to be extremely laborious; because of the low concentration of only 0.01 mg vitamin per liter of blood, chromatographic separation methods had to be employed. The therapeutically useful but not directly active 'vitaminic' form, cyanocobalamin (3.1), was obtained in pure form in 1948 [4].

As the complete determination of the molecular constitution of cobalamins was impossible with chemical means alone, the eventual structural elucidation required X-ray diffraction of single crystals, a method that could only be applied to relatively simple systems in those days (see Section 4.2). With approximately 100 nonhydrogen atoms, vitamin B$_{12}$ and later coenzyme B$_{12}$ posed a formidable crystallographic challenge; for the solution of these problems Dorothy Crowfoot-Hodgkin was awarded the 1964 Nobel prize in chemistry.

The structure of the coenzyme (3.1) as obtained from crystallographic studies shows the cobalt–corrin macrocycle featuring an alkyl-bound 5'-deoxyadenosine group and an N(1)-coordinated 5,6-dimethylbenzimidazole ring as axial ligands (Figure 3.1). The latter is connected to the corrin macrocycle via a long pendant chain so that this corrin derivative effectively functions as a pentadentate chelate ligand. In spite of extensive conjugation of π electrons the unsaturated macrocycle is not completely flat but adopts a slightly bent 'butterfly' or 'saddle' conformation (Figures 2.9 and 3.1). Model studies have demonstrated the relevance of this structural feature for the reactivity in terms of an entatic state structure.

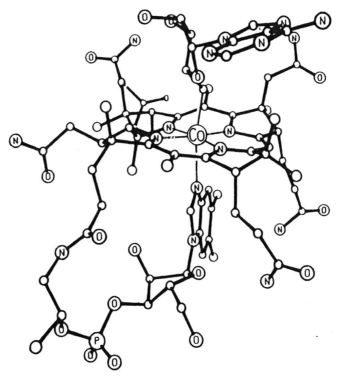

Figure 3.1
Molecular structure of coenzyme B_{12} (from [5])

The nonplanarity results from the fact that the relatively large metal ion (Table 2.7) is encapsulated by a 15-membered instead of a 16-membered macrocycle. In lieu of the 5'-deoxyadenosine moiety a simple methyl group can be bound at the metal center (methylcobalamin, 3.1); exchange with water, hydroxide or cyanide leads to the physiologically not directly active forms aqua-, hydroxo- and cyanocobalamin. The latter is called vitamin B_{12}, although this form represents only an artifact, i.e. the result of the isolation procedure.

The existence of a relatively inert bond between a transition metal center and a primary alkyl ligand is quite remarkable, especially since this true organometallic compound is stable under physiological conditions, i.e. in aqueous solution at pH 7 and in the presence of oxygen. The corrin macrocycle creates a strong ligand field resulting in a low-spin situation (2.9). However, the six-coordinate arrangement shows distinctive tetragonal distortion and a corresponding splitting of the d orbitals according to Figure 2.10 is to be expected.

3.2 Reactions of the Alkylcobalamins

3.2.1 One-Electron Reduction and Oxidation

Starting out from configuration (3.1), the trivalent cobalt (d^6) ion exists as a six-coordinate metal center with the corrin monoanion, an anionic axial group (e.g.

alkyl) and the neutral dimethylbenzimidazole base as ligands; a negatively charged phosphate in the side chain completes the charge balance. Starting from this configuration two one-electron reduction steps are possible, the kind and number of axial ligands determining the actual redox potentials [6]. The stepwise reduction of the metal from a d^6 to a d^8 configuration is accompanied by a tendency toward decreased axial coordination (3.3).

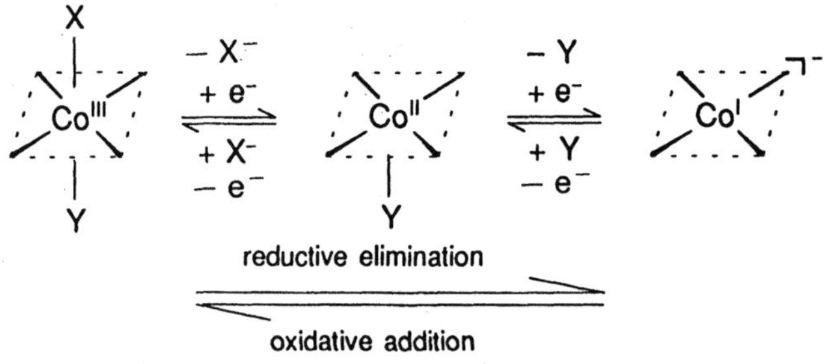

$$\tag{3.3}$$

One-electron reduction of Co^{III}–methylcobalamin thus leads to a more than 50% decreased Co—C bond strength and corresponding rate enhancement for the homolysis [7] due to a half-filled antibonding σ^*(Co—CH$_3$) orbital (d_{z^2} component). Excitation with light can also lead to a population of this σ^*(Co—CH$_3$) orbital and thus to a cleavage of the Co—C bond; however, this is probably not relevant for enzymatic reactions. Nonoccupation of the strongly antibonding $d_{x^2-y^2}$ orbital in a d^8 system such as Co^I favors the sterically less favorable square-planar configuration (see Figure 2.10). The stabilization of the Co^I state is a particular characteristic of the cobalt–corrin system; this state is not reversibly accessible under physiological conditions for cobalt–porphyrin complexes.

3.2.2 Co—C Bond Cleavage

The reactivity of the physiologically relevant alkyl cobalamins is characterized by the fact that the reactive alkyl groups are made available in a *controlled* fashion for follow-up reactions. Three formal alternatives (3.4) are conceivable for a bond cleavage Co┼CH$_2$R which may be induced by the interaction of the coenzymes with apoprotein and substrate.

Heterolytic bond cleavage can lead either to low-spin Co^{III} and a carbanion equivalent, $^-$C$\lessgtr$, involving substitution, e.g. by water, or to Co^I and a carbocationic alkyl moiety, $^+$C$\lessgtr$. In the latter case, a d^8-configurated metal center is formed which behaves as a σ electron-rich 'supernucleophile', i.e. its filled antibonding d_{z^2} orbital results in a high affinity towards σ electrophiles. A typical d^8 metal reactivity is thus the 'oxidative addition', e.g. of organic halogen compounds R—X. The carbanionic or carbocationic alkyl groups would not be produced as free ions but would be transferred in the presence of a reaction partner and the polar reaction medium in the transition state of the reaction.

The third alternative (3.4) is the homolytic bond cleavage which leads to paramagnetic, EPR (electron paramagnetic resonance) spectroscopically detectable Co^{II} with a low-spin d^7 configuration (*one* unpaired electron) and a primary alkyl radical.

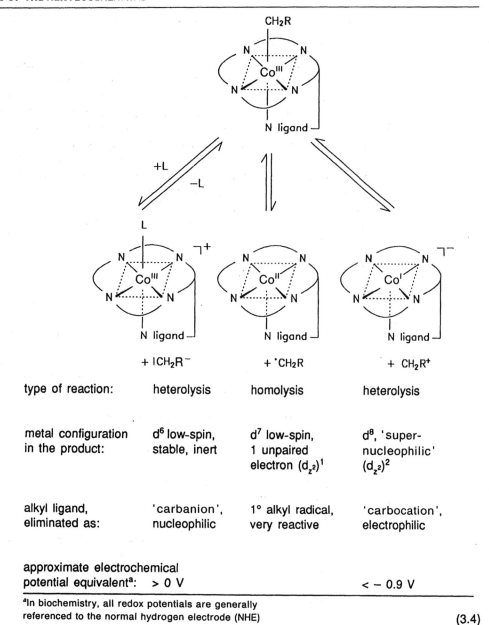

type of reaction:	heterolysis	homolysis	heterolysis
metal configuration in the product:	d^6 low-spin, stable, inert	d^7 low-spin, 1 unpaired electron $(d_{z^2})^1$	d^8, 'super-nucleophilic' $(d_{z^2})^2$
alkyl ligand, eliminated as:	'carbanion', nucleophilic	1° alkyl radical, very reactive	'carbocation', electrophilic
approximate electrochemical potential equivalent[a]:	> 0 V		< − 0.9 V

[a]In biochemistry, all redox potentials are generally referenced to the normal hydrogen electrode (NHE)

(3.4)

All three alternatives (3.4) are possible; axial coordination or noncoordination ('base-off') of the heterocyclic ligand, the nature of the substrate and the redox potential determine the actual reaction pathway. In the absence of a special base in the axial position, the carbanionic mechanism is realized at potentials above 0 V vs. the normal hydrogen electrode (NHE); the Co^I/carbocation cleavage occurs only below approximately −0.9 V. Homolytic bond cleavage is a viable reaction in the physiologically interesting potential range between 0 and −0.4 V, resulting in EPR detectable radical intermediates [8,9].

Electron Paramagnetic Resonance I

Electron paramagnetic resonance (EPR) or electron spin resonance (ESR) is an important spectroscopic method in bioinorganic chemistry [10,11]. The unpaired electron(s) in radicals or in complexes of transition metal centers with only partially filled d orbitals feature a spin, the orientation of which in a magnetic field can give rise (in the most simple case) to two energetically different states (spin as a binary quantum mechanical feature of electrons). The transition from the energetically favorable and more populated state (spin state I) to the high-energy and thus less populated state (spin state II), the 'resonance', can be induced by applying electromagnetic radiation, provided that certain spectroscopical selection rules are observed. With the normally used magnetic fields of a few tenths of one tesla (1 Tesla = 1000 G(auss)), this resonance requires microwave frequencies approximately 10^{10} Hz.

(a) one unpaired electron ($S = 1/2$)

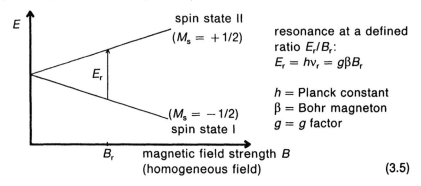

resonance at a defined ratio E_r/B_r:
$E_r = h\nu_r = g\beta B_r$

h = Planck constant
β = Bohr magneton
g = g factor

(3.5)

(b) one unpaired electron (electron spin $S = 1/2$), interacting with one proton (nuclear spin $I = 1/2$)

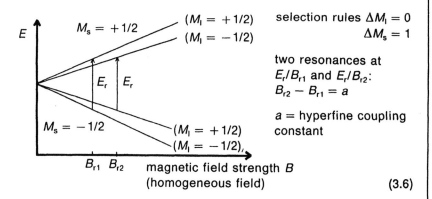

selection rules $\Delta M_I = 0$
$\Delta M_s = 1$

two resonances at E_r/B_{r1} and E_r/B_{r2}:
$B_{r2} - B_{r1} = a$

a = hyperfine coupling constant

(3.6)

Both the magnitude of the resonance field strength at a given frequency (g factor as the proportionality constant) and the hyperfine interaction(s) of the unpaired electron(s) with atomic nuclei that possess a (nuclear) spin are characteristic for the paramagnetic system. With this information it is often possible to determine oxidation and spin states and, under favorable circumstances, even details in the coordination sphere of metal centers. Since EPR spectroscopy is 'blind' to the diamagnetic bulk of a protein, this method is often used for the initial characterization of a newly isolated metalloenzyme. The sensitivity of the EPR technique is such that small effective concentrations suffice to detect electron spin-bearing centers, even in large proteins.

Due to the complexity and asymmetry of many paramagnetic centers in biological systems and because of an inherently high linewidth, the simple EPR method often yields only unresolved signals. In such cases the less sensitive but better resolving double resonance technique ENDOR (electron nuclear double resonance) is increasingly being used [12,13]. In this method the change in EPR intensity under saturation conditions is monitored as a function of the nuclear frequency.

If several unpaired electrons are present, their exchange behavior and the resulting exited states have to be taken into account (see Section 4.3). Consequences of a not spherically symmetrical, i.e. anisotropic distribution of the unpaired electron become evident in the solid state or when the mobility of the paramagnetic center is otherwise restricted (see EPR II in Section 10.1).

The very reactive primary carbon radicals formed after homolysis of RH_2C—Co-(corrin) bonds are short-lived and, therefore, not always detectable by EPR [8]. The remaining low-spin Co^{II} complex (d^7) features an EPR signal for *one* unpaired electron [9]. Interaction (coupling) of the electron spin occurs with the nuclear spin of the metal center (^{59}Co: 100% natural isotopic abundance, nuclear spin $I = 7/2$) and the nuclear spin of *one* nitrogen atom (^{14}N: 99.6% natural abundance, $I = 1$). These findings can be explained only if the unpaired electron is assumed to occupy the d_{z^2} orbital, interacting mainly with the single, axially coordinated nitrogen atom of the benzimidazole ligand. If the $d_{x^2-y^2}$ orbital were occupied by the unpaired electron, all four nitrogen centers of the macrocyclic corrin ligand would contribute with essentially similar nuclear spin/electron spin coupling [11]. The thus determined order of d orbitals corresponds to a relatively small distortion of octahedral symmetry (Figure 2.10) and is in accordance with the observed supernucleophilicity in the axial direction after double occupation of the d_{z^2} orbital (3.7).

radical scavenger: supernucleophile:

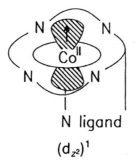

N ligand

$(d_{z^2})^1$

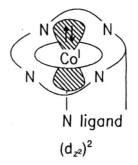

N ligand

$(d_{z^2})^2$ (3.7)

Despite convincing evidence from spectroscopy and mechanistic studies suggesting a radical mechanism, this hypothesis [14] has not remained undisputed [15]; however, radical formation and electron transfer may be combined according to the relation $R^{\bullet} + e^{-} \rightarrow R^{-}$ to resemble carbanion reactivity.

3.2.3 Mutase Activity of Coenzyme B₁₂

According to the reactivity modes mentioned in (3.4), cobalamin-containing enzymes participate in redox reactions, alkylations and in radical-induced rearrangements, especially in highly stereospecific 1,2-shifts (3.8) at saturated hydrocarbon centers. Such 'mutases' are generally large, coenzyme B₁₂-dependent enzymes.

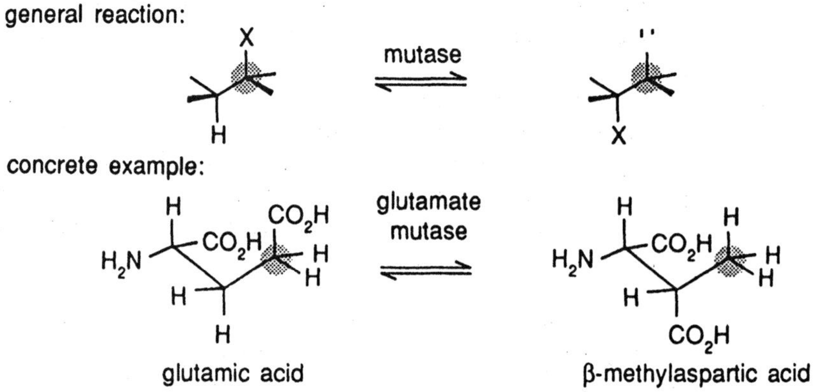

general reaction:

concrete example:

glutamic acid   β-methylaspartic acid

(3.8)

An overview and mechanistic interpretation of mutase activity are given in Table 3.1 and formula (3.10); in a larger context these reactions include the reduction of ribonucleotides to their corresponding deoxy forms in certain bacteria. For the latter,

Table 3.1 Reactions requiring coenzyme B₁₂

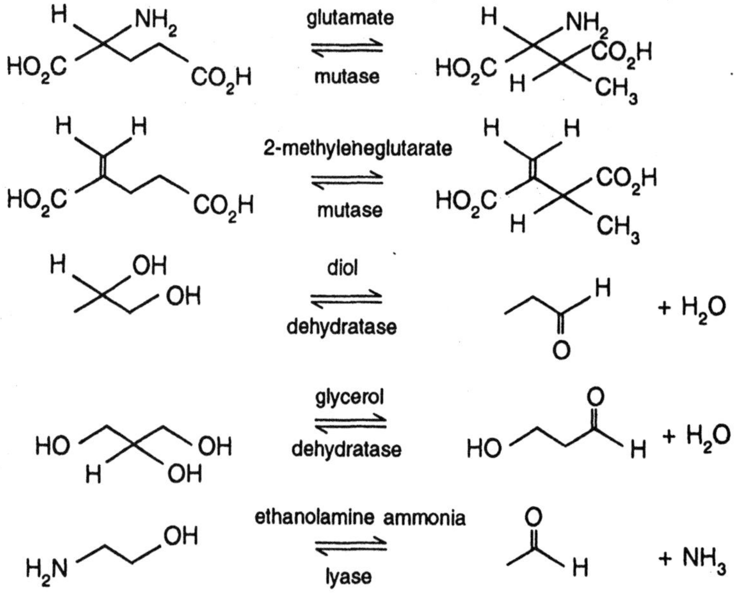

Table 3.1 (*continued*)

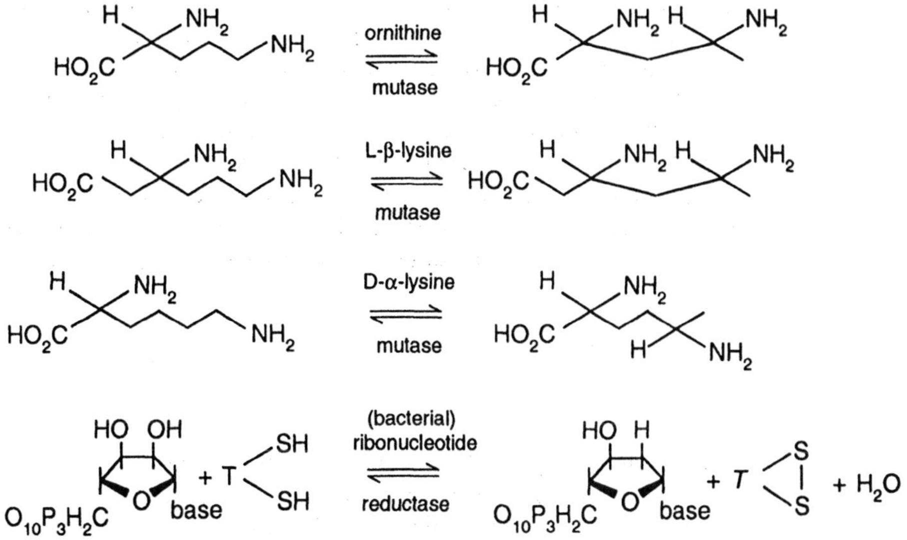

ornithine mutase

L-β-lysine mutase

D-α-lysine mutase

(bacterial) ribonucleotide reductase

$T(SH)_2$: dithiol, e.g. lipoic acid (3.12) or thioredoxin proteins

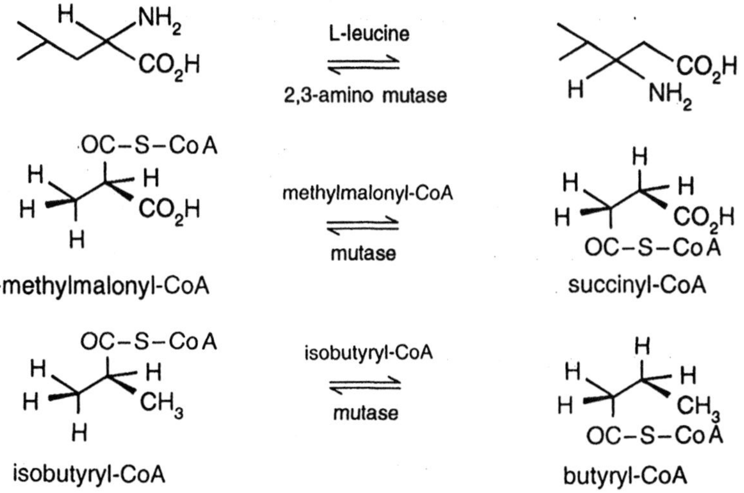

L-leucine 2,3-amino mutase

methylmalonyl-CoA mutase

R-methylmalonyl-CoA

succinyl-CoA

isobutyryl-CoA mutase

isobutyryl-CoA

butyryl-CoA

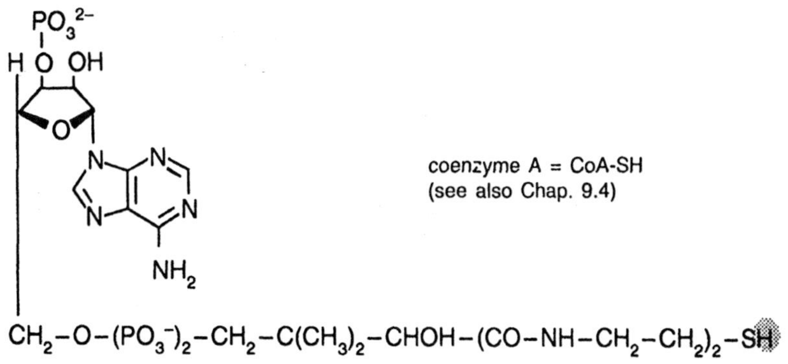

coenzyme A = CoA-SH
(see also Chap. 9.4)

$CH_2-O-(PO_3^-)_2-CH_2-C(CH_3)_2-CHOH-(CO-NH-CH_2-CH_2)_2-SH$

biochemically very important reaction, other organisms and bacteria such as *Escherichia coli* use manganese- or iron-containing ribonucleotide reductases which, however, also require radicals for proper function ([16]; compare Section 7.6.1). Most reactions catalyzed by coenzyme B$_{12}$ are restricted to microorganisms which are also able to synthesize this coenzyme. For mammals, the methylmalonyl-*CoA* mutase is particularly important; it is required for the metabolism of amino acids in the liver and its absence due to genetic defects is lethal [17].

Many dehydratases, desaminases and lyases are coenzyme B$_{12}$-dependent enzymes since a 1,2-shift in 1,2-diols or 2-aminoalcohols leads to geminal (1,1)-isomers which readily lose water or ammonia to form carbonyl compounds (3.9).

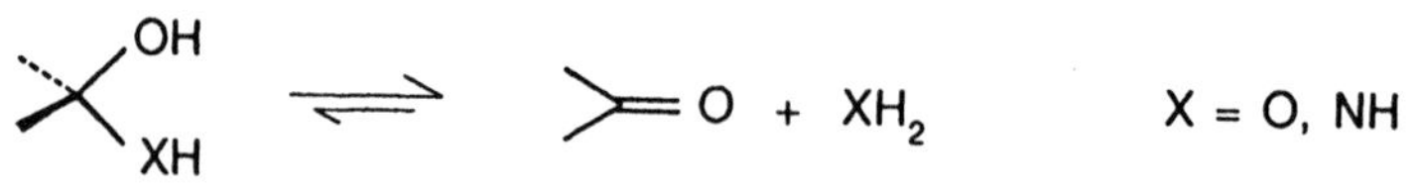

$$X = O, NH$$

$$(3.9)$$

In synthetic organic chemistry the deceptively simple 1,2-shifts (3.8) are not easily accomplished; accordingly, there is much interest in the use of alkylcobalt–corrin complexes and corresponding model compounds for organic synthesis, even if substrate and stereospecificity are not guaranteed in the absence of the apoenzyme [18]. Numerous results from spectroscopic (EPR) and mechanistic studies (e.g. isotopic labeling) now point to the following, radical-based reaction cycle (3.10):

initiation:

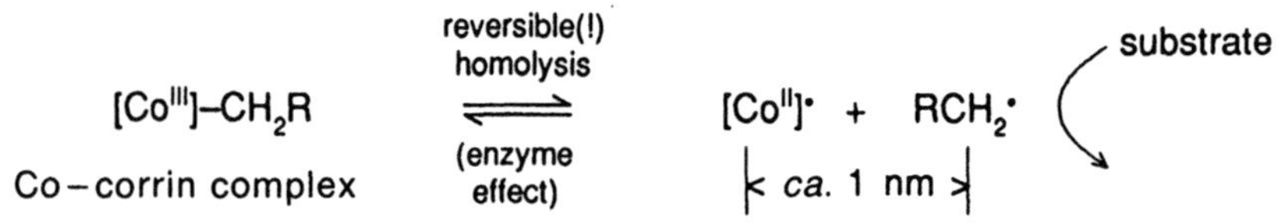

general radical mechanism:

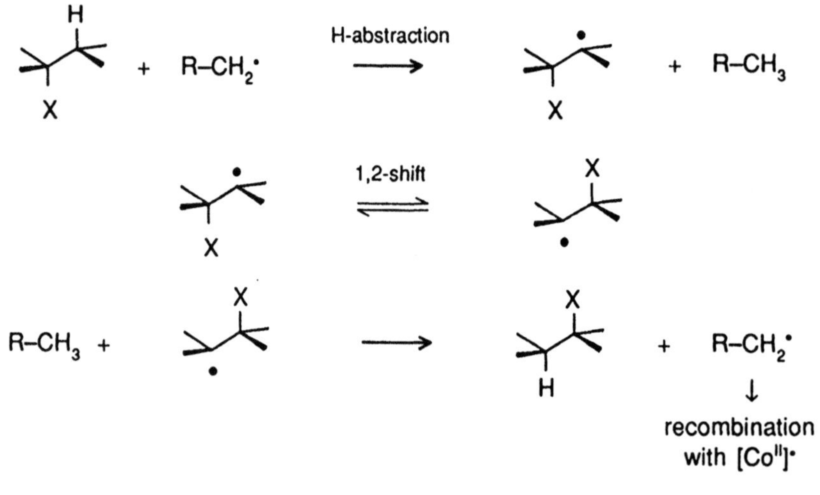

example:

$$R'-CH-CH_2OH \quad \rightarrow \quad R'-CH_2-CHO \ + \ H_2X$$
$$\underset{XH}{|}$$

X = O, NH
R = 5'-deoxyadenosyl

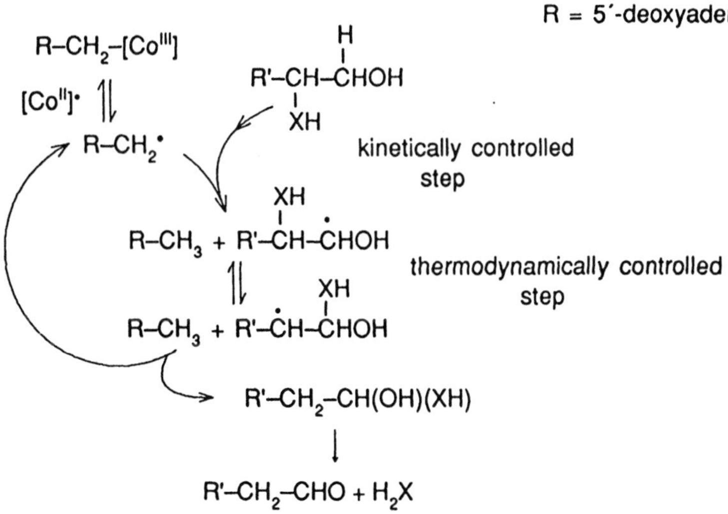

$$(3.10)$$

In a reversible step which requires only little activation for the Co—C bond dissociation (about 100 kJ/mole in the coenzyme and not more than 65 kJ/mole in the actual enzyme) an adenosyl radical is produced homolytically from the alkyl-cobalamin. This primary alkyl radical can *selectively* [19] attack an exposed H—C center in a kinetically, i.e. activation energy-controlled step and abstract a hydrogen atom which is the typical behavior of reactive alkyl radicals. The second step is the actual rearrangement (1,2-shift) which may be determined by the equilibrium position, for instance favoring a secondary alkyl radical over a primary one. The reabstraction of a hydrogen atom from enzyme-bound 5'-deoxyadenosine by such a secondary substrate radical inside the 'radical cage' system would lead to a rearranged reaction product which, in the case of a geminal diol, rapidly loses water and forms the carbonyl compound. Elimination of water, however, can also be conceived for the radical itself; 1,2-dihydroxyalkyl radicals tend to lose OH^- or H_2O (after protonation) under formation of carbonylalkyl radicals. If not immediately used, the 5'-deoxyadenosyl radical can recombine with the Co^{II} system so that alkylcobalamins can be described as *reversibly acting radical carriers* (3.10) [14].

A direct participation of the cobalt complex in the actual rearrangement is not assumed, providing the alkyl radical is its main function [16]. Incidentally, the radical-induced transformation of glycols to aldehydes is not unknown in organic synthesis; such reactions can be triggered by hydroxyl radicals generated from Fenton's reagent:

$$Fe^{2+} \ + \ H_2O_2 \quad \rightarrow \quad Fe^{3+} \ + \ OH^- \ + \ OH^\bullet$$

$$(3.11)$$

The reactions that can be interpreted via the mechanistic hypothesis (3.10) are listed in Table 3.1. The protein in the actual B$_{12}$-dependent enzymes has at least three functions. (i) After substrate binding it effects a drastic attenuation of the Co—C bond energy in an as yet unknown manner (conformational changes, electron transfer?), causing an acceleration of this initial reaction step by a factor of more than 10^{10}. (ii) It protects the reactive primary alkyl radical from the multitude of other, undesired reactants (negative catalysis [20]). (iii) Finally, the protein guarantees a high stereoselectivity of the isomerization.

B$_{12}$-dependent ribonucleotide reductase requires coenzymatic dithiol which is oxidized to disulfide (compare lipoic acid in 3.12 and Section 7.6.1).

Organic Redox Coenzymes

Organic coenzymes, particularly those with a redox function, often interact with inorganic cofactors such as metal ions or their complexes. As already mentioned in the introduction, the modern artificial distinction between organic and inorganic molecules has no meaning for organisms; the important point is the relation between biosynthetic requirements and functional benefits.

Compounds (3.12) are among the most common organic redox coenzymes; they can be roughly divided into *N*-heterocycles, quinonoid systems and sulfur compounds. The 5,6,7,8-tetrahydro form of folic acid (which is related to the flavin systems) is an important carrier of C$_1$ fragments.

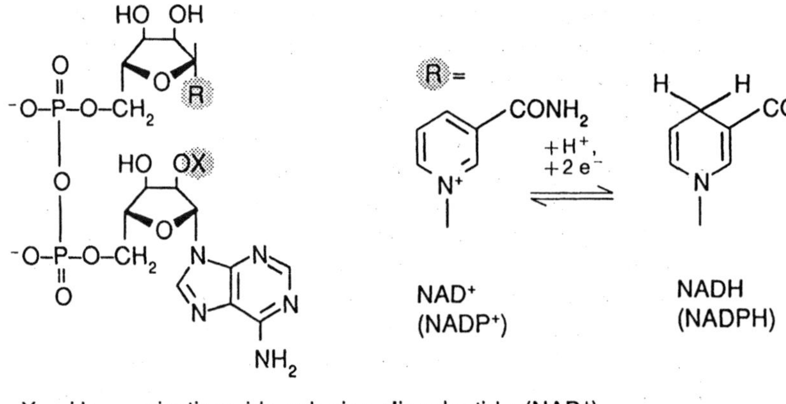

X = H: nicotinamide adenine dinucleotide (NAD$^+$)
X = PO$_3^{2-}$: nicotinamide adenine dinucleotide phosphate (NADP$^+$)

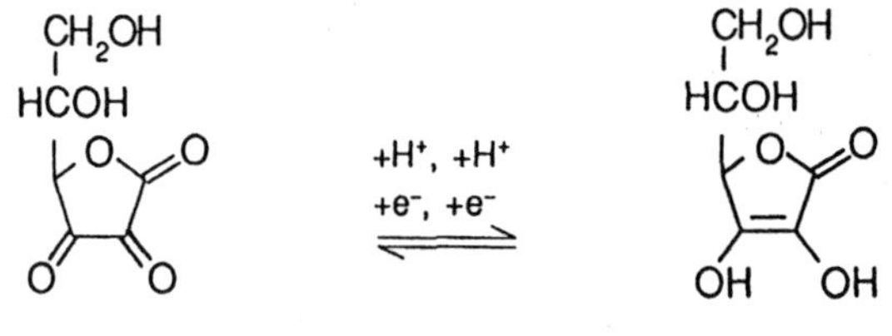

dehydroascorbic acid   ascorbic acid (vitamin C)

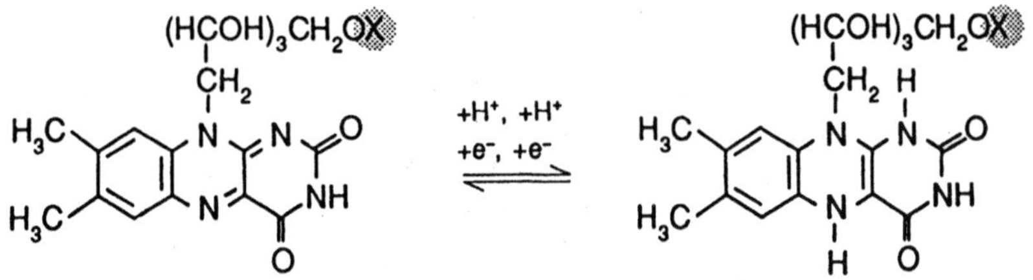

X = H: riboflavin (vitamin B$_2$, see 2.2)
X = PO$_3^{2-}$:   FMN (flavin mononucleotide)
X = adenosine diphosphate:
                  FAD (flavin adenine dinucleotide)

5,10-dihydroriboflavin
(FMNH$_2$)

(FADH$_2$)

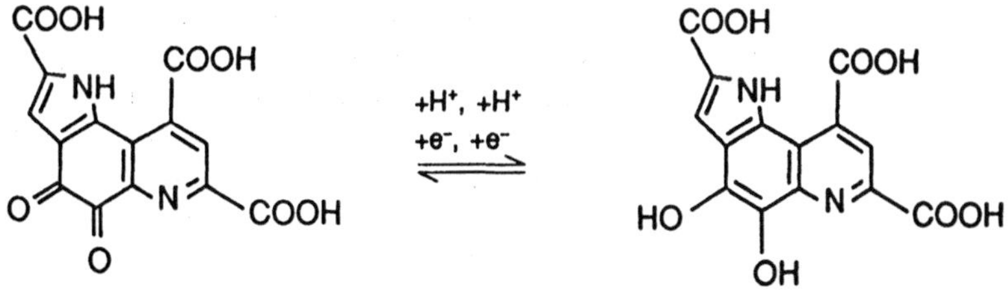

methoxatin (o-quinone form)
cofactor PQQ

(catechol form)

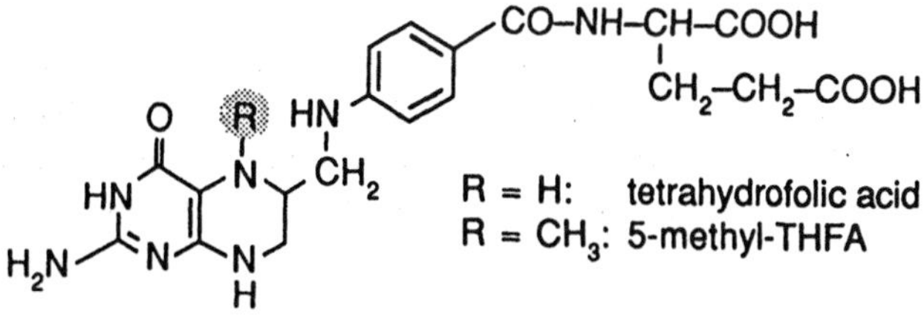

R = H: tetrahydrofolic acid (THFA)
R = CH$_3$: 5-methyl-THFA

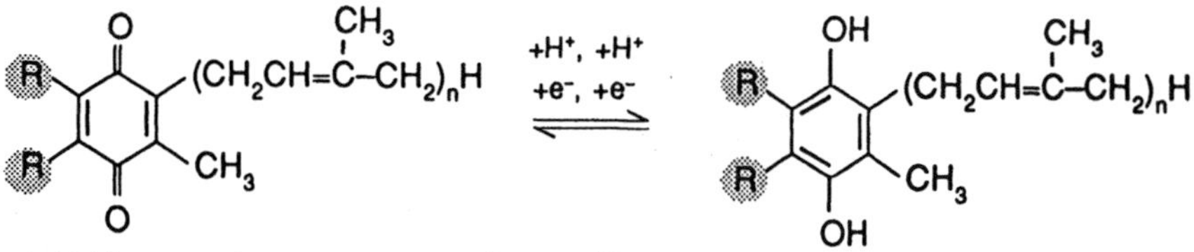

2 R = (CH)$_4$, n = 9:   menaquinone (Q$_a$)
R = OCH$_3$, n = 2–10:   ubiquinones (Q$_b$)
R = CH$_3$, n = 6–10:   plastoquinones (PQ)
R = H, n = 4–7:   vitamin K group

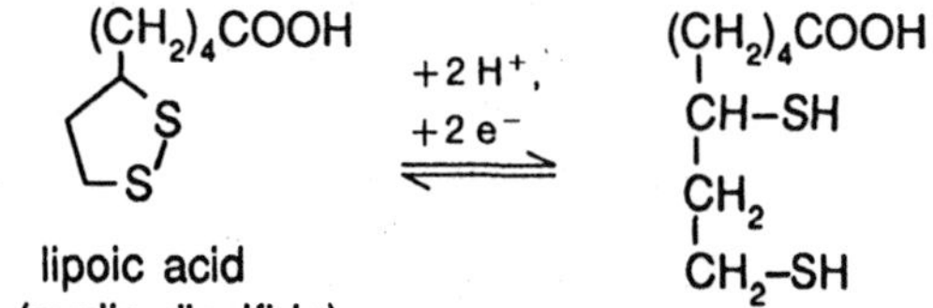

hydroquinone form

lipoic acid
(cyclic disulfide)

(dithiol form)

(3.12)

3.2.4 Alkylation Reactions of Methylcobalamin

Methylcobalamin-induced 'bio'-methylations are of great importance for microbiology, biosynthesis and toxicology (compare Sections 17.3 and 17.5). While methyl groups with an electrophilic character, with a 'positive partial charge', are biochemically available through the sulfonium species S-adenosyl methionine, (adenosyl)-$(CH_3)S^+(CH_2-CH_2-CH(NH_3^+)CO^-)$, or through 5-methyltetrahydrofolic acid (3.12), the methylation of electrophilic *substrates* typically requires an organometallic compound that can react either in a carbanionic fashion (S_N2 reaction) or as a radical, involving a single-electron transfer process (compare 3.4). The methylation of compounds of less electropositive and 'soft' elements like selenium or mercury (3.13) with oxidation potentials $E_0 > 0$ V presumably occurs via a carbanionic mechanism, while less noble elements such as arsenic or tin ($E_0 < 0$ V) are alkylated in their compounds via radical pathways (see Section 17.3). In many instances, very toxic species like the methylmercury cation $(CH_3)Hg^+$ are formed by these reactions (3.13). If sufficiently stable under physiological conditions, such mixed hydrophilic/hydrophobic organometallic cations are able to penetrate the blood–brain barrier and deactivate sulfur-containing enzymes (Section 17.5). In contrast to coenzyme B_{12}, methylcobalamin is found in the circulatory system of mammals rather than in the liver.

$$Hg^{2+} + CH_3-[Co^{III}] \longrightarrow (CH_3)Hg^+ + [Co^{III}]^+ \qquad (3.13)$$

A biologically valuable methylation requiring cobalamin-dependent enzymes has been established for the substrate homocysteine (*homo*: extended by one CH_2 chain link); the essential and often 'limiting' amino acid methionine is biosynthesized in this reaction (methionine synthetase from *E. coli*; 3.14 [21]).

$$
\begin{array}{c}
H_3N^+ \\
| \\
CH-CH_2-CH_2-S^- + \text{5-methyl-THFA} + H^+ \\
| \\
{}^-OOC \quad \text{homocysteine}
\end{array}
\xrightarrow{\begin{array}{c} CH_3 \\ | \\ [Co^{III}]/[Co^I]^- \end{array}}
$$

$$
\begin{array}{c}
H_3N^+ \\
| \\
CH-CH_2-CH_2-S-CH_3 \quad + \quad \text{THFA} \\
| \\
{}^-OOC \quad \text{methionine} \qquad \text{(see 3.12)}
\end{array}
$$

$$(3.14)$$

In microorganisms, especially in 'acetogenic' or 'methanogenic' bacteria which produce acetic acid or methane (see Figure 1.1), methyl-transferring 'corrinoid', i.e. cobalt–corrin-containing enzymes, are of great importance. During bacterial CO_2-fixation they participate in the catalytic formation of acetyl-CoA as 'activated acetic acid' (3.15) [22], a process that involves a nickel enzyme-requiring carbonylation (compare Section 9.4) and a methyl group transfer from a 5-methyltetrahydropterin (Pter-N_5)—CH_3 to coenzyme A (see Table 3.1) via a methylcobalt–corrin intermediate.

$$CO_2 + CO + 6 \; 'H' + HS-CoA \; \rightleftharpoons \; CH_3C(O)S-CoA + 2 \; H_2O$$

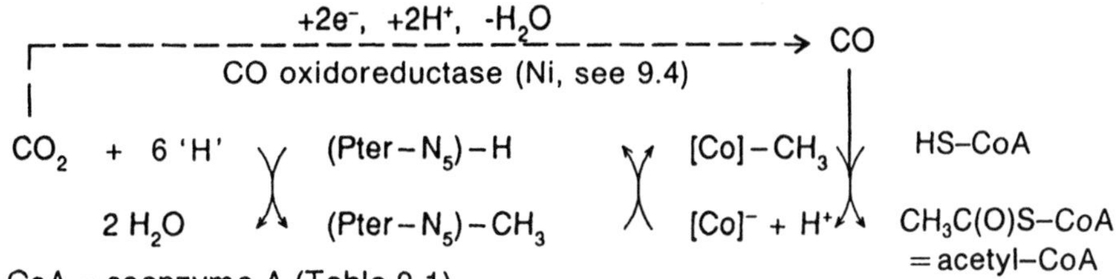

$$(3.15)$$

The nickel-containing porphinoid 'factor 430' is directly involved in the actual methane formation (see Section 9.5); however, cobalt–corrinoid-containing membrane proteins in various coordination and oxidation states were detected in microorganisms [23]. These corrinoid centers are presumably involved in the synthesis of coenzyme M ($H_3CSCH_2CH_2SO_3{}^-$; compare Section 9.5), the last methyl carrier before eventual methane production.

Very little is known about noncorrinoid cobalt enzymes. A possible cobalt–porphinoid factor has been described in the context of decarbonylation of long-chain aldehydes (3.16 [24]).

$$R(CH_2)_nCH_2CHO \longrightarrow R(CH_2)_nCH_3 + CO \qquad (3.16)$$

This transformation which would not be untypical for cobalt- or nickel-containing catalysts (see Section 9.4) is important for the biosynthesis of paraffin hydrocarbons ($\rightarrow$ water proofing, e.g. of leaves or feathers); the long-chain aldehyde precursors can be formed via peroxidase-catalyzed reactions (see 6.13).

3.3 Model Systems and the Role of the Apoenzyme

Suitable model systems of the cobalamins are of considerable interest because of the organic–synthetic attractiveness of coenzyme B_{12}-catalyzed reactions; the actual cobalamins are very sensitive and exhibit only limited solubility. Schrauzer [25,26] realized quite early that bis(diacetyldioxime) complexes ('cobaloximes'; 3.16) represent good models for B_{12} systems; these complexes contain a 'quasi-macrocyclic' chelate ring structure as a consequence of two strong hydrogen bonds. Even better suited with respect to redox properties are the 'Costa complexes' with covalently linked α-diimine moieties [27]. Other model complexes contain the chelate ligand bis(salicyl-aldehyde)ethylenediimine (*salen*; 3.17) which also binds cobalt in a square planar fashion.

All these model complexes as well as the cobalt-containing porphyrins have as disadvantage that the supernucleophilic Co^I state is much less stable under physiological conditions than that of the cobalt–corrin systems. The 'cobester' complexes which lack the nucleotide part of the B_{12} systems fare better in this regard but are also harder to synthesize [6].

cobaloxime complex	a Costa complex	*salen* ligand

B: base   (3.17)

The value of model systems for mechanistic considerations has been evident from comparative studies of cobalt porphyrins and of cobaloximes as cobalt–corrin model compounds with axially coordinated alkyl and organophosphine ligands, PR$_3$, of different size and basicity [28]. The results suggest that the enzymatically relevant activation, e.g. benzyl–cobalt bond cleavage, is influenced by electronic effects such as organophosphine basicity in the porphyrin complexes, but by steric requirements of these axial PR$_3$ ligands in cobaloximes. This observation seems to confirm the significance of the nonplanarity of cobalt–corrin complexes and lent support to the assumption that radical formation in the enzyme is sterically controlled (→ entatic state formation); after all, B$_{12}$-containing enzymes react faster by more than ten orders of magnitude than the corresponding free B$_{12}$ cofactors. Additional activation of Co—C bond cleavage can possibly occur through binding of the nucleotide part of the 5′-deoxyadenosyl group to the enzyme.

Structural data from Kräutler, Keller and Kratky [29] have exhibited an essentially unchanged geometry for Co^{II} binding to the corrin ring in cobalamin as compared to Co^{III} analogues, which enhances the electron transfer rate due to a small reorganization energy and negligible conformational changes (see Section 6.1). However, the bond to the axial dimethylbenzimidazole ligand is shortened in the Co^{II} state, leading to a more pronounced *out-of-plane* situation for the metal center and a corresponding weakening of the Co—C bond. A similar, better-known example for the cooperativity of both axial ligands X and Y (2.6) is the reversible binding of O$_2$ to heme-iron centers (see Chapter 5).

Despite extensive knowledge of the structure and reactivity of the B$_{12}$ coenzymes and their model compounds rather little is known about the conditions in the actual enzymes and about details of the reaction mechanisms [30]; the catalytic cycles shown here (3.10 and 3.15) still have only model character. It seems to be clear, however, that the unique function of the element cobalt in B$_{12}$ systems is its tolerance of a bond from the redox-active transition metal to primary alkyl groups that can be set free through well-defined activation processes and act as specific, reactive, and reversibly transferable species.

References

1. P. J. Toscano and L. G. Marzilli, B$_{12}$ and related organocobalt chemistry: formation and cleavage of cobalt carbon bonds, *Prog. Inorg. Chem.*, **31**, 105 (1984).
2. D. Dolphin (ed.), *B$_{12}$*, Wiley, New York, 1982.
3. Z. Schneider and A. Stroinski, *Comprehensive B$_{12}$*, de Gruyter, Berlin, 1987.

4. K. Folkers, Perspectives from research on vitamins and hormones, *J. Chem. Educ.*, **61**, 747 (1984).

5. P. G. Lenhert, The structure of vitamin B_{12}. VII. The X-ray analysis of the vitamin B_{12} coenzyme, *Proc. Roy. Soc. A*, **303**, 45 (1968).

6. B. Kräutler, Thermodynamic trans-effects of the nucleotide base in the B_{12} coenzymes, *Helv. Chim. Acta*, **70**, 1268 (1987).

7. B. D. Martin and R. G. Finke, Methylcobalamin's full- vs. half-strength cobalt-carbon σ bonds and bond dissociation enthalpies: $A > 10^{15}$ Co—CH_3 homolysis rate enhancement following one-antibonding-electron reduction of methylcobalamin, *J. Am. Chem. Soc.*, **114**, 580 (1992).

8. Y. Zhao, P. Such and J. Retey, Detection of radical intermediates in the coenzyme-B_{12}-dependent methylmalonyl-CoA-mutase reaction by ESR spectroscopy, *Angew. Chem. Int. Ed. Engl.*, **31**, 215 (1992).

9. C. Michel, S. P. J. Albracht and W. Buckel, Adenosyl-cobalamin and cob(II)alamin as prosthetic groups of 2-methyleneglutarate from *Clostridium barkeri*, *Eur. J. Biochem.*, **205**, 767 (1992).

10. J. A. Weil, J. R. Bolton and J. E. Wertz, *Electron Paramagnetic Resonance: Elemental Theory and Practical Applications*, Wiley, New York, 1993.

11. M. Symons, *Chemical and Biochemical Aspects of Electron-Spin Resonance Spectroscopy*, van Nostrand Reinhold, New York, 1978.

12. H. Kurreck, B. Kirste and W. Lubitz, *Electron Nuclear Double Resonance Spectroscopy of Radicals in Solution*, VCH, Weinheim, 1988.

13. B. M. Hoffman, Electron nuclear double resonance (ENDOR) of metalloenzymes, *Acc. Chem. Res.*, **24**, 164 (1991).

14. J. Halpern, Mechanisms of coenzyme B_{12}-dependent rearrangements, *Science*, **227** 869 (1985).

15. P. Dowd, B. Wilk, B. K. Wilk, First hydrogen abstraction-rearrangement model for the coenzyme B_{12}-dependent methylmalonyl-CoA to succinyl-CoA carbon skeleton rearrangement reaction, *J. Am. Chem. Soc.*, **114**, 749 (1992).

16. J. Stubbe, Radicals in biological synthesis, *Biochemistry*, **27**, 3893 (1988).

17. P. Zurer, Error charged in infant's death, mother's jailing, *Chem. Eng. News*, 18, 26 August 1991.

18. G. Pattenden, Cobalt-mediated radical reactions in organic synthesis, *Chem. Soc. Rev.*, **17**, 361 (1988).

19. H. Fischer, Unusual selectivities of radical reactions by internal suppression of fast modes, *J. Am. Chem. Soc.*, **108**, 3925 (1986).

20. J. Retey, Reaction selectivity of enzymes through negative catalysis or how enzymes work with highly-reactive intermediates, *Angew. Chem. Int. Ed. Engl.*, **29**, 355 (1990).

21. R. G. Matthews and J. T. Drummond, Providing one-carbon units for biological methylations, *Chem. Rev.*, **90**, 1275 (1990).

22. G. Diekert, CO_2 reduction to acetate in anaerobic bacteria, *FEMS Microbiol. Rev.*, **87**, 391 (1990).

23. M. D. Wirt, M. Kumar, S. W. Ragsdale and M. R. Chance, X-ray absorption spectroscopy of the corrinoid/iron–sulfur protein involved in acetyl coenzyme A synthesis by *Clostridium thermoaceticum*, *J. Am. Chem. Soc.*, **115**, 2146 (1993).

24. M. Dennis and P. E. Kolattukudy, A cobalt-porphyrin enzyme converts a fatty aldehyde to a hydrocarbon and CO, *Proc. Natl. Acad. Sci. USA*, **89**, 5306 (1992).

25. G. N. Schrauzer, Late developments in the vitamin B_{12} field: reactions of cobalt atoms in corrin derivatives and vitamin B_{12} model, *Angew. Chem. Int. Ed. Engl.*, **15**, 417 (1976).

26. G. N. Schrauzer, Recent developments in the vitamin B_{12} field: enzyme reactions depend on simple corrins and coenzyme B_{12}, *Angew. Chem. Int. Ed. Engl.*, **16**, 233 (1977).

27. G. Costa, G. Mestroni and L. Stefani, Organometallic derivatives of cobalt(III) chelates of bis(salicylaldehyde)ethylenediimine, *J. Organomet. Chem.*, **7**, 493 (1967).

28. M. K. Geno and J. Halpern, Why does nature not use the porphyrin ligand in vitamin B_{12}?, *J. Am. Chem. Soc.*, **109**, 1238 (1987).

29. B. Kräutler, W. Keller and C. Kratky, Coenzyme B_{12} chemistry: the crystal and molecular structure of cob(II)alamin, *J. Am. Chem. Soc.*, **111**, 8936 (1989).

30. B. T. Golding, The B_{12} mystery, *Chem. Br.*, **26**, 950 (1990).

4 Metals at the Center of Photosynthesis: Magnesium and Manganese

In addition to their well-established functions in photosynthesis, the divalent forms of the main group element magnesium and of the transition metal manganese [1] are important as centers of hydrolytic and phosphate-transferring enzymes (Section 14.1). Moreover, higher-oxidized (+III, +IV?) manganese plays a role as redox center in several enzymes [2], including certain forms of ribonucleotide reductase (see Section 7.6.1), catalase and peroxidase (see Section 6.3) and in the particular superoxide dismutase of mitochondria [3] (see Section 10.5). Iron and copper centers are also prominently involved in the overall photosynthetic process, viz. by contributing to directed electron transfer along or across membrane proteins. However, this chapter, which deals with arguably the most important chemical reaction for life on earth, restricts itself to two fundamental parts of photosynthesis: firstly, the absorption of light and the ensuing charge separation originating from magnesium-containing chlorophylls and, secondly, the manganese-catalyzed oxidation of water to oxygen ('dioxygen', O_2) in cyanobacteria, algae and higher plants.

4.1 Volume and Total Efficiency of Photosynthesis

Increasing public awareness of our dependence on fossil fuels and of the steadily rising accumulation of carbon dioxide in the atmosphere due to the combustion of those fuels have been responsible for intensified efforts to understand the molecular workings of photosynthesis. This chemical process, often summarized as in (4.1), is fundamental to the existence of higher forms of life on earth; the production of reduced carbon compounds ('organic' material, including fossil fuels) on one hand and the production of oxygen on the other hand are based on this energy-consuming process.

$$H_2O + CO_2 \underset{\substack{\text{respiration} \\ \text{(downhill catalysis)}}}{\overset{\substack{\text{photosynthesis} \\ \text{(uphill catalysis)}}}{\rightleftharpoons}} 1/n\,(CH_2O)_n + {}^3O_2$$

$$\Delta H = +470 \text{ kJ/mole}$$

(4.1)

Progress in many diverse areas of science such as protein crystallography [4], pico- and femtosecond laser spectroscopy [5,6] or high-resolution magnetic resonance [7] has revealed numerous details of the many single reaction steps that are involved in the highly complex photosynthetic process. Accordingly, recent Nobel prizes in chemistry have been awarded for the structural elucidation of a bacterial photosynthetic reaction center (1988; J. Deisenhofer, R. Huber and H. Michel [4]) and for the theoretical description of the underlying electron transfer processes (1992; R. A. Marcus [8]). For the chemical sciences, an understanding of photosynthesis would be particularly attractive within the context of model systems since valuable (high-energy) substances are produced in these reactions from very simple, low-energy starting materials by using a readily available, innocuous and rather 'diluted' form of energy. However, all hitherto constructed model systems have shown only moderate success in mimicking partial processes of photosynthesis. The reason lies in the very demanding requirements for an 'uphill-catalysis', which also explains the high degree of complexity of the photosynthetical 'apparatus' in biology [9].

Certain bacteria and algae are photosynthetically active as well as green plants. Purple bacteria like *Rhodopseudomonas viridis* possess a comparatively simple photosynthetical apparatus without the ability to oxidize water. These bacteria use photosynthesis primarily to separate charges and thus create a transmembrane proton gradient (pH difference) which is then used to synthesize high-energy adenosine triphosphate (ATP) from adenosine diphosphate (ADP-phosphorylation, 14.2). Other, anaerobic bacteria use the redox equivalents to oxidize substrates like hydrogen sulfide (H_2S) or dihydrogen (H_2) instead of water.

In plants, the primary photosynthetical events take place in the highly folded, disc-shaped thylakoid membrane vesicles inside chloroplasts (see Figure 4.8 [10]), and even in simple bacteria the process is membrane-spanning (compare Figure 4.2 a,b). Since immobilization and a defined orientation of pigments and reaction centers are crucial for the success of photosynthesis, all chlorophyll molecules (which differ slightly with regard to some substituents, 2.5, 4.2) feature a long aliphatic phytyl side chain for anchoring these pigments in the hydrophobic phospholipid membrane with its thickness of about 5 nm (compare Figure 13.5).

The photosynthetic output of green plants in normal sunlight is usually assumed to be about 1 g of glucose per hour per 1 m^2 of leaf surface area. In a global context, the photosynthetically active algae (phytoplankton) also play an important role as the water coverage of the earth is about 71%. Even though the total efficiency of photosynthesis is, on the average, much less than 1% if measured as the production of fuel equivalents in comparison to the available radiation energy, the total global turnover is tremendous: about 200 billion tons of carbohydrate equivalents $(CH_2O)_n$ are produced from CO_2 each year [11]. Incidentally, the efficiency of the primary, 'physical' energy transformation in photosynthesis from incident light to transmembrane redox potential differences (about 20%) is comparable to that of very good photovoltaic elements.

4.2 Primary Processes in Photosynthesis

What are the main requirements for photosynthesis and what are the roles of the inorganic components?

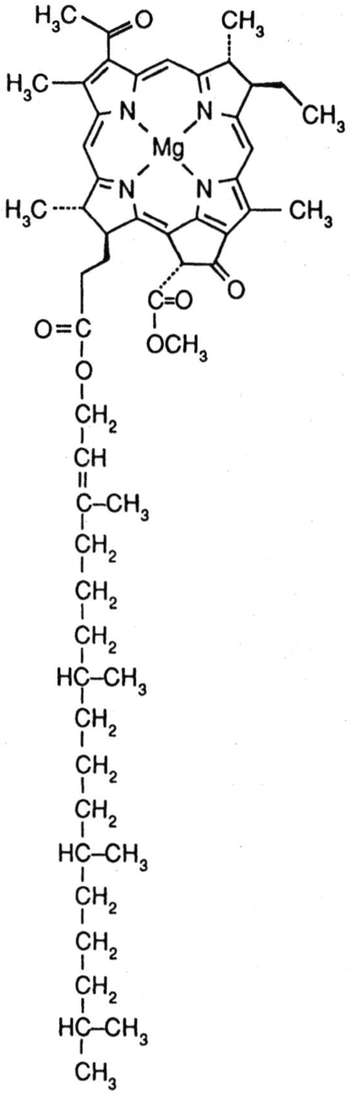

bacteriochlorophyll *a*   (4.2)

4.2.1 Light Absorption (Energy Acquisition)

The sunlight which is available at the earth's surface includes the wavelength range visible to the human eye from about 380 to 750 nm; however, a considerable number of the incoming solar photons also features higher wavelengths in the near infrared region, up to more than 1000 nm. An efficient photosynthetic transformation of light of this energy (1.24–3.26 eV) requires the absorption of as many photons as possible. This condition is fulfilled through the presence of different organic pigments, including chlorophyll molecules, which are positioned in a highly folded membrane with an inherently large inner surface area and therefore a high cross-section for

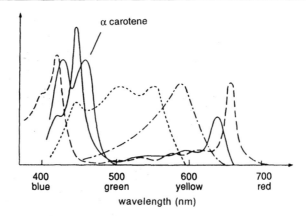

Figure 4.1
Absorption spectra of various pigments from algae and plants (according to [11]): chlorophyll
a (– –), chlorophyll b (——), α-carotene (——), phycocyanin (–·–·), phycoerythrin (– – – –)

photon capture (compare Figure 4.2). Chlorophylls themselves contain a conjugated
tetrapyrrole π system (2.5, 4.2) which shows a high absorptivity with molar extinction
coefficients of about $10^5 \, M^{-1} \, cm^{-1}$ at both the long- and short-wavelength ends
of the visible spectrum. The complementary colors blue (after long-wavelength
absorption) and yellow (after short-wavelength absorption) combine to form the
typical green color, e.g. of fresh leaves. Starting from a completely unsaturated
porphyrin π system (2.4), the partial hydrogenation of pyrrole rings leads to a shift
of photon absorption to longer wavelengths. Bacteriochlorophylls, which contain
two partially hydrogenated pyrrole rings in contrast to the 'normal' chlorophylls
with only one dihydropyrrole ring (2.5), absorb light at particularly low energy in
the near infrared region. Carotenoids and open-chain tetrapyrrole molecules, e.g.
phycobilins, complement the chlorophyll pigments [12] so that a broad spectral
absorption range is covered (Figure 4.1); the separation of these 'leaf pigments'
marked the beginning of chromatography (M. Tswett, 1906). At the end of each
growth period the nongreen carotenoid leaf pigments became visible (autumn colors)
after the disintegration of chlorophyll, which is quite unstable in its free, unprotected
state.

4.2.2 Exciton Transport (Directed Energy Transfer)

The absorption of energy in the form of photons by the pigments requires less than
10^{-15} s and yields short-lived electronically excited (singlet) states which, in principle,
can produce a charge separation. In view of the rather low photon density in diffuse
sunlight (rate of absorption less than one photon per pigment molecule per second)
and the necessarily rapid charge separation process it is more economical to use the
major part (>98%) of the chlorophyll molecules to act as 'antenna' devices and
collect available photons ('light-harvesting'). This means, however, that there has to
be efficient and spatially oriented ('vectorial') transfer of the absorbed energy in the
form of excited states ('excitons') to the actual reaction centers which contain less
than 2% of the total chlorophyll content.

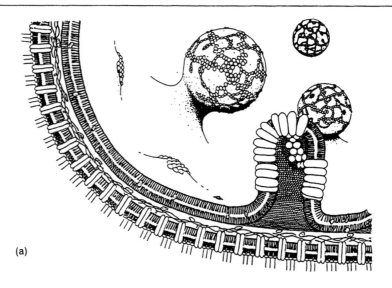

(a)

(b)

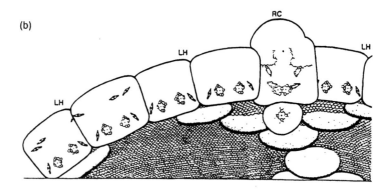

(c)

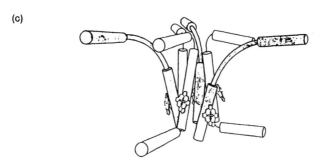

Figure 4.2
(a) Model of the photosynthetic membrane of *Rhodobacter sphaeroides* with invaginations that contain the photosynthetic centers; (b) a model for the arrangement of bacteriochlorophyll molecules in light-harvesting (LH) and reaction center (RC) protein complexes within a membrane indentation; (c) model structure for a typical light-harvesting protein complex (cylindrically depicted protein helices and chlorophyll 'discs'; other pigments not shown) (from [13])

The energy transfer which requires neither mass nor charge movement is made possible by a special arrangement (see Figure 4.2) of many chlorophyll chromophors in a network of 'antenna pigments' [13,14]. These chromophors are arranged in spatial proximity and a certain, well-defined orientation; they are able to 'funnel' the light energy to the actual reaction centers with about 95% efficiency within 10–100 ps. In physical terms this Förster mechanism of 'resonance transfer' proceeds via spectral overlap of emission bands of the exciton source with absorption bands of the exciton acceptor. This kind of mechanism also exists for the exciton transfer from other, higher energies absorbing pigments to the reaction centers (energy transfer along an energy gradient; see Figure 4.3) so that the light-harvesting complexes of the photosynthetic membrane feature a spatially as well as spectrally optimized cross-section for photon capture.

The role of magnesium in chlorophyll is to contribute to the particular arrangement of pigments. The virtually loss-free exciton transfer through a cluster network of antenna pigments requires a high degree of three-dimensional order (Figure 4.2). Therefore, a well-defined spatial orientation of the chlorophyll π chromophors with regard to each other is necessary. This orientation cannot be solely guaranteed by anchoring the chlorophyll molecules in the membrane via the phytyl side chains but must rely also on the coordination of polypeptide side-chain ligands to the two free axial coordination sites at the metal (three-point fixing for defined spatial orientation).

From the point of view of coordination chemistry, the more direct *in vitro* aggregation of the dihydrate of a chlorophyll derivative with an ethyl instead of a phytyl side chain is quite revealing (Figure 4.4) because a one-dimensional coordination polymer is thus formed. The coordinatively doubly unsaturated Lewis-acidic, i.e. electron pair-accepting Mg^{2+} centers can interact via dipolar, hydrogen-bonding

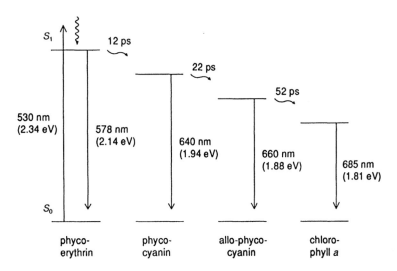

Figure 4.3
Energy-transfer cascade for antenna pigments in light-harvesting complexes of the algae *Porphyridium cruetum* (according to [5]). The data on the vertical arrows indicate absorption (↑) and emission (↓) wavelengths; S_0 denotes the singlet ground states, S_1 the lowest excited singlet states

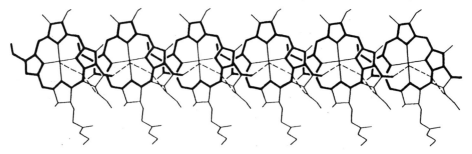

Figure 4.4
Structure of the one-dimensional aggregate occurring in crystals of ethyl chlorophyllide dihydrate [15]. The π electron conjugation is represented by bold lines; hydrogen bond links via water molecules are marked by broken lines

water molecules with the Lewis-basic carbonyl groups in the characteristic cyclo-pentanone ring of adjacent chlorophyll molecules. Such a direct linkage, however, is not assumed to occur in the light-harvesting proteins (Figure 4.2).

Exactly defined spatial orientation, e.g. the arrangement of tetrapyrrole ring planes parallel to the plane of the membrane [13] and the resulting high degree of organization in light-harvesting systems is thus ensured by the presence of a coordinatively doubly unsaturated electrophilic metal center in the macrocycle. Of all the metals with the proper size (Table 2.7), sufficient natural abundance (noncataly-

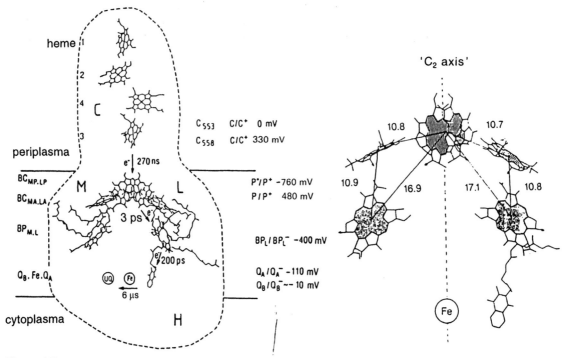

Figure 4.5
Photosynthetic reaction center of *Rhodopseudomonas viridis*. Left: Arrangement of electron-transferring components in the membrane complex (protein subunits C, L, M, H) as well as some potentials of redox pairs. Right: Orientation and distances (in Å = 0.1 nm, measured from center to center) of the main electron-transferring components (according to [12])

tic function) and a strong tendency for hexacoordination, only Mg^{2+} remains of all the main group metal ions. Magnesium is a rather light atom with a small spin-orbit coupling constant. Heavier elements, including main-group metal ions have higher spin-orbit coupling constants and can thus enhance an inter-system crossing (ISC) from very short-lived singlet to considerably longer-lived triplet excited states, thus slowing down the necessarily very rapid primary events in photosynthesis; the result would be a competition of undesired light- or heat-producing processes with actual chemical reactions. The fact that transition metals in particular are not suited as central ions of chlorophylls is related to the next step of photosynthesis: the charge separation.

4.2.3   Charge Separation and Electron Transport

When excitonic energy reaches a photosynthetic 'reaction center', the essential step for the separate production of an electron-rich, i.e. reduced component and an electron-poor, i.e. oxidized species can take place. Since the structural elucidation of bacterial reaction centers via X-ray diffraction of single crystals (Figure 4.5) has been accomplished [4,12,16], there is now a fairly solid basis to discuss the functions of the molecular units involved (see Table 4.1).

The purple bacteria *Rhodopseudomonas viridis* and *Rhodopseudomonas sphaeroides* which contain only one photosystem show similarities as well as differences with regard to their photosynthetic reaction centers. In both cases, the reaction center is situated in a polyprotein complex which *in vivo* spans the membrane. On the

Table 4.1   Active components in photosystems I and II of plants (compare Figure 4.9)

photosystem I (including cytochrome b/f complex)		
about 200	antenna chlorophylls	(Fig. 4.2)
about 50	carotenoids	(compare Fig. 4.1)
1	reaction center P_{700}	(compare Fig. 4.5)
1	chlorophyll *a* (primary acceptor A_0)	(2.5)
1	vitamin K_1 (secondary acceptor A_1)	(3.12)
3	Fe/S-clusters (FeS)	(Chap. 7.1-7.4)
1	bound ferredoxin (Fd)	(Chap. 7.1-7.4)
1	soluble ferredoxin (Fp)	(Chap. 7.1-7.4)
1	plastocyanin (PC, primary donor)	(Chap. 10.1)
1	Rieske Fe/S center	(7.5)
1	cytochrome *f* (cyt *f*)	(Chap. 6.1)
2	cytochromes b_6 (cyt b_6)	(Chap. 6.1)
photosystem II (including OEC)		
about 200	antenna chlorophylls	(Fig. 4.2)
about 50	carotenoids	(compare Fig. 4.1)
1	reaction center P_{680}	(compare Fig. 4.5)
2	chlorophylls	(4.2)
2	pheophytins (primary acceptor)[b]	(Chap. 4.2.3)
2	plastoquinones (PQ)	(3.12)
2	tyrosine residues (primary donor)[a]	(compare Tab. 2.5)
4	manganese centers	(Fig. 4.10)
1	calcium ion Ca^{2+}	(Chap. 14.2)
several	chloride ions Cl^-	
1	cytochrome b_{559}	(Chap. 6.1)

[a] Components of the protein. [b] Metal-free chlorophyll.

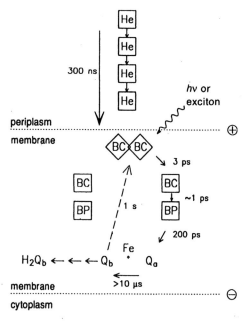

Figure 4.6
Schematic representation of the temporal and spatial sequence of the light-induced charge separation in the reaction center of bacterial photosynthesis (*Rps. viridis*) (according to [4,12]) He: heme systems; BC: bacteriochlorophyll; BP: bacteriopheophytin; $Q_{a,b}$: quinones

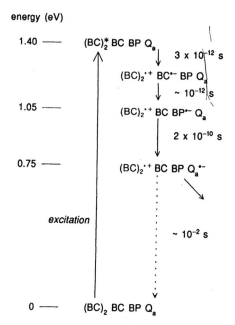

Figure 4.7
Primary events of photosynthetic charge separation: time and energy aspects (according to [5]) (compare also Figure 4.6)

symmetry axis of the nearly C_2-symmetrical reaction centers there is a bacterio-chlorophyll dimer, the 'special pair' BC/BC (Figure 4.6). The asymmetry of these π dimers which feature a dihedral angle of about $15°$ for the chlorin planes is different from a coordinative interaction between acetyl substituents at the tetrapyrrole ring and the metal centers [12]. Because of the significant π/π orbital interaction, the and the metal centers [12]. Because of the significant π/π orbital interaction, the special pair may function as an electron donor; electronic excitation of the dimer, which is also called P960 (pigment with a long-wavelength absorption maximum at 960 nm), leads to a primary charge separation within a very short time. One energetically elevated electron of the electronically excited dimer is transferred to the primary acceptor, a monomeric bacteriochlorophyll (BC) molecule (Figures 4.6 and 4.7).

Structure Determination by X-ray Diffraction

In this context as well as in the description of other physical methods we can only provide very elementary presentations. The main purpose is to point out the principles and the usefulness of these methods for bioinorganic chemistry, and also their limitations.

In structure determination by X-ray diffraction, uniformly crystallizing systems are examined with monochromatic X-rays. The mathematical analysis of the resulting diffraction pattern can then provide an idea of the three-dimensional distribution of electron density in the 'unit cell', i.e. the periodically repeated unit of a single crystal, revealing an illustration of molecular shapes (see Figure 2.7).

A first requirement which is particularly difficult to achieve in the biochemical field is the growing of sufficiently large (mm range) and qualitatively suited single crystals. For this reason alone, the number of structurally well-examined metalloproteins is relatively small [17]. Large proteins have many degrees of freedom and can often be arranged in a single-crystal arrangement only under very specific conditions (temperature, solvent, pH). Hydrophilicity and the propensity to form hydrogen bonds usually lead to the inclusion of water molecules during crystallization. A major problem, the structural determination of proteins that exist only within biological membranes, was solved by using co-crystallizing, membrane-analogous detergent molecules [18]. It was this achievement that allowed for the structural elucidation of a reaction center of bacterial photosynthesis by X-ray diffraction methods which was honored by the 1988 Nobel prize in chemistry to J. Deisenhofer, R. Huber and H. Michel.

The necessary computational effort for a structure determination is much higher for proteins with their very large number of atoms than for low molecular weight compounds. Even with a good set of data, the resulting multiparameter problem can often be solved only to a certain extent, leaving a molecular resolution of 0.2 nm or higher, which is chemically not satisfactory and only provides for a rather diffuse structure, revealing, for example, the protein folding. In such cases, bond angles and bond lengths between small, little diffracting atoms such as C, N or O are not available with sufficient precision. In general, hydrogen atoms cannot be localized with X-ray diffraction methods under the conditions of protein crystallography.

An important and not always easily dismissed objection to the relevance of single-crystal structural analyses concerns the mainly static aspect of crystalline systems. It is possible and has been demonstrated in some cases that important structural features (conformations) of biochemically active compounds are different in the crystal and in solution. Recent studies via high-resolution NMR (nuclear magnetic resonance) spectroscopy in solution and temperature-dependent diffraction studies of crystalline compounds have begun to contribute to an understanding of the molecular dynamics of biochemical systems.

The next step of the charge separation consists in the transfer of negative charge to the secondary acceptor bacteriopheophytin (BP), a bacteriochlorophyll ligand without coordinated metal (Figure 4.7). From the coordination chemistry of porphyrins it is well known that neutral M^{2+} complexes are often harder to reduce than corresponding doubly protonated neutral ligands; the more ionic bond to the metal leaves considerable amounts of negative charge at the ligand. Since the central ion Mg^{2+} is redox-inert and thus not directly involved in electron donation or acceptance, the radical anions of chlorophylls can be regarded as complexes of a divalent metal cation and the radical *trianion* of the macrocyclic ligand: $(Chl/Mg)^{\cdot-} = Chl^{\cdot 3-}/Mg^{2+}$. Correspondingly, radical cations of tetrapyrrole complexes can be formulated as compounds of metal dications with anion radical ligands (compare Sections 6.2 to 6.4).

The third detectable acceptor for the electron which has been generated through light-induced charge separation in the reaction center is a *para*-quinone Q_a, e.g. menaquinone (3.12), which is reduced to a semiquinone radical anion, $Q_a^{\cdot-}$, in this process. After 'quenching' at low temperatures the longer-lived paramagnetic states of the charge separation process can be examined by EPR/ENDOR spectroscopy, as can the *in vitro* generated radicals of the individual components [7]. Reduced Q_a may in turn reduce a different, more labile quinone such as ubiquinone, Q_b (compare 3.12), with a high-spin iron(II) center connecting the two quinones at the axis of the reaction center. The role of this six-coordinate Fe^{II}, which is equatorially bound by four histidine ligands and axially by a carboxylate, is not yet clear; an active redox function in the sense of an *inner-sphere* electron transfer between Q_a and Q_b does not seem likely since the exchange, e.g. with redox-inert Zn^{II}, does not make a significant difference in the electron transfer rate, at least in *Rps. sphaeroides*. Possibly, a divalent metal ion is required to guarantee the controlled electron transfer from reduced bacteriopheophytin to the primary quinone through polarization of the hydrogen bond-forming histidine ligands or simply to maintain the necessary structure. Quinone Q_b or its $2\,e^-/2\,H^+$ reaction product, the hydroquinone H_2Q_b (3.12), is not tightly bound to the protein but exchanges with quinones in the 'quinone pool' of the membrane (compare Figure 4.8) so that electrons can now be transported further outside the protein. In simple bacteria, the electron gradient finally gives rise to a coupled H^+ gradient and photosynthetic phosphorylation takes place (ATP synthesis; compare Figure 4.8). In higher organisms there are further steps, the 'dark reactions', which eventually lead to the production of electron-rich coenzymes NAD(P)H and to CO_2 reduction (the Calvin cycle).

The radical cation of the 'special pair' which remained after the initial charge separation is reduced after a relatively long time through regulated electron flow

(Figure 4.6) via one (*Rps. sphaeroides*) or several (*Rps. viridis*) heme centers of cytochrome proteins (Section 6.1). The electron deficiency or 'hole' is thus further translocated and finally is filled up in purple bacteria through back electron transfer at a different site in the membrane. In higher evolved organisms, the hole created at the special pair may be the starting point for substrate oxidation (Section 4.3), which requires an additional photosynthetic system consisting of a light-harvesting complex, reaction center and an oxidase enzyme.

The capability of an electronically excited state to serve as effective reductant *and* oxidant is illustrated in the orbital energy diagram (4.3). Electronic excitation creates an electron hole in a low-lying orbital which invites electron transfer from an external donor (photooxidation). Simultaneously, the presence of the excited electron in a high-lying, previously unoccupied orbital allows for the photoreduction of an external acceptor. The photosynthetically undesired alternative of a simple radiative or radiationless recombination is clearly obvious from diagram (4.3).

The formation of long-lived, storable oxidation and reduction products as a consequence of light-induced charge separation is very unusual, the normal course of events being a rapid recombination of charges to produce heat or light (emission, luminescence). According to recent results from structural, spectroscopic and magnetic resonance studies, the success of photosynthesis is based on the strong preference for the extremely rapid charge-separating steps in comparison with the slower, energy-releasing recombination processes, the ratio of corresponding rates amounting to about 10^8. The basis for this ratio, which is impossible to reach in 'normal' chemical reaction systems, is the immobilization of the participating components in a special orientation to each other within *nonpolar* regions of proteins anchored in the membrane (Figure 4.5). Only then may a 'vectorial' chemical uphill reaction prevail over the natural tendency for charge recombination which would dominate if free diffusion were possible. In other words, the special arrangement of electron transfer components results in a sizable reduction or even disappearance of the activation energies for the 'forward' reaction process, while the reverse reaction (back electron transfer) falls in an 'inverted region' of electron transfer. This means that the reaction rate *decreases* in spite of an increasing free reaction energy, ΔG^0, i.e. in spite of a more favorable equilibrium [8]. Naturally, the electron transfer rates should be highest for the initial steps (Figures 4.6 and 4.7). In summary, the redox potentials in the ground and excited states, the individual structural changes during the charge separation processes, and the orientations of the participating components with respect to each other have to be finely tuned in order to achieve a useful charge separation. Many functional details, e.g. the asymmetry of the electron transfer

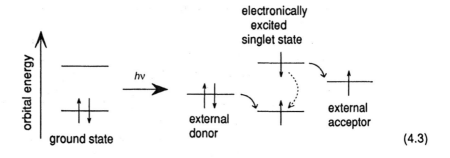

(4.3)

pathway in the actual axially symmetrical reaction centers (Figures 4.5 and 4.6), are not yet understood, even after the structural characterization.

From the absolute necessity to prevent heat- or light-producing back electron transfer reactions it follows that the chlorophyll molecules must not contain a redox-active transition metal. Metal centers like $Fe^{II/III}$ which easily transfer electrons themselves could accept or donate electrons in the ground or electronically excited states of an iron–chlorophyll π system and thus preclude the photosynthesis, which requires a rapid spatial charge separation. An *intra*molecular instead of an *inter*-molecular electron transfer would result without chemical energy storage.

The role of the magnesium ion in chlorophylls is therefore to act as a *light, redox-inert, Lewis-acidic coordination center* which thus contributes to a defined three-dimensional organization in light-harvesting systems and reaction centers. Monomer, π/π dimer and the metal-free ligand each have separate, unique functions in the primary processes of photosynthesis. Redox-active transition metals or heavier metal centers with higher spin-orbit coupling constants would support additional, undesired reaction alternatives; of the remaining bioavailable metal ions only Mg^{2+} fits exactly with regard to size and charge (Table 2.7).

4.3   Manganese-Catalyzed Oxidation of Water to O_2

Although the photosynthetic fixation of CO_2, the carboxylation, requires polarizing Mg^{2+} ions [19], the reductive side (4.4) of photosynthesis which proceeds via the universal 'hydride' carrier NAD(P)H is interesting mainly from the organic biochemical point of view (Calvin cycle). The other side, the more 'inorganic' oxidative part (4.5), e.g. of plant photosynthesis, has recently attracted much attention from coordination chemistry [1,20].

$$4\ e^- + 4\ H^+ + CO_2 \rightarrow \ '1/n\ (CH_2O)_n' + H_2O \tag{4.4}$$

$$2\ H_2O \rightarrow O_2 + 4\ H^+ + 4\ e^- \tag{4.5}$$

The main reason for this interest lies in the crucial but little understood catalytic function of polynuclear manganese complexes in the mechanistically challenging oxidation of water to dioxygen, O_2 [2,9,21–27]. The efficient and long-term stable catalysis of dioxygen formation also poses a problem in the (photo)electrochemical water splitting for the technical production of dihydrogen, H_2, as the promising future energy carrier.

Since the production of NAD(P)H *and* O_2 requires more potential difference than can be generated by *one* photosystem of the type depicted in Figures 4.6 and 4.7, especially when considering energy losses, the photosynthesis in plants and cyanobacteria differs from that of the more simple purple bacteria by featuring *two* separately excitable photosystems (Figure 4.8). These two systems can be connected in a redox potential diagram, the 'Z scheme', which shows several, often metal-containing, electron transfer components (Table 4.1 and Figure 4.9 [28]). In addition to photosystem I (PS I, absorption maximum 700 nm), a photosystem II (PS II,

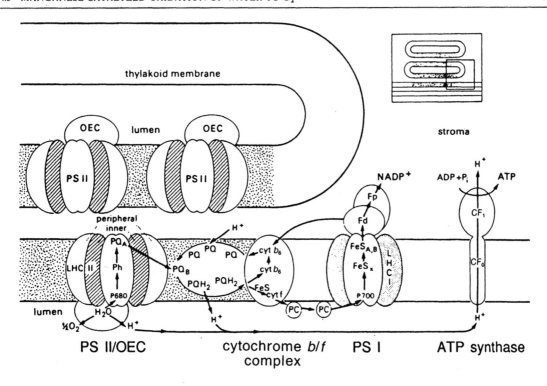

Figure 4.8
Structural organization of the lamellar thylakoid membrane of higher plants with the following components:
two photosystems (PS) and two light-harvesting complexes (LHC), the oxygen-evolving complex (OEC) at PS
II (Section 4.3), the cytochrome b/f complex (Section 6.1), plastoquinones (PQ/PQH$_2$; 3.12) and plastocyanin
(PC; Section 10.1), several iron–sulfur centers (FeS; Sections 7.1 to 7.4), soluble ferredoxin (Fd) and the flavoprotein
Fp (ferredoxin/NADP reductase), and ATP synthase as the center of photosynthetic phosphorylation (according
to [10])

absorption maximum 680 nm) exists which provides electrons for phosphorylation
and for photosystem I. The electron holes remaining in PS II represent a very positive
potential which is used to oxidize two water molecules to dioxygen in an overall
four-electron process (4.5).

Whereas photosystem II (without the oxygen evolving complex, OEC) shows
structural similarity to the reaction centers of purple bacteria (Figures 4.5 and 4.10),
photosystem I exhibits a different arrangement according to low-resolution diffraction studies [30]. Table 4.1 summarizes the nonprotein components for both
photosystems.

Because of the higher excitation energy in comparison to bacterial pigments,
photosystem II of plants can utilize the tyrosine/tyrosine radical cation redox pair
($E_0 = +0.95$ V [25]) to transfer electrons to the OEC; a secondary electron donor
for P_{700} is the copper-containing protein plastocyanin (see Section 10.1).

According to the redox potential scheme (4.6), the removal of one, two or three
instead of four electrons per two H_2O molecules requires rather high potentials and
leads to such reactive high-energy products as the hydroxyl radical, $\cdot OH$, hydrogen
peroxide, H_2O_2, or superoxide, $O_2^{\cdot -}$. All of these are potentially harmful substances for biological membranes, particularly in the presence of transition metal ions
(Sections 10.5 and 16.8). Important functions of water oxidation catalysts are therefore

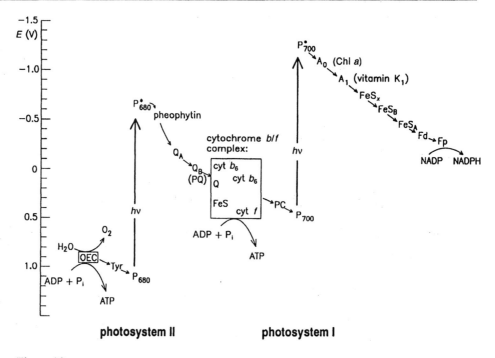

Figure 4.9
Z scheme of electron transfer in plant photosynthesis (see Figure 4.8). Tyr: tyrosine/tyrosine radical cation redox pair; for other abbreviations see Table 4.1 (scheme modified according to [28])

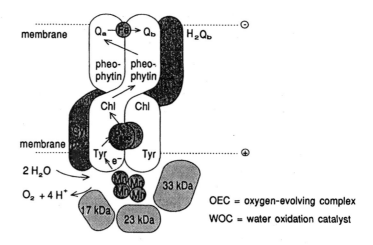

Figure 4.10
Schematic structure of photosystem II, including the oxygen evolving center: protein subunits, electron-transferring components and electron flow (according to [23,29])

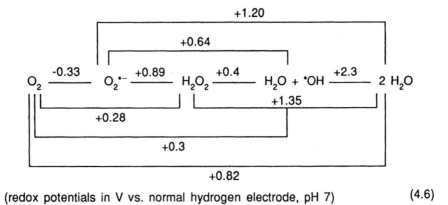

(redox potentials in V vs. normal hydrogen electrode, pH 7) (4.6)

to lower the oxidation potential, to prevent the formation of free reactive intermediates and to guarantee substrate specificity since many molecules are easier to oxidize than H_2O.

The overall reaction in photosystem II can be summed up as follows (4.7), employing quinonoid proton and electron carriers (PQ: plastoquinone, PQH_2: plastohydroquinone, see 3.12):

$$2\ H_2O\ +\ 2\ PQ\ +\ 4\ H^+(out)\ \xrightarrow[\ [Mn]_x\ ]{4h\nu}\ O_2\ +\ 2\ PQH_2\ +\ 4\ H^+(in)$$

(catalyzing) (4.7)

The transport of protons is linked to the charge separation and electron flow across the membrane which eventually results in energy-storing phosphorylation (Figure 4.9); *out* and *in* refer to the location either outside or inside membrane vesicles (Figure 4.8). The required four oxidation equivalents for O_2 production are available only after excitation with at least four photons in photosystem II; measurements of the actual quantum yield for photosynthesis have shown that about eight photons are needed for every molecule of converted CO_2. The binding site of mobile Q_B in photosystem II is the target of many herbicides, an understanding of its mechanism thus being of immediate practical importance [31].

In addition to inorganic Ca^{2+} and several chloride ions, the function of which is still largely obscure [26], effective dioxygen generation requires manganese centers as redox-active components. EPR spectroscopically detectable manganese deficiency thus impairs tree growth and can even be related to forest damage [32]. The catalyzing oligoprotein complex OEC apparently contains a total of four manganese centers in one of the subunits (33 kDa), of which two are rather loosely bound and easily extractable using chelate ligands. Manganese 'clusters' with a nuclearity $N \geq 2$ seem to be absolutely essential for the catalysis of O_2 production and several hypotheses have been proposed recently concerning their individual coordination environment and mutual arrangement (symmetry), followed by detailed mechanistic schemes on the water oxidation as based on spectroscopic measurements and model studies [27]. Since the corresponding, very labile membrane proteins have not yet been crystallized, the elucidation of structural features had to rely on variants of X-ray absorption spectroscopy.

X-ray Absorption Spectroscopy

The crystallization of proteins and their structure determination via X-ray diffraction can often require many years (see Section 4.2.3). A faster, but often less satisfactory determination of a few structural features, especially around unique heteroatoms, is possible via X-ray absorption spectroscopic (XAS) measurements of not (yet) crystallized substances, a technique that has become very popular in recent years [33,34]. The method is based on the element-specific absorption of X-rays, i.e. the ionization from the innermost electron shells. Under more detailed inspection the typical absorption edge and its high-energy tailings show a fine structure which is due to the interference of the ejected photoelectron with back-scattering due to electrons around the absorbing nucleus (Figure 4.11).

In the region close to the absorption edge this method is referred to as XANES spectroscopy (X-ray absorption near edge structure); the high-energy part shows the EXAFS effect (extended X-ray absorption fine structure). Through mathematical Fourier analysis of this fine structure the immediate (coordination) environment of a metal center can be determined, even in a protein that is not characterized further structurally. The oxidation state, symmetry and electronic structure are better estimated from XANES data while EXAFS analysis through mathematical simulation can provide information on the type, number and distance of the atoms in the first coordination spheres, i.e. up to distances of 400–500 pm. However, the difficulty in distinguishing between O and N donor centers and remaining large uncertainties with respect to coordination numbers and symmetry render X-ray absorption spectroscopy more ambiguous as a method for structure determination; frequently, only certain alternatives can be excluded. The intense and monochromatically tunable X-rays required in the K edge regions of heavy metal centers are only available from synchrotron radiation; the measurements are carried out at low temperatures in order to avoid structural damage and follow-up reactions.

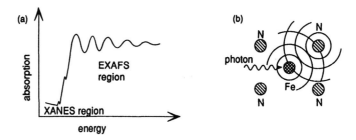

Figure 4.11
(a) Typical X-ray absorption spectrum with XANES and EXAFS region; (b) interference at the absorbing nucleus through back-scattering effects (Fe absorber in a heme environment as an example)

The manganese EXAFS data [2,27] for the more stable of the S states (see below) of the water-oxidizing complex in photosystem II suggest that the metal centers are surrounded by four to six lightweight donor atoms (O and/or N, distance at about 200 pm) but not by stronger scattering sulfur atoms. Signals indicating one or more possible Mn–Mn distances of about 270 pm point to a polynuclear arrangement in which the metal centers are linked by single-atom (oxide) bridges (see the relevant model complexes, 4.14).

According to the EXAFS measurements another Mn–Mn (or Mn–Ca?) distance could exist at about 330 pm (carboxylate bridge?); a further feature at about 430 pm involving a heavy atom was interpreted as an Mn–Ca distance. A direct bond between chloride ions (which are necessary for the catalytic function) and manganese was not observed in the examined material up to a distance of 220 pm.

The charge-separation cascade illustrated in (4.8) clearly shows how single oxidation equivalents become available for the polynuclear manganese cluster via the tyrosine radical cation as a primary acceptor.

temporal and spatial course of the charge separation process in
photosystem II (Ph: pheophytin):

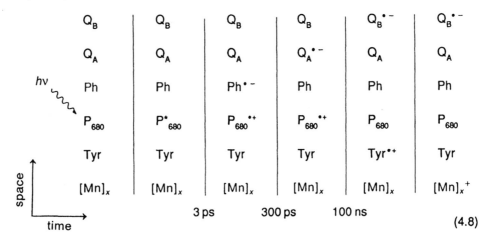

(4.8)

Until two molecules of water have been oxidized *stepwise* to dioxygen through successive light excitation/charge separation events, five exactly tuned (oxidation) states of PS II, referred to as S_0 to S_4, are observed in the millisecond range by flash-photolysis techniques [24]. Scheme (4.9) shows the charge-induced coupling of electron and proton flow as well as the different lifetimes of those five states under physiological conditions; the structural changes at the manganese centers are small, as demonstrated by EXAFS spectroscopy of frozen material, in agreement with a low activation energy for effective redox catalysis.

The position of the manganese XAS edge which is characteristic for the oxidation state changes from S_0 via S_1 to S_2, corresponding to an oxidation of the metal; however, there are conflicting reports about the change from S_2 to S_3 [35], which may or may not involve the oxidation of a histidine ligand instead of a metal site [23]. Of special interest is the relatively long-lived intermediate state S_2 which exhibits an EPR signal at about $g = 4$ and, most conspicuously, a highly structured 'multiline'

to the reaction center via
tyrosine radical cation

$$2\,H_2O \longrightarrow S_0 \underset{hv}{\overset{e^-}{\rightleftarrows}} S_1 \underset{hv}{\overset{e^-}{\rightleftarrows}} S_2 \underset{hv}{\overset{e^-}{\rightleftarrows}} S_3 \underset{hv}{\overset{e^-}{\rightleftarrows}} S_4 \longrightarrow O_2$$

$$H^+ \qquad\qquad H^+ \quad 2H^+$$

40 ms   100 ms   250 ms   ~ 1 ms

to the aqueous phase   (4.9)

EPR spectrum around $g = 2$. This latter feature is characteristic for antiferro-magnetically coupled (see below) polynuclear manganese complexes with an odd number of electrons and an $S = 1/2$ ground state. Naturally occurring manganese contains exclusively the isotope ^{55}Mn which has a nuclear spin of $I = 5/2$ and a large nuclear magnetic moment; a spin–spin hyperfine interaction of *one* unpaired electron with *two or more* different ^{55}Mn centers can thus be assumed for the S_2 state.

Such a situation would be the natural consequence of the stepwise process (4.9) in the course of which there are odd-electron *mixed-valent* states, i.e. paramagnetic polynuclear complexes with different oxidation states of the two or more transition metal centers involved. The spectroscopic results (EPR, EXAFS) may be interpreted using several combinations of oxidation states [23]; diagram (4.10) shows one of the alternatives. Tetravalent manganese and an odd total number of electrons are postulated for the extremely short-lived O_2-evolving S_4 state. S_4 contains two electrons less than S_2, which definitely contains an odd number of electrons based on its EPR characteristics [1].

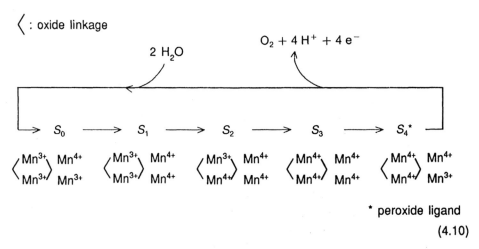

$\langle$: oxide linkage

$2\,H_2O \qquad\qquad O_2 + 4\,H^+ + 4\,e^-$

$$S_0 \longrightarrow S_1 \longrightarrow S_2 \longrightarrow S_3 \longrightarrow S_4^*$$

$$\left\langle{Mn^{3+} \atop Mn^{3+}}\right\rangle Mn^{4+} \atop Mn^{3+} \quad \left\langle{Mn^{3+} \atop Mn^{3+}}\right\rangle Mn^{4+} \atop Mn^{4+} \quad \left\langle{Mn^{3+} \atop Mn^{4+}}\right\rangle Mn^{4+} \atop Mn^{4+} \quad \left\langle{Mn^{4+} \atop Mn^{4+}}\right\rangle Mn^{4+} \atop Mn^{4+} \quad \left\langle{Mn^{4+} \atop Mn^{4+}}\right\rangle Mn^{4+} \atop Mn^{3+}$$

* peroxide ligand

(4.10)

What are the properties that make manganese centers particularly well suited for a catalysis of water oxidation and for the rapid evolution of dioxygen? In this context it should be remembered that freshly precipitated manganese dioxide, a typically non-stoichiometric mixed-valent ($+$IV, $+$III) system of the composition $MnO_{2-x} \cdot nH_2O$, may act as a good heterogeneous catalyst for the decomposition of hydrogen peroxide to dioxygen and water. Mn^{III,IV} oxides or hydroxides certainly were available in sea water under the conditions of developing photosynthesis about 3×10^9 years ago; oxidic manganese nodules with about 20% Mn content are quite abundant on the sea floor. Also, the importance of manganese for the O$_2$ metabolism is not restricted to photosynthesis; further established examples include a manganese-containing superoxide dismutase (Section 10.5 [3]), an azide-insensitive catalase and other peroxidases (see Section 6.3 [1]).

The remarkable features of manganese are:

(a) a large variety of stable or at least metastable oxidation states ($+$II, $+$III, $+$IV, $+$VI, $+$VII),
(b) the often very labile binding of ligands and
(c) a pronounced preference for high-spin states because of inherently small d orbital splitting (2.9), resulting in an often complex magnetic behavior.

Spin–Spin Coupling

When two or more centers with unpaired electrons ($\uparrow$) interact, the result may be a parallel 'ferromagnetic' ($\uparrow\uparrow$) or an antiparallel 'antiferromagnetic' ($\uparrow\downarrow$) coupling of the electron spins [36,37].

If the orbital interaction is small, e.g. because of orthogonal arrangement of p or d orbitals, Hund's rule requiring maximal multiplicity in order to avoid

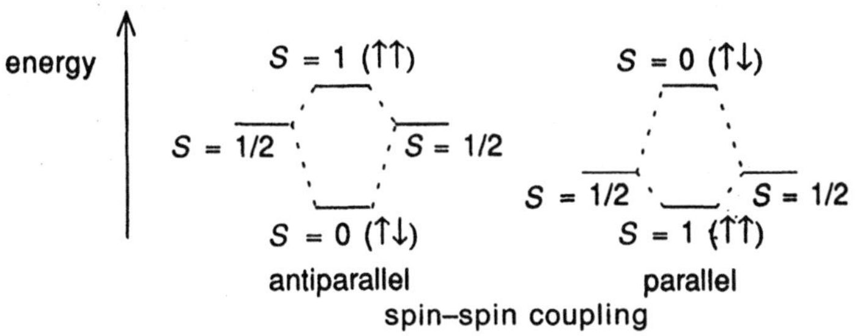

$$(4.11)$$

the spin-pairing energy favors a parallel spin–spin coupled situation. The more frequent case, however, is the antiparallel (antiferromagnetic) coupling in which the energy gain from possibly only indirect orbital interactions ('super-exchange') compensates for the spin pairing.

In the case of a dimer containing high-spin Mn^{III} (d^4, $S = 2$) and high-spin Mn^{IV} (d^3, $S = 3/2$), ferromagnetic coupling would lead to a ground state with a total electron spin of $S = 7/2$ ($\uparrow\uparrow\uparrow\uparrow + \uparrow\uparrow\uparrow \rightarrow \uparrow\uparrow\uparrow\uparrow\uparrow\uparrow\uparrow$) and partial spin pairing could occur only in thermally accessible excited states. With antiferromagnetic coupling the resulting ground state would have a total electron spin of $S = 1/2$ ($\uparrow\uparrow\uparrow\uparrow + \downarrow\downarrow\downarrow \rightarrow \uparrow\uparrow\downarrow\uparrow\downarrow\uparrow\downarrow$); in this case, there are thermally accessible magnetically excited states with higher multiplicity. Ground states with an intermediate electron spin are possible in polynuclear systems with different extents and signs of the coupling interactions.

Magnetic states can be examined by measuring the paramagnetic component of the magnetic susceptibility, e.g. via the effects experienced by a substance in an inhomogeneous magnetic field (Faraday balance; SQUID susceptometer). Theoretical models help to interpret the observed data; in particular, the simulation of the temperature dependence of the susceptibility provides information on the type and extent of couplings between electron spins at various centers. According to the Curie law, a higher susceptibility of paramagnetic systems should result at low temperatures due to the diminished averaging through thermal motion of the particles; however, this effect can be (over)compensated by antiferromagnetic behavior, i.e. by the tendency for spin pairing.

Due to at least partial antiferromagnetic coupling between the individual high-spin metal centers the total spin of the manganese cluster in the oxygen-evolving complex is distinctly smaller than that of some synthetic polymanganese compounds with $S = 10$–14 ground states. Nevertheless, the water-oxidizing centers have *several odd-electron states* at their disposal. This fact, as well as the already mentioned availability of fairly high oxidation states and the high lability regarding coordinated ligands, makes manganese centers uniquely suited to catalyze the generation and *release* of the molecule $^3O_2$ in its triplet ground state, i.e. with an *even* number of unpaired electrons (compare Section 5.1). To appreciate this behavior one must remember that the reaction of transition metals with dioxygen normally involves an irreversible *binding*.

It would require a 'flipping' of electron spins to make the odd-electron catalyst and even-electron systems such as O_2 compatible; however, spin-flipping processes during chemical reactions often implicate a high activation energy because of their low probability ($\rightarrow$ statistical aspect of the reaction rate). The best-known example for a spin-inhibited reaction is the H_2/O_2 mixture which reacts to give water only after activation by a bond-breaking catalyst or after ignition in a radical chain reaction (4.12).

$$
\text{H–H} \; + \; \text{H–H} \; + \; \text{O=O} \; \xrightarrow{\substack{\text{(spin-} \\ \text{inhibited)}}} \; \text{H}\diagdown\!\!\overset{O}{}\!\!\diagup\text{H} \; + \; \text{H}\diagdown\!\!\overset{O}{}\!\!\diagup\text{H} \qquad \text{(atom balance)}
$$

$$
\quad\; \uparrow\downarrow \qquad\quad \uparrow\downarrow \qquad\; \uparrow\downarrow \; \uparrow\uparrow \qquad\qquad\quad \uparrow\downarrow \; \uparrow\downarrow \quad\;\; \uparrow\downarrow \; \uparrow\downarrow \qquad \text{(spin balance)}
$$

$$
\qquad\qquad\qquad S = 1 \qquad\qquad\qquad\qquad\qquad S = 0 \qquad\qquad\qquad (4.12)
$$

The hypothetical spin balance (4.13) shows a possible function of catalytic metal centers with variable spin quantum numbers $S = n/2$.

$$
2\,H_2O \; + (\text{Mn—Mn})^{n+} \rightarrow \; ^3O_2 \; + (\text{Mn—Mn})^{(n-4)+} + 4\,H^+
$$

$$
\qquad\quad \uparrow\downarrow \qquad\qquad \downarrow \qquad\qquad \uparrow\uparrow \qquad\quad \downarrow\downarrow\downarrow
$$

$$
\qquad\; S = 0 \quad\;\; S = 1/2 \qquad\; S = 1 \quad\; S = 3/2 \qquad\qquad (4.13)
$$

Numerous di- and polynuclear complexes of Mn^{II-IV} with biologically relevant oxygen donor ligands such as carboxylates, oxo or hydroxo groups and nitrogen donor-containing bi- or terdentate chelate ligands to mimic the protein environment were synthesized as models for the active sites of Mn-containing enzymes and characterized by extensive EPR, XAS and magnetic susceptibility measurements (4.14 [1,20]).

dinuclear manganese complexes with model character:

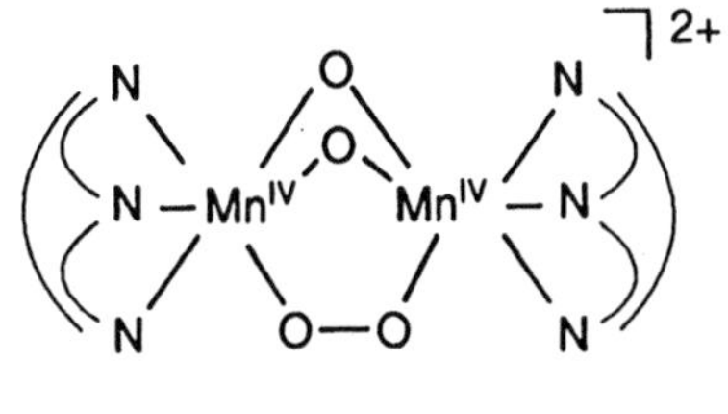

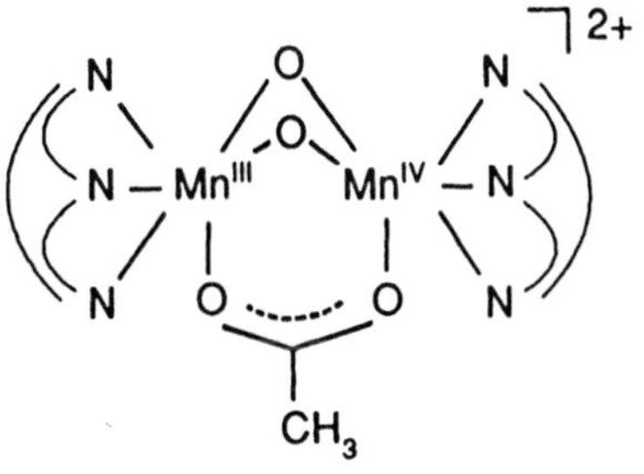

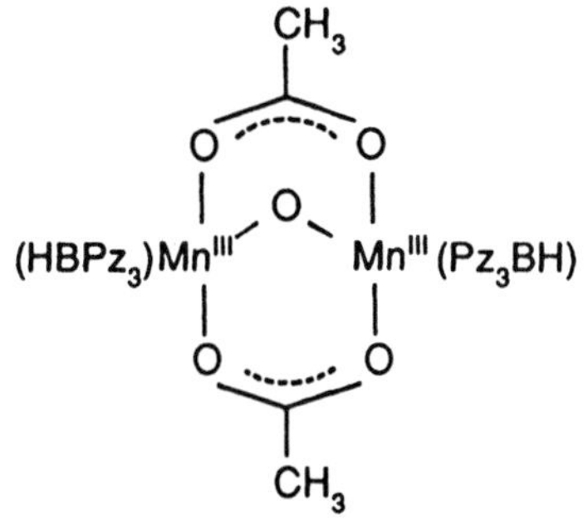

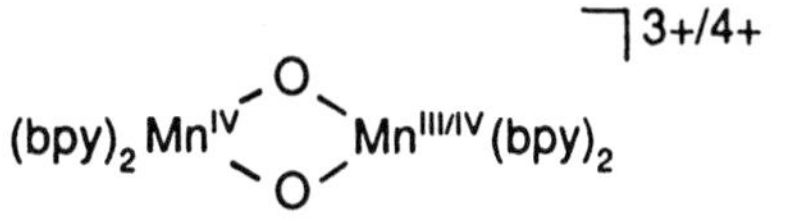

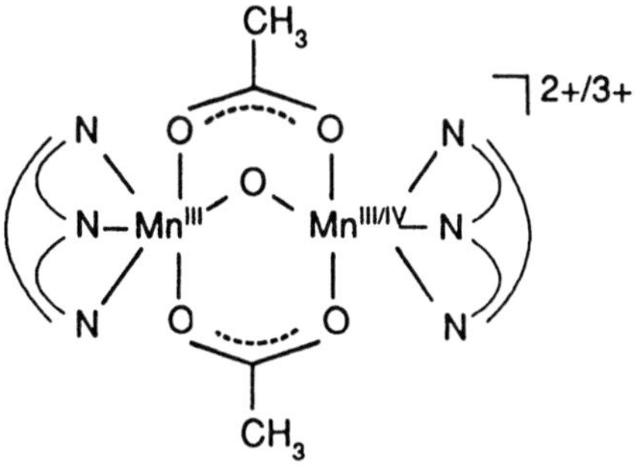

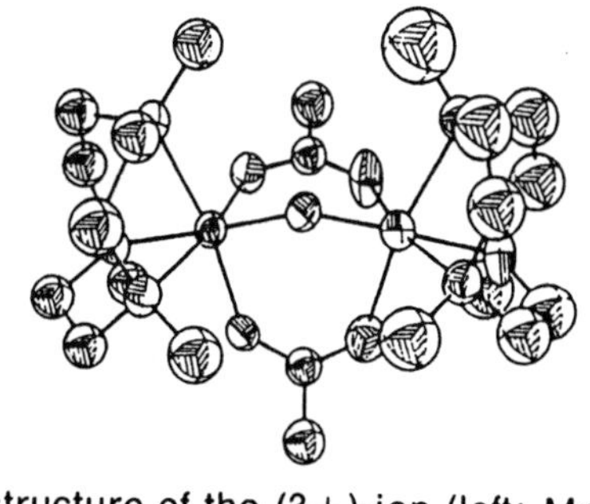

structure of the (3+)-ion (left; Mn^{III}/Mn^{IV}) in the crystal [40]

= 1,4,7-trimethyl-1,4,7-triazacyclononane :

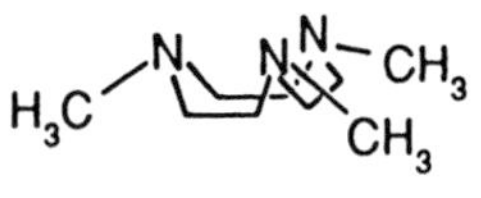

bpy = 2,2'-bipyridine :

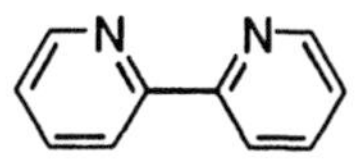

Pz$_3$BH$^-$ = tris(pyrazolyl)borate :

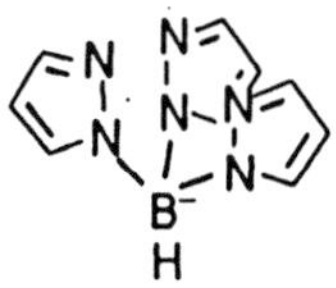

(4.14)

It became clear that both the magnitude and the sign of the magnetic coupling are strongly dependent on the configuration of the ligands. Therefore, magnetic data or EPR spectra alone are not sufficient to make a definite statement regarding the arrangement of metal centers in the actual enzyme [1,38]. The already mentioned lability of manganese-to-ligand bonds and the integration of the whole water-oxidizing protein complex within a membrane (Figure 4.10) have so far precluded more detailed structural studies. Based on the available experimental and theoretical results [27], an association of two oxide-bridged dimers ('dimer of dimers' model) is being considered as most likely.

While the number, arrangement (symmetry) and distribution of oxidation states of the manganese centers of the OEC have not yet been unambiguously established, there is even more uncertainty regarding the actual mechanism of dioxygen production, starting from two water molecules. A main problem of pertinent studies is the fact that enzymatic substrate and the solvent are identical in this case. It is generally assumed that two molecules of water have to be coordinated close to each other before the individual steps of electron removal (mainly metal-centered oxidations) with simultaneous deprotonation of coordinated water to OH^- or beyond can take place: $S_0 \rightarrow S_4$ (Figure 4.9). It is still unclear at which stage the crucial O—O bonds are actually formed (S_3, S_4?); there are polynuclear metal complexes with peroxide, O_2^{2-}, as well as superoxo ligands, $O_2^{\cdot-}$, as pointed out in the following chapter in Section 5.1. An O_2-producing peroxodimanganese(IV) complex (4.14, upper left) as a model for the $S_0 \rightarrow S_4$ reaction was first described by Bossek et al. [39]. A hypothetical minimal mechanism involving a dimer [1] is depicted in (4.15).

hypothetical mechanism of cluster-catalyzed water oxidation:

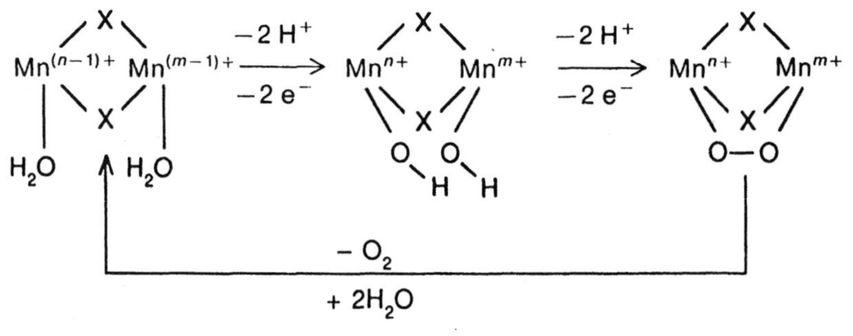

$$X = O^{2-}, OH^-, OR^- \text{ or } RCOO^- \tag{4.15}$$

In the four-electron oxidation of water to O_2 the polymanganese system acts (i) as an electron reservoir, accumulating charge in an exactly controlled fashion at physiologically high redox potential and (ii) as a non-$^3O_2$-retaining catalyst [41].

The apparently essential role of Ca^{2+} and Cl^- ions for dioxygen production [26] is still unclear. For calcium, a structure-stabilizing/regulatory function is assumed in accord with its otherwise known biochemical characteristics (compare Section 14.2). The role of chloride ions could be that of a temporary substitute for water or hydroxide ligands which are to be oxidized by this particular 'water oxidase' enzyme. This role of a stand-in ligand for the actual substrate in the active center of enzymes is usually assumed by the weakly coordinating water (see, for example, Chapter 12). In this case, however, water itself is the substrate so that the weakly coordinating chloride ions which are present in relatively high concentrations could fulfil this task and prevent possible oxidation reactions of the peptide without running the risk of being oxidized themselves ($E_0 = +1.36$ V for the oxidation to Cl_2).

With regard to synthetic polymanganese complexes, the emphasis has mainly been on *spectroscopic* and *structural* model compounds [42]; functional model studies, e.g. of coupled H^+/e^- transfer reactions, have only recently begun to emerge [43]. For ruthenium, which is related to manganese via a diagonal relationship in the periodic system, there are functional models available. In attempts to approach artificial photosynthesis [44], oxide-bridged ruthenium dimers (4.16) were thus synthesized which contain the strongly bound π electron acceptor 2,2'-bipyridine as an electron reservoir component [45].

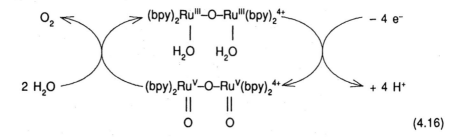

$$(4.16)$$

References

1. K. Wieghardt, The active centers in manganese-containing metalloproteins and inorganic model complexes, *Angew. Chem. Int. Ed. Engl.*, **28**, 1153 (1989).
2. V. L. Pecoraro (ed.), *Manganese Redox Enzymes*, VCH, New York, 1992.
3. G. E. O. Borgstahl, H. E. Parge, M. J. Hickey, W. F. Beyer, Jr, R. A. Hallewell and J. A. Tainer, The structure of human mitochondrial manganese superoxide dismutase reveals a novel tetrameric interface of two 4-helix bundles, *Cell*, **71**, 107 (1992).
4. J. Deisenhofer and H. Michel, The photosynthetic reaction center of the purple bacterium *Rhodopseudomonas viridis* (Nobel lecture), *Angew. Chem. Int. Ed. Engl.*, **28**, 829 (1989).
5. G. S. Beddard, Some applications of picosecond spectroscopy, *Eur. Spectrosc. News*, **65**, 10 (1986).
6. W. Holzapfel, U. Finkele, W. Kaiser, D. Oesterhelt, H. Scheer, H. U. Stilz and W. Zinth, Observation of a bacteriochlorophyll anion radical during the primary charge separation in a reaction center, *Chem. Phys. Lett.*, **160**, 1 (1989).
7. G. Feher, Identification and characterization of the primary donor in bacterial photosynthesis: a chronological account of an EPR/ENDOR investigation, *J. Chem. Soc. Perkin Trans. 2*, 1861 (1992).
8. R. A. Marcus, Electron transfers reactions in chemistry—theory and experiment, *Angew. Chem. Int. Ed. Engl.*, **32**, 1111 (1993).

9. H. T. Witt, Examples for the cooperation of photons, excitons, electrons, electric fields and protons in the photosynthetic membrane, *Nouv. J. Chim.*, **11**, 91 (1987).
10. J. M. Anderson and B. Andersson, The dynamic photosynthetic membrane and regulation of solar energy conversion, *Trends Biochem. Sci.*, **13**, 351 (1987).
11. J. D. Coyle, R. R. Hill and D. R. Roberts (eds.), *Light, Chemical Change and Life*, The Open University Press, Milton Keynes (England), 1982.
12. R. Huber, A structural basis for the transmission of light energy and electrons in biology (Nobel address), *Angew. Chem. Int. Ed. Engl.*, **28**, 848 (1989).
13. C. N. Hunter, R. van Grondelle and J. D. Olsen, Photosynthetic antenna proteins: 100 ps before photochemistry starts, *Trends Biochem. Sci.*, **14**, 72 (1989).
14. H. Zuber, Structure of light-harvesting antenna complexes of photosynthetic bacteria, cyanobacteria and red algae, *Trends Biochem. Sci.*, **11**, 414 (1986).
15. H. C. Chow, R. Serlin and C. E. Strouse, The crystal and molecular structure and absolute configuration of ethyl chlorophyllide *a* dihydrate, *J. Am. Chem. Soc.*, **97**, 7230 (1975).
16. J. P. Allen, G. Feher, T. O. Yeates, H. Komiya and D. C. Rees, Structure of the reaction center from *Rhodobacter sphaeroides* R-26: the cofactors, *Proc. Natl. Acad. Sci. USA*, **84**, 5730 (1987).
17. W. H. Armstrong, Metalloprotein crystallography: survey of recent results and relationships to model studies, in *Metal Clusters in Proteins* (ed. L. Que, Jr), ACS Symposium Series 372, 1988, p. 1.
18. H. Michel, Three-dimensional crystals of a membrane protein complex. The photosynthetic reaction center from *Rhodopseudomonas viridis*, *J. Mol. Biol.*, **158**, 567 (1982).
19. I. Andersson, S. Knight, G. Schneider, Y. Lindqvist, T. Lundqvist, C.-I. Bränden and G. H. Lorimer, Crystal structure of the active site of ribulose-bisphosphate carboxylase, *Nature (London)*, **337**, 229 (1989).
20. G. W. Brudvig, H. H. Thorp and R. H. Crabtree, Probing the mechanism of water oxidation in photosystem II, *Acc. Chem. Res.*, **24**, 311 (1991).
21. G. Renger, Biological solar energy utilization by photosynthetic water splitting, *Angew. Chem. Int. Ed. Engl.*, **26**, 643 (1987).
22. G. Renger, On the mechanism of photosynthetic water oxidation to dioxygen, *Chem. Scr.*, **28A**, 105 (1988).
23. A. W. Rutherford, Photosystem II, the water-splitting enzyme, *Trends Biochem. Sci.*, **14**, 227 (1989).
24. Govindjee and W. J. Coleman, How plants make oxygen, *Sci. Am.*, **262**(2), 42 (1990).
25. G. T. Babcock, B. A. Barry, R. J. Debus, C. W. Hoganson, M. Atamaian, L. McIntosh, I. Sithole and C. F. Yocum, Water oxidation in photosystem II: from radical chemistry to multielectron chemistry, *Biochemistry*, **28**, 9557 (1989).
26. R. J. Debus, The manganese and calcium ions of photosynthetic oxygen evolution, *Biochim. Biophys. Acta*, **1102**, 269 (1992).
27. D. M. Proserpio, R. Hoffmann and G. C. Dismukes, Molecular mechanism of photosynthetic oxygen evolution: a theoretical approach, *J. Am. Chem. Soc.*, **114**, 4374 (1992).
28. R. E. Blankenship and R. C. Prince, Excited-state redox potentials and the Z scheme of photosynthesis, *Trends Biochem. Sci.*, **10**, 382 (1985).
29. J. Barber, Photosynthetic reaction centres: a common link, *Trends Biochem. Sci.*, **12**, 321 (1987).
30. N. Krauss, W. Hinrichs, I. Witt, P. Fromme, W. Pritzkow, Z. Dauter, C. Betzel, K. S. Wilson, H. T. Witt and W. Saenger, Three-dimensional structure of system I of photosynthesis at 6 Å resolution, *Nature (London)*, **361**, 326 (1993).
31. W. Draber, J. F. Kluth, K. Tietjen and A. Trebst, Herbicides in photosynthesis research, *Angew. Chem. Int. Ed. Engl.*, **30**, 1621 (1991).
32. P. Laggner, R. Mandl, A. Schuster, M. Zechner and D. Grill, Rapid determination of manganese deficiency in conifer needles by ESR spectroscopy, *Angew. Chem. Int. Ed. Engl.*, **27**, 1722 (1988).
33. J. E. Penner-Hahn, X-ray absorption spectroscopy for characterizing metal clusters in proteins: possibilities and limitations, in *Metal Clusters in Proteins* (ed. L. Que, Jr), ACS Symposium Series 372, 1988, p. 28.

34. C. D. Garner, X-ray absorption spectroscopy and the structures of transition metal centers in proteins, *Adv. Inorg. Chem.*, **36**, 303 (1991).

35. T. Ono, T. Noguchi, Y. Inoue, M. Kusunoki, T. Matsushita and H. Oyanagi, X-ray detection of the period-four cycling of the manganese cluster in photosynthetic water oxidizing enzyme, *Science*, **2587**, 1335 (1992).

36. R. L. Carlin, *Magnetochemistry*, Springer-Verlag, Berlin, 1986.

37. G. Blondin and J.-J. Girerd, Interplay of electron exchange and electron transfer in metal polynuclear complexes in proteins or chemical models, *Chem. Rev.*, **90**, 1359 (1990).

38. M. Sivaraja, J. S. Philo, J. Lary and G. C. Dismukes, Photosynthetic oxygen evolution: changes in magnetism of the water-oxidizing enzyme, *J. Am. Chem. Soc.*, **111**, 3221 (1989).

39. U. Bossek, T. Weyhermüller, K. Wieghardt, B. Nuber and J. Weiss, $[L_2Mn_2(\mu\text{-}O_2)](ClO_4)_2$: the first binuclear ($\mu$-peroxo)dimanganese(IV) complex (L = 1,4,7-trimethyl-1,4,7-triazacyclononane). A model for the $S_4 \rightarrow S_0$ transformation in the oxygen-evolving complex in photosynthesis, *J. Am. Chem. Soc.*, **112**, 6387 (1990).

40. K. Wieghardt, U. Bossek, J. Bonvoisin, P. Beauvillain, J. J. Girerd, B. Nuber, J. Weiss and J. Heinze, Binuclear manganese(II,III,IV) model complexes for the active centers of metalloprotein photosystems. II. Preparation, magnetism, and crystal structure of $[LMn(II)(\mu\text{-})O)\text{-}(\mu\text{-}OAc)_2Mn(IV)L][ClO_4]_3$ (L = *N,N',N''*-trimethyl-1,4,7-triazacyclononane), *Angew. Chem. Int. Ed. Engl.*, **25**, 1030 (1986).

41. R. J. P. Williams, The where, the when, the how and the why of biological oxygen reactions, *Chem. Scr.*, **28A**, 5 (1988).

42. M. L. Kirk, M. K. Chan, W. H. Armstrong and E. I. Solomon, Ground-state electronic structure of the dimer-of-dimers complex $[(Mn_2O_2)_2(tphpn)_2]^{4+}$: potential relevance to the photosystem II water oxidation catalyst, *J. Am. Chem. Soc.*, **114**, 10432 (1992).

43. H. H. Thorp, J. E. Sarneski, G. W. Brudvig and R. H. Crabtree, Proton-coupled electron transfer in $[(bpy)_2Mn(O)_2Mn(bpy)_2]^{3+}$, *J. Am. Chem. Soc.*, **111**, 9249 (1989).

44. T. J. Meyer, Chemical approaches to artificial photosynthesis, *Acc. Chem. Res.*, **22**, 163 (1989).

45. J. A. Gilbert, D. S. Eggleston, W. R. Murphy, D. A. Geselowitz, S. W. Gersten, D. J. Hodgson and T. J. Meyer, Structure and redox properties of the water-oxidation catalyst $[(bpy)_2(OH_2)RuORu(OH_2)(bpy)_2]^{4+}$, *J. Am. Chem. Soc.*, **107**, 3855 (1985).

5 The Dioxygen Molecule, O_2: Uptake, Transport and Storage of an Inorganic Natural Product

5.1 Molecular and Chemical Properties of Dioxygen, O_2

The rather high concentration of potentially reactive dioxygen, O_2, in the earth's atmosphere (about 21 vol%) is the result of continuous photosynthesis of 'higher' organisms. Thus, O_2 is a *natural product*, i.e. a secondary metabolic product, just like, for example, alkaloids or terpenes; initially, it must have been classified as an exclusively toxic waste product. Studies of the atmospheres of planets and moons have generally shown O_2 contents of far less than 1 vol% and the primeval atmosphere of the earth until about 2.5 billion years ago is assumed to have been very similar. Because of the continuous growth of organisms and a corresponding need for reduced carbon compounds synthesized from fixated CO_2, the amount of photosynthesis increased to such an extent that the simultaneously produced oxidation equivalents could no longer be scavenged by auxiliary substrates such as sulfur($-$II) or iron($+$II) compounds; eventually, a (Mn-)catalyzed oxidation (Section 4.3) of the surrounding water to dioxygen developed. For a while, the 'environmentally' extremely harmful O_2 could be deactivated by reaction with reduced compounds, particularly soluble ions such as Fe^{2+} or Mn^{2+}, to form massive oxidic sediments, e.g. the 'banded iron formations' (5.1). However, about 2 billion years ago the concentration of O_2 in the atmosphere started to increase; at least since about 400 million years ago the obviously very stable equilibrium between biogenic production and general, biogenic and nonbiogenic consumption of O_2 has been reached with the well-known stationary concentration.

$$4 \text{ Fe}^{2+} + O_2 + 2 \text{ H}_2\text{O} + 8 \text{ OH}^- \rightarrow 4 \text{ Fe(OH)}_3 \rightarrow 2 \text{ Fe}_2\text{O}_3 + 6 \text{ H}_2\text{O} \qquad (5.1)$$

The appearance of free dioxygen, O_2, as a *toxic waste product of an energy-producing process* was a true 'ecological catastrophe'. Most organisms living at that time probably perished as a consequence of this biogenic 'self-poisoning'; however, some may have survived in O_2-free niches as today's *anaerobic* organisms. In addition to the strongly oxidizing character of O_2, partly reduced and highly reactive species (4.6) are easily formed through catalysis by transition metals such as Fe^{II}. Detoxification with regard to these species has made it necessary to develop a multitude of biological antioxidants (see Table 16.1). Therefore, only those *aerobic* organisms have survived in contact with the atmosphere that were able to develop protective mechanisms against O_2 *and* the very toxic intermediates (often radicals) resulting from its partial reduction (4.6, 5.2 [1,2]).

$$O_2 \quad \text{(dioxygen)}$$

$$-e^- \Big\Updownarrow +e^-$$

$$O_2^{\bullet-} \underset{-H^+}{\overset{+H^+}{\rightleftharpoons}} HO_2^{\bullet} \qquad\qquad (pK_a \approx 4.7)$$

superoxide
(radical anion)

$$-e^- \Big\Updownarrow +e^-$$

$$O_2^{2-} \underset{-H^+}{\overset{+H^+}{\rightleftharpoons}} HO_2^- \underset{-H^+}{\overset{+H^+}{\rightleftharpoons}} H_2O_2 \qquad\qquad \begin{array}{l} (pK_a \approx 11.6) \\ \text{(first stage)} \end{array}$$

peroxide   hydro-peroxide   hydrogen peroxide

$$-e^- \Big\Updownarrow +e^-$$

$$[O^{2-} + O^{\bullet-}] \overset{+3H^+}{\rightleftharpoons} H_2O + {}^{\bullet}OH \qquad\qquad (pK_a \approx 10)$$

water   hydroxyl radical

$$-e^- \Big\Updownarrow +e^-$$

$$2\,O^{2-} \underset{-2H^+}{\overset{+2H^+}{\rightleftharpoons}} 2\,OH^- \underset{-2H^+}{\overset{+2H^+}{\rightleftharpoons}} 2\,H_2O \qquad\qquad \begin{array}{l} (pK_a \approx 15.7) \\ \text{(first stage)} \end{array}$$

oxide   hydroxide   water   (5.2)

Through the process of evolution organisms have developed which were able to use the reverse process of photosynthesis (4.1) in an oxygen-containing atmosphere for a much more efficient and only indirectly light-dependent metabolic energy conversion and thus for a more dynamic form of life. This so-called 'respiration' may be viewed as a 'cold', i.e. controlled and carefully catalyzed combustion of reduced substrates, also known as 'food'. This remarkable development required not only a successful and eventually even biosynthetically useful degradation of partially reduced oxygen intermediates such as H_2O_2, $O_2^{\bullet-}$ or ${}^{\bullet}OH$ but also a coping with a drastically changed bioavailability of some elements and their compounds [3]. The following changes resulted from the formation of an oxidized atmosphere at pH 7:

$$Fe^{II}\,(\text{soluble}) \rightarrow Fe^{III}\,(\text{insoluble})$$
$$Cu^{I}\,(\text{insoluble}) \rightarrow Cu^{II}\,(\text{soluble})$$
$$S^{-II}\,(\text{insoluble}) \rightarrow SO_4^{2-}\,(\text{soluble})$$
$$Se^{-II}\,(\text{insoluble}) \rightarrow SeO_3^{2-}\,(\text{soluble})$$
$$MoS_x\,(\text{insoluble}) \rightarrow MoO_4^{2-}\,(\text{soluble})$$
$$CH_4 \rightarrow CO_2$$
$$H_2 \rightarrow H_2O$$
$$NH_3 \rightarrow NO_3^-,\ NO_2^-,\ NO$$

Incidentally, as a consequence of O_2 production the ozone layer in the stratosphere has formed which may have contributed to a more controlled development of organisms by protecting them from high-energy components of solar radiation.

An understanding of why the presence of free O_2 is, at the same time, extraordinary hazardous but also a great opportunity for organisms requires a discussion of the molecular properties and coordination behavior of dioxygen [4,5]. According to its position in the periodic table and the resulting second highest electronegativity of all elements, O_2 is a very strong oxidizing agent. Many substances react very exothermically with dioxygen although activation is often required. This frequently observed and quite characteristic energy barrier for many reactions involving O_2 can be rationalized considering the triplet ground state of the O_2 molecule with two unpaired electrons. This situation, which is quite unusual for a small, stable molecule, is a consequence of the molecular orbital scheme (5.3) in connection with Hund's rule (compare 2.8). When filling up degenerate orbitals such as $\pi^*(2p)$, the state with maximum multiplicity is favored, in the case of the dioxygen triplet $^3O_2$ over the singlet $^1O_2$.

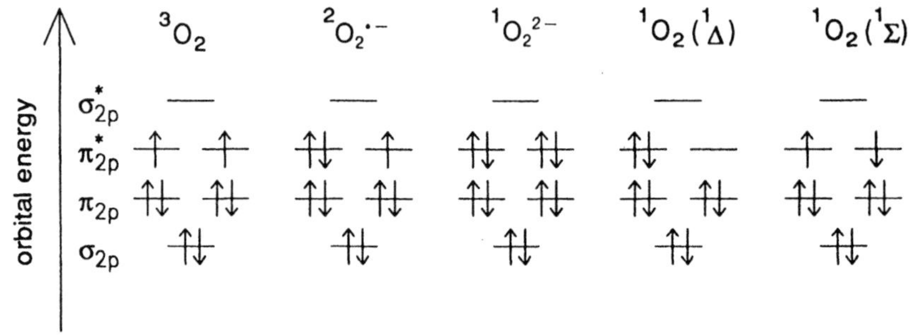

orbital energy		$^3O_2$	$^2O_2^{\cdot-}$	$^1O_2^{2-}$	$^1O_2(^1\Delta)$	$^1O_2(^1\Sigma)$
	σ^*_{2p}	—	—	—	—	—
	π^*_{2p}	↑ ↑	↑↓ ↑	↑↓ ↑↓	↑↓ —	↑ ↓
	π_{2p}	↑↓ ↑↓	↑↓ ↑↓	↑↓ ↑↓	↑↓ ↑↓	↑↓ ↑↓
	σ_{2p}	↑↓	↑↓	↑↓	↑↓	↑↓

bond order	2	1.5	1
bond length (pm)	121	≈ 128	≈ 149
vibrational frequency (cm^{-1})	1560	1150–1100	850–740

$$(5.3)$$

In fact, $^3O_2$ with two unpaired electrons in the ground state is favored over both singlet states $^1O_2$ ($^1\Delta$) and $^1O_2$ ($^1\Sigma$) by more than 90 and 150 kJ/mole, respectively. The reactions of $^3O_2$ with most normal singlet molecules are thus inhibited because of the necessity for a statistically less likely 'flipping' of spins (compare 4.12). In fact, this phenomenon is responsible for the present *metastable* situation with the simultaneous presence of combustibles (wood, fossil fuels, carbohydrates, etc.) and an oxygen-rich atmosphere without instant formation of the lowest energy products CO_2 and water. Therefore, chemically or photogenerated singlet dioxygen $^1O_2$ represents another toxic form of oxygen; the diamagnetic elemental modification ozone, O_3, also shows uninhibited and therefore uncontrolled oxidation behavior toward biomolecules.

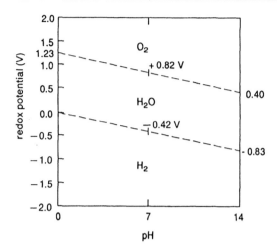

Figure 5.1
Stability diagram of water (- - - -: equilibrium lines)

The 'spin-verbot' is not valid for reaction partners of $^3O_2$ which may easily undergo single-electron transfer reactions [6] or which already contain unpaired electrons. The latter include:

(a) radicals ($S = 1/2$), either as stable free species or as intermediates produced, for example, in an ignition,
(b) compounds with photochemically produced excited triplet states ($S = 1$) and
(c) paramagnetic transition metal centers ($S \geq 1/2$).

Almost all reactions between O$_2$ and metal complexes proceed irreversibly, as illustrated by equations (5.1) or (5.12). In most of these reactions oxygen is eventually reduced to the ($-$II) state so that oxide, hydroxide or water ligands involving a cleavage of the oxygen–oxygen bond result. During the first two one-electron reduction steps (compare 4.6, 5.2) to superoxide radical anion ($O_2^{\cdot -}$, $S = 1/2$) and peroxide dianion (O_2^{2-}, $S = 0$), the oxygen–oxygen bond remains intact while the bond order is stepwise reduced from a double to a single bond. This corresponds to the placement of additional electrons into the degenerate, weakly antibonding $\pi^*(2p)$ molecular orbital (5.3) until this is completely filled. The redox potentials for the reduction of O$_2$ decrease in the presence of electrophiles such as metal ions or H$^+$. The biochemically important stability diagram of water, illustrating the thermodynamic equilibrium between H$_2$O and H$_2$ or O$_2$, respectively, as a function of potential and pH is shown in Figure 5.1; all other states are metastable.

Inorganic and organometallic compounds (5.4) of the metals cobalt, rhodium and iridium from group 9 of the periodic table have long been known to form rather simple *reversibly* dioxygen-coordinating complexes [5]. Vaska's iridium complex shows reversible uptake and release of 'side-on' (η^2) coordinated O$_2$ while pentacyano-, *salen-* (compare 3.16) or pentaammine–cobalt(II) complex fragments can coordinate 'end-on' η^1-O$_2$ (5.4). Further possibilities of coordination which are also discussed in biological O$_2$ utilization are summarized in (5.6).

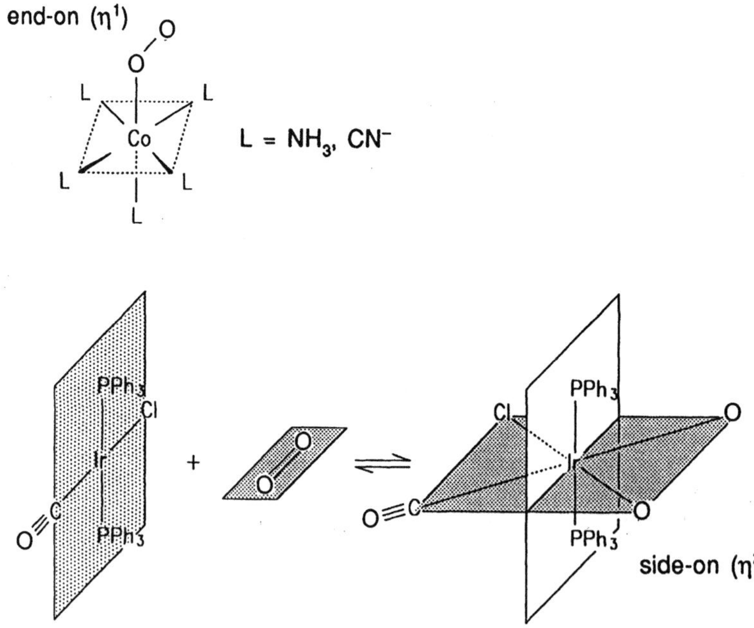

(5.4)

Structural and spectroscopic studies (EPR) suggest that the redox-active 'non-innocent' dioxygen ligand is mainly bound in singly or doubly reduced form (5.5). The ability of superoxide and peroxide as well as of oxide and hydroxide to frequently act as bridging (μ-)ligands between metal centers contributes to the irreversibility of 'O_2' coordination (compare 5.5, 5.6 and 5.12).

The unsaturated molecule O_2 is a σ or π donor/π acceptor ligand. Electron density is transferred to the electropositive metal center via the free electron pairs or the double bond while high-energy, partly filled d orbitals of electron-rich metal centers can contribute to back-bonding via π back-donation into one of the partly filled π^* orbitals of O_2 (5.7).

After intramolecular electron rearrangement the (formal) oxidation states can be reassigned according to criteria (5.3) (see also 5.5).

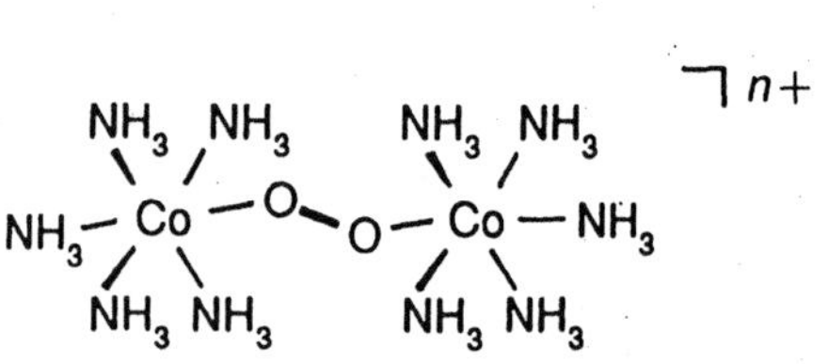

$n = 5, \quad d(O—O) = 131$ pm, $O_2^{\cdot -}$ ligand

$n = 4, \quad d(O—O) = 147$ pm, O_2^{2-} ligand

(5.5)

metal coordination of O_2^{n-}

type of structure:		mode of O_2^{n-} coordination:	example:
	η^1	end-on	$[(CN)_5Co(O_2)]^{3-}$
	η^2	side-on	$(Ph_3P)_2(CO)(Cl)Ir(O_2)$ $(HBPz_3^*)Co(O_2)$: [7]
	$\mu-\eta^1{:}\eta^1$	end-on bridging	$[(H_3N)_5Co(O_2)Co(NH_3)_5]^{5+}$
	$\mu-\eta^2{:}\eta^2$	side-on bridging	$[(Cl_3O_2U)_2(O_2)]^{4-}$ $[(HBPz_3^*)Cu]_2(O_2)$: [8]
	$\mu-\eta^1{:}\eta^2$	end-on/side-on bridging	$[(Ph_3P)_2ClRh]_2(O_2)$
	μ_4-	end-on, fourfold bridging	$Fe_6O_2(O_2)(OCPh)_{12}(OH)_2$: [9]

$HBPz_3^*$: tris(3,5-diisopropylpyrazolyl)borate :

(5.6)

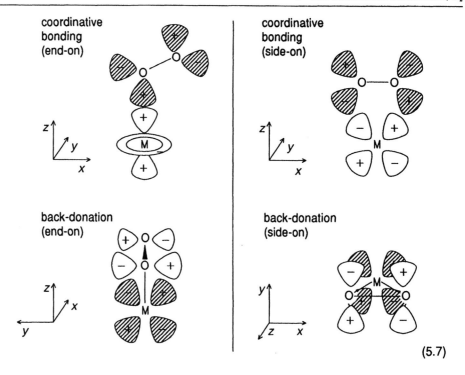

$$(5.7)$$

5.2 Oxygen Transport and Storage through Hemoglobin and Myoglobin

What functions are required by organisms for a controlled utilization of O_2? Before dioxygen can be metabolized, it has to be taken up (reversibly) from the atmosphere and transported to oxygen-depleted tissue where it must be stored until actual use. Certain groups of molluscs, crustaceans, spiders and worms, on one hand, and the majority of higher organisms, particularly vertebrates, on the other hand, differ through their strategies in O_2 coordination. While the formerly mentioned groups of invertebrates contain binuclear metal arrangements with amino acid coordination, viz. the copper protein hemocyanin (Hc, Section 10.2) or the iron protein hemerythrin (Hr, Section 5.3), the higher breathing organisms use the heme system, i.e. monoiron complexes of a certain porphyrin macrocycle, protoporphyrin IX (2.5, 5.8; see also Chapter 6). The corresponding proteins are the tetramer hemoglobin (Hb, O_2 uptake in the lungs and transport in the blood stream) and the monomer myoglobin (Mb, O_2 storage and transport within muscle tissue).

At this point, the versatile role of iron in general (Chapters 5 to 8) and that of the heme group for the biochemistry of humans in particular (Table 5.1) should be specified. Since dioxygen transport is not a catalytic but a 'stoichiometric' function, about 65% of the iron present in a human body is confined to the transport protein hemoglobin (Hb) alone; the contents of the oxygen-storage protein myoglobin is roughly 6%. After all, the share of O_2 in air is only about 21 vol% and a sufficient level has to be maintained in the tissue even under unfavorable circumstances, e.g. above 2000 m sea level; human blood has an approximately 30 times higher

Table 5.1 Distribution of the major iron-containing proteins in an adult human (modified from [10])

protein	molecular mass of the protein (kDa)	amount of iron (g)	% of total body iron	type of iron: heme (h) or non-heme (nh)	number of iron atoms per molecule	function	mentioned in Section
hemoglobin	64.5	2.60	65	h	4	O_2 transport in blood	5.2
myoglobin	17.8	0.13	6	h	1	O_2 storage in muscle	5.2
transferrin	76	0.007	0.2	nh	2	iron transport	8.4.1
ferritin	444	0.52	13	nh	up to 4500	iron storage in cells	8.4.2
hemosiderin		0.48	12	nh		iron storage	8.4.3
catalase	260	0.004	0.1	h	4	metabolism of H_2O_2	6.3
peroxidases	variable	small	small	h	1	metabolism of H_2O_2	6.3
cytochrome c	12.5	0.004	0.1	h	1	electron transfer	6.1
cytochrome c oxidase	>100	<0.02	<0.5	h	2	terminal oxidation ($O_2 \rightarrow H_2O$)	10.4
flavoprotein oxygenases (e.g. P-450 system)	about 50	small	small	h	1	incorporation of molecular oxygen	6.2
iron–sulfur proteins	variable	about 0.04	about 1	nh	2-8	electron transfer	7.1-7.4
ribonucleotide reductase	260 (E. coli)	small	small	nh	4	transformation of ribonucleic acids to deoxyribonucleic acids	7.6

'solubility' for O_2 than water. Metal-storage proteins like ferritin (Section 8.4.2) contain most of the rest of the iron in the human body; the catalytically active enzymes are present in only minute amounts.

Many but not all (see Chapter 7) of the redox catalytically active iron enzymes contain the heme group; peroxidases, cytochromes, cytochrome c oxidase and the P-450 system (Chapter 6 and Section 10.4) belong to the group of heme proteins. Scheme (5.8) illustrates what determining role the protein environment can play for the quite varying biological functions of a tetrapyrrol complex.

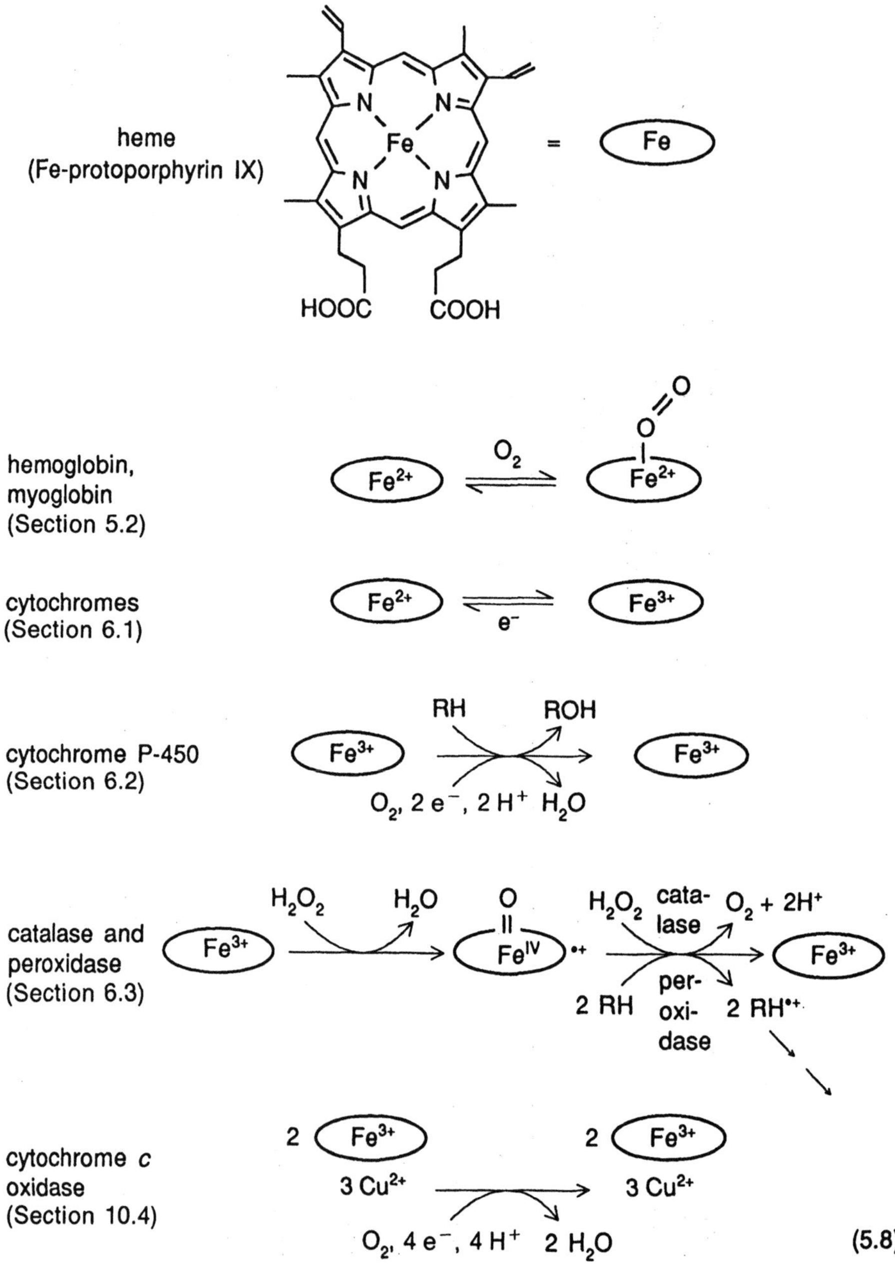

heme
(Fe-protoporphyrin IX)

hemoglobin,
myoglobin
(Section 5.2)

cytochromes
(Section 6.1)

cytochrome P-450
(Section 6.2)

catalase and
peroxidase
(Section 6.3)

cytochrome c
oxidase
(Section 10.4)

(5.8)

Figure 5.2
Schematic structures of myoglobin (left) and the tetrameric protein hemoglobin (right), each with protein folding and indicated heme-'disc' (from [11])

The transport system for O_2 has to take up this molecule as effectively as possible in its ground state form, $^3O_2$, from the gas phase, to transport it in specialized blood cells, the erythrocytes, via the circulatory system to an intermediate storage site, and to release it there completely. This task is not easily solved since both supply and demand of O_2 may change considerably; the storage system always has to have a higher affinity for O_2 than the nonetheless efficient transport system. In the case of hemoglobin (Figure 5.2) with its total of four heme sites, this efficiency is guaranteed by the cooperative effect [12]. In the course of the loading with four molecules O_2, corresponding to 1 ml O_2 per 1 g Hb, the oxygen affinity increases as graphically illustrated by a sigmoidal, nonhyperbolical saturation curve (Figure 5.3).

According to Figure 5.3, the *cooperative effect* guarantees an efficient transfer of O_2 to the storage site: the less O_2 is present in the transport system, the more completely it is released into storage. The biological reason for such a system with

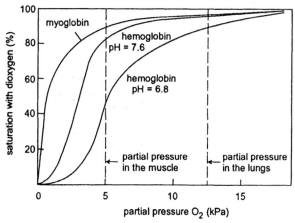

Figure 5.3
Oxygen saturation curves of myoglobin and hemoglobin at different pH values

pH regulated (compare 5.11 and Figure 5.3) 'all or nothing' functionality is directly evident; its molecular realization, however, requires a complex interaction of several hemoprotein subunits (see below), which is still not understood in full detail [12,13].

This is all the more astonishing as the structural determination of myoglobin and hemoglobin (the latter was the first protein to be crystallized in 1849) had been accomplished quite early by the groups of J. C. Kendrew and M. F. Perutz (Nobel prize 1962) and some first mechanistic hypotheses also originated from that time. In simple terms, Hb (2 × 141 and 2 × 146 amino acids, 64.5 kDa) is a tetramer of the monomeric hemoprotein myoglobin Mb (molecular mass 17.8 kDa); two α- and two β-peptide chains form a well-defined quarternary structure in Hb (Figure 5.2). The deoxy-heme systems inside the protein, which are often depicted as disc-shaped, are actually not completely flat but slightly domed due to interactions with the surrounding protein. Of two free axial coordination sites at each iron center one is occupied by the five-membered imidazole ring of the *proximal* histidine. The other axial position remains essentially free for O_2 coordination (Figure 5.4); however, there is a biologically meaningful [14] amino acid arrangement in the form of a *distal* histidine with its ability to form hydrogen bonds and in the form of a valine side chain containing an isopropyl group (Figure 5.5).

Before the molecular basis for the cooperative effect in Hb is discussed, the inorganic chemical questions regarding oxidation and spin states of the metal center before and after O_2 coordination, i.e. in the deoxy and oxy forms, have to be answered.

The presence of high-spin Fe^{II} in the deoxy forms of Hb and Mb is firmly established and an $S = 2$ ground state with four unpaired electrons (2.8, 2.9) is observed. The presence of an even number of unpaired electrons is favorable for a rapid, not spin-inhibited binding of $^3O_2$ with its $S = 1$ ground state (compare Section 4.3). For the diamagnetic oxy forms ($S = 0$) with end-on coordinated O_2 in a nonlinear arrangement with an Fe—O—O angle of about 120° (compare 5.10 or

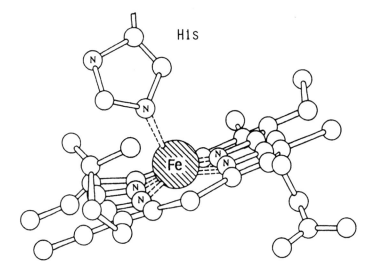

Figure 5.4
Simplified structure of the deoxy heme unit in Mb and Hb (neglecting the doming of the porphyrin ring; from [11])

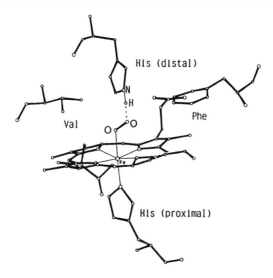

Figure 5.5
Arrangement of proximal and distal amino acid residues with regard to oxy-myoglobin (according to [14,15])

Figure 5.5), the oxidation state assignment is less unambiguous because of the noninnocent nature of coordinated dioxygen; two alternatives (5.9) have been proposed.

In 1936, Pauling and Coryell [16] explained the observed diamagnetism invoking a combination of low-spin Fe^{II} and coordinated singlet dioxygen. Both components would each be diamagnetic and binding would be achieved through the σ acceptor/π donor character of the reduced metal and the σ donor/π acceptor character of the unsaturated ligand (π back-bonding component; compare 5.7). The alternative was formulated by Weiss [17]. Following a single-electron transfer from the metal to the approaching ligand in the ground state, low-spin Fe^{III} with $S = 1/2$ and a superoxide radical anion $O_2^{\cdot-}$, also with $S = 1/2$, should be formed; the observed diamagnetism at room temperature would then be due to strong antiferromagnetic coupling. Some experimental results such as the vibrational frequency of the O—O bond at about $1100\,cm^{-1}$ (characteristic for $O_2^{\cdot-}$; compare 5.3), data from Mössbauer spectra (see Section 5.3) and some aspects of the chemical reactivity seem to favor the formulation of Weiss. In substitution reactions superoxide, $O_2^{\cdot-}$, behaves as a 'pseudo halide' similar, for example, to the azide anion, N_3^-, and, in fact, Mb-bound dioxygen is rather easily replaced by chloride. When iron is substituted by cobalt (hemoglobin → coboglobin), its neighbor in the periodic table with one additional electron, EPR spectroscopy shows that the additional unpaired electron resides mainly at the dioxygen ligand, corresponding to a formulation $O_2^{\cdot-}$ ($S = 1/2$)/low-spin Co^{III} ($S = 0$). However, this result for a model system provides only indirect evidence in favor of the Weiss formulation. Both alternatives have been studied under MO theoretical aspects (5.9) [18,19], but an unambiguous assignment has not yet been possible. It should be remembered here that oxidation states are not measurable quantities but are based on conventions regarding a formal bond heterolysis.

oxidation and spin states in the heme–O₂ system
(d orbital splitting for approximately octahedral
symmetry; see 2.9)

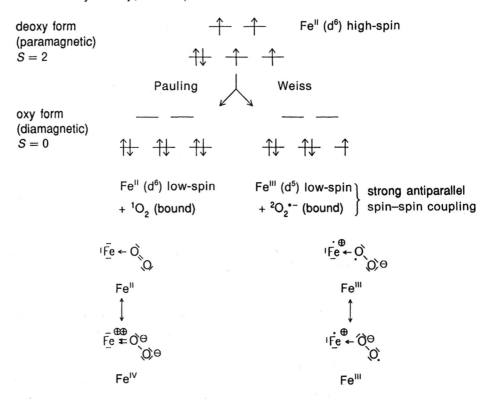

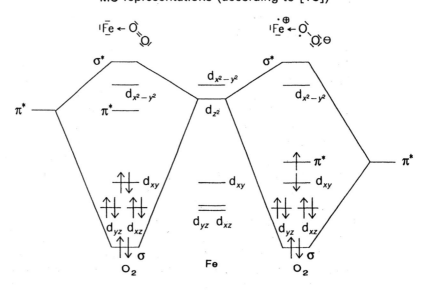

formulation according to Pauling   formulation according to Weiss (5.9)

An argument in favor of Pauling's model with divalent low-spin iron (which would fit better into the porphyrin plane; see Table 2.7 and also Figure 5.6) is the observation that carbon monoxide and other π acceptor ligands such as NO which predominantly coordinate to lower-valent metal centers may effectively replace dioxygen. To a small extent, an endogenous CO production takes place even during the degradation of porphyrin systems in aged erythrocytes. However, some mechanisms exist that at least partially counteract this unwanted competition of poisonous carbon monoxide [14]. According to the valence-bond description (5.10), $\eta^1(C)$-coordinating CO in protein-free model systems prefers a linear arrangement whereas end-on coordinated O_2 prefers a bent situation with an Fe—O—O angle of about $120°$ [15], due to the remaining free electron pair at the coordinating oxygen atom. In myoglobin, spatial restrictions through the protein environment as well as the possibility of hydrogen bond formation with the distal histidine (5.10, Figure 5.5) cause a more favorable equilibrium situation for binding of the definitely weaker π acceptor O_2 ($K_{CO}/K_{O2} = 200$ vs. 25 000 in protein-free heme model systems); nevertheless, it is well known that only small concentrations of CO in air can be tolerated.

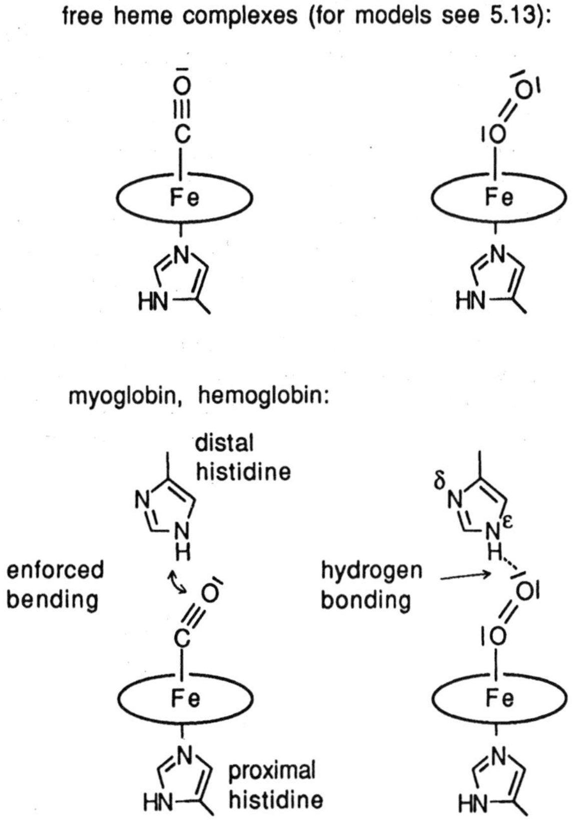

free heme complexes (for models see 5.13):

myoglobin, hemoglobin:

distal histidine

enforced bending   hydrogen bonding

Fe   Fe

proximal histidine   (5.10)

The imidazole ring of the distal histidine blocks an unhindered access to the sixth, 'free' coordination site at the iron center so that a controlled, rapid binding of small molecules may result only as a consequence of side-chain dynamics of the globin protein. Modifications of the distal histidine or of the valine site (Figure 5.5) led to a less favorable binding ratio O_2/CO [14]. Histidine is valuable as the distal amino acid residue because of its basic nature, thus keeping protons away from the coordinated O_2 or binding them via N_ε and liberating them on the other side via N_δ (histidine as the proton shuttle; compare Table 2.6). Protons act as electrophilic competitors in relation to the coordinating iron, weakening its bond to O_2 and thus favoring deleterious autoxidation processes.

It is remarkable that the coordination of an additional 'weak' ligand, either O_2 or $O_2^{\cdot-}$ (the latter after inner-sphere electron transfer), induces a change from a high-spin to a low-spin state of the metal ('spin crossover'). The high-spin situation which is nevertheless essential for the activation of $^3O_2$ is rather uncommon for tetrapyrrole complexes; its occurrence in the deoxy form is the result of a weakened metal–ligand interaction because of the incomplete fit into the cavity of the significantly domed macrocycle (out-of-plane situation; Figure 5.4 and Table 2.7). For this apparent entatic state situation, the relatively light 'pull' from coordinating O_2 or CO is sufficient to effect at least a partial charge transfer from the metal to the incoming ligand and a spin crossover of the metal center. The thus effected contraction of the metal (Table 2.7) and relative motion of about 20 pm toward the now better ('more strongly') coordinating macrocycle (Figure 5.6) are probably essential factors for the cooperative effect [12]. Another significant structural change during O_2 coordination concerns the straightening of the Fe—N bond to the proximal histidine with respect to the porphyrin plane (compare Figure 5.8).

The (positive) cooperative effect of the prototypical 'allosteric' protein hemoglobin as evident from the sigmoidal saturation characteristics (Figure 5.3) is due to the variable interaction between the four heme-containing units of tetrameric Hb, the changes being triggered by O_2 coordination to individual heme sites. Each of the four protein chains which are linked to each other by electrostatic interactions ('salt bridges') shows a geometrical change upon dioxygen coordination (Figure 5.6). This transformation is communicated ('transduced' [21]) to the other units by a process which, in the most simple concept, can be visualized by a 'spring-tension' model involving a conformational change of the quarternary structure. A two-state model (Figure 5.7) has been formulated with a tense (T, low O_2 affinity) and a relaxed (R, high O_2 affinity) state of the tetrameric protein.

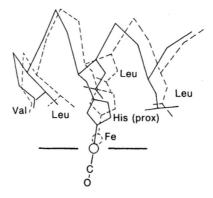

Figure 5.6
Structural changes in the transition from the deoxy (- - - -) to the carbon monoxy (——) form of hemoglobin (according to [20]; stereo images of the deoxy/oxy changes in [12])

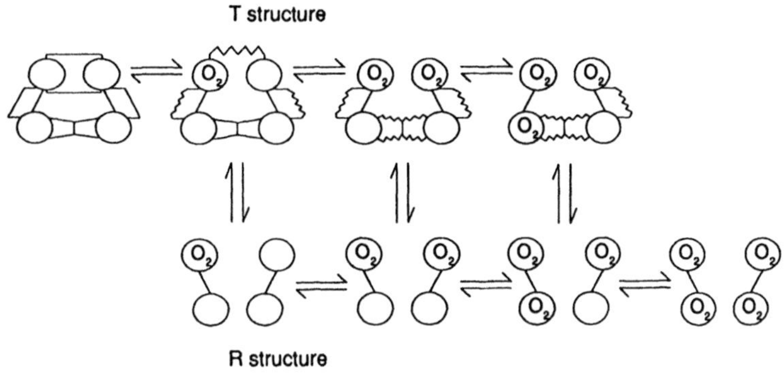

T structure

R structure

Figure 5.7
Functional model for the positive cooperativity of the four subunits in hemoglobin (according
to [12]): increasing O_2 coordination favors the transition to the more O_2-affine R structure;
after complete O_2 transfer to myoglobin with its even higher O_2 affinity the less compact T
structure is formed through multiple interprotein linkages

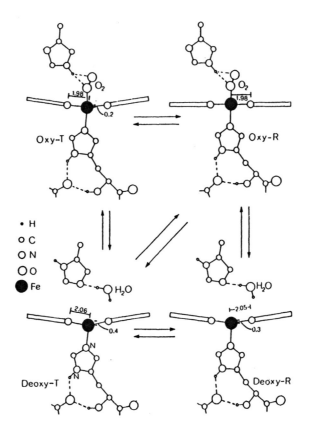

Figure 5.8
Schematic representation of the geometric differences for the four states of the heme environment in the α subunit of Hb (from [12]). Featured are the Fe—N bond lengths and the
distances from the iron center to the plane of the four pyrrole nitrogen atoms (in Å = 0.1 nm)

According to this scheme there are four different states for the individual heme system (oxy or deoxy form, T or R state), their geometrical features being depicted in Figure 5.8 [12]. The T form in particular features a steric interaction of the proximal histidine with the porphyrin ring, thus favoring a porphyrin bending as it is found in the nonplanar deoxy form [22].

Initial binding of O_2 to the significantly domed T form leads to the not yet completely planar oxy T form with a decrease of the Fe—N(porphyrin) distance before a relaxation takes place to a planar oxy R state with a straightened Fe—N bond. Release of O_2 from the oxy R form results in doming of the porphyrin system (deoxy R state). The distal amino acid residues are influenced by these geometrical changes (Figure 5.5), thus being able to effect allosteric interprotein interaction in the Hb tetramer [12]. When the R form is completely loaded with dioxygen, it shows a more compact quaternary structure than the T form and a higher degree of hydration (5.11 [23]).

$$(H^+)_2 Hb \; + \; 4 \, O_2 \; + \; ca. \; 60 \; H_2O \; \rightleftharpoons \; Hb(O_2)_4 \; + \; 2 \, H^+ \qquad (5.11)$$

The phenomenon of pH dependence of the sigmoidal saturation curve (Figure 5.3) is called the 'Bohr effect'. It originates from the fact that the binding of CO_2 (which is formed as the respiratory end product) to a terminal amino group of hemoglobin reduces its capability to bind oxygen, i.e. oxy-Hb is a stronger acid than deoxy-Hb (5.11). Hb thus acts as a O_2 *and* CO_2 transport protein.

With the release of O_2, Hb thus takes up protons and helps to convert 'carbonic acid, H_2CO_3,' to bicarbonate, HCO_3^- (Section 12.2 and Figure 13.13). The sigmoidal character of O_2-binding is more pronounced in (carbonic) acidic solution and the biological relevance for this is immediately obvious. In the case of an increased CO_2 production the CO_2/O_2 gas exchange is promoted in the sense that the transport protein becomes completely loaded with O_2.

Studies on hemoglobin are of continuous interest in biology and medicine [13]. For instance, special adaptations of animal Hb in deep-diving marine mammals or high-flying birds and the extracellular hemoglobins of worms are being investigated in biology, whereas in medicine the fetal Hb with its special dioxygen affinity, the stabilization of Hb outside of erythrocytes, the problem of sickle cell anemia, caused by a single altered amino acid [24] and the therapy of malaria which involves parasitic digestion of Hb [25] are worth mentioning. More than 200 different hemoglobin variants have been found for humans alone. The Hb-degrading *Plasmodium* parasites responsible for malaria have an interesting strategy to deactivate heme, which is quite toxic in the free state. The detoxification by the parasite has to occur through enzymatic coordination polymerization via iron–carboxylate binding because the parasite lacks the enzyme heme oxygenase. The antimalaria treatment with substances of the quinine type presumably effects the blocking of such heme polymerase enzymes [25].

In contrast to Hb and Mb, most simple iron(II)–porphyrin complexes which are not protected by a protein environment react irreversibly with O_2, eventually forming oxo-bridged dimers (5.12) via peroxo intermediates. Therefore, the deliberate synthesis of *reversibly* O_2-coordinating model compounds has posed a chemical challenge for quite some time [26,27].

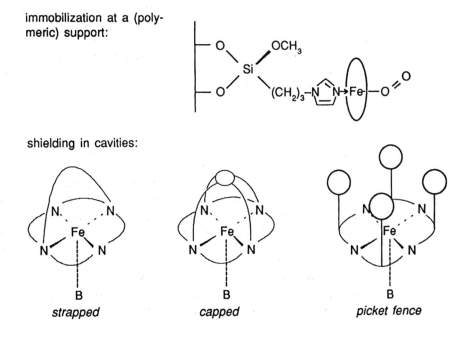

The unwanted dimerization of metalloporphyrins as shown in (5.12) can be avoided using different strategies (5.13). Among the available approaches, Collman's [28] 'picket fence porphyrins', in which space-demanding vertical 'pickets' on the porphyrin ring leave a cavity for O_2 binding but prohibit an $O_{(2)}$ connection between two metalloporphyrin systems, have to be mentioned. Immobilization of the heme analogue, e.g. on the surface of a silica support, also prevents irreversible dimerization.

strategies to prevent reaction (5.12):

immobilization at a (polymeric) support:

shielding in cavities:

strapped   *capped*   *picket fence* an oxygenated 'picket fence' porphyrin complex (from [27]):

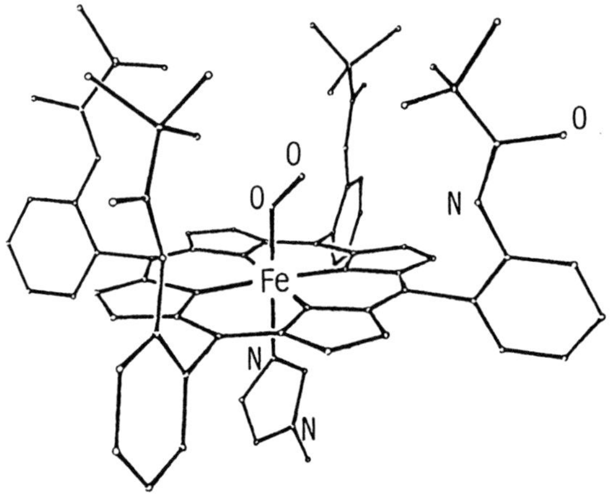

(5.13)

5.3   Alternative Oxygen Transport in Some Lower Animals: Hemerythrin and Hemocyanin

Several groups of invertebrates like crustaceans, molluscs (e.g. snails), arthropods (e.g. spiders) or marine worms possess nonporphinoid metalloproteins for reversible O_2 uptake. While iron-containing hemerythrin (Hr) which does, contrary to its name, *not* contain a heme system according to (5.8) ($\alpha\tilde{\iota}\mu\alpha$: blood) occurs mainly as an octameric protein with a molecular mass of 8×13.5 kDa, the even more complex copper-containing hemocyanins (Hc, Section 10.2) feature molecular masses of more than 1 MDa. For hemerythrin, many of the fundamental characteristics of the active center have by now been verified structurally [29] (compare 5.17 and Figure 5.9); nevertheless, the contributions of various physical methods to an initially *indirect* structural elucidation of the O_2-coordinating centers will be exemplarily summarized in the following [30,31].

5.3.1   Magnetism

Magnetic measurements of the virtually colorless deoxy form of hemerythrin indicate the presence of high-spin iron(II) with four unpaired electrons (2.9) at each center, just as in deoxy Hb and Mb. A weak antiparallel spin-spin coupling (4.11) between two apparently neighboring centers was observed. For the red-violet oxy form with one dioxygen bound per Fe dimer, the susceptibility measurements indicate the presence of two strongly antiferromagnetically coupled ($S = 1/2$) centers which led to the conclusion that each monomeric oxyprotein contains *two* effectively interacting low-spin iron(III) centers (compare 5.9). The strength of the spin–spin coupling can be inferred from the temperature dependence of the magnetic susceptibility: the stronger this coupling, the more thermal energy is needed to observe normal paramagnetic behavior, i.e. uncoupled spins.

5.3.2 Light Absorption

The absence of strong light absorption in the deoxy form suggests protein-bound metal centers because a pronounced color would be expected if a porphyrin π system were present. The color of the oxy form in the apparent absence of π conjugated macrocyclic ligands can be attributed to a charge-transfer transition which is 'allowed' (compare 5.7) due to compatible symmetry and a considerable orbital overlap (ligand-to-metal charge transfer, LMCT). This charge transfer in an electronically excited state occurs from an electron-rich peroxide ligand with a doubly occupied $\pi^*(2p)$ orbital (5.3) to electron-deficient, oxidized iron(III) with only partly filled d orbitals (5.14 [32]).

$$\text{Fe}^{III}(O_2^{2-}) \xrightarrow{\quad h\nu \quad} {}^{\cdot}[\text{Fe}^{II}(O_2^{\cdot-})] \qquad (5.14)$$

$$\text{ground state} \qquad\qquad\qquad \text{LMCT excited state}$$

Many peroxo complexes of transition metals, particularly in higher oxidation states, are colored for that very reason; for instance Ti^{IV}, V^{V} or Cr^{VI} may be analytically detected by color-forming reactions with H_2O_2. In contrast, the light absorption of the deoxy form is weaker by several orders of magnitude since electronic transitions in the visible region can only occur between d orbitals at the *same* metal center. By definition, these d orbitals are of different symmetry (2.7) and the corresponding 'ligand-field' transitions are therefore very weak.

5.3.3 Vibrational Spectroscopy

If a 'resonance' Raman experiment is carried out in the wavelength range of the LMCT absorption band of oxyhemerythrin, a resonance-enhanced vibrational band at 848 cm^{-1} is observed which is characteristic for peroxides (5.3). When the isotopic combination $^{16}O-^{18}O$ is used for the bound dioxygen species, *two* signals are obtained for the O—O stretching vibration, suggesting a strongly asymmetric coordination, e.g. end-on (Figure 5.5).

Resonance Raman Spectroscopy

In this method, molecular vibrations are being excited in a scattering experiment (Raman effect) using an *absorption wavelength* [33,34]. Through coupling of electronic and vibrational transitions, a selective enhancement, the resonance, results for a limited number of vibrational bands which are associated with the chromophor, e.g. a tetrapyrrole macrocycle. Those parts of the molecule experiencing a major geometry change following the electronic excitation respond particularly. This highly selective kind of vibrational spectroscopy is thus suitable even for large proteins where normal infrared or Raman spectra do not provide very useful information because of the large number of atoms.

5.3.4 Mössbauer Spectroscopy

Mössbauer spectroscopy of oxyhemerythrin shows two distinctly different Fe^{III} resonance signals whereas the Fe^{II} centers of the deoxy form cannot be distinguished.

Mössbauer Spectroscopy

Mössbauer spectroscopy is a nuclear absorption/nuclear emission spectroscopy (nuclear fluorescence) involving γ radiation. The strong line-broadening expected at such high energies of several keV is circumvented by embedding the absorbing nucleus in a solid-state matrix at low temperatures; the resulting distribution of recoil energy over many atoms makes resonance detection possible (5.15) [35,36].

Mössbauer spectroscopy
schematic setup:

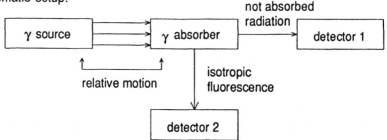

energy level diagram (nuclear energy levels)

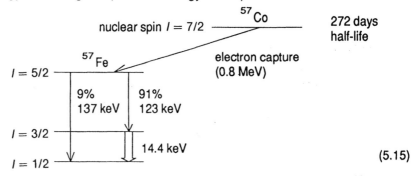

$$(5.15)$$

The very small *relative* linewidth of γ photons (ratio of about 10^{-13} between linewidth and energy) allows a detection of minute effects in the chemical environment (electron shell) of the absorbing nucleus. The actual measurement makes use of the Doppler effect which occurs when the γ emitter and the absorber are moved relative to each other with a stepwise change of constant velocity. The resonance signal observed exhibits an 'isomer shift' and a quadrupolar splitting which reflect the chemical environment of the nucleus (symmetry, ligand field) and, in particular, the *spin and ionization state*. Unfortunately, only very few nuclei are well suited for this physical method; by far the most important isotope in the area of bioinorganic chemistry is ^{57}Fe which is formed in a nuclear excited state during the radioactive decay of ^{57}Co, yielding a γ radiation of 14.4 keV energy.

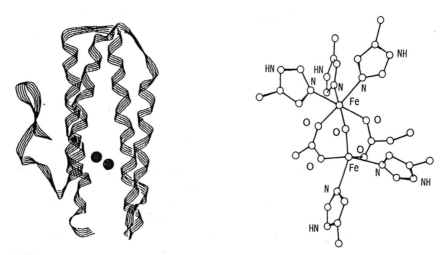

$$(5.16)$$

5.3.5. Structure

Of the conceivable alternatives (5.16) for the $O_2(^{2-})$ coordination by the dinuclear metal arrangement in hemerythrin [30] only the situation 2 or strongly distorted arrangements 4 or 5 remain, considering the above-mentioned results. Scheme (5.17) shows the actual coordination environments for both forms [29] as suggested by crystal structure analyses of various derivatives (Figure 5.9 [38]).

Figure 5.9
Protein folding in monomeric hemerythrin with the positions of both iron atoms (left) and structure of the iron dimer center (right) in the 'met'-form ($Fe^{III}Fe^{III}$ without HO_2^-, i.e. oxidized, but not oxygenated; according to [37])

deoxyhemerythrin   oxyhemerythrin (5.17)

The doubly $\eta^1:\eta^1$ carboxylate- and singly hydroxide-bridged high-spin iron(II) centers of the deoxy forms are coordinatively saturated with five histidine ligands except for one position at Fe_A. When dioxygen is bound there, *both* centers are oxidized to Fe^{III} under simultaneous reduction of the substrate to the peroxide state and the metal centers are more tightly linked by formation of an oxo bridge ('super-exchange' $\rightarrow$ antiferromagnetic coupling). The O_2 taken up presumably exists as hydroperoxo ligand, HOO^-, allowing for a hydrogen bond to be formed with the bridging oxide ion (5.17). While this kind of reactivity between O_2 and Fe^{II} is not surprising (compare 5.1, 5.12), it is astonishing that the O_2 binding in the protein is *reversible*. Structural model systems for hemerythrin have been synthesized with components (5.18 [39]) comparable to those used in modeling dinuclear manganese species (4.14) and several physical data of the protein could be reproduced with these models [29,40].

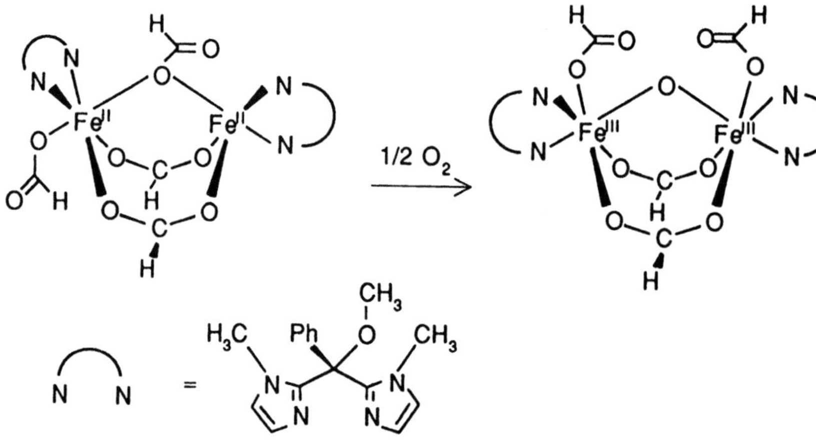

(5.18)

The copper-containing protein hemocyanin will be described in detail in Section 10.2 because of its structural relationship to copper-dependent oxygenases. However, some results of physical examinations parallel those of hemerythrin; e.g. a dimeric arrangement of histidine-coordinated metal centers is assumed for each protein subunit (see 10.6). The diamagnetic deoxy form features two copper(I) centers with formally closed, i.e. completely filled 3d shells; the blue oxy form of hemo*cyanin* contains antiferromagnetically interacting Cu^{II} dimers (d^9 configuration, $2 \times S = 1/2$) and an obviously $\mu\text{-}\eta^2:\eta^2$-coordinated peroxide ligand, O_2^{2-}. In both non-heme systems, Hr and Hc, the cooperative effect between the corresponding protein subunits is smaller than that of hemoglobin of higher animals.

In conclusion, the analogies and differences in oxygen coordination by heme systems (Hb, Mb) and by heme-free metal dimers (Hr, Hc) may be summarized as follows:

(a) The presence of high-spin Fe^{II} with four unpaired electrons as the $^3O_2$-coordinating center and its spin crossover to a low-spin system, formulated either as Fe^{II} or Fe^{III}, is a common feature for Hb, Mb and Hr; the special alternative of the Cu system Hc will be discussed in Section 10.2.

(b) Dioxygen *reversibly* bound to an iron center always shows *end-on* coordination (η^1) which seems better for the necessary rapid exchange than *side-on* coordination or bridging.

(c) The differences lie in the obvious two-electron transfer through the metal dimers to create peroxo ligands versus a smaller extent of electron transfer in the oxy-heme species which involve superoxo or dioxygen ligands, depending on the model used.

(d) The electron 'buffer capacity' necessary for the coordination of small, centrosymmetrical, unsaturated molecules is realized through metal–metal cooperation (cluster effect) in nonheme systems whereas the heme-containing Hb and Mb exhibit interaction between only one redox-active iron center and the redox-active π system of the porphyrin ligand.

Both types of composite systems are apparently better suited for reversible O_2-coordination than simple, isolated metal centers; the hemoproteins seem to represent the more elegant, more flexible and more efficient system, in particular with regard to cooperativity.

References

1. W. Ando and Y. Moro-oka (eds.), *The Role of Oxygen in Chemistry and Biochemistry*, Springer-Verlag, Berlin, 1988.
2. D. T. Sawyer, *Oxygen Chemistry*, Oxford University Press, Oxford, 1991.
3. J. J. R. Fraústo da Silva and R. J. P. Williams, *The Biological Chemistry of the Elements*, Clarendon Press, Oxford, 1991.
4. H. Taube, Interaction of dioxygen species and metal ions—equilibrium aspects, *Prog. Inorg. Chem.*, **34**, 607 (1986).
5. E. C. Niederhoffer, J. H. Timmons and A. E. Martell, Thermodynamics of oxygen binding in natural and synthetic dioxygen complexes, *Chem. Rev.*, **84**, 137 (1984).
6. M. Chanon, M. Julliard, J. Santamaria and F. Chanon, Role of single electron transfer in dioxygen activation. Swing activation in photochemistry, electrochemistry, thermal chemistry, *New. J. Chem.*, **16**, 171 (1992).
7. J. W. Egan, B. S. Haggerty, A. L. Rheingold, S. C. Sendlinger and K. H. Theopold, Crystal structure of a side-on superoxo complex of cobalt and hydrogen abstraction by a reactive terminal oxo ligand, *J. Am. Chem. Soc.*, **112**, 2445 (1990).
8. N. Kitajima, K. Fujisawa and Y. Moro-oka, μ–η^2: η^2-Peroxo binuclear copper complex, $[Cu(HB(3,5-iPr_2pz)_3)]_2(O_2)$, *J. Am. Chem. Soc.*, **111**, 8975 (1989).
9. W. Micklitz, S. G. Bott, J. G. Bentsen and S. J. Lippard, Characterization of a novel μ_4-peroxide tetrairon unit of possible relevance to intermediates in metal-catalyzed oxidations of water to dioxygen, *J. Am. Chem. Soc.*, **111**, 372 (1989).
10. F. A. Cotton and G. Wilkinson, *Advanced Inorganic Chemistry*, 5th edn, Wiley, New York, 1988, p. 1337.
11. J. E. Huheey, *Inorganic Chemistry, Principles of Structure and Reactivity*, 3rd edn, Harper & Row, New York, 1983, pp. 870, 875.
12. M. F. Perutz, G. Fermi, B. Luisi, B. Shaanan and R. C. Liddington, Stereochemistry of cooperative mechanisms in hemoglobin, *Acc. Chem. Res.*, **20**, 309 (1987).
13. R. E. Dickerson and I. Geis, *Hemoglobin: Structure, Function, Evolution and Pathology*, Benjamin Cummings, Menlo Park, Calif., 1983.
14. M. F. Perutz, Myoglobin and haemoglobin: role of distal residues in reactions with haem ligands, *Trends Biochem. Sci.*, **14**, 42 (1989).

15. S. E. V. Phillips and B. P. Schoenborn, Neutron diffraction reveals oxygen–histidine hydrogen bond in oxymyoglobin, *Nature (London)*, **292**, 81 (1981).
16. L. Pauling and C. D. Coryell, Magnetic properties and structure of hemoglobin, oxyhemoglobin and carbon monoxyhemoglobin, *Proc. Natl. Acad. Sci. USA*, **22**, 210 (1936).
17. J. J. Weiss, Nature of the iron–oxygen bond in oxyhemoglobin, *Nature (London)*, **202**, 83 (1964).
18. K. Gersonde, Reversible dioxygen binding, in *Biological Oxidations* (eds. H. Sund and V. Ullrich), Springer-Verlag, Berlin, 1983, p. 170.
19. J. E. Newton and M. B. Hall, Generalized molecular orbital calculations on transition-metal dioxygen complexes. Models for iron and cobalt porphyrins, *Inorg. Chem.*, **23**, 4627 (1984).
20. J. Baldwin and C. Chothia, Hemoglobin: the structural changes related to ligand binding and its allosteric mechanism, *J. Mol. Biol.*, **129**, 175 (1979).
21. G. K. Ackers and J. H. Hazzard, Transduction of binding energy into hemoglobin co-operativity, *Trends Biochem. Sci.*, **18**, 385 (1993).
22. V. Srajer, L. Reinisch and P. M. Champion, Protein fluctuations, distributed coupling, and the binding of ligands to heme proteins, *J. Am. Chem. Soc.*, **110**, 6656 (1988).
23. M. F. Colombo, D. C. Rau and V. A. Parsegian, Protein solvation in allosteric regulation: a water effect on hemoglobin, *Science*, **256**, 655 (1992).
24. L. F. Stryer, *Biochemistry*, 3rd edn, Freeman, New York, 1988, p. 143.
25. A. F. G. Slater and A. Cerami, Inhibition by chloroquine of a novel haem polymerase enzyme activity in malaria trophozoites, *Nature (London)*, **355**, 167 (1992).
26. T. G. Traylor, Synthetic model compounds for hemoproteins, *Acc. Chem. Res.*, **14**, 102 (1981).
27. K. S. Suslick and T. J. Reinert, The synthetic analogs of O_2-binding heme proteins, *J. Chem. Educ.*, **62**, 974 (1985).
28. J. P. Collman, Synthetic models for the oxygen-binding hemoproteins, *Acc. Chem. Res.*, **10**, 265 (1977).
29. S. J. Lippard, Oxo-bridged polyiron centers in biology and chemistry, *Angew. Chem. Int. Ed. Engl.*, **27**, 344 (1988).
30. I. M. Klotz and D. M. Kurtz, Binuclear oxygen carriers: hemerythrin, *Acc. Chem. Res.*, **17**, 16 (1984).
31. D. M. Kurtz, Oxo- and hydroxo-bridged diiron complexes: a chemical perspective on a biological unit, *Chem. Rev.*, **90**, 585 (1990).
32. R. C. Reem, J. M. McCormick, D. E. Richardson, F. J. Devlin, P. J. Stephens, R. L. Musselman and E. I. Solomon, Spectroscopic studies of the coupled binuclear ferric active site in methemerythrins and oxyhemerythrins, *J. Am. Chem. Soc.*, **111**, 4688 (1989).
33. R. J. H. Clark and T. J. Dines, Resonance Raman spectroscopy and its application to inorganic chemistry, *Angew. Chem. Int. Ed. Engl.*, **25**, 131 (1986).
34. J. Sanders-Loehr, Resonance Raman spectroscopy of iron–oxo and iron–sulfur clusters in proteins, in *Metal Clusters in Proteins* (ed. L. Que, Jr), ACS Symposium Series 372, 1988, p. 49.
35. E. Fluck, The Mössbauer effect and its application in chemistry, *Adv. Inorg. Chem. Radiochem.*, **6**, 433 (1964).
36. P. Gütlich, R. Link and A. Trautwein, *Mössbauer Spectroscopy and Transition Metal Chemistry*, Springer-Verlag, Berlin, 1978.
37. R. E. Stenkamp, L. C. Sieker and L. H. Jensen, Binuclear iron complexes in methemerythrin and azidomethemerythrin at 2.0 Å resolution, *J. Am. Chem. Soc.*, **106**, 618 (1984).
38. M. A. Holmes, I. Le Trong, S. Turley, L. C. Sieker and R. Stenkamp, Structures of deoxy and oxy hemerythrin at 2.0 Å resolution, *J. Mol. Biol.*, **218**, 583 (1991).
39. W. B. Tolman, A. Bino and S. J. Lippard, Self-assembly and dioxygen reactivity of an asymmetric, triply bridged diiron(II) complex with imidazole ligands and an open co-ordination site, *J. Am. Chem. Soc.*, **111**, 8522 (1989).
40. L. Que and R. C. Scarrow, Active sites of binuclear iron-oxo proteins, in *Metal Clusters in Proteins* (ed. L. Que, Jr), ACS Symposium Series 372, 1988, p. 152.

6 Catalysis through Hemoproteins: Electron Transfer, Oxygen Activation and Metabolism of Inorganic Intermediates

Iron–porphyrin complexes are not only involved in stoichiometric dioxygen transport but also in a variety of catalytic biochemical processes. In addition to the actual heme system (5.8) there are iron complexes with partly reduced porphyrin ligands such as bacterial heme d or siroheme (see 6.20). Heme-containing proteins participate in electron transport and electron accumulation, in the controlled reaction of oxygen-containing intermediates such as O_2^{2-}, NO, NO_2^- or SO_3^{2-}, and in complex redox processes where they interact with other prosthetic groups (cf. cytochrome c oxidase, Section 10.4).

Hemoproteins do not only feature prominently in the initial stages of the respiratory process. The reversible uptake and storage of O_2 is only the first, not yet energy-producing step in its utilization and, furthermore, the toxic intermediates resulting from incomplete reduction of dioxygen (compare 4.6 or 5.2) have to be eliminated.

At the center of respiration lies the oxidative phosphorylation, i.e. the exergonic but catalytically controlled transformation of oxidizing O_2 and partially reduced carbon compounds to thermodynamically stable low-energy products CO_2 and H_2O (4.1). This process is often referred to as 'cold combustion'; like its counterpart photosynthesis, it proceeds in a stepwise fashion involving the 'respiratory chain'. Overall, the production of three equivalents of ATP from ADP (14.2) is coupled to this process through the necessary proton gradients (Figure 6.1). The respiratory chain features a multitude of organic and inorganic components which function as electron transfer agents according to their redox potentials [1]. These compounds are found especially in the mitochondrial membrane of higher organisms (Figure 6.2); they include *organic* molecules such as the nicotine adenine dinucleotide coenzyme $NAD(P)H/NAD(P)^+$, the ubiquinone/ubihydroquinone system and flavoenzymes such as $FMN/FMNH_2$ or $FAD/FADH_2$ (3.12). Copper proteins (at high potentials; Chapter 10), iron–sulfur proteins (at low potentials; Sections 7.1 to 7.4) and certain hemoproteins, the cytochromes, are the main *inorganic* constituents.

Just as in photosynthesis, most processes essential for respiration occur in membranes since the vectorial (directed) stepwise electron transport is coupled to the likewise vectorial, membrane-spanning proton transport for ATP synthesis ('chemiosmotic' theory according to Mitchell [2]). A further analogy with photosynthesis

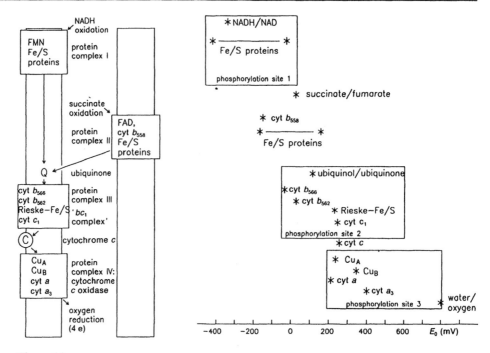

Figure 6.1
Schematic overview over the electron-transferring components in the respiratory chain.
Representation of the functional assembly in the membranes (left: see also Figure 6.2) and of
the redox potentials of individual components (*, right: modified according to [1])

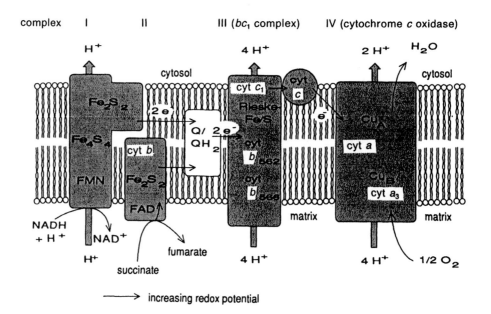

Figure 6.2
Schematic representation of interacting protein complexes in the mitochondrial membrane;
the proton back-transport via ATP synthase is not shown here

consists in the charge transport between functional membrane protein complexes via electron transferring molecules, in particular quinones and heme-containing cyto-chromes.

6.1 Cytochromes

Cytochromes are hemoproteins which exert their electron-transfer function not only in the respiratory chain but also in photosynthesis (Figures 4.5 to 4.7) and other complex biological processes. Structure–reactivity relationships will be discussed using the example of the thoroughly studied and relatively small cytochrome c [3,4].

The more than 50 known cytochromes ('cell pigments') can be divided into different groups depending on their structural constitution and physical properties [5]. The cytochrome a type shows very high redox potentials; these species are important in the reduction of oxygen to water in the cytochrome c oxidase complex (see Section 10.4, Figure 10.5). Cytochromes of the b and c types (subscripts either serve for numbering or indicate the characteristic absorption maximum in nm) contain two tightly bound amino acid residues, histidine/histidine or histidine/methionine, as ligands at the heme iron (Figure 6.3). The heme group of cytochromes c is covalently bound via porphyrin/cysteine bonds. Histidine/lysine axial coordination is assumed for cytochrome f (Figure 4.8 and Table 4.1) while methionine/methionine axial coordination occurs in bacterial ferritins (Section 8.4.2 [6]). The coordinatively *saturated* iron centers show quite variable redox potentials for the Fe^{II}/Fe^{III} transition, depending on the axial ligands and on the coordination environment (hydrogen bonding, electrostatic charge distribution, geometrical distortion); comparable effects of ligands and reaction media are well known from the 'normal' coordination chemistry of iron (Table 6.1).

Table 6.1 Redox potentials for some chemical and biochemical Fe^{II}/Fe^{III} pairs

compound	E_0' (mV)	
hexaaquairon(II/III) $[(H_2O)_6Fe]^{2+/3+}$	771	
tris(2,2'-bipyridine)iron(II/III) $[(bpy)_3Fe]^{2+/3+}$	960	
hexacyanoferrate(II/III) $[(NC)_6Fe]^{4-/3-}$	358	
trisoxalatoiron(II/III) $[(C_2O_4)_3Fe]^{4-/3-}$	20	
		axial amino acid
hemoprotein iron(II/III)		ligands
hemoglobin	170	His/ -
myoglobin	46	His/ -
horseradish peroxidase (HRP)	-170	His/ -
cytochrome a_3	400	His/ -
cytochrome c	260	His/Met
cytochrome b_5	20	His/His
cytochrome P-450	-400	Cys^-/ -

Cytochrome proteins may contain several heme groups as, for example, in the mitochondrial complex III (Figure 6.2), in a nitrite reductase (Section 6.5), in cytochrome c oxidase (complex IV, Section 10.4) or in the subunit which participates in the bacterial photosynthesis of *Rhodopseudomonas viridis* (Figure 4.5). The 'cytochrome P-450' (=pigment with an absorption maximum of the carbonyl complex at 450 nm) differs markedly from other cytochromes by its catalytic function; it shows monooxygenase activity and will be discussed separately in Section 6.2.

The best-examined cytochrome is cytochrome c [3], shown in Figure 6.3, which is typically obtained from heart muscle tissue of tuna or horses. With only about 100 amino acids and a relatively small molecular mass of about 12 kDa it is a rather small protein; structural determinations by protein crystallography could thus reveal several details. The amino acid sequences and tertiary structures of cytochrome c proteins isolated from very different organisms differ only slightly, suggesting that this is a very old protein in evolutionary terms and that its composition and structure have been well optimized. The protein is usually found at the outside of membranes (compare Figure 6.2) and, for that purpose, features a hydrophilic 'surface'.

Although the transfer of an electron is one of the simplest chemical reactions, at least three variables have to be taken into account, regarding (i) energy, i.e. the redox potential, (ii) space, i.e. the directionality, and (iii) time, i.e. the rate of the reaction. Thus, some of the most important questions with respect to electron-transfer proteins are still not sufficiently understood [8]:

(a) What are the requirements on the molecular level for a rapid (→enzymatic catalysis) but nevertheless potential-controlled intramolecular or intermolecular electron transfer?

(b) How is it possible to transfer electrons between redox centers which can be separated by more than 2 nm through an apparently inert protein environment?

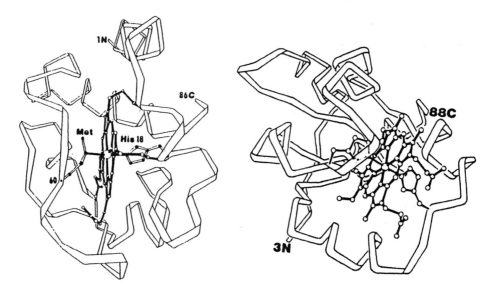

Figure 6.3
Schematic representations of protein folding, heme position and iron coordination in cytochromes c (left) and b_5 (right; from [7])

The main obstacle for rapid electron transfer are the geometrical differences between the oxidized and reduced forms of a redox pair; in general, a 'reorganization energy' has to be provided. Model studies on iron–porphyrin complexes with axial thioether ligands have shown that the Fe—S bond lengths change only little during the Fe^{II}/Fe^{III} transition. This can be explained by the σ donor/π acceptor character of thioether sulfur centers toward transition metals (π back-bonding [4]), the geometric change being small through electron balancing according to $Fe \underset{\sigma}{\overset{\pi}{\rightleftarrows}} S$. Unsaturated nitrogen ligands like the extended porphyrin π system also show such an 'electron buffer' capacity. According to the theory of the entatic state, the enzymatic ground state is then situated *between* the typical structures for each of the individual redox states, i.e. it is close to the transition state. In fact, the holoprotein of cytochrome c does not show a significant conformational change during electron transfer and the same is true for other electron-transfer proteins such as the Fe/S and blue copper systems (see Chapters 7 and 10).

The question of a possible directional dependence of long-range electron transfer has been the subject of numerous research projects [9–11]. It is quite surprising at first that an intramolecular or even intermolecular electron transfer between redox-active centers should be possible across more than 2 nm of apparently inert protein material. However, recent results from molecularly fixated model systems (6.1) consisting of an electron donor, an electron acceptor and, for example, saturated hydrocarbons as 'spacers' have shown the possibility of a fairly rapid long-range directed electron transfer in nonbiological material [12].

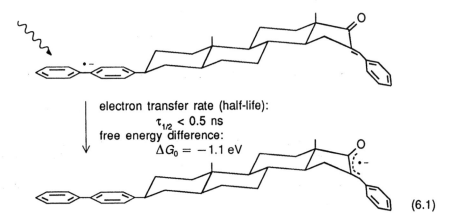

electron transfer rate (half-life):
$$\tau_{1/2} < 0.5 \text{ ns}$$
free energy difference:
$$\Delta G_0 = -1.1 \text{ eV}$$

(6.1)

In this context, the tunnel effect is frequently being invoked which enables particles such as the electron to 'overcome' energy barriers with a certain probability by 'tunneling' according to quantum-theoretical concepts. However, in order to account for the frequently remarkable directionality of this process, suitably interacting molecular orbitals probably play an important role ('through-bond' electron transfer). Although completely saturated spacers with a σ framework (see 6.1) can be effective to some extent, the intensely studied cytochrome c features particularly aromatic and thus potentially redox-active [13] amino acid residues such as tryptophane (Trp) as invariant components of assumed 'electron-transfer pathways'. However, the crystal structure determined for a specifically aligned protein/protein complex between cytochrome c and the likewise heme-containing partner enzyme cytochrome c peroxidase (CCP [14]) has indicated electron transfer between two heme groups via an amino acid framework (Ala-)Ala-Gly-Trp with predominantly nonaromatic amino acid residues [15]. Particularly illustrative examples of an electron transfer 'pathway' are the four heme groups in the cytochrome subunit of the photosynthetic reaction center of *Rps. viridis* (Figure 4.5). Here, the cytochrome is also located at the outside of the membrane (Figure 4.8); four heme groups that are not coincidentally oriented almost orthogonally provide for a controlled electron transfer to fill the electron hole left in the oxidized special pair of bacterio-chlorophyll after light-induced charge separation (Figures 4.6 and 4.7). Except for one heme group with (His, His) axial ligation and a relatively low potential (second closest to the special pair), the heme-iron centers are each coordinated by one methionine and one histidine group as axial ligands [16]. The still poorly understood fine tuning of potential differences, distances and respective spatial orientations is obviously crucial for controlled and effective electron transfer.

In general, normal cytochromes are relatively simple structured proteins with clearly measurable activities, i.e. the redox potential and the electron-transfer rate; therefore they have become attractive objects for deliberate enzyme modification via site-directed mutagenesis. In particular, they may feature different spin states of the iron depending on the coordination environment, because porphyrin complexes of Fe^{II} and Fe^{III} often lie close to the spin-crossover region [17]. For instance, 'engineered' cytochrome b_5 where basic His39 has been exchanged for less basic Met exhibits high-spin ($S = 5/2$) iron(III) instead of a low-spin iron center in its oxidized state. That high-spin center is also characterized by a negatively shifted redox potential and additional substrate-catalytic (here: N-demethylase) activity [18].

6.2   Cytochrome P-450: Oxygen Transfer from O_2 to Non-Activated Substrates

Sulfur ligation and catalytic (enzymatic) activity also go together in a special, ubiquitous kind of 'cytochrome' proteins which are referred to as P-450 because the CO complexes have a characteristic absorption maximum at 450 nm. This remarkable heme system is a constituent of hydroxylases (monooxygenases) which, in cooperation with reducing agents (NADH, flavins, iron–sulfur proteins) and a P-450 reductase, produce oxygenated products from often rather inert substrates and dioxygen (6.2 [19–23]).

$$R-H \text{ or } \bigvee\!\!\!\bigwedge \quad + \; O_2 \; + \; 2 \; e^- \; + \; 2 \; H^+ \quad \xrightarrow{\text{P-450}} \quad R-OH \text{ or } O\!\!-\!\!\bigvee\!\!\!\bigwedge \quad + \; H_2O \qquad (6.2)$$

Several variants exist within the P-450 enzyme 'super family' which play an essential role in the metabolism of endogeneous substances (e.g. steroids) as well as in the transformation of external, 'xenobiotic' substances in animals, plants and microorganisms. For instance, P-450-dependent monooxygenases are active as typical detoxification enzymes [24] in the microsomes of the human liver, where they often show relatively little substrate selectivity; on the other hand, fatty acids, amino acids and hormones (steroids, prostaglandines) are metabolized by more specific P-450 systems in a stereospecific fashion [25]. Of great importance in pharmacology, medicine and toxicology are the monooxygenation reactions of xenobiotic substances to the physiologically active metabolites, e.g. leading from calciferols (vitamin D group) to active 1,25-dihydroxycalciferols (6.3) or from β-naphtylamine to the carcinogenic α-hydroxy-β-aminonaphthalene (6.4). Drugs like morphine and numerous other well-known pharmaceuticals (Table 6.2), as well as chlorinated and

Table 6.2 Some examples of the oxidative metabolism of pharmaceuticals by P-450 monooxygenation (according to Mutschler [26])

reaction type	equation	examples
oxidation of aliphatic chains	$R\text{-}CH_2\text{-}CH_3 \rightarrow R\text{-}\overset{\displaystyle OH}{\underset{\displaystyle }{CH}}\text{-}CH_3$ and $R\text{-}CH_2\text{-}COOH$	barbiturates
oxidative N-dealkylation	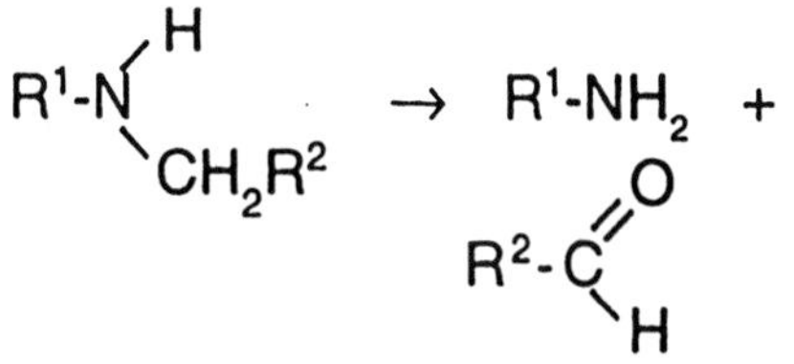 $R^1\text{-}N\overset{\displaystyle H}{\underset{\displaystyle CH_2R^2}{}} \rightarrow R^1\text{-}NH_2 + R^2\text{-}\overset{O}{C}{-}H$	ephedrine
oxidative deamination	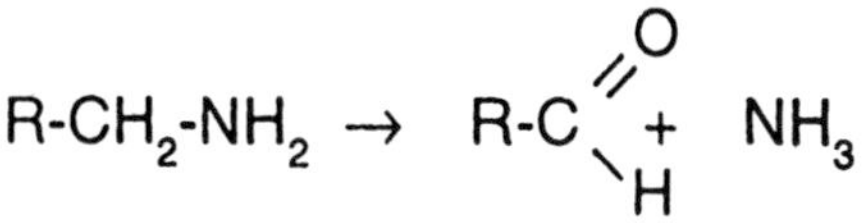 $R\text{-}CH_2\text{-}NH_2 \rightarrow R\text{-}\overset{O}{C}{-}H + NH_3$	histamine norepinephrine mescaline
oxidative O-dealkylation	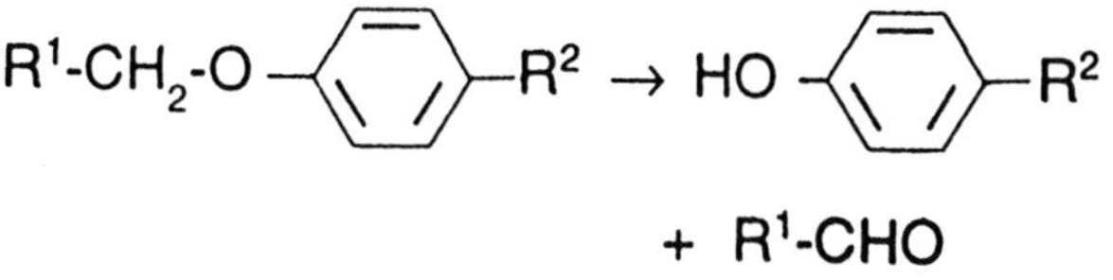 $R^1\text{-}CH_2\text{-}O\text{-}\langle\rangle\text{-}R^2 \rightarrow HO\text{-}\langle\rangle\text{-}R^2 + R^1\text{-}CHO$	phenacetin codeine mescaline
para-hydroxylation of aromatic compounds	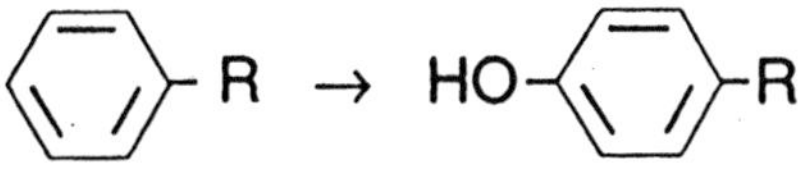 $\langle\rangle\text{-}R \rightarrow HO\text{-}\langle\rangle\text{-}R$	phenobarbital chlorpromazine
oxidation of aromatic amines	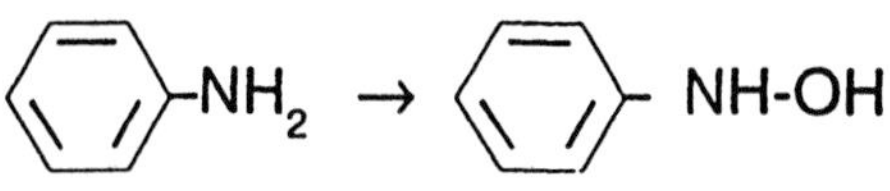 $\langle\rangle\text{-}NH_2 \rightarrow \langle\rangle\text{-}NH\text{-}OH$	aniline derivatives
S-oxidation	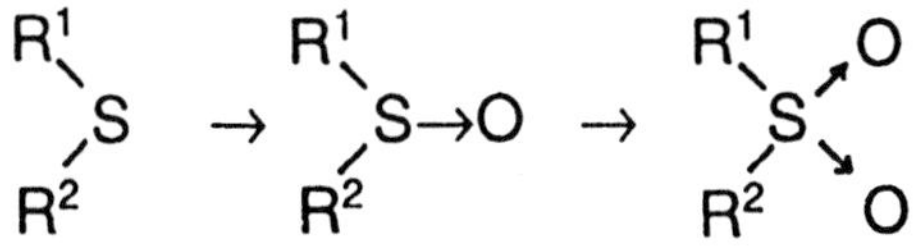 $\overset{R^1}{\underset{R^2}{}}S \rightarrow \overset{R^1}{\underset{R^2}{}}S{\rightarrow}O \rightarrow \overset{R^1}{\underset{R^2}{}}S\overset{O}{\underset{O}{}}$	phenothiazine

nonchlorinated hydrocarbons, are also transformed by P-450 monooxygenases; in the absence of oxidizable aliphatic side chains, the cytochrome P-450 enzymes catalyze the epoxidation of benzene or benzo[a]pyrene to yield the mutagenic derivatives (6.5, 6.6). Nitrosamines and polychlorinated methane derivatives are transformed by P-450 enzymes to reactive radicals and carbocations, and the large amounts of acetaldehyde formed in the oxidation of excess ethanol (see Section 12.5) may cause liver damage. The high reactivity and often low specificity of many P-450-dependent detoxification enzymes requires a precise control of their activity. One reason for the diffuse toxicity of certain polychlorinated dibenzo-1,4-dioxines and biphenyls (PCBs) is probably the interaction of these substances with a receptor which stimulates the expression of P-450-dependent enzymes and thus triggers immunotoxic reactions without being metabolized rapidly enough. The oxidative degradation of such compounds is blocked in particular if the most reactive sites, e.g. the para positions of aromatic molecules or the 2,3,7,8 positions in dibenzo-1,4-dioxine heterocycles are substituted by Cl instead of H; the C—Cl bond is not oxygenated by P-450 and Cl^+ would be a poor leaving group.

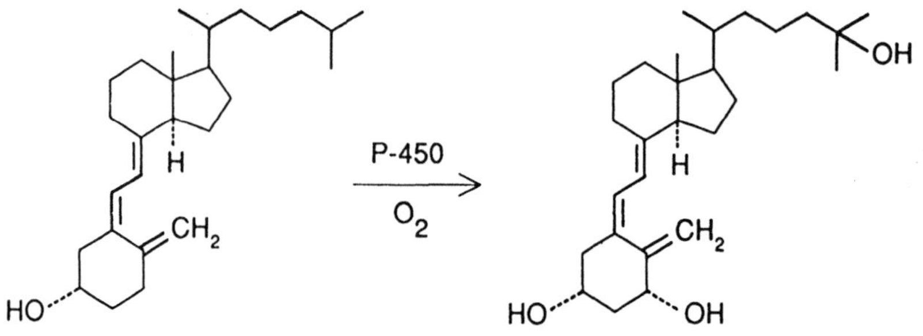

colecalciferol (vitamin D_3)   1,25-dihydroxy-colecalciferol   (6.3)

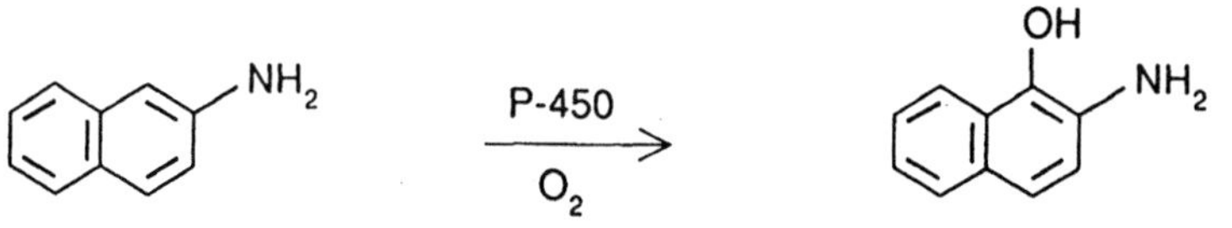

β-naphthylamine   α-hydroxy-β-aminonaphthalene
                                                            (carcinogenic)   (6.4)

The interest in cytochrome P-450 arises not only from a desire to understand its physiological function; the controlled transfer of oxygen atoms from freely available O_2 to not otherwise activated organic chemical substrates, particularly hydrocarbons, continues to pose a formidable challenge for synthetic chemistry [27,28]. This is even more true since studies on the reaction mechanism of P-450 have shown a pattern which is also characteristic for many other metal-catalyzed processes. Obviously, this catalysis involves the *controlled reaction of two ligands in the coordination sphere of the metal*. The enzyme thus functions both as a substrate-selective 'template', bringing the reactants together in a spatially defined orientation (stereospecificity), and as an electronic activator, providing a new reaction pathway with lower activation energy. The activation of bound O_2 is presumably influenced by the axial ligand (2.6) which

toxic: less toxic:

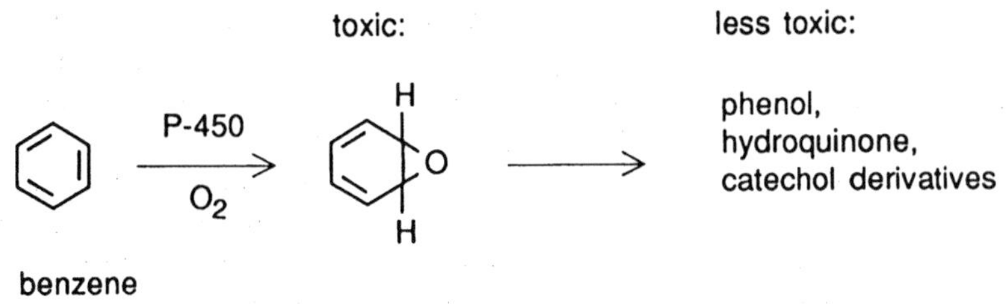

benzene   (6.5)

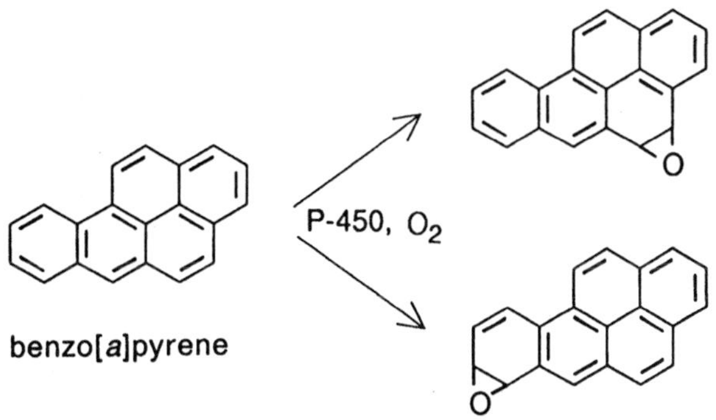

benzo[a]pyrene   (6.6)

has been identified structurally [14,29,30] and via model studies as a cysteinate anion. In contrast to π accepting thioethers such as methionine, the thiolates are strong σ *and* π electron donors and can therefore stabilize high oxidation states of metal centers.

Many P-450-dependent enzymes with their typical molecular mass of about 50 kDa have been analyzed in terms of their amino acid sequences [22]. Some forms were characterized with respect to the reaction mechanism and structurally, both with and without substrate [29] (compare Figure 6.4). According to those studies, the following

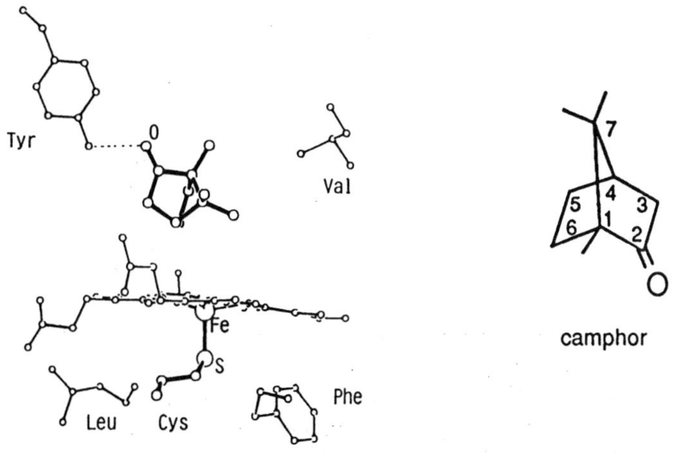

Figure 6.4
Side view of the active center in a P-450-dependent enzyme of *Pseudomonas putida* with a bound camphor molecule $C_{10}H_{16}O$ as substrate: hydrogen bonding between the keto group and a tyrosine side chain, monooxygenation in the 5-position of camphor (according to [14])

mechanism is assumed for dioxygen activation and monooxygen transfer (6.7; an abbreviated version appears on the cover). Starting from the predominantly low-spin iron(III) state (1) with six-coordinate metal (porphyrin, cysteinate, water), the binding of the organic substrate occurs through hydrophobic and other (see Figure 6.4) interactions with the protein inside a cavity and close to the axial coordination site of the heme system, thereby causing a transition to the high-spin (h.s.) iron(III) form (2). After the concomitant loss. of the water ligand, form (2) thus features an open coordination site and an *out-of-plane* structure with a domed porphyrin ring (Figure 6.4).

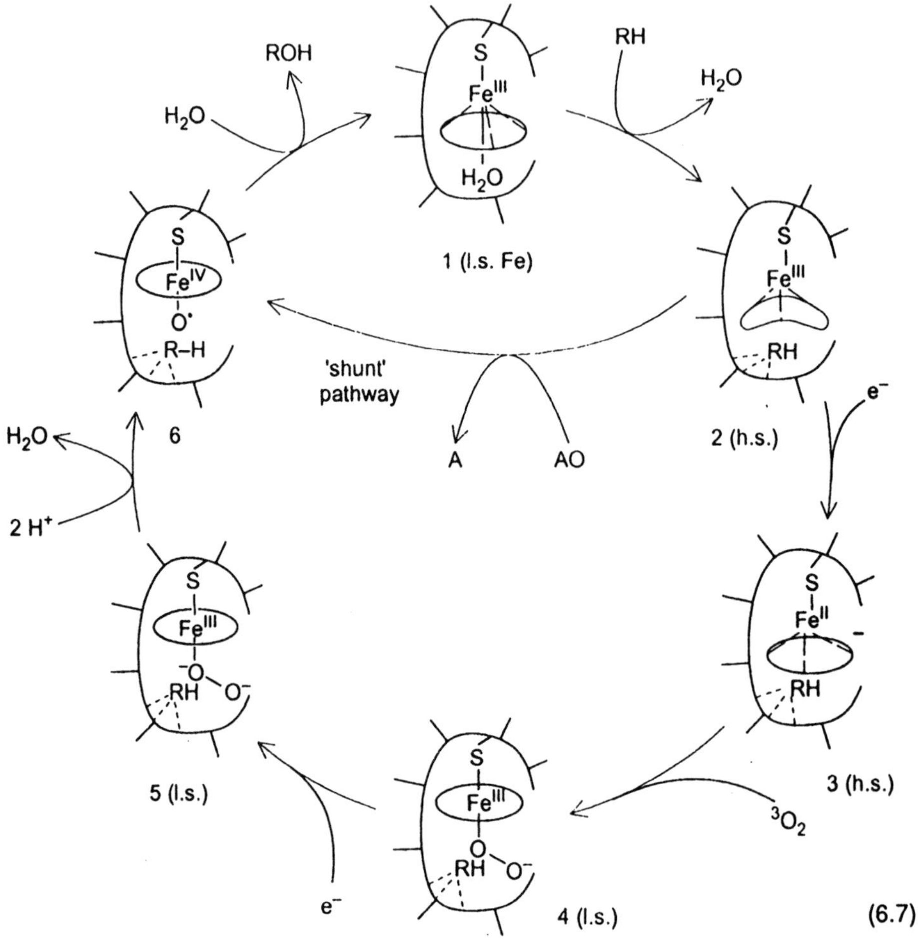

$$(6.7)$$

The next step is a one-electron reduction (via $FADH_2$, see 3.12) to give a high-spin iron(II) complex (3) which is predestined for $^3O_2$ binding because of its ($S = 2$) spin state and the pronounced *out-of-plane* situation (compare Section 5.2). In the presence of external oxygenation agents, AO, such as peracids, this iron(II) state (3) may directly yield compound (6), the highly oxidized productive complex via a 'shunt' pathway. The physiological reaction, however, requires the uptake of dioxygen to form the coordinatively saturated low-spin oxy form (4) with a hydrogen bond to a threonine side chain [31] and a possible $Fe^{III}/O_2^{\cdot-}$ oxidation state formulation (compare hemoglobin and myoglobin, 5.9). After a second one-electron reduction to form a very labile low-spin peroxo iron(III) complex (5), this species adds two protons and releases one water molecule, thereby cleaving the O—O bond [32]. This cleavage, however, requires two oxidation equivalents which have to be made available intramolecularly; the result is a reactive complex (6) which can be formulated with $Fe^V{=}O^{-II}$ or, better, with $Fe^{IV}{-}O^{\cdot-I}$ oxidation states and which collapses to the product and back to the initial state of the catalytic heme, involving presumably the transfer of a monooxygen radical to the substrate (6.8; for the overall reaction see 6.2).

$$(6.8)$$

The formally pentavalent oxo-iron center of the reactive heme group is certainly being stabilized by the electron-donating thiolate group; however, it also has to be regarded in context with the surrounding porphyrin π system. As has been shown for the chlorophylls in Section 4.2, the porphyrin system itself may add or release single electrons to form radical ions; in the presence of a dicationic oxo-metal ion $[Fe^{IV}{=}O]^{2+}$, which was detected here by resonance Raman spectroscopy, the formation of a radical cation complex via single-electron oxidation of the dianionic porphyrin ligand to a radical anion does not seem unreasonable. For the highest oxidized states of the oxo-heme systems in peroxidases (see below) and also in the P-450 enzymes such an electronic configuration is widely accepted [21,33]. A porphyrin radical anion, $Por^{\cdot-}$ ($S = 1/2$), would then coordinate to an oxoferryl(IV) fragment, $[Fe^{IV}{=}O]^{2+}$, with tetravalent iron (6.9).

$$[Fe^{II}(Por^{2-})]^0 \xrightarrow[- H_2O]{+O_2,\ 2\,H^+,\ 1\,e^-} [O^{-II}{=}Fe^{IV}(Por^{\cdot-})]^{\cdot+} \qquad (6.9)$$

The four d electrons of this rare [34] but biochemically very important oxidation state $+ IV$ of iron are not completely paired. From the ligand field splitting of an approximately D_{4h}-symmetrical complex (tetragonal compression; see the correlation diagram in Figure 6.5) with oxide and cysteinate as strongest ligands one can deduce that an occupation with four electrons should lead to a triplet ground state ($S = 1$), in agreement with Hund's rule.

Interpretation (6.9) is based on magnetic measurements which indicate a weak antiferromagnetic coupling between the assumed porphyrin radical anion ($S = 1/2$) and the iron center ($S = 1$). At very low temperatures there is only *one*, predominantly metal-centered unpaired electron observable.

Even considering the highly activated nature of compound (6), it remains remarkable that this last state of the P-450 center in cycle (6.7) is able to transfer the iron-bound oxygen atom (6.8) to a substrate which is little or not at all activated; after release of the oxygenated substrate, the initial state of coordinately unsaturated

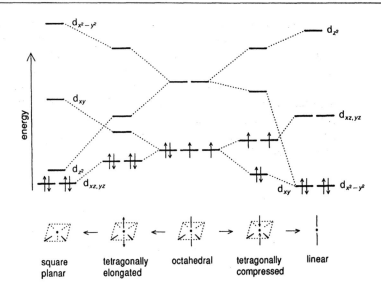

Figure 6.5
Correlation diagram of d orbitals for the tetragonal distortion (compression and elongation) of an originally octahedral d^4 metal complex

low-spin iron(III) is formed again. Formula (6.9) implies that the coordinated monooxygen exists as oxide ligand (O^{2-}) and that the oxidation number should thus be $-II$. This usually unquestioned convention, however, has been challenged by Sawyer [35], in particular with regard to the typical P-450 reactivity; a terminal monooxygen ligand can also be present as radical anion ($O^{\cdot-}$) or, in analogy to carbenes and nitrenes, even as neutral 'oxene' ligand (compare resonance structures 6.10).

$$
\underset{\text{oxide}}{\overset{-II}{O}{=}\overset{+IV}{Fe}}\Big]^{2+} \quad \leftrightarrow \quad \overset{-I}{\cdot}\overset{}{\underline{O}}{-}\overset{+III}{\underset{\bullet}{Fe}}\Big]^{2+} \quad \leftrightarrow \quad \underset{\text{oxene}}{|\overset{0}{\underline{O}}{-}\overset{+II}{Fe}}\Big]^{2+} \tag{6.10}
$$
$$\curvearrowleft \; (\textit{back-donation})$$

In fact, X-ray spectroscopic [35] and mechanistic results suggest that in many 'oxo' complexes of transition metals and, particularly, in the reactive state (6) of the P-450 system [36] (compare 6.14) the formulation with a weakly bound radical oxygen atom as reactive, hydrogen-abstracting ligand contributes significantly to the actual electronic structure. Adding the possible porphinate oxidation as an alternative (6.10), there is then a multitude of conceivable resonance structures [37]. 'Genuine' oxide (O^{2-}) ligands which might tolerate even organic ligands in the metal coordination sphere [38] are found mainly in metal complexes with extremely stabilized high oxidation states, such as Re^{VII}. In the P-450 system, the interactions of stabilizing cysteinate and reactivity-enhancing distal ligands of the Fe—O(—O) moiety aid in controlling the reaction [31], especially preventing an autoxygenation alternative which prevails in the case of the heme-degrading enzyme heme dioxygenase.

6.3 Peroxidases: Detoxification and Utilization of Doubly Reduced Dioxygen

Heme-containing peroxidases and catalases are closely related to cytochrome P-450; however, there are also manganese- or vanadium-containing (Section 11.4) and even metal-free peroxidases [39,40] as well as manganese-dependent catalases with polynuclear Mn^{II-IV} centers. Contrary to cytochrome P-450, peroxidases use the doubly reduced peroxidic form of O_2 to oxidize substrates of the type AH_2 to radical cations and their reaction products (6.11–6.13). The peroxide oxidation state of dioxygen can be produced as an undesired intermediate in the course of photosynthetic water oxidation or via incomplete oxygen reduction during respiration (4.6, 5.2); only about 80% of the dioxygen taken up by breathing is *completely* reduced. Therefore, peroxidases may be regarded at least partly as detoxification enzymes. This is especially true for the catalases since their second substrate is also hydrogen peroxide (6.12); overall, the resulting reaction is the enzymatically catalyzed disproportionation of metastable H_2O_2, the equilibrium constant for (6.12) being about 10^{36}!

$$H_2O_2 \ + \ AH_2 \ \underset{}{\overset{\text{peroxidases}}{\rightleftharpoons}} \ 2\,H_2O \ + \ A \qquad (6.11)$$

$$H_2O_2 \ + \ H_2O_2 \ \underset{}{\overset{\text{catalases}}{\rightleftharpoons}} \ 2\,H_2O \ + \ O_2 \qquad (6.12)$$

There are numerous, not easily oxidized compounds like fatty acids, amines, phenols, chloride or xenobiotic substances (toxins) that can serve as substrates for peroxidases. For example, the controlled α-oxidations of fatty acids during plant growth yield an α-carbonyl carboxylic acid intermediate which loses CO_2 (decarboxylation) to form an aldehyde with one CH_2 group less; its oxidation product is the correspondingly shorter fatty acid (6.13).

$$R-CH_2-COOH \ + \ 2\,H_2O_2 \ \xrightarrow{\substack{\text{fatty acid} \\ \text{peroxidase}}} \ 3\,H_2O \ + \ R-CHO \ + \ CO_2 \quad (6.13)$$
$$\downarrow \text{oxidation}$$
$$R-COOH$$

Other important reactions of heme peroxidases [39] concern the coupling of tyrosines and their iodination to the thyroid hormones by thyreoperoxidases [41] (see Section 16.7), the oxidation of cytochrome c by cytochrome c peroxidase (CCP [14,15]), the oxidation of chloride to bactericidal hypochlorite, ^-OCl, by myeloperoxidase with cysteinate-coordinated iron, or the oxidative degradation of lignin by lignin peroxidase [42]. A remarkable curiosity is the utilization of H_2O_2 and hydroquinone by certain beetles (*Brachymus*) to effect an explosive burst of O_2 and aggressive, oxidizing quinone in a peroxidase-catalyzed reaction.

The most thoroughly studied enzyme in the peroxidase group is horseradish peroxidase (HRP) which has a molecular mass of about 40 kDa [43] and which has been investigated since the days of R. Willstätter (1892–1942). Its low molecular mass distinguishes it from the classical heme-containing catalases which are associated proteins (tetramers) with total masses of 260 kDa and partially tyrosine-coordinated heme iron [14]. Under physiological conditions, the resting state of most heme peroxidases contains high-spin iron(III) ($S = 5/2$, half-filled d shell, *out-of-plane* structure) and, in contrast to the P-450 system, an imidazole base from histidine as axial ligand. Oxidation, i.e. the protein-induced [44] internally catalyzed mono-oxygen transfer from H_2O_2 to the iron center under release of water (6.14), adds two oxidation equivalents and leads formally to an iron(V) species, again, most probably, involving a dicationic oxoferryl(IV) center with a coordinated porphyrin radical anion. This first, very electron-deficient intermediate ('HRP I', $E_0 > 1$ V) can react with the substrate in a one-electron oxidation step; in fact, the follow-up products observed when phenols or certain strained hydrocarbons are used as substrates are otherwise known only from secondary reactions of chemically or electrochemically generated substrate radical cations. The resulting second enzymatic intermediate shows only one more oxidation equivalent than the resting state but can still undergo a second one-electron oxidation reaction; according to physical measurements, this 'HRP II' state contains an oxoferryl(IV) center ($S = 1$) coordinated to a normal, i.e. dianionic porphyrin ligand (compare 6.9 [33]).

$$[Fe^{III}(Por^{2-})]^{\bullet+} \rightarrow [O=Fe^{IV}(Por^{\bullet-})]^{\bullet+} \rightarrow O=Fe^{IV}(Por^{2-}) \rightarrow [Fe^{III}(Por^{2-})]^{\bullet+}$$

$$H_2O_2 \quad H_2O \qquad\qquad AH_2 \quad AH_2^{\bullet+} \qquad\quad AH_2 \quad AH_2^{\bullet+}$$
$$2\,H^+ \quad H_2O$$

$$\text{HRP I} \qquad\qquad\qquad \text{HRP II}$$
$$\text{green} \qquad\qquad\qquad\quad \text{red} \qquad\qquad\qquad\qquad\qquad (6.14)$$

In cytochrome c peroxidase [14] the ligand oxidation by H_2O_2 apparently involves a (tryptophane) amino acid residue of the peptide chain instead of the porphyrin. In all heme peroxidases, the hydrogen bonds from distal histidine and carboxylate groups must play an important role in tuning the reactivity [45] since, at a first glance, the not-peroxidase-active myoglobin features a heme coordination similar to that of HRP.

6.4 Controlling the Reaction Mechanism of the Oxyheme Group — Generation and Function of Organic Free Radicals

Cytochrome P-450 as well as the heme peroxidases go through reactive intermediate states with unusually high oxidation levels of the iron center; for instance, there are very few examples of Fe^{IV} oxidation states in coordination chemistry [34]. What are the reasons for the rather different reactivities [36,43], viz. for monooxygenase activity (monooxygen transfer, one O from O_2) in one case and direct electron withdrawal, i.e. formation of a substrate radical cation, in the other? Contrary to P-450, most peroxidase iron centers feature a neutral histidine as axial ligand which, however, may become deprotonated. Compared to the anionic thiolate ligand of the P-450 systems neutral histidine is a weaker electron donor which possibly effects a shift of the radical activity (spin density) from the iron-bound oxygen to the porphyrin π system (6.15). Electrophilic attack is then no longer connected to the oxygen transfer

(6.7, 6.8) but consists of the extraction of one electron from the substrate and the conversion of the peroxidic oxygen to water (6.14). In particular, the different protein environments must be responsible for such different reactivities: in cytochrome P-450, the possibly generated radicals rapidly combine to yield the oxygenated products ('cage reaction') while in peroxidases, a dissociation of the reactants may lead to the typical 'escape' products of free radicals (6.16). For example, the peroxidases do not catalyze the formation of dihydroxyaromatic compounds from phenolic substrates, as do the P-450 systems, but rather of aryl coupling products which are typical for phenoxyl radicals [21,46]. This observation is important in as much as the controlled oxidation of phenols to catechols may also be catalyzed by copper-dependent enzymes (see Sections 10.2 and 10.3).

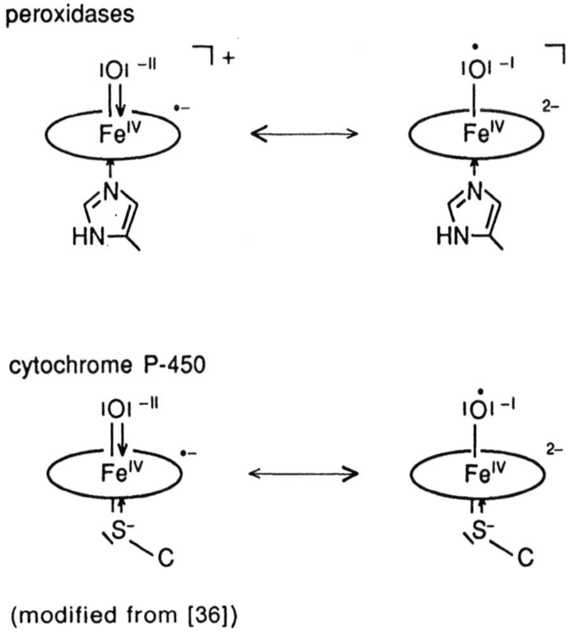

peroxidases

cytochrome P-450

(modified from [36])   (6.15)

'cage'

+ Su

+ 2 H$^+$

'escape'

+ SuO

+ H$_2$O + Su$^{\bullet +}$ → → → 1/2 Su$_2$

Su: substrate, e.g. phenol $\langle\!\!\!\!\bigcirc\!\!\!\!\rangle$— OH;

SuO: e.g.   (OH)$_2$

Su$_2$: e.g. HO—$\langle\!\!\!\bigcirc\!\!\!\rangle$—$\langle\!\!\!\bigcirc\!\!\!\rangle$—OH

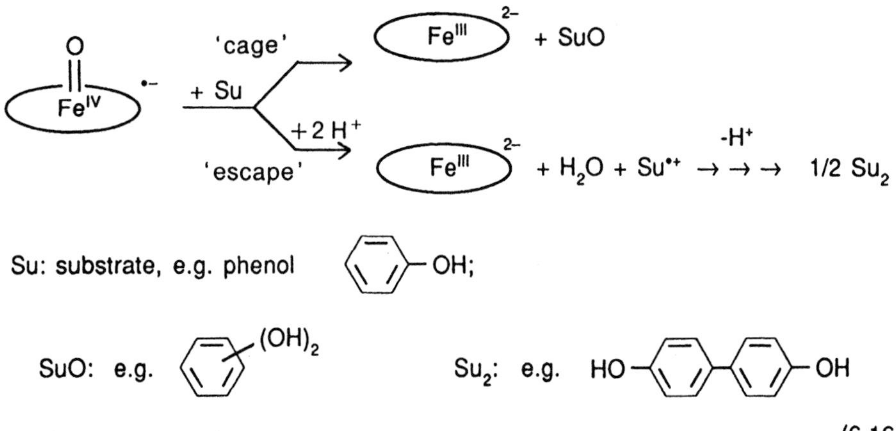

(6.16)

With their unusually high redox potentials of about $+1.5$ V the reactive radical forms of oxidized peroxidases play an important role in the metabolism and degradation of the polymeric lignin of higher plants (about 25% of the world's biomass). Arylether bonds are formed and broken with the help of lignin peroxidases which prefer rather low pH values [42]. Peroxidase-utilizing microorganisms are thus also being tested for the degradation (detoxification!) of aromatic toxins such as chlorinated polyaryls, phenols, dioxines and furanes which can relatively easily be oxidized to unstable radical cations [47,48].

On the other hand, the potentially high reactivity of oxyheme iron centers with regard to substrates has to be avoided at all cost in myoglobin and hemoglobin. Otherwise, an autoxidation of these exclusively O_2-transporting and -storing systems would result, a condition that only occurs in a pathological context. With this in mind, the requirements for the protein environment.to inhibit such thermodynamically favorable oxidative processes in Hb and Mb have to be particularly appreciated in retrospect (Section 5.2). The multiple functions of the heme group (5.8) are thus summarized in (6.17) and correlated with their electronic structure (according to [21]).

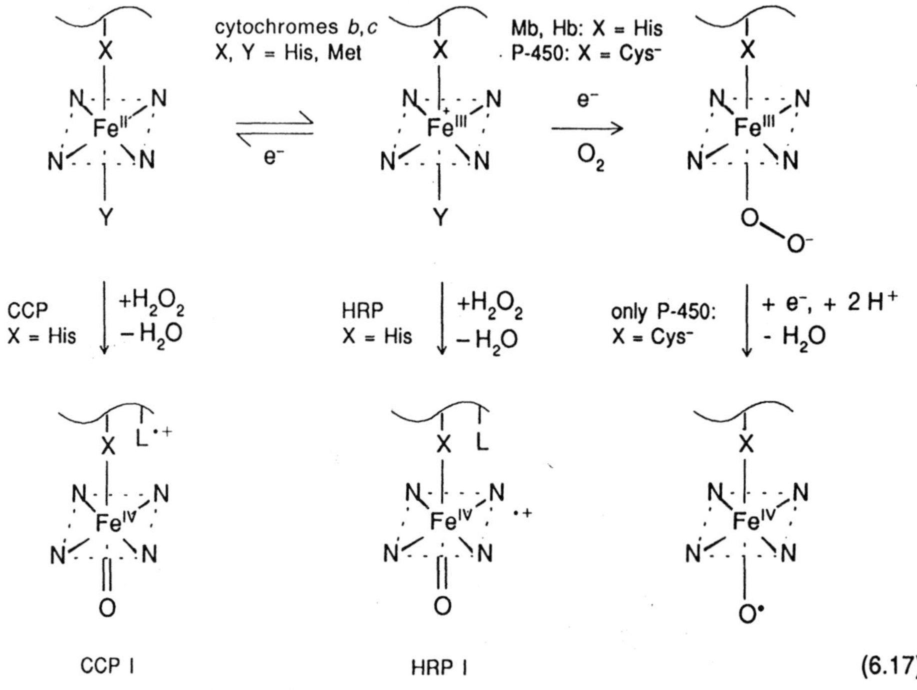

$$(6.17)$$

6.5 Hemoproteins in the Catalytic Transformation of Partially Reduced Nitrogen and Sulfur Compounds

The heme group with its open coordination sites can not only bind small unsaturated molecules via free oxygen (O_2) and carbon electron pairs (CO); partially reduced nitrogen oxo and sulfur oxo compounds, such as nitrogen monoxide (nitric oxide, nitrosyl radical), NO, nitrite, NO_2^-, or sulfite, SO_3^{2-}, have also become known as substrates for hemoprotein catalysis [49–53]. In each case, binding to iron is possible through electron pair coordination and π back-donation from the divalent metal into low-lying orbitals of the unsaturated ligands (6.18).

$$(6.18)$$

It has long been known that organic nitrites and nitrates as well as the nitroprusside dianion, $[Fe(CN)_5(NO)]^{2-}$, cause muscle relaxation in vertebrates as 'nitrovasodilators' and may thus be used against high blood pressure and angina pectoris [49,51]. Since muscle relaxation may be caused by an accumulation of the cyclic nucleotide guanosine monophosphate (cGMP; see Section 14.2) it is now assumed that the guanylate cyclase enzyme is stimulated through binding of its heme component with endogeneous or exogeneous NO. While the binding to iron most probably involves the nitrosonium form, NO^+, which is isoelectronic with the CO and CN^- ligands, the transport mode (NO·, NO^+, thiol-bound form?) is not yet clear. Similar to O_2, the biological binding of NO, e.g. in the bacterial nitrogen cycle, can be effected not only through heme iron but also through polynuclear copper centers [52] (see Section 10.3). The increasingly recognized essential role of NO ('molecule of the year 1992' [54]) not only as a vasodilatory or cytotoxic agent [55] but also as a common neurotransmitter in the brain, muscular relaxation and penile erection [49,50,56] requires its rapid and well-controlled synthesis from arginine whereby, in the crucial step paralleling the P-450 reactivity, one oxygen from O_2 is inserted into an N—H bond. Not surprisingly, the calmodulin-dependent (Section 14.2) NO synthases with molecular masses greater than 130 kDa are heme proteins [57]. The involvement of 'NO' both in normal neurotransmitting and neuropathological disorders has invited speculations on the different roles of facile interconverted $^{+II}NO·$ and $^{+III}NO^+$ forms [58].

Incidentally, guanylate cyclase can also be activated by CO, another potential neurotransmitter. Carbon monoxide is formed during the degradation of the heme group by hemoxygenase [59] (see Section 5.2).

The complete, i.e. six-electron reduction (6.19) of nitrite, NO_2^-, another form of nitrogen(+III) which can be generated from nitrate, NO_3^-, via molybdoenzyme action (see Section 11.1), is part of the microbial denitrification and involves polyheme-containing nitrite reductases. These relatively large proteins contain several, partly unique (i.e. catalytic) heme centers [53].

$$NO_2^- + 6\ e^- + 8\ H^+ \xrightarrow[\text{nitrite reductase}]{\text{denitrification}} NH_4^+ + 2\ H_2O \qquad (6.19)$$

The four-electron oxidation of hydroxylamine, NH_2OH, to nitrite within the microbial nitrogen cycle (see Figure 11.1) is also catalyzed by a complex multiheme enzyme [60].

Another special heme group, heme d_1, is catalytically active in a dissimilatory bacterial nitrite reductase which reduces nitrite to NO and features an unusual porphyrin ligand with two carbonyl groups in two pyrrole rings [61] (6.20). Dissimilatorily nitrite-reducing bacteria may also contain copper enzymes; some assimilatorily NO_2^--utilizing microorganisms contain siroheme with a doubly hydrogenated porphyrin ring (6.20).

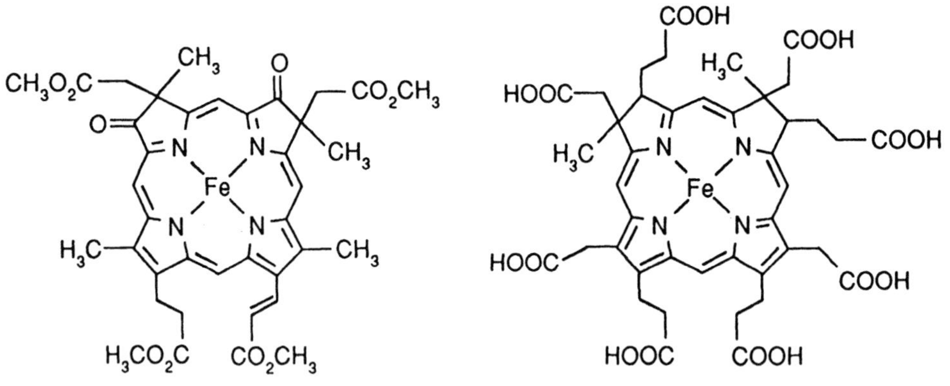

heme d_1 siroheme

(6.20)

In case of the sulfite reductases, which most probably coordinate the partly reduced sulfur($+IV$) atom of SO_3^{2-} there is also a distinction possible between dissimilatory and assimilatory enzymes. The dissimilatory enzymes are found in simple bacteria where SO_3^{2-} functions only as terminal electron acceptor whereas the assimilatory enzymes of, for example, *E. coli* serve to supply (hydrogen) sulfide for biosynthesis. Thermodynamically metastable sulfite can be formed from sulfate by molybdenum-containing oxotransferase enzymes (see Section 11.1.1).

$$SO_3^{2-} + 6 \, e^- + 7 \, H^+ \xrightarrow{\hspace{2cm}} HS^- + 3 \, H_2O \qquad (6.21)$$
$$\text{sulfite reductase}$$

The problem of reaction (6.21) is the catalysis of a six-electron process via several possible intermediates [52], using the normally available one-electron equivalents, a problem which is similar to that of reaction (6.19). Many-electron processes such as the transformation $2 \, H_2O/O_2$ ($4 \, e^-$) or the nitrogen fixation $8 \, H^+ + N_2/2 \, NH_3 + H_2$ ($8 \, e^-$) require the combined action of several redox-active, often inorganic centers (see Sections 4.3, 10.4 and 11.2) in order to utilize favorable potentials and to avoid the formation of undesired free reactive intermediates. Sulfite reductases are therefore complex $\alpha_n\beta_m$ oligomers in which the α subunits are flavoproteins while the β subunits contain Fe/S clusters (see Chapter 7) and a siroheme group with high-spin iron. A direct bridging of both types of iron centers (distance about 0.44 nm) via a μ-cysteinate sulfur center was inferred both from spectroscopic and preliminary structural studies (6.22) [62]. As an alternative, an unconventional Fe/S cluster with an $S = 9/2$ spin state and not necessarily strong interactions with the heme group has been postulated [63] (see Section 7.4).

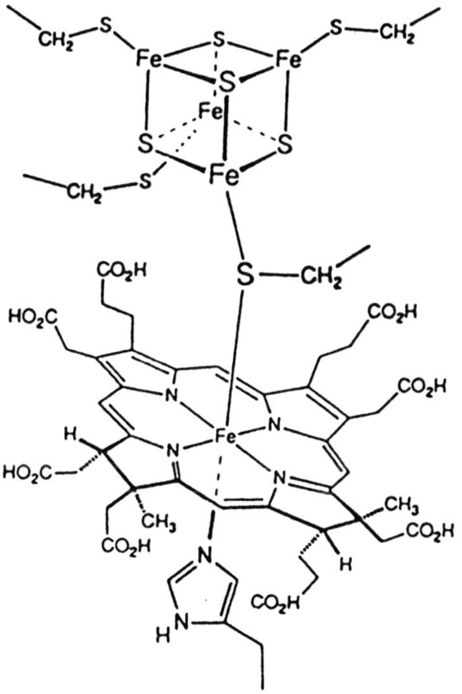

$$(6.22)$$

References

1. G. von Jagow and W. D. Engel, Structure and function of the energy converting systems of mitochondria, *Angew. Chem. Int. Ed. Engl.*, **19**, 659 (1980).
2. P. Mitchell, Keilin's respiratory chain concept and its chemiosmotic consequences, *Science*, **206**, 1148 (1979).
3. C. Greenwood, Cytochromes *c* and cytochrome *c* containing enzymes, in *Metalloproteins* (ed. P. Harrison), Part 1, Verlag Chemie, Weinheim, 1985, p. 43.
4. G. R. Moore and G. W. Pettigrew, *Cytochromes c*, Springer-Verlag, Berlin, 1990.
5. G. Palmer and J. Reedijk, Nomenclature of electron-transfer proteins, *Eur. J. Biochem.*, **200**, 599 (1991).
6. G. N. George, T. Richards, R. E. Bare, Y. Gea, R. C. Prince, E. I. Stiefel and G. D. Watts, Direct observation of bis-sulfur ligation to the heme of bacterioferritin, *J. Am. Chem. Soc.*, **115**, 7716 (1993).
7. F. R. Salemme, Structure and function of cytochromes *c*, *Annu. Rev. Biochem.*, **46**, 299 (1977).
8. R. J. P. Williams, Overview of biological electron transfer, in *Electron Transfer in Biology* (eds. M. K. Johnson *et al.*), Advances in Chemistry Series 226, 1990, p. 3.
9. H. B. Gray and B. G. Malmström, Long-range electron transfer in multisite metalloproteins, *Biochemistry*, **28**, 7499 (1989).
10. G. Palmer (ed.), *Long-Range Electron Transfer in Biology*, Springer, Berlin, 1991.
11. H. Sigel and A. Sigel (eds.), Electron Transfer Reactions in Metalloproteins, in *Metal Ions in Biological Systems*, Vol. 27, Marcel Dekker, New York, 1991.
12. J. R. Miller, Controlling charge separation through effects of energy, distance and molecular structure on electron transfer rates, *Nouv. J. Chim.*, **11**, 83 (1987).
13. R. C. Prince and G. N. George, Tryptophane radicals, *Trends Biochem. Sci.*, **15**, 170 (1990).
14. T. L. Poulos, Heme enzyme crystal structures, in *Advances in Inorganic Biochemistry* (eds. G. L. Eichhorn and L. G. Marzilli), Vol. 7, Elsevier, Amsterdam, 1988, p. 1.

15. H. Pelletier and J. Kraut, Crystal structure of a complex between electron transfer partners, cytochrome *c* peroxidase and cytochrome *c*, *Science*, **258** 1748 (1992).
16. J. Deisenhofer and H. Michel, The photosynthetic reaction center of the purple bacterium *Rhodopseudomonas viridis* (Nobel address), *Angew. Chem. Int. Ed. Engl.*, **28**, 829 (1989).
17. W. R. Scheidt and C. A. Reed, Spin-state/stereochemical relationships in iron porphyrins: implications for the hemoproteins, *Chem. Rev.*, **81**, 543 (1981).
18. S. G. Sligar, K. D. Egeberg, J. T. Sage, D. Morikis and P. M. Champion, Alteration of heme axial ligands by site-directed mutagenesis: a cytochrome becomes a catalytic demethylase, *J. Am. Chem. Soc.*, **109**, 7896 (1987).
19. J. T. Groves, Key elements of the chemistry of cytochrome P-450, *J. Chem. Educ.*, **62**, 928 (1985).
20. R. I. Murray, M. T. Fisher, P. G. Debrunner and S. G. Sligar, Structure and chemistry of cytochrome P-450, in *Metalloproteins* (ed. P. Harrison), Part 1, Verlag Chemie, Weinheim, 1985, p. 157.
21. P. R. Ortiz de Montellano, Control of the catalytic activity of prosthetic heme by the structure of hemoproteins, *Acc. Chem. Res.*, **20**, 289 (1987).
22. F. P. Guengerich, Reactions and significance of cytochrome P-450 enzymes, *J. Biol. Chem.*, **266**, 10019 (1991).
23. T. D. Porter and M. H. Coon, Cytochrome P-450, *J. Biol. Chem.*, **266**, 13469 (1991).
24. W. B. Jakoby and D. M. Ziegler, The enzymes of detoxification, *J. Biol. Chem.*, **265**, 20715 (1990).
25. J. A. Fruetel, J. R. Collins, D. L. Camper, G. H. Loew and P. R. Ortiz de Montellano, Calculated and experimental absolute stereochemistry of the styrene and β-methylstyrene epoxides formed by cytochrome P-450$_{cam}$, *J. Am. Chem. Soc.*, **114**, 6987 (1992).
26. E. Mutschler, *Arzneimittelwirkungen, Lehrbuch der Pharmakologie und Toxikologie*, 6th edn, WVG, Stuttgart, 1991.
27. D. Mansuy, P. Battioni and J. P. Battioni, Chemical model systems for drug-metabolizing cytochrome-P-450-dependent monooxygenases, *Eur. J. Biochem.*, **184**, 267 (1989).
28. H. Patzelt and W. D. Woggon, O-Insertion into nonactivated C—H bonds: the first observation of O_2 cleavage by a P-450 enzyme model in the presence of a thiolate ligand, *Helv. Chim. Acta*, **75**, 523 (1992).
29. R. Raag and T. L. Poulos, The structural basis for substrate-induced changes in redox potential and spin equilibrium in cytochrome P-450CAM, *Biochemistry*, **28**, 917 (1989).
30. K. G. Ravichandran, S. S. Boddupalli, C. A. Hasemann, J. A. Peterson and J. Deisenhofer, Crystal structure of hemoprotein domain of P450BM-3, a prototype for microsomal P450's, *Science*, **261**, 731 (1993).
31. N. C. Gerber and S. G. Sligar, Catalytic mechanism of cytochrome P-450: evidence for a distal charge relay, *J. Am. Chem. Soc.*, **114**, 8742 (1992).
32. D. Mandon, R. Weiss, M. Franke, E. Bill and A. X. Trautwein, High-valent oxoiron porphyrin complexes. Preparation via solvent-dependent protonation of a peroxyiron porphyrinate, *Angew. Chem. Int. Ed. Engl.*, **28**, 1709 (1989).
33. J. T. Groves and Y. Watanabe, Reactive iron porphyrin derivatives related to the catalytic cycles of cytochrome P-450 and peroxidase, *J. Am. Chem. Soc.*, **110**, 8443 (1988).
34. K. L. Kostka, B. G. Fox, M. P. Hendrich, T. J. Collins, C. E. F. Richard, L. J. Wright and E. Münck, High-valent transition metal chemistry. Mössbauer and EPR studies of high-spin ($S = 2$) iron(IV) and intermediate-spin ($S = 3/2$) iron(III) complexes with a macrocyclic tetraamido-N ligand, *J. Am. Chem. Soc.*, **115**, 6746 (1993).
35. D. T. Sawyer, The nature of the bonding and valency for oxygen in its metal compounds, *Comments Inorg. Chem.*, **6**, 103 (1987).
36. P. M. Champion, Elementary electronic excitations and the mechanism of cytochrome P450, *J. Am. Chem. Soc.*, **111**, 3433 (1989).
37. H. Sugimoto, H. C. Tung and D. T. Sawyer, Formation, characterization, and reactivity of the oxene adduct of [tetrakis(2,6-dichlorophenyl)porphinato]iron(III) perchlorate in acetonitrile. Model for the reactive intermediate of cytochrome P-450, *J. Am. Chem. Soc.*, **110**, 2465 (1988).
38. W. A. Herrmann, Zufallsentdeckung am Beispiel Rhenium: Oxo-Komplexe in hohen und niedrigen Oxidationsstufen, *J. Organomet. Chem.*, **300**, 111 (1986).

39. J. Everse, K. E. Everse and M. B. Grisham (eds.), *Peroxidases in Chemistry and Biology*, Vol. 2, CRC Press, Boca Raton, 1990.

40. T. Haag, F. Lingens and K.-H. van Peé, A metal ion and cofactor independent enzymatic redox reaction. Halogenation by bacterial non-heme haloperoxidases, *Angew. Chem. Int. Ed. Engl.*, **33**, 1487 (1991).

41. S. Hashimoto, R. Nakajima, I. Yamazaki, T. Kotani, S. Ohtaki, T. Kitagawa, Resonance Raman characterization of hog thyroid peroxidase, *FEBS Lett.*, **248**, 205 (1989).

42. H. E. Schoemaker, On the chemistry of lignin biodegradation, *Recl. Trav. Chim. Pays-Bas*, **109**, 255 (1990).

43. J. H. Dawson, Probing structure-function relations in heme-containing oxygenases and peroxidases, *Science*, **240**, 433 (1988).

44. K. Yamaguchi, Y. Watanabe and I. Morishima, Direct observation of the push effect on the O—O bond cleavage of acylperoxoiron(III) porphyrin complexes, *J. Am. Chem. Soc.*, **115**, 4058 (1993).

45. J. E. Erman, L. B. Vitello, M. A. Miller and J. Kraut, Active-site mutations in cytochrome *c* peroxidase: a critical role for histidine-52 in the rate of formation of compound I, *J. Am. Chem. Soc.*, **114**, 6592 (1992).

46. M. G. Peter, Chemical modification of biopolymers with quinones and 'quinone' methides, *Angew. Chem. Int. Ed. Engl.*, **28**, 555 (1989).

47. K. E. Hammel, Oxidation of aromatic pollutants by lignin-degrading fungi and their extracellular peroxidases, in *Metal Ions in Biological Systems* (ed. H. Sigel), Vol. 28, Marcel Dekker, New York, 1992, p. 41.

48. G. Winkelmann (ed.), *Microbial Degradation of Natural Products*, VCH Publishers, New York, 1992.

49. A. R. Butler and D. L. H. Williams, The physiological role of nitric oxide, *Chem. Soc. Rev.*, 233 (1993).

50. J. S. Stamler, D. J. Singel and J. Loscalzo, Biochemistry of nitric oxide and its redox-activated forms, *Science*, **258**, 1898 (1992).

51. M. J. Clarke and J. B. Gaul, Chemistry relevant to the biological effects of nitric oxide and metallonitrosyls, *Struct. Bonding (Berlin)*, **81**, 147 (1993).

52. P. M. H. Kroneck, J. Beuerle and W. Schumacher, Metal-dependent conversion of inorganic nitrogen and sulfur compounds, in *Metal Ions in Biological Systems* (ed. H. Sigel), Vol. 28, Marcel Dekker, New York, 1992, p. 455.

53. T. Brittain, R. Blackmore, C. Greenwood and A. J. Thomson, Bacterial nitrite-reducing enzymes, *Eur. J. Biochem.*, **209**, 793 (1992).

54. E. Culotta and D. E. Koshland, NO news is good news, *Science*, **258**, 1862 (1992).

55. M. A. Tayeh and M. A. Marletta, Macrophage oxidation of L-arginine to nitric oxide, nitrite and nitrate, *J. Biol. Chem.*, **264**, 19654 (1989).

56. A. L. Burnett, C. J. Lowenstein, D. S. Bredt, T. S. K. Chang and S. H. Snyder, Nitric oxide: a physiologic mediator of penile erection, *Science*, **257**, 401 (1992).

57. M. A. Marletta, Nitric oxide synthase structure and mechanism, *J. Biol. Chem.*, **268**, 12231 (1993).

58. S. A. Lipton *et al.*, A redox-based mechanism for the neuroprotective and neurodestructive effects of nitric oxide and related nitroso-compounds, *Nature (London)*, **364**, 626 (1993).

59. A. Verma, D. J. Hirsch, C. E. Glatt, G. V. Ronnett and S. H. Snyder, Carbon monoxide. A putative neural messenger?, *Science*, **259**, 381 (1993).

60. R. C. Prince and A. B. Hooper, Resolution of the hemes of hydroxylamine oxidoreductase by redox potentiometry and electron spin resonance spectroscopy, *Biochemistry*, **26**, 970 (1987).

61. C. K. Chang, R. Timkovich and W. Wu, Evidence that heme d_1 is a 1,3-porphyrindione, *Biochemistry*, **25**, 8447 (1986).

62. D. E. McRee, D. C. Richardson, J. S. Richardson and L. M. Siegel, The heme and Fe_4S_4 cluster in the crystallographic structure of *Escherichia coli* sulfite reductase, *J. Biol. Chem.*, **261**, 10277 (1986).

63. A. Pierik and W. R. Hagen, $S = 9/2$ EPR signals are evidence against coupling between the siroheme and the Fe/S cluster prosthetic groups in *Desulfovibrio vulgaris* (Hildenborough) dissimilatory sulfite reductase, *Eur. J. Biochem.*, **195**, 505 (1991).

7 Iron–Sulfur and Other Nonheme Iron Proteins

Three main groups of iron-containing proteins can be distinguished according to the ligation of the metal. Iron centers which are exclusively coordinated by amino acid residues, components of water (H_2O, HO^-, O^{2-}) or oxoanions occur in photosynthetic reaction centers (Figures 4.5 to 4.7), in hemerythrin (Section 5.3), in nonheme iron enzymes (Section 7.6) and in iron transport and storage proteins (see Chapter 8). In addition to these often polynuclear systems and the heme species (5.8) in which the iron centers are chelated through a porphinoid macrocycle (see Sections 5.2 and Chapter 6), the *iron–sulfur (Fe/S) centers* represent another major and important class of iron species in proteins [1–3].

7.1 Biological Relevance of the Element Combination Iron/Sulfur

The majority of the ubiquitous iron–sulfur centers in proteins is involved in electron transfer at typically negative redox potentials (Tables 7.1 and 7.2).

Fe/S centers have essential functions in photosynthesis (Figure 4.9), cell respiration (Figures 6.1 and 6.2), nitrogen fixation (see Section 11.2) and in the metabolism of H_2 (hydrogenases with and without nickel; Section 9.3), NO_2^- or SO_3^{2-} (sulfite oxidation and reduction; Sections 6.5 and 11.1.2). In addition to the electron transfer function, Fe/S centers have been recognized as sites for redox and nonredox catalysis; examples include the exclusively Fe/S-dependent hydrogenases and nitrogenases as well as (de)hydrolase/isomerase enzymes of the aconitase type (see 7.10; [4]). A further role of coordinatively open Fe/S centers would be to serve as iron sensors [5], whereas the [4Fe–4S] center found in the DNA repair enzyme endonuclease III [6] seems to have 'only' a structural/polarizing function.

Iron–sulfur centers in proteins sometimes occur as isolated clusters, e.g. in the small, electron-transferring 'ferredoxins' [Fd]; however, they frequently show close interactions with other prosthetic groups such as other metal centers (Ni, Mo, V, heme-Fe) or flavins [7]. A characteristic feature of iron–sulfur proteins is the coordination of iron ions with protein-bound cysteinate sulfur (RS^-) and, in polynuclear Fe/S centers, with 'inorganic' acid-labile sulfide (S^{2-}). Sulfide and iron ions are often extractable and, in many cases, the remaining apoenzymes can be reconstituted with external S^{2-} and $Fe^{2+/3+}$ ions (Figure 7.1).

Approximately 1% of the iron content of mammals is present in the form of Fe/S proteins (compare Table 5.1). The facile formation and thermal robustness of such proteins, their distribution in nearly all organisms, particularly in evolutionary 'old' species, and the remarkable conservation of critical amino acid sequences (7.7) suggest that they might have played an important role very early in evolution, i.e. in the

Table 7.1 Some reactions that are catalyzed by Fe/S center-containing redox enzymes

enzyme	catalyzed reaction	further explanations in chapter
hydrogenases	$2 H^+ + 2 e^- \rightleftharpoons H_2$	9.3
nitrogenases	$N_2 + 10 H^+ + 8 e^- \rightleftharpoons 2 NH_4^+ + H_2$	11.2
sulfite reductase	$SO_3^{2-} + 7 H^+ + 6 e^- \rightleftharpoons HS^- + 3 H_2O$	6.5
aldehyde oxidase	$R\text{-}CHO + 2 OH^- \rightleftharpoons R\text{-}COOH + H_2O + 2 e^-$	12.5
xanthine oxidase	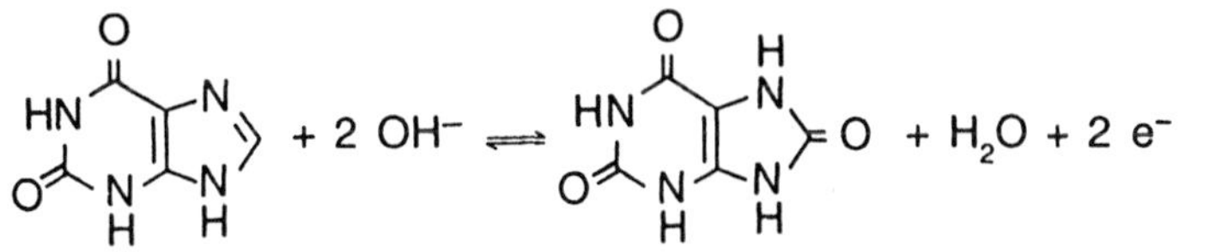	11.1.1
NADP oxidoreductase	$NADP^+ + H^+ + 2 e^- \rightleftharpoons NADPH$	(Fig. 4.8)

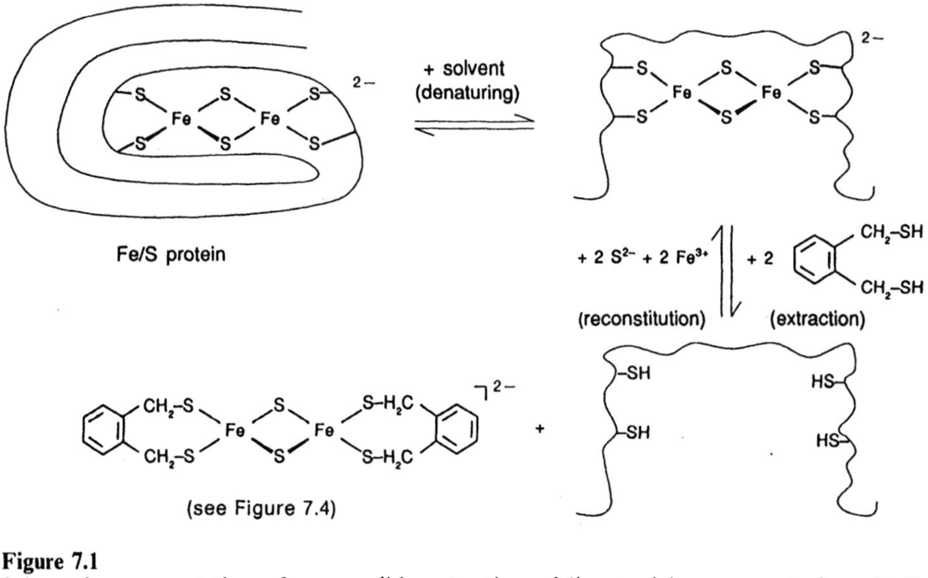

Figure 7.1
Schematic representation of a reversible extraction of 'inorganic' components from Fe/S proteins (according to [8])

Table 7.2   Redox potentials of representative iron–sulfur proteins (according to [1]; see also Table 6.1)

protein	typical origin	type of Fe/S center	molecular mass (kDa)	E (mV)
rubredoxin	*Clostridium pasteurianum*	$[Rd]^{2+;3+}$	6	-60
2Fe ferredoxin	spinach	$[2Fe-2S]^{1+;2+}$	10.5	-420
adrenodoxin	adrenal mitochondria	$[2Fe-2S]^{1+;2+}$	12	-270
Rieske center	adrenal mitochondria	$[2Fe-2S]^{1+;2+}$	250 (bc$_1$ complex)	+280
4Fe ferredoxin	*Bacillus stearothermophilus*	$[4Fe-4S]^{1+;2+}$	9.1	-280
8Fe ferredoxin	*Cl. pasteurianum*	$2[4Fe-4S]^{1+;2+}$	6	-400
High Potential Iron-Sulfur Protein (HiPIP)	*Chromatium vinosum*	$[4Fe-4S]^{2+;3+}$	9.5	+350
ferredoxin II	*Desulfovibrio gigas*	$[3Fe-4S]^{n+}$	24 (tetramer)	-130
ferredoxin I	*Azotobacter vinelandii*	$[3Fe-4S]^{n+}$ $[4Fe-4S]^{n+}$	14	-460

absence of free O_2 [9,10]. The typically negative redox potentials (Table 7.2), the occurrence in highly temperature-resistant ($> 100\,°C$) 'hyperthermophilic' micro-organisms [11] and the general oxygen sensitivity of the reduced states also support this hypothesis.

There are experimentally substantiated [12] theories according to which inorganic iron sulfides, FeS or the disulfide (S_2^{2-})-containing pyrite (FeS_2), might have been involved in the beginning of chemoautotrophic metabolism through reduction of CO_2 according to (7.1a) or (7.1b) and thus possibly in the evolution of life [10,13].

$$H_2S \;+\; 2\,FeS \;+\; CO_2 \;\rightarrow\; 2\,FeS_2 \;+\; HCOOH \tag{7.1a}$$

$$H_2O \;+\; 2\,FeS \;+\; CO_2 \;\xrightarrow{h\nu}\; 2\,FeO \;+\; 1/n\,(CHOH)_n \;+\; 2\,S \tag{7.1b}$$

Robust 'chemolithotrophic' sulfur bacteria which obtain their energy from the transformation of inorganic compounds have become important in geobiotechnology as the organisms essential for the process of 'bacterial leaching' [14,15]. This leaching process of ores or slag heaps with the help of the ubiquitous bacteria *Thiobacillus thiooxidans* and *T. ferrooxidans* is directed at hardly soluble sulfides such as CuS or $CuFeS_2$ or at oxides like UO_2 which become transformed into soluble sulfates (7.2), at the same time releasing enclosed noble metals such as gold.

$$S + 1.5\ O_2 + H_2O \xrightarrow{\text{\textit{T. thiooxidans}}} H_2SO_4$$

$$2\ FeSO_4 + 0.5\ O_2 + H_2SO_4 \xrightarrow{\text{\textit{T. ferrooxidans}}} Fe_2(SO_4)_3 + H_2O$$

$$MS + Fe_2(SO_4)_3 \longrightarrow MSO_4 + 2\ FeSO_4 + S$$

overall reaction:
$$MS\ (\text{insoluble}) + 2\ O_2 \longrightarrow MSO_4\ (\text{soluble})$$

M: e.g. Cu, Zn, Ni, Co   (7.2)

The enzymes in the above-mentioned bacteria catalyze the oxidation of iron(II) and elemental sulfur to such an extent that metal-poor ores, slag heaps of primary ore mining and even metal-contaminated industrial wastes and effluents can be treated using this ecologically and economically advantageous process with excellent yields. Worldwide, more than 25% of the produced copper is already obtained via microbially supported leaching; in a single mine the corresponding yield can amount to 50 tons of copper per day. The pH optimum of reaction (7.2) at pH 2–3, the temperature resistance and the tolerance of thiobacilli with respect to heavy metal concentrations are quite remarkable. Biotechnological modifications to increase and transfer this temperature stability and heavy metal tolerance are being attempted with regard to possible applications in metal decontamination and recycling. In nature, reactions like (7.2) proceed reversibly depending on the external conditions; read from right to left such processes are referred to as 'bioweathering' or 'geo-chemical biomineralization'.

Returning to the Fe/S proteins, four general kinds of Fe/S centers can be distinguished, according to the degree of aggregation in 'clusters' (7.3, Figure 7.2); with few exceptions, the metal atoms in these clusters are surrounded by four sulfur atoms in a distorted tetrahedral fashion. This arrangement, implicating a relatively small coordination number, is enforced by the steric requirements of the large sulfur donor atoms, thus differing significantly from the usual hexacoordination of 'bio-logical' iron when it is bound to O or N ligands, i.e. smaller donor atoms from the first full period. As an important consequence, the iron atoms in the Fe/S centers are exclusively high-spin because the ligand field splitting in tetrahedral symmetry is less than half of what it would be in an octahedral arrangement (see 12.4).

Although some enzymes such as, for example, fumarate reductase [17] may contain several different Fe/S centers, the individual established centers will be introduced consecutively in the following.

Structurally confirmed Fe/S centers (nomenclature and charges):

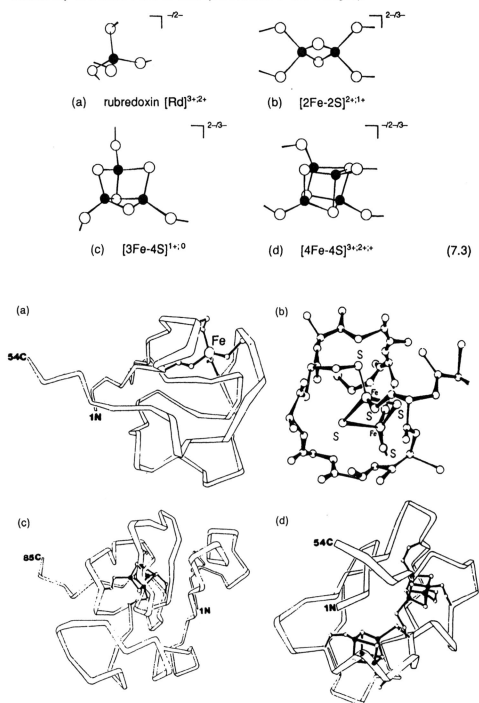

(a)   rubredoxin [Rd]$^{3+;2+}$ (b)   [2Fe-2S]$^{2+;1+}$

(c)   [3Fe-4S]$^{1+;0}$ (d)   [4Fe-4S]$^{3+;2+;+}$ (7.3)

Figure 7.2
Structures of Fe/S centers in proteins. (a), (c), (d) Ribbon representations of the protein folding
(from [2]); (b) detailed close-up of the immediate protein environment of a [2Fe–2S] center
(according to [16])

7.2 Rubredoxins

Rubredoxins are small redox proteins which occur in certain bacteria and contain just one iron center. Four cysteinate ligands from two amino acid sequences —Cys—X_2—Cys— ligate the iron center in a distorted tetrahedral fashion (Figure 7.2a). The transition between the nearly colorless iron(II) state ($S = 2$) and the red iron(III) form occurs without a major change in Fe–S distances; larger changes and, therefore, slower electron transfer (compare Section 6.1) were observed for simple model complexes (see Section 7.5). The intensely red color responsible for the name of these proteins results from a ligand-to-metal charge-transfer transition (7.4) from the σ and π electron-rich thiolate ligands to the oxidized, i.e. electron-poor iron(III) center [18]. Comparably intense colors are known from the analytically important thiocyanate (NCS^-) complexes of iron(III).

$$Fe^{III}(^-S-R) \xrightarrow{h\nu} [Fe^{II}(^{\cdot}S-R)]^* \qquad (7.4)$$

An important function of the protein is to stabilize the trivalent form with respect to an otherwise conceivable intramolecular redox reaction to give Fe^{II} and disulfide. Rubredoxin-type centers, [Rd], also occur in more complex proteins such as ruberythrin which, in addition, contains nonheme diiron centers similar to those of hemerythrin or ribonucleotide reductase [19].

7.3 [2Fe–2S] Centers

In the [2Fe–2S] centers of 2Fe ferredoxins [16] or more complex enzymes [7] two iron centers are each coordinated by two cysteinate side chains of the protein and by two shared, i.e. bridging (μ-)sulfide dianions (7.3b, Figure 7.2b). The [2Fe–2S] centers are particularly common in chloroplasts (Figure 4.8), the 2Fe ferredoxin obtained from spinach leaves having become particularly well known ($\rightarrow$Popeye). Because of their different protein environment and thus resulting electrostatic and structural asymmetry, the two iron centers are not equivalent; however, the question is whether or not the same is true for their redox behavior. The biologically relevant electron transfer consists in a transition from the Fe^{III}/Fe^{III} state with compensating, i.e. strongly antiferromagnetically coupled spins, to a one-electron reduced mixed-valent form. Does the unsymmetrical formulation Fe^{II}/Fe^{III}, implicating localization, or rather the symmetrical description $Fe^{2.5}/Fe^{2.5}$, corresponding to delocalization in the ground state, conform with experimental results? Mössbauer spectroscopy (Section 5.3) and other physical methods suggest a localized description with fixed valences Fe^{II} and Fe^{III} for reduced [2Fe–2S] centers in proteins and in some model complexes (Section 7.5). Despite antiferromagnetic coupling of the high-spin centers via the sulfide ions (which are capable of effecting superexchange) one EPR-detectable unpaired electron remains for the reduced form.

Within the group of [2Fe–2S] proteins there are some systems that contain centers with unusual spectroscopic properties and relatively high redox potentials (Table 7.2). These 'Rieske centers' are found in cytochrome-containing membrane protein complexes of mitochondria ('bc_1-complexes'; compare Figures 6.1 and 6.2) and in chloroplasts (b/f-complexes compare Figure 4.8). The Rieske proteins contain two

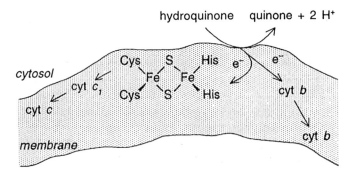

(7.5)

markedly different iron centers; an unsymmetrical coordination as in (7.5) involving the neutral *non*sulfur ligand histidine is considered most likely [20].

In combination with a cytochrome b, the function of the Rieske centers is to guarantee a splitting (bifurcation) of the electron flow in the intramembrane electron transport chain [22]. Starting from the two-electron donating hydroquinones there is one electron pathway *along* the membrane at high potential and another pathway across the membrane at low potential (7.5).

7.4 Polynuclear Fe/S Clusters: Relevance of the Protein Environment and Catalytic Activity

The most common and most stable iron–sulfur centers are of the [4Fe–4S] type. These 'clusters' are found in numerous complex enzymes and in electron-transfer proteins such as the 4Fe, 7Fe and 8Fe ferredoxins, the latter containing two rather isolated [4Fe–4S] centers within the protein (distance > 1 nm; compare Figure 7.2d). The 4Fe and 8Fe ferredoxins with exclusively S-coordinated iron are relatively small ubiquitous electron-transfer proteins. As the term [4Fe–4S] suggests, four iron centers and four sulfide ions are arranged in what may be described as a distorted cuboidal arrangement with approximate D_2 symmetry, each four-coordinate iron center being anchored in the protein by one cysteinate residue (7.3, Figure 7.2c,d). To a first approximation, the four iron centers form a tetrahedron with the μ_3-sulfide ions above the triangular planes so that these, too, are tetrahedrally arranged. [4Fe–4S] centers participate in nearly all complex biological redox reactions such as photosynthesis, respiration or N_2 fixation, when they act mainly as electron-transfer centers at negative potentials; however, they may also have nonredox catalytic or noncatalytic functions [6].

Mössbauer spectroscopy indicates that there are two pairs of iron dimers present in the diamagnetic (2–) charged state of the system ([4Fe–4S]$^{2+}$ nucleus and 4 Cys$^-$); both pairs show approximately the same isomer shift, corresponding to an oxidation state of + 2.5, but different quadrupole splitting. The fact that two mixed-valent Fe^{II}/Fe^{III} pairs do not show any resulting paramagnetism is not self-evident despite the even number of electrons. Extensive electron delocalization, including the sulfur ligands, and effective spin coupling seem to be operating in this configuration. Starting from this (2–) state, paramagnetic EPR-active forms are created in the 'normal' 4Fe ferredoxins through one-electron reduction to 3Fe^{II}/1Fe^{III} species with spin ground states of $S = 1/2$ or higher. However, there are also protein types in which the diamagnetic [4Fe–4S] center can be reversibly oxidized at relatively high potentials to a paramagnetic ($S = 1/2$) form with a $1Fe^{II}/3Fe^{III}$ oxidation state distribution; these proteins are referred to as high potential iron–sulfur proteins (HiPIP; Table 7.2, Figure 7.2c).

A 'three-state hypothesis' has been applied to describe the fact that HiPIP and normal 4Fe ferredoxins are distinguished by different stabilities of their one-electron oxidized or reduced forms, respectively (7.6 [23]). Although it is possible to obtain the respective unphysiological 'superreduced' or 'superoxidized' states under denaturation of the protein, the protein environments of the intact species allow only the biologically intended redox behavior.

HiPIP	normal ferredoxin	n in $[4Fe–4S]^{n+}$	EPR
oxidized HiPIP $\uparrow\downarrow$ + 350 mV reduced HiPIP	super- oxidized ferredoxin $\uparrow\downarrow$ −50 mV oxidized ferredoxin	3	active
		2	inactive
$\uparrow\downarrow$ −600 mV super- reduced HiPIP	$\uparrow\downarrow$ −400 mV reduced Ferredoxin	1	active

$$(7.6)$$

In accord with the explanation given for the other two classes of 'inorganic' electron-transfer proteins, i.e. the cytochromes (Section 6.1) and blue copper proteins (see Section 10.1), the geometries of the Fe/S cluster and of the whole protein change very little during electron transfer. For example, the reduction of the $[4Fe–4S]^{3+}$ form in HiPIP causes only a small elongation of Fe—S bonds, an expansion and a slightly more pronounced distortion of the cluster. In non-HIPIP ferredoxins, it is also the diamagnetic form that seems to exhibit a somewhat higher deviation from the ideally tetrahedral arrangement of the iron atoms [1]; a geometrically more distorted diamagnetic cluster would conform with the entatic state concept (Section 2.3.1). Apparently small changes in the protein framework, like a larger number of hydrophobic amino acid residues around the HiPIP cluster and a thus diminished access of water determine the redox potential and stability of individual oxidation states. The expansion of the clusters upon reduction is due to the extra electron being transferred to nonbonding or antibonding cluster molecular orbitals; in any case, the smaller metal cations are closer to the cluster center than the larger sulfide anions (compare 7.3d).

Experimental and theoretical studies on proteins, modified proteins and model complexes have shown that the electronic structure of [4Fe–4S] clusters is very sensitive toward minute geometrical changes in bond lengths, bond angles and torsion angles. Not only redox potentials and stabilities but also spin–spin coupling patterns depend on the conformation, implicating, for example, the accessibility of higher spin states than $S = 1/2$ for odd electron forms. Like the central diamagnetic $(2+)$ state, the neighboring paramagnetic states of the $[4Fe–4S]^{n+}$ system also contain two antiferromagnetically coupled iron dimers, one mixed-valent pair (II,III) and one homo-valent pair, each with predominantly parallel internal spin–spin coupling [24]. In contrast to the $[2Fe–2S]^{+}$ systems with their localized state, the higher degree of delocalization (resonance) found in the $[4Fe–4S]^{n+}$ clusters can be attributed to the structurally determined orthogonality of metal orbitals which interact via super-exchanging sulfide bridges. Following theoretical analyses, this favors parallel spin–spin interactions with higher resonance energies according to Hund's rule [25]. Thus, the electron delocalization as such is less susceptible to perturbation from external asymmetries as induced from the protein. Nevertheless, these asymmetries exist and can be detected by sensitive EPR/ENDOR [24] and NMR spectroscopical methods [26]. The participation of the cysteinate ligands in the accommodation of additional electrons is evident from a strengthening of hydrogen bond interactions $X—H\cdots{}^{-}S(Cys)$ [27] suggesting an increase in effective negative charge at the cysteinate sulfur centers.

The amino acid sequence does not only determine whether a HiPIP center or the 'normal' ferredoxin form of the [4Fe–4S] system occurs, but is primarily responsible for the alternative between a 4Fe or a 2Fe ferredoxin formed from a cysteine-containing protein and enzymatically introduced iron (Chapter 8) and sulfide (via sulfide reductase and thiosulfate sulfur transferase). Some representative amino acid positions for the invariant cysteines in Fe/S cluster-forming proteins are summarized in (7.7) [3,6,23,28].

[Rd]	:	- Cys - X_2 - Cys - X_n - Cys - X_2 - Cys -
[2Fe–2S]	:	- Cys - X_4 - Cys - X_2 - Cys - X_{29} - Cys -
[3Fe–4S]	:	- Cys - $X_{5,7}$ - Cys - X_n - Cys -
[4Fe–4S] or 'normal Fd'	:	- Cys - X_2 - Cys - X_2 - Cys - X_n - Cys -
HiPIP	:	- Cys - X_2 - Cys - X_{16} - Cys - X_{13} - Cys -
nonredox active (endonuclease III)	:	- Cys - X_6 - Cys - X_2 - Cys - X_5 - Cys -
nitrogenase Fe protein (dimer)	:	- Cys - X_{34} - Cys - - Cys - X_{34} - Cys -

$$(7.7)$$

Different amino acid sequences are obviously responsible for the formation of relatively new kinds of iron–sulfur proteins which were first detected by Mössbauer spectroscopy and, in addition to [4Fe–4S] moieties, contain special [3Fe–4S] centers (3Fe and 7Fe ferredoxins [29]). After suitable modification, these [3Fe–4S] centers may be converted to [4Fe–4S] centers (7.8 [3,17,21]). Structurally, they can be derived from [4Fe–4S] analogues by removal of a labile, *not cysteinate-coordinated* iron atom from the distorted cuboidal arrangement [30]; another [3Fe–4S] form with *linearly* arranged metal centers has been detected albeit under unphysiological (basic) conditions (7.3c, 7.8) (see Figure 7.3).

In addition to their occurrence in ferredoxins of microorganisms (compare Table 7.2), the [3Fe–4S] systems have been found in equilibrium with a labile [4Fe–4S] form (7.8) as a component of the enzyme aconitase (aconitate hydratase/isomerase),

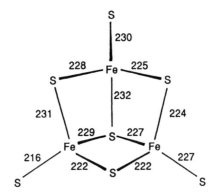

Figure 7.3
Structure and some geometrical data (bond lengths in pm) of the [3Fe–4S] cluster from *Desulfovibrio gigas* ferredoxin (according to [30])

Cys–S / Fe / S / Cys–S / Fe — S / S — Fe / S–Cys
[3Fe–4S]

$+Fe_a^{n+}$
$-Fe_a^{n+}$

Cys–S / S — Fe_a / Cys–S / Fe — S / S — Fe / S–Cys
[4Fe–4S]

pH > 9.5

$+Fe^{n+}$

Cys–S / S / S / S–Cys
Cys–S / Fe / Fe / Fe / S–Cys
S / S / S–Cys

'linear' [3Fe–4S] cluster   (7.8)

e.g. in mitochondria. This iron–sulfur enzyme catalyzes the equilibrium (7.9) within the Calvin cycle in what is clearly a nonredox reaction.

$$\underset{\substack{\text{citrate}\\(90\%\text{ in equilibrium})}}{\begin{array}{c} H \\ HC-COO^- \\ | \\ HO-C-COO^- \\ | \\ CH_2-COO^- \end{array}} \overset{-H^+}{\underset{-OH^-}{\rightleftharpoons}} \underset{\substack{\text{Z-aconitate}\\(4\%)}}{\begin{array}{c} H \quad COO^- \\ \diagdown\diagup \\ C \\ \| \\ C \\ \diagup\diagdown \\ {}^-OOC-H_2C \quad COO^- \end{array}} \overset{+H^+}{\underset{+OH^-}{\rightleftharpoons}} \underset{\substack{\text{isocitrate}\\(6\%)}}{\begin{array}{c} H \\ HO-C-COO^- \\ | \\ HC-COO^- \\ | \\ CH_2-COO^- \end{array}} \quad (7.9)$$

The following reaction mechanism has been suggested for aconitase [31,32]:

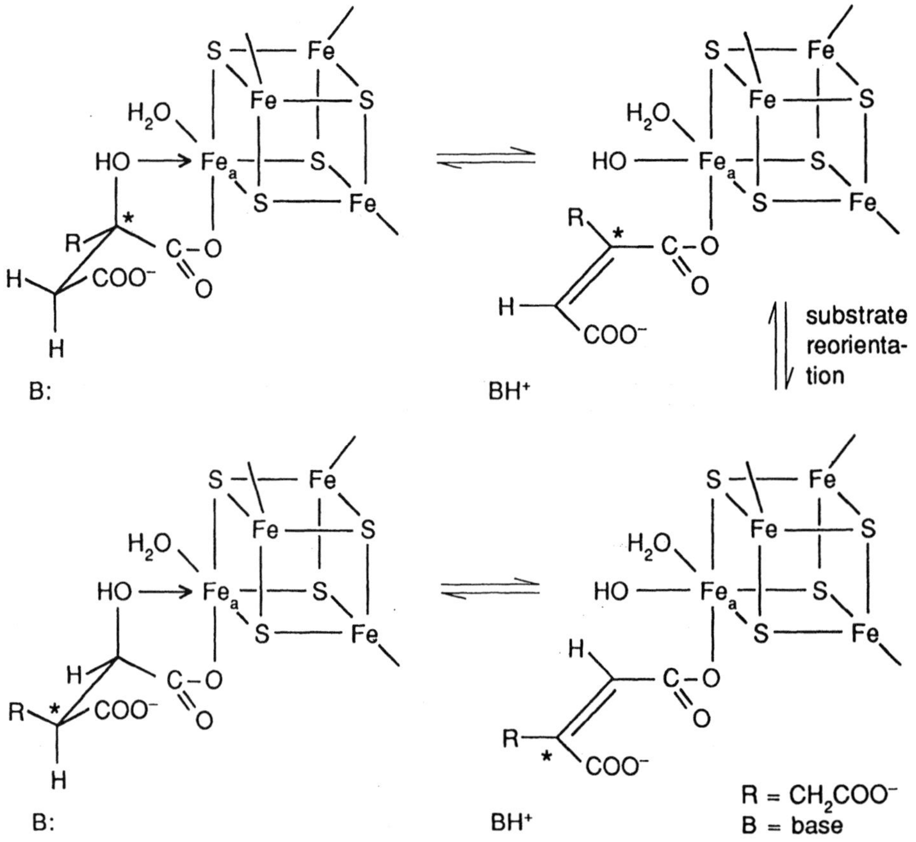

$$R = CH_2COO^-$$
$$B = base$$

(7.10)

In the active state, the labile iron center, Fe_a, of the [4Fe–4S] form of aconitase is not coordinated by cysteinate but by water molecules. After substitution of H_2O and coordination of the chelating substrate (five-membered chelate ring, coordination number five or six at Fe_a), a sequence (7.10) of (HO)—C* bond cleavage/C—H deprotonation, rotation of the nonchelating Z-aconitate ligand and hydroxide–C(olefin) bond formation/C*(olefin) protonation leads to a rapid equilibration (7.9).

The electronic structure of the [3Fe–4S] centers is interesting both from a coordination chemical and spectroscopical point of view. The ground state of the oxidized form shows *one* unpaired electron ($S = 1/2$) corresponding to a formulation with three almost equally antiferromagnetically coupled high-spin Fe^{III} centers. The reduced form may be described as the combination of $1Fe^{II}$ with $2Fe^{III}$, featuring an $S = 2$ ground state. According to Mössbauer data, this situation results from antiparallel spin–spin interactions between a high-spin Fe^{III} center ($S = 5/2$) and a Fe^{III}/Fe^{II} mixed-valent pair with $S = 9/2$, i.e. which shows parallel internal spin–spin coupling. In some cases there are even excited spin states readily accessible; according to their 'open', coordinatively unsaturated structure, the [3Fe–4S] centers are predestined for chemical catalytic activity [21], including iron-sensor functions.

The natural inventory of Fe/S centers is not exhausted with the 3Fe clusters; preliminary studies on biochemical material [3,11,33,34] as well as model studies [35] suggest that a variety of other possible structures with higher numbers of iron atoms than four ($\rightarrow$six, eight) or with nonthiolate ligands can be found in the system $Fe^{II/III}/S^{-II}$/protein.

Among such more complex $[xFe–yS]$ centers are the sulfide-bridged double-cubane 'P clusters' in nitrogenase (7.11 [34]; see Section 11.2) with their very negative redox potential of $-470\,mV$ and an $S = 5/2$ or $S = 7/2$ ground state in the oxidized form, and the (6Fe) 'H clusters' in nickel-free hydrogenases [11] which catalyze the equilibrium $2\ H^+ + 2\ e^- \leftrightarrows H_2$. The molecular mechanism of H_2 activation by hydrogenase Fe/S centers is not yet known; a side-on coordination of H_2 to the metal (η^2-ligation) with ensuing base-assisted heterolysis to give H^+ and enzymatically bound hydride is being discussed [11] (compare 9.9).

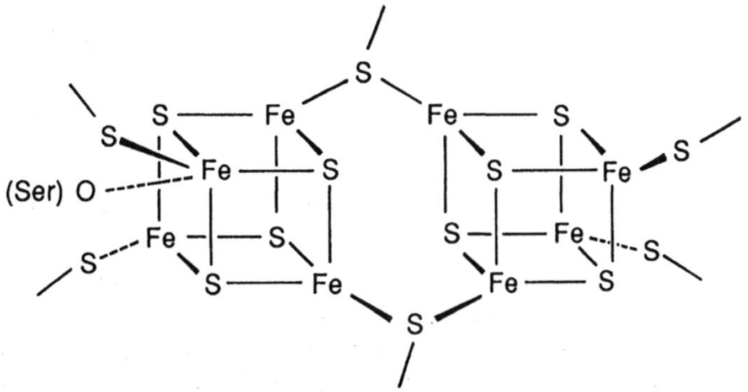

$$(7.11)$$

The [4Fe–4S] center of the DNA repair enzyme endonuclease III seems to have 'only' a polarizing and structure-determining function, directed at recognition. The negatively charged cluster binds positively polarized amino acids, which in turn interact with the negatively charged phosphate backbone of DNA [6].

7.5 Model Systems for Iron-Sulfur Proteins

Several model complexes for Fe/S centers in proteins can be prepared surprisingly easily via 'spontaneous self-assembly' reactions [36,37]. For example, [4Fe–4S] cluster ions are obtained from thiol, hydrogen sulfide, iron(III) and base under reducing conditions in polar aprotic solvents such as dimethylsulfoxide (DMSO), as follows:

$$6\ RSH + 4\ NaHS + 4\ FeCl_3 + 10\ NaOR\ \rightarrow$$

$$Na_2[Fe_4S_4(RS)_4] + 10\ ROH + 12\ NaCl + RSSR \quad (7.12)$$

Whereas the stable [4Fe–4S] systems are typically formed with sterically unhindered thiols as model ligands for cysteine, conformational constraints as imposed by preferentially chelate-forming dithiols such as o-xylene-α,α'-dithiol (Figure 7.1)

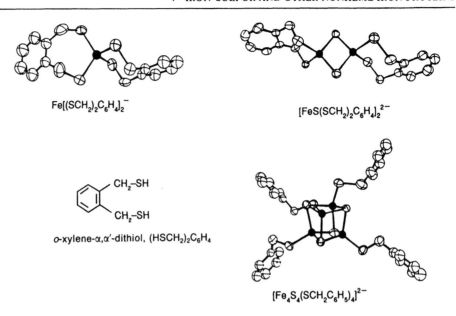

Fe[(SCH₂)₂C₆H₄]₂⁻

[FeS(SCH₂)₂C₆H₄]₂²⁻

$Fe[(SCH_2)_2C_6H_4]_2^-$

$[FeS(SCH_2)_2C_6H_4]_2^{2-}$

o-xylene-α,α'-dithiol, $(HSCH_2)_2C_6H_4$

$[Fe_4S_4(SCH_2C_6H_5)_4]^{2-}$

Figure 7.4
Molecular structures of model complex anions (according to [23])

lead to models of [2Fe–2S] dimers or, in the absence of sulfide, to models of rubredoxin (Figure 7.1 and 7.4).

Modelling of [3Fe–4S] centers and of unsymmetrically (3 + 1)-substituted clusters requires more effort, particularly the synthesis of specially designed polythiolate ligands (Figure 7.5 [21]).

As expected, most model systems show lower reaction rates for electron transfer than the natural Fe/S proteins because of more pronounced geometrical changes during the process; in addition, the models exhibit significantly lower and thus unphysiological redox potentials [8]. This can be attributed to the absence of the

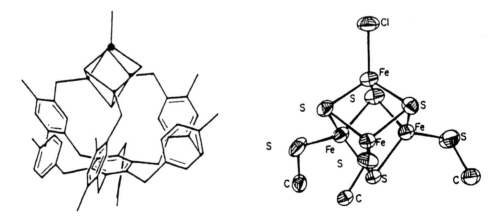

Figure 7.5
Molecular structure of a [4Fe–4S] cluster, functionalized at only one metal center (left: overall structure, right: cluster arrangement) (according to [21])

protein environment, where amine–sulfide hydrogen bond interactions $HN \cdots S$ [27], electrostatic effects in the peptide backbone and small but effective distortions of the cluster may be important. In view of the rather simple structure and function of Fe/S proteins, some 'artificial proteins' of this kind, i.e. peptides with short chain lengths, have already been synthesized, based on known amino acid sequences (7.7), and tested, for example, for hydrogenase activity [38]. The possibility of heteroatom substitution in the clusters has been observed both in the course of chemical synthesis and in important biological materials such as Ni-containing hydrogenases (see Section 9.3) and Mo- or V-containing nitrogenases (see Sections 11.2 and 11.3).

7.6 Iron-Containing Enzymes without Porphyrin or Sulfide Ligands

After introduction of the heme–iron proteins in Section 5.2 and Chapter 6 and of the iron–sulfur proteins described above, the most important iron-containing enzymes which apparently manage without such 'special' ligands will be discussed in the following. The O_2-transporting and, therefore, nonenzymatic protein hemery-thrin with its dimeric iron center has already been described in Section 5.3; several other iron proteins containing neither heme nor sulfide similarly resort to an (indirect) metal–metal interaction in polynuclear metalloproteins [39–41] in order to achieve sufficient electronic flexibility. However, there are also numerous mono-nuclear nonheme iron enzymes with very important biological functions (Section 7.6.4).

7.6.1 Iron-Containing Ribonucleotide Reductase (RR)

A coenzyme B_{12}-dependent form of ribonucleotide reductase found in *Rhizobium* bacteria and lactobacilli has already been introduced in Section 3.2.3. However, in most organisms, particularly higher ones, an iron-containing form of RR is respon-sible for the catalysis of the biologically essential reaction (7.13) [42,43]. Anaerobic microorganisms feature an Fe/S cluster-containing ribonucleotide reductase [43] and still another dimanganese-containing form of bacterial RR was described by Willing, Follmann and Auling [44]. All known ribonucleotide reductases contain metal centers and organic radicals; they catalyze the deoxygenation (reduction) of the ribose ring to yield 2'-deoxyribose in nucleotides and thus promote the first step in *de novo* DNA biosynthesis. The required electrons are made available by dithiols, e.g. peptides of the thioredoxin type, which can themselves be oxidized to a disulfide form (7.13).

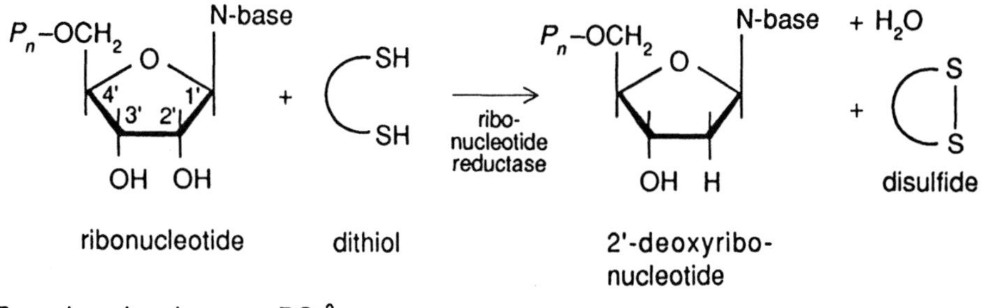

P : phosphoryl group, PO_3^{2-}
(compare 14.2)

(7.13)

The iron-containing ribonucleotide reductases are quite large proteins with molecular masses exceeding 200 kDa; this complexity is necessitated *inter alia* by the requirements for an exact control of the reaction (feedback, allosteric regulation). The most thoroughly studied enzyme of this type is the RR isolated from *E. coli* which consists of two different dimeric subunits ($\alpha_2\beta_2$) with molecular masses of about 170 (α_2) and 87 kDa (β_2 protein). According to X-ray crystal structure determinations [45], the active center is located at the interface between the two different subunits; whereas the larger protein contains thiol groups as the assumed direct electron donors, the smaller protein features one tyrosyl radical as well as one diiron site per subunit. The iron-containing RR was the first enzyme for which a stable free radical in the protein was established as an essential component [46]; previously, it had only been assumed that the function of the coenzyme B_{12}-dependent RR was based on radical reactivity (Section 3.2.3). As in the related hemerythrin protein (Section 5.3), numerous spectroscopic and magnetic measurements had already provided information concerning the approximate structure of the diiron center and its magnetic interaction with the tyrosyl radical before details became available from a crystal structure analysis of the smaller protein subunit [45].

In the oxidized state, the trivalent high-spin iron centers in each subunit are bridged by one μ-oxo and one μ-η^1:η^1-glutamate ligand and are otherwise coordinated rather unsymmetrically (7.14): one metal center binds to two η^1-glutamate groups, one histidine ligand (N_δ coordination) and a water molecule, while the other iron ion coordinates to one H_2O molecule, one histidine ligand (N_δ) and a chelating (η^2-)aspartate in a less regular octahedral configuration. The μ-oxo/μ-carboxylato-bridged structure of the dimetal center as also found in oxyhemerythrin is quite characteristic and is also assumed for some dimanganese-containing enzymes (see 4.14). Studies on synthetic models [39,47,48] suggest that this arrangement has a high tendency of formation ('self assembly' process), it favors an efficient metal–metal interaction at a distance of about 0.33 nm (RR) and thus a certain degree of electronic flexibility.

$$(7.14)$$

The presence of labile water ligands at the iron centers facilitates interactions with the tyrosyl radical situated at a distance of 0.53 nm. A magnetic interaction between antiferromagnetically coupled high-spin Fe^{III} centers and the reactive radical situated approximately along the Fe–Fe axis has been established [46].

The reduced Fe^{II}/Fe^{II} form of the iron dimer presumably features two $\mu\text{-}\eta^1:\eta^1$-glutamate ligands without an oxo bridge [49]. This state can possibly interact with O_2 [50] to create the neutral tyrosyl radical; additional conceivable functions of the diiron moiety are to stabilize, protect or even activate the tyrosyl radical in a controlled way. However, according to the results from the structure analysis [45], the situation of the tyrosyl radical at least 1 nm away from the protein surface prohibits a direct interaction, e.g. in form of a hydrogen abstraction, between tyrosyl radical and substrate. A possible role of the tyrosyl radical and the iron dimer is to participate in electron-transfer processes involving a tryptophane which is situated at the protein surface and which also has the potential to form a reactive radical [51]. Dioxygen species such as O_2 or O_2^{2-}, further activated by the iron centers, could contribute to the oxidative formation of the tyrosyl radical [52]. Starting from the tryptophyl radical, the deoxygenation of ribose may take place via hydrogen abstraction, loss of OH^- from a radical intermediate and reduction of the resulting radical cation species via enzymatic sulfhydryl groups; their reduction, in turn, proceeds through the external dithiol (7.13). The deoxygenation by RR enzymes may thus be viewed as the reverse reaction of the O-insertion which is catalyzed by P-450 enzymes or the methane monooxygenases described in the following.

7.6.2 Soluble Methane Monooxygenase (MMO)

This enzyme, a multiprotein complex with a molecular mass of about 300 kDa [40], is employed by methano*trophic* microorganisms which use CH_4 as a source of carbon and energy, the latter provided through the first step (7.15) of methane oxidation. (Metalloenzymes of methano*genic*, i.e. CH_4-producing organisms, are discussed elsewhere; compare Sections 3.2.3 and 9.5 as well as Figure 1.1.)

$$\text{NADH} + \text{CH}_4 + \text{O}_2 + \text{H}^+ \xrightarrow{\substack{\text{methane} \\ \text{monooxygenase}}} \text{NAD}^+ + \text{CH}_3\text{OH} + \text{H}_2\text{O}$$

$$(7.15)$$

The very large (251 kDa, $\alpha_2\beta_2\gamma_2$) hydroxylase component of the enzyme [53] contains two diiron centers which have been characterized in several oxidation state combinations. Typical for the Fe^{III}/Fe^{III} form are the antiparallel spin–spin coupling of two high-spin Fe^{III} centers and the absence of an intense charge-transfer absorption in the visible spectrum [54]. In accord with spectroscopic data, a μ-hydroxo bridge has been found between the two coordinatively not fully saturated metal centers, in addition to a bidentate glutamate bridge ($\mu\text{-}\eta^1:\eta^1$) and monodentate non-bridging glutamate and histidine ligands [53]. Only the high-spin Fe^{II}/Fe^{II} form is active with respect to dioxygen activation; presumably, the diiron site is the catalytic center. Other metal ions, organic cofactors or stable radicals have not been found.

The mechanism of reaction (7.15) apparently involves the formation of an oxygenated dimetal center with oxoferryl(IV) groups [55] (see Sections 6.3 and 6.4) which then effect monooxygen insertion into the C—H bond of CH_4; other hydrocarbons are also oxygenated with a certain stereospecificity [56]. The reactivity and assumed catalytic mechanism found for MMO thus shows many parallels to the P-450 system (Section 6.2).

7.6.3 Purple Acid Phosphatases (Fe/Fe and Fe/Zn)

Polyphosphate- and phosphate ester-cleaving enzymes typically require a nonredox-active divalent metal ion such as Mg^{2+} (Section 14.1) or Zn^{2+}, as, for example, in alkaline phosphatase (Section 12.3 [57]). However, there are also iron-containing phosphatases which are distinguished by their intense color and their activity optimum in the acidic region. Such enzymes may be obtained from a variety of plants, microorganisms or animals (uteroferrin, bovine spleen phosphatase); they have molecular masses of about 40 kDa and contain dimetal centers. Detailed studies of the diiron species have shown easily distinguishable colors and absorption spectra for the inactive purple Fe^{III}/Fe^{III} state and for the enzymatically active pink Fe^{II}/Fe^{III} mixed-valent form, the latter exhibiting weak antiferromagnetic spin–spin coupling [58].

The iron centers are presumably bridged by a μ-hydroxo ligand and by a μ-η^1:η^1-carboxylate. A tyrosinate ligand binds to the Fe^{III} center in either of the two states, the coordination of this π electron-rich ligand to an oxidized metal center being responsible for intense low-energy ligand-to-metal charge-transfer (LMCT) absorptions in the visible region. Both metals are probably coordinated by one histidine ligand each [59]. There are different arrangements possible for the binding of phosphate ester substrates, either η^1- or μ-η^1:η^1-coordination [40]; model complexes have not yet provided clear answers [59,60]. A plausible mechanism comprises the attack of hydroxide activated by binding to Fe^{II} (see 12.3 [61]) on an Fe^{III}-coordinated phosphate derivative; variable metal–metal distances would favor such a differentiated reaction mechanism.

The native form of a phosphatase from kidney beans obviously contains zinc(II) and iron(III) in a similarly bridged arrangement as in Fe^{II}/Fe^{III} phosphatases [62]. For the mechanism, this again implies a division of functions between divalent and trivalent metal centers, viz. through water activation by the (+ II) ion and phosphate binding by the trivalent ion.

7.6.4 Mononuclear Nonheme Iron Enzymes

This group of enzymes with only one, apparently simple iron center includes mainly monooxygenases and dioxygenases which are important, for example, in the metabolism of fatty acids or aromatic compounds, and in the synthesis of amino acids, neurotransmitters, β-lactams or leukotrienes [52].

For instance, iron-containing and tetrahydropterin-dependent monooxygenases (hydroxylases) catalyze the transformation of L-phenylalanine to L-tyrosine, the tetrahydropterin (compare 3.12 or 11.8) being oxidized to its dihydro form (see 11.8) and dioxygen being reduced to one molecule of water (*mono*oxygenation [52,63]. Obviously, both the high-spin Fe^{III} or Fe^{II} centers and the pterin redox system (11.8) cooperate in oxygen activation and electron transfer.

Nonheme iron-containing dioxygenases are quite common, participating *inter alia* in the oxidative cleavage of aromatic compounds (compare 7.16). In their oxidized state, catechol 1,2-dioxygenases and related enzymes contain high-spin iron(III) with tyrosinate ligands [64], giving rise to an LMCT transition in the visible region. According to mechanistic hypotheses [52], the binding of π electron-rich catecholate to the π electron-deficient and Lewis-acidic high-spin Fe^{III} center in 'intradiol-cleaving' enzymes results in a ligand-to-metal electron transfer with corresponding weakening of the intradiol (O)C—C(O) bond of the thus formed o-semiquinone radical. $^3O_2$ activation may then proceed in a spin-allowed fashion via high-spin Fe^{II} (compare Section 5.2) before the final product is formed via peroxidic intermediates (7.16). There are also extradiol-cleaving nonheme iron enzymes [52].

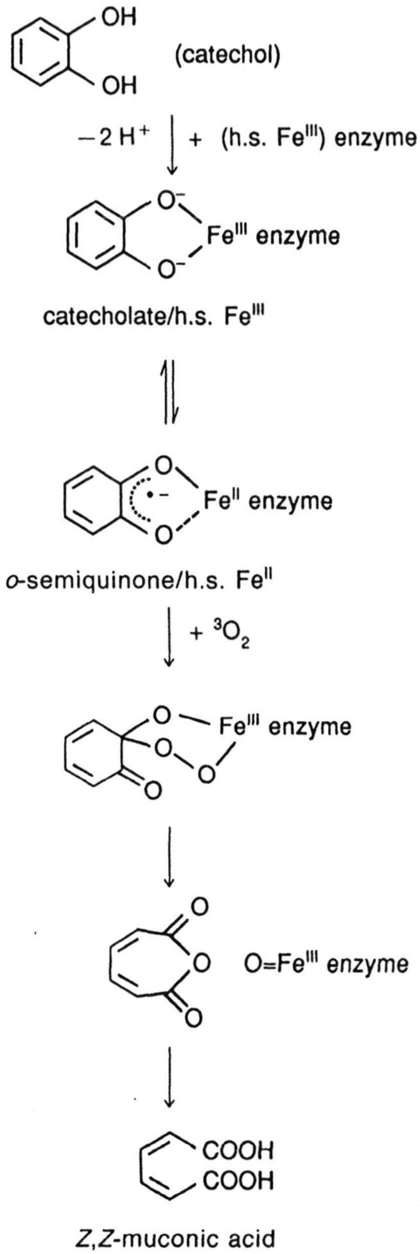

$$\tag{7.16}$$

Lipoxygenases are among those iron-containing enzymes which, if only indirectly, catalyze reactions of dioxygen with organic substrates. In the metabolism of 1,4-diene-containing fatty acids they stereospecifically catalyze autoxidation or 'peroxidation' reactions, i.e. the O_2 insertion into C—H bonds, as a first step in leukotriene synthesis.

Crystal structure analyses of soy bean lipoxygenase [65,66] exhibit coordinatively unsaturated high-spin iron ligated by three histidine residues and the (isoleucine) carboxy terminus, perhaps also by weakly coordinating carboxamide oxygen of an asparagine side chain [66]. The reaction mechanism involves oxidative deprotonation of the 1,4-diene part of the coordinated substrate by Fe^{III} as well as $^3O_2$ binding and activation by the resulting Fe^{II} ion [65].

No oxygenation occurs during the ring-forming reaction (7.17) to give isopenicillin N which is catalyzed by another high-spin iron(II) enzyme, isopenicillin-N synthase; *both* oxygen atoms of O_2 are being reduced to water. For the active state of the metal center, three histidine residues and a water molecule have been postulated as ligands in addition to the coordinated reactants [52,67].

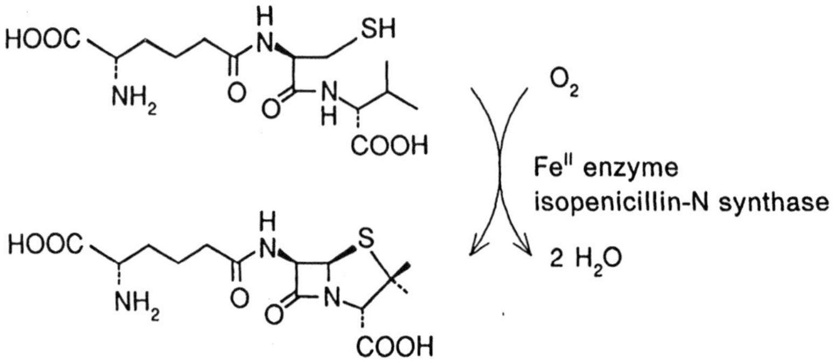

isopenicillin N   (7.17)

The unusual low-spin iron(III) center in bacterial nitrile hydratase does apparently not have a biological redox activity. This enzyme is interesting from a biotechnological point of view since it also catalyzes the formation of acrylamide for synthetic fiber production from acrylonitrile (7.18).

$$R\text{–}CN + H_2O \xrightarrow{\text{nitrile hydratase}} R\text{–}C(O)NH_2 \tag{7.18}$$

EXAFS and ENDOR data and the long-wavelength LMCT absorption of the enzyme suggest six-coordinated iron with two (cysteine) thiolates and three histidine side chains as protein-based ligands [68].

References

1. A. J. Thomson, Iron–sulphur proteins, in *Metalloproteins* (ed. P. Harrison), Part 1, Verlag Chemie, Weinheim, 1985, p. 79.

2. F. R. Salemme, Structure and function of cytochromes *c*, *Annu. Rev. Biochem.*, **46**, 299 (1977).
3. R. Cammack, Iron–sulfur clusters in enzymes: themes and variations, *Adv. Inorg. Chem.*, **38**, 281 (1992).
4. R. Grabowski, A. E. M. Hofmeister and W. Buckel, Bacterial L-serine dehydratases: a new family of enzymes containing iron–sulfur clusters, *Trends Biochem. Sci.*, **18**, 297 (1993).
5. T. V. O'Halloran, Transition metals in control of gene expression, *Science*, **261**, 715 (1993).
6. C.-F. Kuo, D. E. McRee, C. L. Fisher, S. F. O'Handley, R. P. Cunningham and J. A. Tainer, Atomic structure of the DNA repair [4Fe–4S] enzyme endonuclease III, *Science*, **258**, 434 (1992).
7. C. C. Correll, C. J. Batie, D. P. Ballou and M. L. Ludwig, Phthalate dioxygenase reductase: a modular structure for electron transfer from pyridine nucleotides to [2Fe–2S], *Science*, **258**, 1604 (1992).
8. B. Averill and W. H. Orme-Johnson, Iron–sulfur proteins and synthetic analogs, in *Metal Ions in Biological Systems* (ed. H. Sigel), Vol. 7, Marcel Dekker, New York, 1978, p. 127.
9. A. Müller, N. Schladerbeck, Systematik der Bildung von Elektronentransfer-Clusterzentren $[Fe_nS_n]^{m+}$ mit Relevanz zur Evolution von Ferredoxinen, *Chimia*, **39**, 23 (1985).
10. G. Wächtershäuser, Pyrite formation, the first energy source for life: a hypothesis, *System. Appl. Microbiol.*, **10**, 207 (1988).
11. M. W. W. Adams, Novel iron–sulfur centers in metalloenzymes and redox proteins from extremely thermophilic bacteria, *Adv. Inorg. Chem.*, **38**, 341 (1992).
12. E. Blöchl, M. Keller, G. Wächtershäuser and K. O. Stetter, Reactions depending on iron sulfide and linking geochemistry with biochemistry, *Proc. Natl. Acad. Sci. USA*, **89**, 8117 (1992).
13. R. J. P. Williams, Iron and the origin of life, *Nature (London)*, **343**, 213 (1990).
14. D. K. Ewart and M. N. Hughes, The extraction of metals from ores using bacteria, *Adv. Inorg. Chem.*, **36**, 103 (1991).
15. M. N. Hughes and R. K. Poole, *Metals and Micro-organisms*, Chapman & Hall, London, 1989.
16. T. Tsukihara, K. Fukuyama, H. Tahara, Y. Katsube, Y. Matsuura, N. Tanaka, M. Kakudo, K. Wada and H. Matsubara, X-ray analysis of ferredoxin from *Spirulina platensis*. II. Chelate structure of active center, *J. Biochem.*, **84**, 1646 (1978).
17. A. Manodori, G. Cecchini, I. Schröder, R. P. Gunsalus, M. T. Werth and M. K. Johnson, [3Fe–4S] to [4Fe–4S] cluster conversion in *Escherichia coli* fumarate reductase by site-directed mutagenesis, *Biochemistry*, **31**, 2703 (1992).
18. M. S. Gebhard, J. C. Deaton, S. A. Koch, M. Millar, E. I. Solomon, Single-crystal spectral studies of $Fe(SR)_4^-$ [R = 2,3,5,6-$(Me)_4C_6H$]: the electronic structure of the ferric tetrathiolate active site, *J. Am. Chem. Soc.*, **112**, 2217 (1990).
19. B. C. Prickril, D. M. Kurtz, Jr, J. LeGall and G. Voordouw, Cloning and sequencing of the gene for ruberythrin from *Desulfovibrio vulgaris* (Hildenborough), *Biochemistry*, **30**, 11118 (1991).
20. J. A. Fee, K. L. Findling, T. Yoshida, R. Hille, G. E. Tarr, D. O. Hearshen, W. R. Dunham, E. P. Day, T. A. Kent and E. Münck, Purification and characterization of the Rieske iron–sulfur protein from *Thermus thermophilus*, *J. Biol. Chem.*, **259**, 124 (1984).
21. R. H. Holm, Trinuclear cuboidal and heterometallic cubane-type iron–sulfur clusters: new structural and reactivity themes in chemistry and biology, *Adv. Inorg. Chem.*, **38**, 1 (1992).
22. T. A. Link, H. Schägger and G. von Jagow, Analysis of the structures of the subunits of the cytochrome *bc* I complex from beef heart mitochondria, *FEBS Lett.*, **204**, 9 (1986).
23. D. O. Hall, R. Cammack and K. K. Rao, Chemie und Biologie der Eisen-Schwefel-Proteine, *Chem. Unserer Zeit*, **11**, 165 (1977).
24. J.-M. Mouesca, G. Rius and B. Lamotte, Single-crystal proton ENDOR studies of the $[Fe_4S_4]^{3+}$ cluster: determination of the spin population distribution and proposal of a model to interpret the ^1H NMR paramagnetic shifts in high-potential ferredoxins, *J. Am. Chem. Soc.*, **115**, 4714 (1993).
25. L. Noodleman, J. G. Norman, J. H. Osborne, A. Aizman and D. A. Case, Models for ferredoxins: electronic structures of iron–sulfur clusters with one, two, and four iron atoms, *J. Am. Chem. Soc.*, **107**, 3418 (1985).

26. I. Bertini, F. Capozzi, S. Ciurli, C. Luchinat, L. Messori and M. Piccioli, Identification of the iron ions of high potential iron protein from *Chromatium vinosum* within the protein frame through two-dimensional NMR experiments, *J. Am. Chem. Soc.*, **114**, 3332 (1992).

27. G. Backes, Y. Mino, T. M. Loehr, T. E. Meyer, M. A. Cusanovich, W. V. Sweeney, E. T. Adman and J. Sanders-Loehr, The environment of Fe_4S_4 clusters in ferredoxins and high-potential iron proteins. New information from X-ray crystallography and resonance Raman spectroscopy, *J. Am. Chem. Soc.*, **113**, 2055 (1991).

28. M. M. Georgiadis, H. Komiya, P. Chakrabarti, D. Woo, J. J. Kornuc and D. C. Rees, Crystallographic structure of the nitrogenase iron protein from *Azotobacter vinelandii*, *Science*, **257**, 1653 (1992).

29. G. N. George and S. J. George, X-ray crystallography and the spectroscopic imperative: the story of the [3Fe–4S] clusters, *Trends Biochem. Sci.*, **13**, 369 (1988).

30. C. R. Kissinger, E. T. Adman, L. C. Sieker and L. H. Jensen, Structure of the 3Fe–4S cluster in *Desulfovibrio gigas* ferredoxin II, *J. Am. Chem. Soc.*, **110**, 8721 (1988).

31. H. Beinert and M. C. Kennedy, Engineering of protein bound iron–sulfur clusters, *Eur. J. Biochem.*, **186**, 5 (1989).

32. M. H. Emptage, Aconitase: evolution of the active-site picture, in *Metal Clusters in Proteins* (ed. L. Que, Jr), ACS Symposium Series 372, 1988, p. 343.

33. A. J. Pierik, R. B. G. Wolbert, P. H. A. Mutsaers, W. R. Hagen and C. Veeger, Purification and biochemical characterization of a putative [6Fe–6S] prismane-cluster-containing protein from *Desulfovibrio vulgaris* (Hildenborough), *Eur. J. Biochem.*, **206** 697 (1992).

34. J. Kim and D. C. Rees, Structural models of the metal centers in the nitrogenase molybdenum-iron protein, *Science*, **257**, 1677 (1992).

35. M. S. Reynolds and R. H. Holm, Iron–sulfur–thiolate basket clusters, *Inorg. Chem.*, **27**, 4494 (1988).

36. A. Müller, N. H. Schladerbeck and H. Bögge, $[Fe_4S_4(SH)_4]^{2-}$, the simplest synthetic analogue for a ferredoxin, *J. Chem. Soc., Chem. Commun.*, 35 (1987).

37. A. Müller and N. H. Schladerbeck, Einfache aerobe Bildung eines $[Fe_4S_4]^{2+}$ Clusterzentrums, *Naturwiss.*, **73**, 669 (1986).

38. A. Nakamura and N. Ueyama, Importance of peptide sequence in electron-transfer reactions of iron–sulfur clusters, in *Metal Clusters in Proteins* (ed. L. Que, Jr), ACS Symposium Series 372, 1988, p. 292.

39. S. J. Lippard, Oxo-bridged polyiron centers in biology and chemistry, *Angew. Chem. Int. Ed. Engl.*, **27**, 344 (1988).

40. J. B. Vincent, G. L. Olivier-Lilley and B. A. Averill, Proteins containing oxo-bridged dinuclear iron centers: a bioinorganic perspective, *Chem. Rev.*, **90**, 1447 (1990).

41. R. G. Wilkins, Binuclear iron centres in proteins, *Chem. Soc. Rev.*, 171 (1992).

42. M. Lammers and H. Follmann, The ribonucleotide reductases: a unique group of metalloenzymes essential for cell proliferation, *Struct. Bonding (Berlin)*, **54**, 27 (1983).

43. P. Reichard, From RNA to DNA, why so many ribonucleotide reductases? *Science*, **260**, 1773 (1993).

44. A. Willing, H. Follmann and G. Auling, Ribonucleotide reductase of *Brevibacterium ammoniagenes* is a manganese enzyme, *Eur. J. Biochem.*, **170**, 603 (1988).

45. P. Nordlund, B.-M. Sjöberg and H. Eklund, Three-dimensional structure of the free radical protein of ribonucleotide reductase, *Nature (London)*, **345**, 593 (1990).

46. A. Ehrenberg, Magnetic interaction in ribonucleotide reductase, *Chem. Scr.*, **28A**, 27 (1988).

47. K. Wieghardt, The active centers in manganese-containing metalloproteins and inorganic model complexes, *Angew. Chem. Int. Ed. Engl.*, **28**, 1153 (1989).

48. D. M. Kurtz, Oxo- and hydroxo-bridged diiron complexes: a chemical perspective on a biological unit, *Chem. Rev.*, **90**, 585 (1990).

49. M. Atta, C. Scheer, P. H. Fries, M. Fontecave and J. M. Latour, Multified saturation magnetization measurements on oxidized and reduced ribonucleotide reductase form *Escherichia coli*, *Angew. Chem. Int. Ed. Engl.*, **31**, 1513 (1992).

50. J. M. Bollinger, Jr, D. E. Edmondson, B. H. Huynh, J. Filley, J. R. Norton and J. Stubbe, Mechanism of assembly of the tyrosyl radical–dinuclear iron cluster cofactor of ribonucleotide reductase, *Science*, **253**, 229 (1991).

51. R. C. Prince and G. N. George, Tryptophan radicals, *Trends Biochem. Sci.*, **15**, 170 (1990).

52. L. Que, Jr, Oxygen activation at nonheme iron centers, in *Bioinorganic Catalysis* (ed. J. Reedijk), Marcel Dekker, New York, 1993, p. 347.

53. A. C. Rosenzweig, C. A. Frederick, S. J. Lippard and P. Nordlund, Crystal structure of a bacterial non-haem iron hydroxylase that catalyses the biological oxidation of methane, *Nature (London)*, **366**, 537 (1993).

54. B. G. Fox, M. P. Hendrich, K. K. Surerus, K. K. Andersson, W. A. Froland, J. D. Lipscomb and E. Münck, Mössbauer, EPR, and ENDOR studies of the hydroxylase and reductase components of methane monooxygenase from *Methylosinus trichosporium* OB3b, *J. Am. Chem. Soc.*, **115**, 3688 (1993).

55. S.-K. Lee, B. G. Fox, W. A. Froland, J. D. Lipscomb and E. Münck, A transient intermediate of the methane monooxygenase catalytic cycle containing an $Fe^{IV}Fe^{IV}$ cluster, *J. Am. Chem. Soc.*, **115**, 6450 (1993).

56. K. E. Liu, C. C. Johnson, M. Newcomb and S. J. Lippard, Radical clock substrate probes and kinetic isotope effect studies of the hydroxylation of hydrocarbons by methane monooxygenase, *J. Am. Chem. Soc.*, **115**, 939 (1993).

57. J. B. Vincent, M. W. Crowder and B. A. Averill, Hydrolysis of phosphate monoesters: a biological problem with multiple chemical solutions, *Trends Biochem. Sci.*, **17**, 105 (1992).

58. K. Doi, B. C. Antanaitis and P. Aisen, The binuclear iron centers of uteroferrin and the purple acid phosphatases, *Struct. Bonding (Berlin)*, **70**, 1 (1988).

59. A. E. True, R. C. Scarrow, C. R. Randall, R. C. Holz and L. Que, Jr, EXAFS studies of uteroferrin and its anion complexes, *J. Am. Chem. Soc.*, **115**, 4246 (1993).

60. B. Bremer, K. Schepers, P. Fleischhauer, W. Haase, G. Henkel and B. Krebs, The first binuclear iron(III) complex with a terminally coordinated phosphato ligand—a model compound for the oxidized form of purple acid phosphatase from beef spleen, *J. Chem. Soc., Chem. Commun.*, 510 (1991).

61. E. G. Mueller, M. W. Crowder, B. A. Averill and J. R. Knowles, Purple acid phosphatase: a diiron enzyme that catalyzes a direct phopho group transfer to water, *J. Am. Chem. Soc.*, **115**, 2974 (1993).

62. J. L. Beck, J. de Jersey, B. Zerner, M. P. Hendrich and P. G. Debrunner, Properties of the Fe(II)–Fe(III) derivative of red kidney bean purple phosphatase. Evidence for a binuclear Zn–Fe center in the native enzyme, *J. Am. Chem. Soc.*, **110**, 3317 (1988).

63. T. A. Dix and S. J. Benkovic, Mechanism of oxygen activation by pteridine-dependent monooxygenases, *Acc. Chem. Res.*, **21**, 101 (1988).

64. D. H. Ohlendorf, J. D. Lipscomb and P. C. Weber, Structure and assembly of protocatecholate 3,4-dioxygenase, *Nature (London)*, **336**, 403 (1988).

65. J. C. Boyington, B. J. Gaffney and L. M. Amzel, The three-dimensional structure of an arachidonic acid 15-lipoxygenase, *Science*, **260**, 1482 (1993).

66. W. Minor, J. Steczko, J. T. Bolin, Z. Otwinowski and B. Axelrod, Crystallographic determination of the active site iron and its ligands' in soybean lipoxygenase L-1, *Biochemistry*, **32**, 6320 (1993).

67. L. J. Ming, L. Que, A. Kriauciunas, C. A. Frolik and V. J. Chen, Coordination chemistry of the metal binding site of isopenicillin N synthase, *Inorg. Chem.*, **29**, 1111 (1990).

68. H. Jin, I. M. Turner, Jr, M. J. Nelson, R. J. Gurbiel, P. E. Doan and B. M. Hoffman, Coordination sphere of the ferric ion in nitrile hydratase, *J. Am. Chem. Soc.*, **115**, 5290 (1993).

8 Uptake, Transport and Storage of an Essential Element as Exemplified by Iron

> Despite their fundamental role in processes of signaling, homeostasis, and cytotoxicity little detailed information is available on the mechanisms whereby metal ions enter eukaryotic cells. Exceptions include the uptake of Fe ... and the permeation of Ca^{2+} through Ca channels.
> D. M. Templeton, *J. Biol. Chem.*, **265** 21764 (1990).

The description of structures and physiological functions of metal centers in enzymes or proteins has always played an important role in bioinorganic chemistry. However, there are complex mechanisms operating between the overall abundance of an element in the organism (compare Section 2.1) and its specific function, e.g. in an enzyme. These necessarily selective and controlled processes for uptake, transport, storage and directed transfer of an element, e.g. to the designated apoprotein, have to occur under temporally and spatially well-defined conditions. This intricate aspect of the space and time dependence of 'bioinorganic processes' in actual organisms has been repeatedly pointed out by Williams [1,2].

The compounds which are responsible for transport and storage of iron, the physiologically most abundant and also most versatile transition metal, have been very thoroughly studied. Much less factual information is available for the other elements; however, comparable mechanisms to those of iron transport and storage are probably operating there as well.

8.1 The Problem of Iron Mobilization — Oxidation States, Solubility, and Medical Relevance

Iron is an essential trace element for almost all organisms with the exception of lactobacilli; its distribution in an adult human has been summarized in Table 5.1. Complex regulatory mechanisms serve in the uptake, transport and storage of iron [3–7]. These processes may be divided into several single steps:

(a) active or passive resorption during food ingestion, involving, for example, dissolution, redox reactions or complexation,
(b) selective transport of the iron ions through membranes into cells,
(c) processes within the cells, e.g. incorporation into a protein, and
(d) elimination from the metabolism either through excretion or temporary storage.

An excess of free iron, in particular high-spin Fe^{II}, is dangerous for any organism since radicals can be generated in the presence of dioxygen or peroxide (compare 4.6 or 5.2) according to equations (8.1) and (8.2) [8]:

$$\text{high-spin } Fe^{II} + {}^3O_2 \rightarrow Fe^{III} + O_2^{\bullet-} \tag{8.1}$$

$$Fe^{II} + H_2O_2 \rightarrow Fe^{III} + OH^- + OH^\bullet \tag{8.2}$$

Potentially pathogenic microorganisms need a continuous supply of iron as a growth-determining factor for their reproduction. Accordingly, the availability of iron to invading bacteria in a multicellular organism plays an important role for infectious diseases such as tuberculosis, leprosy or cholera, during which a decrease of the iron content in the human blood plasma is symptomatic [9]. Since microorganisms cannot activate tightly bound iron in the blood serum they have to utilize 'free' iron species; effectively iron-scavenging and membrane-diffusive complexing ligands can thus exhibit antibiotic properties. The iron of heme-type porphyrin complexes is very tightly bound and can be liberated only within cells through enzymatic action.

The metabolism of iron in the human organism has general medical relevance; of the recommended 10–20 mg iron in the daily food supply (see Table 2.3) only about 10% are resorbed on the average. Both iron deficiency which is common in developing nations ($\rightarrow$anemia [10]) and excessive iron intake as a suspected long-term risk factor for coronary diseases may cause severe pathological symptoms. Blood infusions result in an acute iron enrichment in the body since a human being can only excrete about 1–2 mg iron per day. Complexing agents such as polychelating deferrioxamine (8.8) can bind iron even if it is coordinated by the transport protein transferrin (Section 8.4.1) or by storage proteins, but they cannot release iron from heme. The resulting ferrioxamine complex can be excreted from the body via the kidneys. Complex formation or, more exactly, substitution (recomplexation) is rather slow, making high and continually supplied doses of the drug necessary for a successful therapy. Careful studies of naturally occurring iron-complexing ligands, the 'siderophores', and of their synthetic analoga are therefore of great interest in medicine, where such molecules are not only used to treat primary or secondary hemochromatosis (iron poisoning) but also as antibiotics or drug-delivery agents [11]. The aim is to find or develop chelating systems which can be applied orally and which are therapeutically active in small doses without being degraded in the gastrointestinal tract, in the circulatory system or in the liver.

In contrast to substitutionally labile Fe^{II}, Fe^{III} is insoluble at pH 7 if strong complexing agents are absent (5.1). According to the solubility product of 10^{-38} M^4 for $Fe(OH)_3$ the theoretical concentration of Fe^{3+} would be only about 10^{-17} M; however, the formation of aqua/hydroxo complexes contributes to a significantly higher if still very small solubility. Thus, even those microorganisms that have survived the development of an oxidizing atmosphere feature special low molecular weight compounds to accumulate iron(III) from their environment, if necessary, even from solid phases such as iron oxide containing particles [5]. These special soluble chelating ligands have an extremely high affinity for iron(III) and are referred to as siderophores ('iron carriers'). They provide iron for further transport, storage or incorporation into proteins and act via reduction/protonation processes, e.g. at membrane receptors [3].

Plants also need iron for growth, particularly for chlorophyll biosynthesis. Iron-efficient plants are capable of extracting iron in sufficient amounts even from lime-rich soil with high pH values and thus very low concentrations of soluble iron. Iron-complexing phytosiderophores have been isolated from the roots of such plants. The relatively simple iron transport systems of microorganisms as well as the more complex mechanisms of the iron metabolism in higher organisms will be discussed in the following.

8.2   Siderophores: Iron Uptake by Microorganisms

More than 200 different siderophores are currently known as isolated from bacteria, yeast and fungi. All are naturally occurring chelate ligands with low molecular masses between 500 and 1000 Da and a high affinity for iron(III). Their biosynthesis is regulated by the supply of iron and a DNA-binding regulatory protein, FUR (Fe uptake regulation), which is activated by Fe^{II} and plays an essential role in the feedback mechanism [7]. All siderophores form very stable chelate complexes with approximately octahedrally coordinated high-spin Fe^{III} [12]. A quantitative measure for the 'stability' of the complex formed between siderophore and iron(III) is the complex formation constant, K_f:

$$Fe^{3+} + Sid^n \rightleftharpoons FeSid^{(3-n)} \tag{8.3}$$

$$K_f = \frac{[FeSid^{(3-n)}]}{[Fe^{3+}][Sid^{n-}]}$$

[ ]: molar concentrations
Fe^{3+}: hydrated iron(III)
Sid^{n-}: anionic siderophore ligand   (8.4)

Although generally very large, the constants K_f vary over a wide range—from 10^{23} for the aerobactin complex to about 10^{49} for the Fe^{III}/enterobactin complex (Table 8.1). The constants for the corresponding Fe^{II} complexes are much smaller due to the smaller charge and a larger ionic radius of that ion (see Table 2.7); thus, a release

Table 8.1   Stability constants and redox potentials for iron complexes of naturally occurring siderophores [12, 13]

siderophore ligand	log K_f (Fe^{III} complex)	E_o (mV) (at pH 7)	ligand type
coprogen	30.2	-447	hydroxamate
deferrioxamine B	30.5	-468	hydroxamate
ferrichrome A	32.0	-448	hydroxamate
aerobactin	22.5	-336	hydroxamate, carboxylate
enterobactin	49	-790 (pH 7.4)	catecholate
mugineic acid (phytosiderophore)	18.1	-102	carboxylate, amino-N

mechanism for iron can be effected by reduction, coupled to proton transfer. According to the Nernst equation for the concentration dependence of the redox potential, scheme (8.5) shows the correlation between the stability constant, K_f, the redox potential of the Fe^{II}/Fe^{III} transition and the acid dissociation constant, K_a, of the siderophore ligand. Keeping K_a constant, an increasing dissociative stability of the Fe^{III} complex thus correlates directly with a more negative redox potential (Table 8.1).

$$Fe^{3+}(Sid^{3-}) + 3\ H^+ + e^- \rightleftharpoons Fe^{2+} + H_3Sid$$

$$E = E_0 + 0.059\ V \cdot \log \frac{[Fe^{3+}(Sid^{3-})]\ [H^+]^3}{[Fe^{2+}][H_3Sid]}$$

$$= E_0 + 0.059\ V \cdot \log \frac{[Fe^{3+}]}{[Fe^{2+}]} \cdot K_s \cdot K_f$$

$$\text{with} \qquad 3\ H^+ + Sid^{3-} \overset{K_s}{\rightleftharpoons} H_3Sid$$

$$\text{and} \qquad K_s = \frac{[H_3Sid]}{[H^+]^3\ [Sid^{3-}]} \qquad (8.5)$$

The majority of effective siderophores can be divided into two groups: hydroxamates and catecholates.

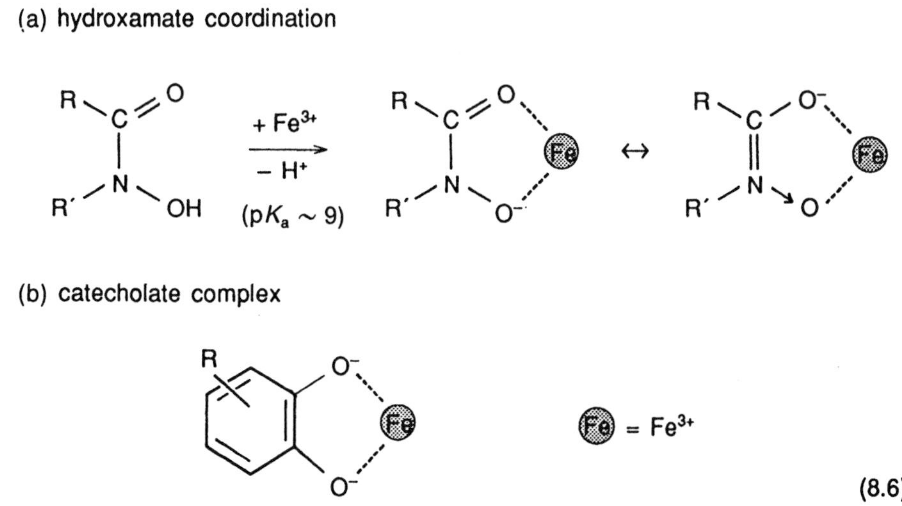

(a) hydroxamate coordination

(b) catecholate complex

$$(8.6)$$

In both cases, the ligands are able to form unstrained and unsaturated five-membered ring chelate systems with negatively charged oxygen atoms as coordination centers. This situation results in a high stability of corresponding complexes with the 'hard', highly charged Lewis acid Fe^{3+}. Ferrichromes (8.7) and ferrioxamines (2.1, 8.8) as well as complexes based on rhodotorulic acid (8.12) or aerobactin (8.13) are typical representatives of hydroxamate siderophore complexes.

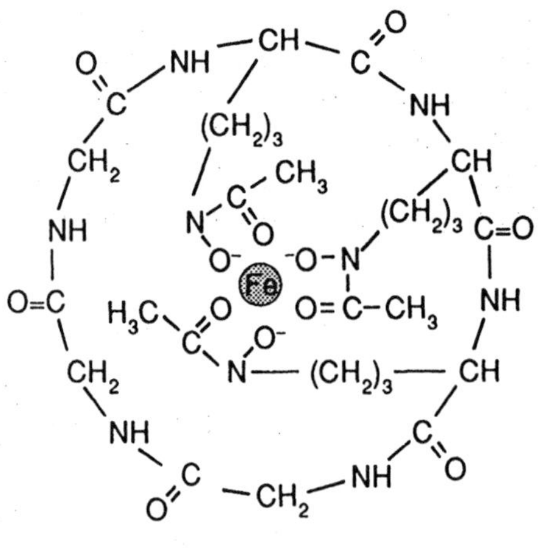

ferrichrome   (8.7)

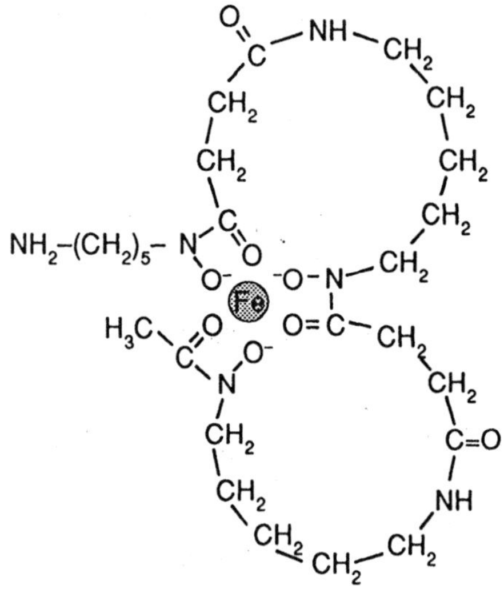

ferrioxamine B

without Fe: deferrioxamine B   (8.8)

Ferrichromes contain cyclic hexapeptides based on glycine and *N*-hydroxy-L-ornithine in which the three hydroxamine groups are derivatized with acetyl functions. They can be synthesized by fungi and used also by bacteria. *In vivo*, ferrichrome adopts a Λ configuration around the iron atom (see below) while the enantiomeric Δ form is much less effective with regard to bacterial iron transport.

A prominent member of the ferrioxamine ligand family is deferrioxamine B (2.1, 8.8). This 'linear' molecule features three nonequivalent hydroxamate groups as part of the chain. It is synthesized *in vivo* by *Streptomyces* species and is used in the prevention of iron poisoning, e.g. after blood transfusions (see Section 8.1).

Optical Isomerism in Octahedral Complexes

Even strictly 'octahedral' complexes with $90°$ bond angles may lose mirror planes and the center of inversion through the ligation of three identical chelate ligands ($O_h \rightarrow C_3$ symmetry). They will then potentially occur in enantiomeric forms, i.e. as mirror images of each other, with the metal functioning as the chirality center. If the octahedron is viewed lying on one of its triangular faces, this isomerism can also be described involving a left- or right-turning helix; the former arrangement is referred to as Λ, the latter as Δ stereoisomer (8.10). Provided that the metal-to-ligand bonds are sufficiently inert (kinetic 'stability'), the two optical isomers can be distinguished by different interactions with enantiomeric reactants such as biological receptors or polarized light ('specific rotation', circular dichroism (CD) in absorption spectroscopy).

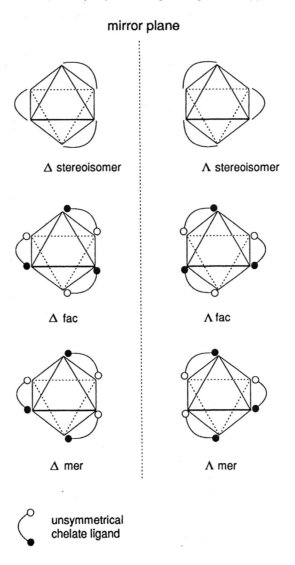

$$(8.9)$$

Another kind of (positional) isomerism, the *fac/mer* alternative, can occur in tris-chelated complexes if the coordinating sites in one chelate ring are not equivalent (○, ●), as is the case with the hydroxamates. In the *fac*-arrangement (derived from facial, often also referred to as 'cis'), equivalent coordinating sites from different chelate rings are situated on one of the triangular faces of the octahedron, whereas in the *mer*-arrangement (derived from meridional, often called 'trans'), such equivalent sites are found lying on a 'meridian' of the octahedron.

The combination of three different, each unsymmetrical chelate arrangements as it occurs in the siderophore complexes, either in a chain (8.8) or in macrocyclic fashion, leads to an increase in the number of possible isomers [14]. The example (8.10) illustrates all possible combinations for Δ-ferrioxamine; for the specific nomenclature the reader is referred to [12]. As a result of spatial restrictions due to chain length or ring conformations only a few isomers are actually realized; ideally, one form is highly preferred [15].

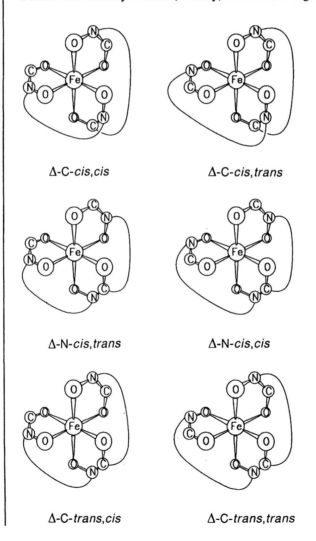

Δ-C-*cis,cis*   Δ-C-*cis,trans*

Δ-N-*cis,trans*   Δ-N-*cis,cis*

Δ-C-*trans,cis*   Δ-C-*trans,trans*

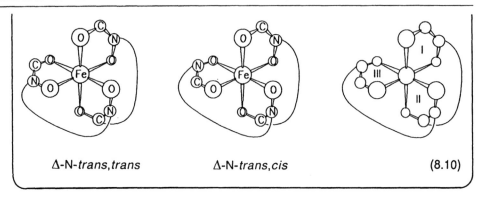

Δ-N-*trans,trans* Δ-N-*trans,cis*   (8.10)

Rhodotorulic acid (8.11) is a dipeptide of *N*-acetyl-*N*-hydroxy-L-ornithine and is synthesized by the yeast organism *Rhodotorula pilimanea*. In contrast to most other siderophores which form 1:1 complexes with iron, rhodotorulic acid features a 3:2 stoichiometry. In addition to iron(III), chromium(III) may also be tightly bound.

rhodotorulic acid   (8.11)

Aerobactin (8.12) is a derivative of citric acid in which the peripheral carboxylic acid moieties have been replaced by hydroxamic acid groups. Some strains of *E. coli* as well as bacteria like *Aerobacter aerogenes* synthesize this chelating agent.

aerobactin   (8.12)

Some siderophores contain several of the coordination features mentioned above. Mycobactin (8.13), for instance, contains two hydroxamic acid groups, one phenolate function and an oxazoline group which donates an (imine) nitrogen atom for coordination to the iron.

Similar to rhodotorulic acid, coprogenes (8.14) as isolated from molds contain one diketopiperazine ring.

While hydroxamate siderophores are mostly found in higher microorganisms such as fungi and yeast, the catecholate siderophores are synthesized mainly by bacteria. Enterobactin (8.15), parabactin and agrobactin (8.16) are the most prominent representatives of siderophores from the catecholate group; in contrast to the hydroxamates they form negatively charged complexes with Fe^{III}. Of all the naturally

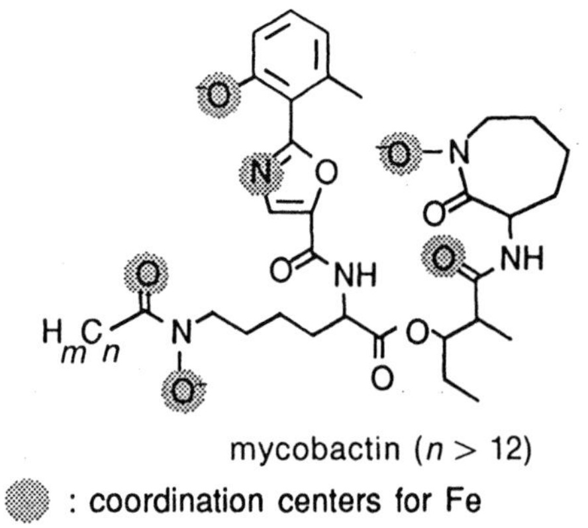

mycobactin ($n > 12$)

⬤ : coordination centers for Fe   (8.13)

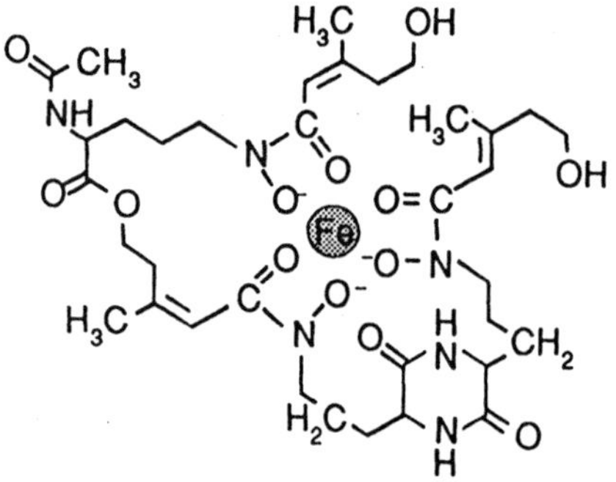

coprogen complex   (8.14)

occurring substances examined, enterobactin (8.15, also called enterochelin) is by far the strongest chelator for iron; it may be isolated, for example, from *Salmonella thyphimurium* and *E. coli* [16]. Related effective siderophores have been isolated recently from a marine bacterium [17], the surface ocean water having a very low concentration of soluble iron (see Figure 2.1). Despite its extremely high affinity for iron ($K_f \approx 10^{49}$) which may even result in the 'extraction' of Fe^{III} from glass,

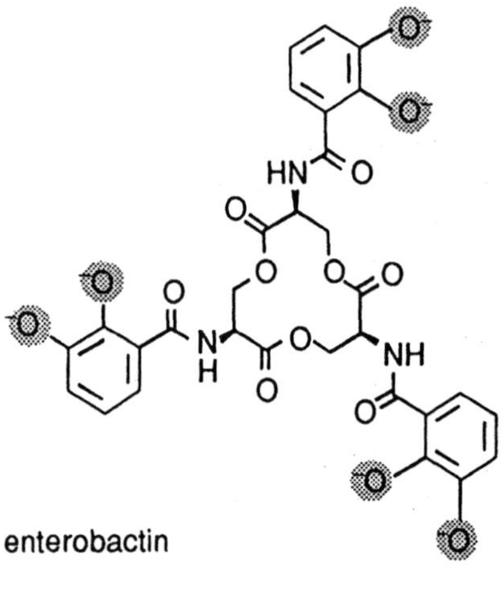

enterobactin   (8.15) enterobactin is not suitable for a therapy of iron poisoning. One reason is that enterobactin contains a hydrolytically labile triester ring and catecholate moieties which are sensitive toward oxidation, being transformed to o-semiquinones and further to o-quinones (see Sections 7.6 and 10.3). In addition, the free ligand is quite insoluble in aqueous solution. Finally, the iron complex of enterobactin supports the growth of higher bacteria and may thus lead to their propagation in the human body (→infection risk). Nevertheless, chemical derivatives of enterobactin are being tested as promising chelating agents for iron therapy [16].

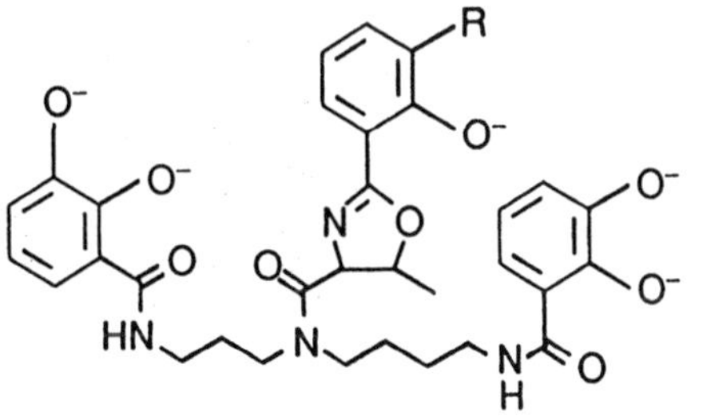

(a) agrobactin (R = OH)
(b) parabactin (R = H)   (8.16)

Synthetic analogues of enterobactin generally contain three catecholate moieties which have been affixed to frameworks with approximately threefold symmetry, such as mesitylene, triamine or cyclododecane derivatives (8.17a–c; compare also Figure 7.5).

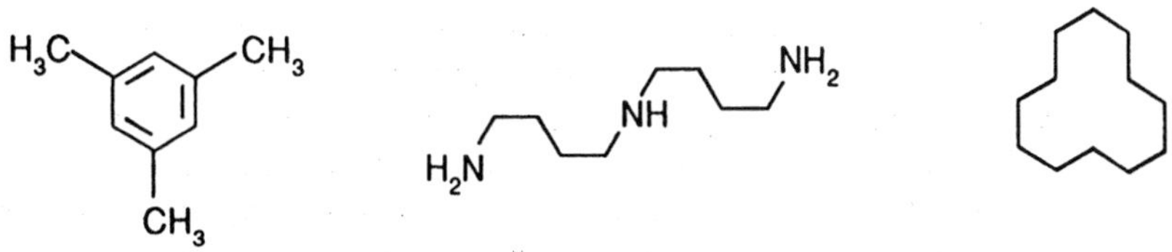

(a) mesitylene   (b) a triamine   (c) cyclododecane

(8.17)

Although a crystal structure analysis of the Fe^{III} complex of enterobactin has not yet been accomplished, absorption (CD) spectroscopy of an isomorphous Cr^{III} complex and the structural analysis of the V^{IV} analogue (Figure 8.1) point towards a Δ configuration. Optical isomers of the naturally occurring Fe^{III} complexes are biologically ineffective. The extraordinarily high stability of the enterobactin complex is caused *inter alia* by multiple intramolecular hydrogen bonds between the amide–NH groups and coordinating catecholate oxygen atoms. Furthermore, these amide functions are probably crucial for the recognition at the membrane receptor and for the proton-induced reductive release of complexed iron [18].

As the coordination chemistry of In^{III} and Ga^{III} is very similar to that of Fe^{III}, there are efforts to make siderophore chelating agents available for radiodiagnostics (Section 18.3). By varying the lipophilicity of the ligands, a relatively organ-specific distribution of the radioisotopes ^{111}In or ^{67}Ga can be achieved (see Table 18.1).

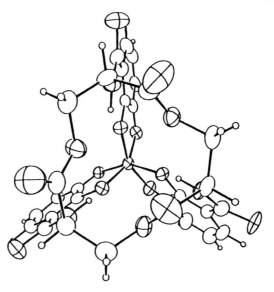

Figure 8.1
Structure of the vanadium(IV) complex of enterobactin (from [13])

8.3   Phytosiderophores: Iron Uptake by Plants

Higher plants require iron for numerous components of the photosynthetic apparatus (Figure 4.9, Table 4.1) and also for the biosynthesis of chlorophyll. Considering the widely varying supply of iron in the soil and the different tolerances towards iron (e.g. rice is very sensitive to iron deficiency), that 'mineral' component has to be extracted from oxidic material via the roots and made available to the plant. Symbiotic microorganisms can often synthesize siderophores which then also mobilize iron for the host plant. The group of actual phytosiderophore ligands includes low molecular weight compounds such as the amino acids mugineic acid and nicotianamine (8.18).

$$
\begin{array}{ccc}
COO^- & COO^- & COOH
\end{array}
$$

X = OH, Y = OH: mugineic acid

X = H, Y = NH$_2$: nicotianamine   (8.18)

These compounds with four chirality centers in *S*-configuration contain a four-membered azetidine ring. Figure 8.2 shows the molecular structure of the Co^{III} complex of mugineic acid; carboxylate as well as amino groups of the hexadentate ligand are involved in the coordination to the metal.

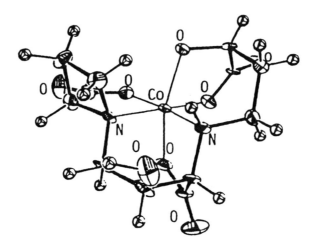

Figure 8.2
Structure of the Co^{III} complex of mugineic acid (according to [19])

8.4 Transport and Storage of Iron

In complex organisms such as human beings, the problem of the selective transport of iron to very different types of cells is of particular importance. To begin with, the uptake of iron from food should be effective; for instance, the potentially chelating reductant ascorbate (vitamin C, see 3.12) enhances a rapid resorption of iron (see 8.5) while the nonreducing but strongly Fe^{III}-binding phosphates may counteract such a resorption from food. In higher animals, the scavenging and transport of iron is not effected by low molecular weight siderophores but rather by fairly large nonheme iron proteins, the transferrins [4,7,20–22]. When iron is set free from its transport system and released, e.g. within the cell, it has to be either directly used or stored because of the potential hazardous character of free iron according to equations (8.1) and (8.2). Highly specialized nonheme iron proteins, in particular ferritin and the insoluble hemosiderin, serve to store that unused iron [7,23–26]. Storage as well as transport systems have to function rapidly and be completely reversible under physiological conditions in order to preclude local excess or deficiency symptoms. Such transport and storage systems have not only been found in animals; the iron-storage protein phytoferritin was detected in plants while certain bacteria were shown to have heme-containing bacterioferritins.

Figure 8.3 shows a simplified overview specifying the roles of transferrin and ferritin in the mammalian (human) metabolism. Iron(III) is liberated from the transferrin transport system in the blood-forming cells of bone marrow and is then taken up by ferritin, presumably after intermediary reduction to more labile Fe^{II}. The uptake by the ferritin storage system eventually involves reoxidation to Fe^{III}. While iron incorporation into the porphyrin macrocycle during heme biosynthesis occurs via a ferrochelatase enzyme (see Section 17.2) the hemoglobin-rich erythrocytes have only a finite life expectancy and are degraded, e.g. in the spleen, where the iron released

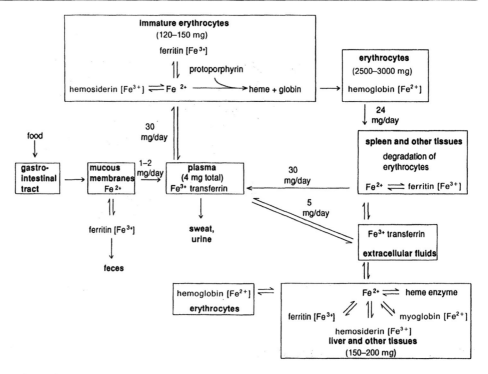

Figure 8.3
Simplified flow chart for the iron metabolism in the human body (according to [6])

in this process is again stored as part of ferritin complexes. The transport form, transferrin, also delivers iron to liver or muscle cells where it can be utilized in the biosynthesis of enzymes or myoglobin. If required, Fe^{II} can be resorbed via the mucous membranes of the intestines where the iron saturation of ferritin inside the membrane regulates the uptake.

8.4.1 Transferrin

The iron transport protein ovotransferrin was first isolated from egg white in 1900 under the name of conalbumin; in 1946, the serum transferrin was obtained from human blood. All transferrins, including the lactoferrin from exocrine secretions (milk, mucosal tissue), show antibacterial properties which are strongly reduced in the presence of excess iron. It has long been known that one molecule of transferrin can tightly bind two iron(III) ions together with stoichiometric amounts of carbonate [20–22].

Quantitatively, the main function of the transferrins in the human body is to transport iron from places of resorption, storage or degradation of erythrocytes to the blood-forming cells in the bone marrow. The major part of the iron is then utilized by the precursor cells of new erythrocytes to form hemoglobin (compare Table 5.1); during pregnancy, considerable amounts of iron are being delivered to the placenta and from there on to the fetus.

A second, more indirect but nonetheless very essential, function of transferrin and related proteins is the protection against infectious diseases. The protein binds iron with such high affinity ($\log K_f \approx 20$) that it is no longer available to parasitic microorganisms, their development thus being inhibited by this nonspecific immune system. The antibacterial and antiinflammatory properties of lactoferrin (from milk and other secretions) or ovotransferrin thus serve to protect very sensitive bodily fluids, mucous membranes and the developing embryo. The particularly strong Fe^{III} binding to lactoferrin presumably plays an important role in the development of cells that are directly involved in immune reactions of the body; therefore, efforts are being made to obtain human lactoferrin from accordingly modified 'transgenic' organisms.

The transferrins (ovotransferrin, serum transferrin and lactoferrin) are glycoproteins with molecular masses of about 80 kDa. The single polypeptide chain is folded to form two lobes, each of which contains one binding site for iron (see Figure 8.4).

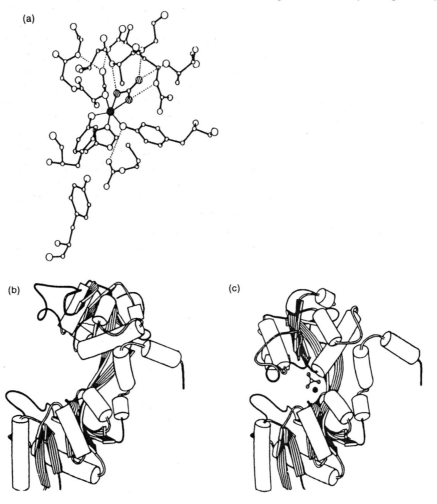

(a)

(b)   (c)

Figure 8.4
(a) Coordination environment of Fe^{3+} (●) and CO_3^{2-} (oxygen atoms hatched; ···· hydrogen bonds) and reversible conformational change (b ↔ c) upon Fe^{3+}/CO_3^{2-} binding by human lactoferrin (N-terminal region) (according to [21])

The amino acid sequences for the C- and N-termini of human serum transferrin show 42% homology. The size of the carbohydrate portion of transferrins varies among different species and may even vary for different regions of the same organism. The polysaccharide chains do not contribute to the iron transport; their presumed function is to engage in specific interactions with receptors.

All transferrins can bind two high-spin Fe^{3+} ions per molecule but also other metal ions such as Cr^{3+}, Al^{3+} (see Section 17.6), Cu^{2+}, Mn^{2+}, Co^{3+}, Co^{2+}, Cd^{2+}, Zn^{2+}, VO^{2+}, Sc^{2+}, Ga^{3+}, Ni^{2+} or trivalent lanthanoid ions. The transport of Al^{3+} by transferrins into the central nervous system (contrary to Fe^{3+}, Al^{3+} cannot be remobilized by reduction) comprises an important aspect of aluminum toxicity [27] (see Section 17.6). While the iron-free apoenzyme is colorless, a red-brown color indicates the coordination of Fe^{3+}. The binding of the metal ion to either one of the two nearly but not completely equivalent coordination sites in the C- and N-terminal region involves a simultaneous 'synergistic' binding of one carbonate ion and the release of three protons (charge effect). The protons can either be produced through hydrolysis of bound aqua ligands or can originate from the protein. The two coordination sites for iron are not completely equivalent, the C-terminus showing a slightly higher affinity for Fe^{3+} ions which results in complex formation at lower pH values. For iron-saturated lactoferrin, the following ligands were found in the coordination sphere of each of the two approximately octahedrally configured metal centers (Figure 8.4): two tyrosinate residues which are responsible for the intense color of the Fe^{III} form (ligand-to-metal charge transfer), one η^1-aspartate, one histidine ligand and η^2-coordinated (chelating) carbonate which links the metal to the protein via hydrogen bonds. The reversible transformation of the 'open' conformation of the bilobal protein to a closed, compact structure after binding of Fe^{3+} and the 'synergistic anion' CO_3^{2-} (Figure 8.4) is an essential feature of its efficient transport function ('iron shuttle').

In vitro studies have shown that Fe^{2+} as well as Fe^{3+} may be taken up by transferrins. However, Fe^{2+} is only weakly bound and, therefore, has to be

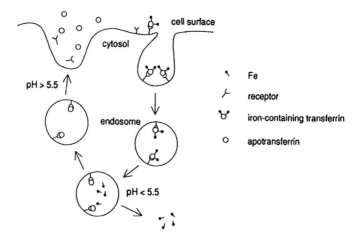

Figure 8.5
Transferrin recognition and demetallation by example of a human tumor cell (according to [28])

oxidized within the protein. Fe^{3+} forms very stable complexes with effective stability constants K_f in the range of 10^{20} M^{-1} (compare Table 8.1). It can be shown from *in vitro* experiments that the stability of the complexes decreases strongly with decreasing pH values (compare 8.5). At pH = 4.5 the stability constant is already lower than that of the corresponding citrate complex and it thus becomes possible to release iron from transferrin by addition of chelating citrate, $^-OOC-CH_2-C(OH)(COO^-)-CH_2-COO^-$. It is yet unknown in which form the iron is supplied to the transferrin *in vivo*, but there are some hypotheses (Figure 8.5) on how the transferrin-bound iron is being made available to the cell. Three phases can be distinguished in such a process. After the binding of the iron-containing transferrin complex to a specific receptor at the outside of the cell membrane one may assume endocytosis, the iron then being released into the cytoplasma from transferrin in a chelate-assisted step involving reduction and, possibly, protonation. Finally, the free apotransferrin is transferred back to the cell surface and into the plasma. Such a cycle can explain how the small total amount of transferrin (see Table 8.1) is able to transport an average of 40 mg of iron per day while the individual capacity of human transferrin is only about 7 mg iron.

8.4.2 Ferritin

Iron which has been released into the cell as described in Figure 8.5 must either be used immediately for biosynthesis or stored in a safe form. Particularly after the advent of a biogenically O_2-enriched atmosphere there had to be a system which served (i) as a storage for iron, i.e. an element that was increasingly less bioavailable because of its precipitation as Fe^{III} hydroxide/oxide, and (ii) as a safekeeping site, inhibiting the uncontrolled radical-producing reactions of reduced free iron with O_2 or its metabolic products (8.1, 8.2). Two very large proteins, ferritin and hemosiderin, serve as such iron stores, the soluble, much better defined ferritin having been thoroughly studied [23–25]. As early as 1935 Laufberger isolated and crystallized ferritin and suggested a role as iron-store because of its high iron content of up to 20% Fe by weight. Ferritin is found in animals as well as in plants. Approximately 13% of the iron in the human body is present as ferritin (Table 5.1); it is mainly found in liver, spleen and bone marrow. The very large molecule consists of an inorganic core of variable size and composition which is surrounded and held together by a protein shell.

Apoferritin, the iron-free protein component of ferritin, has an average molecular mass of 440 kDa. It can be obtained from iron-containing ferritin by reduction with sodium dithionite or ascorbate in the presence of suitable chelating agents. The water-soluble protein consists of 24 equivalent subunits which arrange to form a hollow sphere with an outer diameter of about 12 nm (Figure 8.6). The empty inside has a diameter of approximately 7.5 nm; in holoferritin, this space is filled with inorganic material [29].

The crystal structure of apoferritin was determined with a resolution of 0.28 nm (Figure 8.6). According to this structure analysis, each subunit consists of four long α helices, forming a bundle, and a fifth short α helix arranged perpendicularly. Two of the long helices are connected by a loop and two of these loops from neighboring subunits arrange to form a β sheet. A total of 24 subunits forms the apoferritin molecule with its high symmetry (rhombic dodecahedron, cubic space group F432,

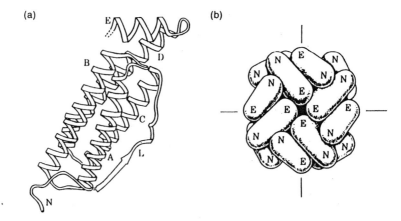

Figure 8.6
(a) Ribbon diagram of the α-helical framework of an apoferritin subunit. N: N-terminus; A, B, C, D: long α helices; E: short α helix; L: loop. (b) Schematic representation of the subunit arrangement in the apoferritin molecule (hollow sphere); view along the fourfold axis. N and E represent the relative position of the N-terminus and the short α helix, respectively (from [23])

$a \approx 18.5$ nm; Figure 8.6). The particular spatial arrangement allows for the formation of channels along the threefold and fourfold symmetry axes. These channels play an important role in the deposition and release of iron. The six channels with fourfold symmetry are hydrophobic because of twelve lining leucine residues, whereas the eight channels with threefold symmetry are rather hydrophilic due to the presence of aspartate and glutamate and, to a lesser extent, histidine and tyrosine residues. During crystal structure analysis these channels were filled with Cd^{2+} ions.

The data obtained seem to indicate that the ferritins of mammals are largely isomorphous. However, ferritins isolated from bacteria like *E. coli* differ a great deal, even in their amino acid sequences; bacterioferritins may contain additional heme groups with axial bis(methionine) coordination (Section 6.1).

In the center of the hollow sphere formed by mammalian apoferritin there is space for the inorganic core which may contain a maximum of 4500 iron centers in oxidic form, a typical filling involving approximately 1200 iron centers. The inorganic nucleus of ferritin represents such a large space of high electron density that it is clearly visible in the electron microscope without further modification. From a stoichiometrical aspect, the iron(III) is present as $Fe_9O_9(OH)_8(H_2PO_4)$; however, the variable amount of phosphate does not determine the actual structure. EXAFS data have suggested that every iron center is surrounded by 6.4 ± 0.6 oxygen atoms at a distance of 195 ± 2 pm and by 7 ± 1 iron atoms at a distance of 329 ± 5 pm. The structure is very similar to that of the metastable mineral ferrihydrite, $5 \ Fe_2O_3 \cdot 9 \ H_2O = Fe^{III}_{10}O_6(OH)_{18}$ [26]. This mineral has a layered structure with a hexagonally close-packed lattice of oxygen centers (O^{2-}, OH^-) and iron(III) in half of the tetrahedral and octahedral interstices [30] (Figure 8.7). There are some well-characterized synthetic $Fe^{III}/OH^-/O^{2-}$-containing oligomers which can serve as models for the iron oxide core of ferritin [31]. The phosphate anions in the ferritin core seem to be bound at the edges of the layers, the approximate ratio being nine Fe^{III} centers per phosphate group. Most of the phosphate is not essential for the formation

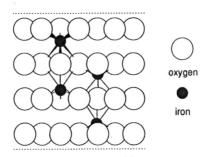

oxygen

iron

Figure 8.7
Representative section of the ferrihydrite structure

of ferritin, as is apparent from its variable amount. However, a general importance of phosphates lies in the linkage of organic polymers such as proteins with inorganic particles (compare Figures 15.3 and 18.2). Formation of such a linkage can also be assumed, for example, between the inorganic layers and the protein hull of ferritin.

The magnetic behavior of the ferritin core has been studied in great detail. The results are best explained locally by assuming antiferromagnetic coupling between oxide-bridged iron centers (compare with ribonucleotide reductase, Section 7.6.1). However, the effective magnetic moment per metal center, μ_{eff}, was determined with 3.85 Bohr magnetons and thus does not fit the ground state of free Fe^{3+} ions $(S = 5/2)$ but rather a state with a spin quantum number of $S = 3/2$; a superexchange process between the iron atoms has been invoked as explanation. The Mössbauer spectrum shows another characteristic behavior of ferritin iron [32]: while the expected six-line spectrum for the well-ordered magnetic state was obtained at low temperatures, these lines coalesce at high temperatures to form a doublet (arising from quadrupolar splitting). This effect is caused by ferromagnetic spin–spin coupling within small 'nanoparticles' $(d < 20\,\text{nm})$ and their temperature-dependent orientation with respect to each other ('superparamagnetism', [33]) — a phenomenon that has parallels in various extremely finely ground metal oxides. It is thus only consequential that nanoparticles with special material properties are now being generated synthetically inside the ferritin shell [33,34].

There are two basically different possibilities for the formation of ferritin from apoferritin: if iron(III) were present as polymeric oxide/hydroxide aggregate, the apoferritin shell, in equilibrium with its subunits, could assemble the ferritin framework around the already existing iron core (template effect); alternatively, the formation of ferritin can also be envisaged (8.19) as a redox process where Fe^{II} is oxidized to Fe^{III} in the presence of apoferritin and an electron acceptor (ultimately O_2) *during* deposition into the apoferritin. Since the biosynthesis of apoferritin precedes that of ferritin and a dissociation into subunits requires fairly unphysiological conditions, the second alternative (8.19) is now well accepted.

$$H^+ + Fe^{2+} + O_2 + \text{apoferritin} \rightarrow \text{ferritin} \; (\equiv \text{apoferritin} \bullet {}'Fe^{III}O(OH)')$$

(8.19)

It is assumed that Fe^{2+} ions (Fe^{3+} is inactive in this respect) can enter the apoferritin through the hydrophilic channels with a rate of about 3000 Fe/s and that they are then catalytically oxidized at active 'ferroxidase' centers. Carboxylate residues such as glutamate or aspartate are abundant at the inner surface of the apoferritin; their complete derivatization in the form of esters blocks the iron loading. Carboxylate-bound and already oxidized iron centers can then serve as nuclei for the deposition and oxidation of further Fe^{II} and thus for the growth of the iron core; mixed-valent Fe^{II}/Fe^{III} centers have been postulated as preferred nucleation sites for the ferritin core. Eventually, O_2 is the oxidizing agent for this process because, after all, it was its bioproduction which necessitated such a soluble, vesicular iron store and protection system. However, not all 100% of iron have to be stored as Fe^{III} in ferritin, small amounts of Fe^{II} may be important for quick mobilization of the mineral. In this case, the proton flux inevitable for the oxidation reaction (8.19) would also be smaller.

According to the typical condensation polymerization pattern of aqua and hydroxo complexes [31,35,36] (8.20), the iron oxide/hydroxide forms more or less well-ordered 'quasi-crystalline' clusters which grow until the inside of the apoferritin is filled.

schematic process of iron hydrolysis
(polycondensation; according to [35]):

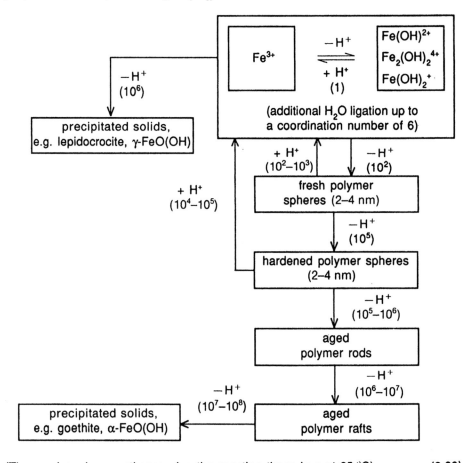

(The numbers in parentheses give the reaction times in s at 25 °C)   (8.20)

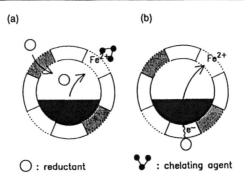

Figure 8.8
Schematic representation of the different possibilities for reductive mobilization of iron from ferritin (see text for details) (according to [37])

The form in which iron(II) is supplied to the apoferritin as well as the receptor system for the reductively released Fe^{2+} ions are still unknown. Furthermore, it is not yet clear what species serve as immediate electron acceptors in the ferritin reductase and whether other electron-transfer reactants such as quinones do participate.

A probable alternative of physiological iron mobilization from ferritin involves electron transfer through the protein shell. It was demonstrated by *in vitro* experiments that ferritin-bound iron could be removed from the core as Fe^{II} using nonphysiological reductants like sodium dithionite. Furthermore, it could be shown that a dynamic equilibrium exists between iron(III) ions inside and outside ferritin.

The different possibilities of iron mobilization are summarized in Figure 8.8. Representation (a) shows the hydrophilic channels with threefold symmetry for the transport of iron(II) *out* of the ferritin and the hydrophobic channels with fourfold symmetry (shaded) for the *import* of small reductants. Iron(II) chelators may shift the equilibrium and thus simplify the release of iron. The second possibility (b) is an electron transfer through the protein shell upon which iron(III) is reduced and transported out through the hydrophilic channels as Fe^{II}.

8.4.3 Hemosiderin

In addition to ferritin, hemosiderin is another storage form for iron in organisms, in particular during iron overload [7]. Hemosiderin was first isolated from horse spleen in 1929. Although the structures of the iron cores of ferritin and hemosiderin are similar, the protein component of hemosiderin is largely unknown. With approximately 35% Fe, the iron/protein ratio is even higher in the extremely large hemosiderin (typically 4 MDa) than in ferritin and it is assumed that this insoluble species is formed via lysosomal decomposition of ferritin. According to this hypothesis, proteases in the lysosome contribute to the degradation of the protein shell of ferritin. The released iron core dissociates and reforms as the amorphous hemosiderin.

References

1. R. J. P. Williams, Inorganic elements in biological space and time, *Pure Appl. Chem.*, **55**, 1089 (1983).
2. R. J. P. Williams, Bio-inorganic chemistry: its conceptual evolution, *Coord. Chem. Rev.*, **100**, 573 (1990).
3. G. Winkelmann, D. van der Helm and J. B. Neilands (eds.), *Iron Transport in Microbes, Plants and Animals*, VCH, Weinheim, 1987.
4. T. M. Loehr (ed.), *Iron Carriers and Iron Proteins*, VCH, Weinheim, 1989.
5. W. Schneider, Iron hydrolysis and the biochemistry of iron—the interplay of hydroxide and biogenic ligands, *Chimia*, **42**, 9 (1988).
6. R. R. Crichton, *Inorganic Biochemistry of Iron Metabolism*, Ellis Horwood, New York, 1991.
7. R. R. Crichton and R. J. Ward, Iron metabolism—new perspectives in view, *Biochemistry*, **31**, 11255 (1992).
8. B. Halliwell and J. M. C. Gutteridge, Iron and free radical reactions: two aspects of antioxidant protection, *Trends Biochem. Sci.*, **11**, 372 (1986).
9. E. D. Letendre, The importance of iron in the pathogenesis of infection and neoplasia, *Trends Biochem. Sci.*, **10**, 166 (1985).
10. N. S. Scrimshaw, Iron deficiency, *Sci. Am.*, **265**(4), 24 (1991).
11. M. J. Miller and F. Malouin, Microbial iron chelators as drug delivery agents: the rational design and synthesis of siderophore-drug conjugates, *Acc. Chem. Res.*, **26**, 241 (1993).
12. K. N. Raymond, G. Müller and B. F. Matzanke, Complexation of iron by siderophores. A review of their solution and structural chemistry and biological functions, *Top. Curr. Chem.*, **123**, 49 (1984).
13. T. B. Karpishin, T. M. Dewey and K. N. Raymond, The vanadium(IV) enterobactin complex: structural, spectroscopic, and electrochemical characterization, *J. Am. Chem. Soc.*, **115**, 1842 (1993).
14. H. Bickel, G. E. Hall, W. Keller-Schierlein, V. Prelog, E. Vischer and A. Wettstein, Über die Konstitution von Ferrioxamin B, *Helv. Chim. Acta*, **43**, 2129 (1960).
15. F. Vögtle, *Supramolecular Chemistry*, Wiley, New York, 1993.
16. K. N. Raymond, M. E. Cass and S. L. Evans, Metal sequestering agents in bioinorganic chemistry: enterobactin mediated iron transport in *E. coli* and biomimetic applications, *Pure Appl. Chem.*, **59**, 771 (1987).
17. R. T. Reid, D. H. Live, D. J. Faulkner and A. Butler, A siderophore from a marine bacterium with an exceptional ferric ion affinity constant, *Nature (London)*, **366**, 455 (1993).
18. C.-W. Lee, D. J. Ecker and K. N. Raymond, The pH-dependent reduction of ferric enterobactin probed by electrochemical methods and its implication for microbial iron transport, *J. Am. Chem. Soc.*, **107**, 6920 (1985).
19. Y. Mino, T. Ishida, N. Ota, M. Inoue, K. Nomoto, T. Takemoto, H. Tanaka and Y. Sugiura, Mugineic acid–iron(III) complex and its structurally analogous cobalt(III) complex: characterization and implication for absorption and transport of iron in gramineous plants, *J. Am. Chem. Soc.*, **105**, 4671 (1983).
20. B. F. Anderson, H. M. Baker, G. E. Norris, S. V. Rumball and E. N. Baker, Apolactoferrin structure demonstrates ligand-induced conformational change in transferrins, *Nature (London)*, **344**, 784 (1990).
21. B. F. Anderson, H. M. Baker, G. E. Norris, D. W. Rice and E. N. Baker, Structure of human lactoferrin: crystallographic structure analysis and refinement at 2.8 Å resolution, *J. Mol. Biol.*, **209**, 711 (1989).
22. E. N. Baker and P. F. Lindley, New perspectives on the structure and function of transferrins, *J. Inorg. Biochem.*, **47**, 147 (1992).
23. G. C. Ford, P. M. Harrison, D. W. Rice, J. M. A. Smith, A. Treffry, J. L. White and J. Yariv, Ferritin: design and formation of an iron-storage molecule, *Phil. Trans. Roy. Soc. London*, **B304**, 551 (1984).
24. E. C. Theil, Ferritin: structure, gene regulation, and cellular function in animals, plants, and microorganisms, *Annu. Rev. Biochem.*, **56**, 289 (1987).

25. E. C. Theil, Ferritin: a general view of the protein, the iron–protein interface, and the iron core, *Metal Clusters in Proteins* (ed. L. Que, Jr), ACS Symposium Series 372, 1988, p. 259.
26. T. G. St. Pierre, J. Webb and S. Mann, Ferritin and hemosiderin: structural and magnetic studies of the iron core, in *Biomineralization*, (eds. S. Mann, J. Webb and R. J. P. Williams), VCH, Weinheim, 1989.
27. S. J. A. Fatemi, F. H. A. Kadir, D. J. Williamson and G. R. Moore, The uptake, storage, and mobilization of iron and aluminum in biology, *Adv. Inorg. Chem.*, **36**, 67 (1991).
28. N. D. Chasteen, C. P. Thompson and D. M. Martin, The release of iron from transferrin. An overview, in *Frontiers in Bioinorganic Chemistry* (ed. A. V. Xavier), VCH Verlagsgesellschaft mbH, Weinheim, 1986, p. 278.
29. J. M. A. Smith, R. F. D. Stansfield, G. C. Ford, J. L. White and P. M. Harrison, A molecular model for the quarternary structure of ferritin, *J. Chem. Educ.*, **65**, 1083 (1988).
30. R. A. Eggleton and R. W. Fitzpatrick, New data and a revised structural model for ferrihydrite, *Clays Clay Miner.*, **36**, 111 (1988).
31. K. S. Hagen, Model compounds for iron–oxygen aggregation and biomineralization, *Angew. Chem. Int. Ed. Engl.*, **31**, 1010 (1992).
32. T. G. St. Pierre, K.-S. Kim, J. Webb, S. Mann and D. P. E. Dickson, Biomineralization of iron: Mössbauer spectroscopy and electron microscopy of ferritin cores from the chiton *Acanthopleura hirtosa* and the limpet *Patella laticostata*, *Inorg. Chem.*, **29**, 1870 (1990).
33. R. Dagani, Nanostructured materials promise the advance range of technologies, *Chem. Eng. News*, 23, 18 November 1992.
34. F. Meldrum, B. R. Heywood and S. Mann, Magnetoferritin: *in vitro* synthesis of a novel magnetic protein, *Science*, **257**, 522 (1992).
35. C. M. Flynn, Hydrolysis of inorganic iron (III) salts, *Chem. Rev.*, **84**, 31 (1984).
36. M. Henry, J. P. Jolivet and J. Livage, Aqueous chemistry of metal cations: hydrolysis, condensation and complexation, *Struct. Bonding (Berlin)*, **77**, 153 (1992).
37. P. M. Harrison, G. C. Ford, D. W. Rice, J. M. A. Smith, A. Treffry and J. L. White, The three-dimensional structure of apoferritin: a framework controlling ferritin's iron storage and release, in *Frontiers in Bioinorganic Chemistry* (ed. A. V. Xavier), VCH Verlagsgesellschaft mbH, Weinheim, 1986, p. 268.

9 Nickel-Containing Enzymes: The Remarkable Career of a Long Overlooked Biometal

9.1 Overview

For a long time, nickel has been the only element of the 'late' 3d transition metals for which a biological role could not be definitely established. The reasons for this 'oversight' were manifold: nickel ions do not exhibit a very characteristic light absorption in the presence of physiologically relevant ligands, Mössbauer effects are not generally accessible for nickel isotopes, and even paramagnetic Ni^I (d^9) or Ni^{III} (d^7) cannot always be unambiguously detected by EPR spectroscopy due to a lack of metal isotope hyperfine coupling (the natural abundance of ^{61}Ni with $I = 3/2$ is only 1.25%). In addition, it has now been shown that Ni is often only one of several components of complex enzymes which may otherwise contain several coenzymes as well as additional inorganic material. For instance, nickel centers remained undetected for a long time due to their frequent association with Fe/S clusters. However, applying more sensitive detection methods in atomic absorption or emission spectroscopy (AAS or AES), magnetic measuring (SQUID susceptometry), or EPR spectroscopy using ^{61}Ni-enriched material, some nickel-containing enzymes of plants and microorganisms have now been established and partly characterized (compare Figure 1.1). Nickel is not particularly rare in the lithosphere or in sea-water where it is soluble as Ni^{2+}, in view of its very low potential requirement as ultratrace element [1] no real deficiency symptoms have been reported for human beings. (Ni^{2+}-specific antibodies are responsible for the not uncommon 'nickel allergy' [2]). Nevertheless, even the nickel content of stainless steel has been mobilized by certain microorganisms (Figure 1.1) to satisfy their need for corresponding essential enzymes. Incidentally, one of the many hypotheses for the extinction of dinosaurs and many other creatures at the end of the cretaceous involves a global nickel poisoning caused by meteorite material [3].

Only in the late 1960s was nickel beginning to be considered as a necessary component for the growth of some anaerobic bacteria; in 1975 the metal was reported to occur in plant urease. According to current knowledge [4–6] nickel is present in four essentially different kinds of enzymes, with rather different oxidation states and coordination environments. Whereas the urease enzymes of bacteria and plants contain five- or six-coordinate nickel(II) bound to N,O-ligands, both the hydrogenases of many bacteria, e.g. 'Knallgas' bacteria or sulfate-reducing strains, and the CO dehydrogenase (= bacterial acetyl–coenzyme A synthase) of anaerobic bacteria contain nickel which is mainly coordinated by sulfur ligands. The methyl–coenzyme M reductase of methanogenic bacteria features a nickel tetrapyrrole complex as the prosthetic group, the coenzyme F430 (compare 2.5). Another tetrapyrrole complex with an as yet unspecified function, the nickel-containing tunichlorin, was isolated from a tunicate species [7]. With respect to biosynthetic utilization, nickel as well as the chemically related cobalt are largely confined to evolutionary 'old' organisms.

9.2 Urease

Urease enzymes which may be isolated, for example, from bacteria or plant products such as jack beans (*Canavalia ensiformis*) have an interesting history [8]. Contrary to the opinion of Richard Willstätter, it was the first enzyme to be prepared in pure crystalline form (J. Sumner, 1926). However, only about fifty years later was the nickel content of this enzyme determined [9].

The 'classical' urease (urea amidohydrolase) catalyzes the degradation of urea to carbon dioxide and ammonia (9.1). Urea is a very stable molecule which normally hydrolyzes very slowly to give isocyanic acid and ammonia (9.2), the half-life value of the uncatalyzed reaction being 3.6 years at 38 °C.

$$H_2N-CO-NH_2 \; + \; H_2O \; \xrightarrow{\text{urease}} \; [H_2N-COO^- + NH_4^+] \; \longrightarrow \; 2\,NH_3 \; + \; CO_2 \quad (9.1)$$

$$H_2N-CO-NH_2 \; + \; H_2O \; \longrightarrow \; NH_3 + H_2O + H-N{=}C{=}O \quad (9.2)$$

The catalytic activity of the enzyme increases the rate of complete hydrolysis by a factor of about 10^{14}. Such an astonishing acceleration can only be explained by a change in the reaction mechanism (see Figure 2.8). While the uncatalyzed reaction involves a direct elimination of ammonia, the enzyme presumably catalyzes a hydrolysis reaction with carbamate, H_2N-COO^-, as the first intermediate. A metal-to-substrate binding as depicted in (9.3) would facilitate that latter reaction mechanism. Additional support for substrate binding to the metal comes from the fact that phosphate derivatives binding strongly to nickel inhibit the activity of the enzyme.

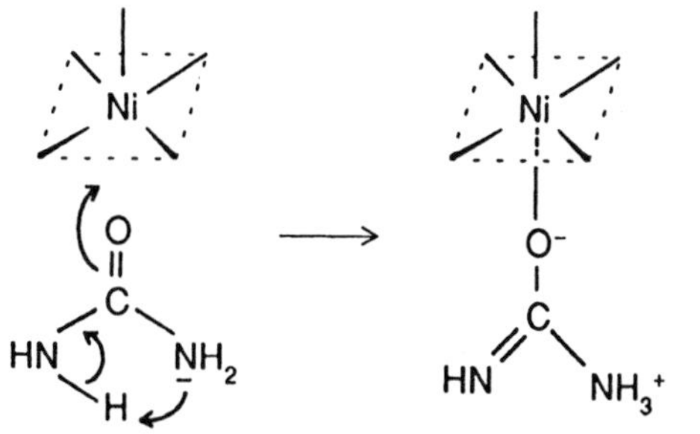

$$(9.3)$$

While amino acid sequences are known for the various ureases there is yet little structural information available regarding the nickel centers in the active site. The holoenzyme of jack bean urease consists of six equivalent subunits; each subunit (91 kDa) contains two close but apparently different nickel ions [10]. EXAFS measurements have shown only nitrogen and oxygen ligands in the first coordination spheres and have suggested coordination numbers of five and/or six, in agreement with magnetic and absorption spectroscopic studies. Measurements of the magnetic susceptibility point to an equilibrium between high-spin ($S = 1$) and low-spin ($S = 0$) forms [10], which would be typical for five-coordinate or distorted six-coordinate Ni^{II} due to small and variable energy differences between d_{z^2} and $d_{x^2-y^2}$ orbitals.

The following reaction mechanism (9.4) [11] was proposed in agreement with model studies and the above-mentioned experimental data. It involves an electrophilic attack of one of the nickel centers on the carbonyl oxygen atom and a nucleophilic attack of a nickel hydroxo species on the carbonyl carbon center (push–pull mechanism; see also Section 12.2).

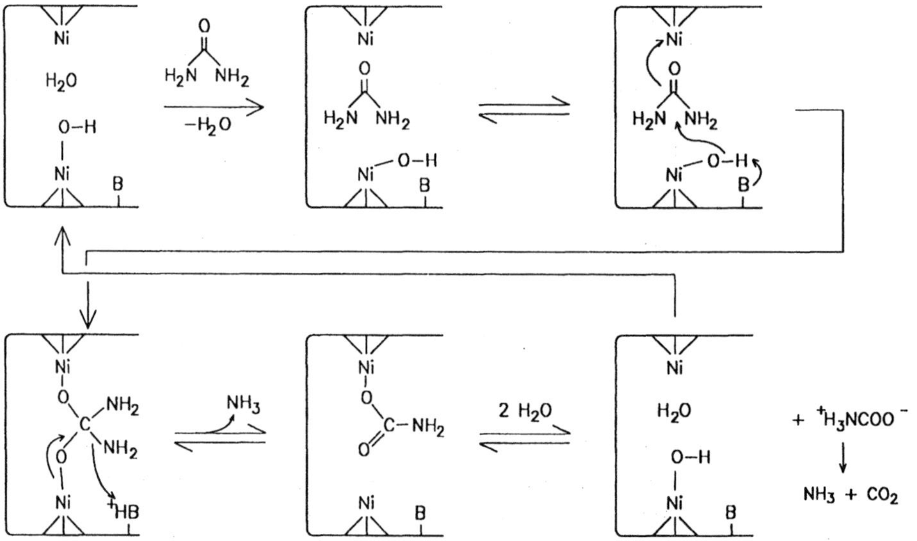

B: base   (9.4)

Similar mechanistical hypotheses exist for the function of zinc-containing hydrolytic enzymes (Section 12.2) where protons often function as the electrophilic species. Hydrolytic zinc enzymes thus require only one metal center to provide the hydroxide nucleophile in an H_2O activation step. It is not entirely obvious why hydrolytic nickel centers are present only in ureases; e.g. it is possible to substitute nickel for zinc in well-characterized zinc-containing enzymes such as carboxypeptidase A (CPA) and carboanhydrase (CA; see Chapter 12). Perhaps the less stringent stereochemical requirements regarding selectivity and the larger number of potentially coordinating heteroatoms in urea favor nickel(II) centers over typically lower-coordinate zinc(II). In agreement with this notion, a urea-related substrate, the guanidinium group of arginine, requires dinuclear Mn^{II} centers for enzymatic hydrolysis [12].

9.3 Hydrogenases

Hydrogenases ('H_2ases') are enzymes that catalyze the reversible two-electron oxidation (9.5) of molecular hydrogen ('dihydrogen').

$$\text{H}_2 \; \xrightleftharpoons{\text{hydrogenase}} \; 2 \, \text{H}^+(\text{aq}) \; + \; 2 \, e^- \qquad\qquad (9.5)$$

This reaction plays a major role, for example, in the course of dinitrogen 'fixation' (Section 11.2), in microbial phosphorylation or in the fermentation of biological substances to methane. Anaerobic and also some aerobic microorganisms contain hydrogenase enzymes; H_2 can either serve as an energy source instead of NADH or it can occur as the product of reductive processes [13]. Most but not all [14] hydrogenases contain iron–sulfur clusters (see Section 7.4); the 'Fe-only' enzymes contain conventional [4Fe–4S] and special catalytic 'H'-clusters (6Fe [15]) and do not require other metals or coenzymes. Depending on the presence of other prosthetic groups, e.g. flavins or unusual 'inorganic' elements, the hydrogenases are further divided into Ni/Fe hydrogenases, the most common form, and Ni/Fe/Se hydrogenases. The Ni/Fe hydrogenases contain a nickel center in addition to separate iron–sulfur clusters; the Ni/Fe/Se enzymes feature nickel and selenium (as Ni-coordinating selenocysteinate) in an equimolar ratio as well as Fe/S clusters. Nitrogenases also show a distinct hydrogenase activity, particularly in the absence of the primary substrate N_2 (see Section 11.2). Catalysis of reaction (9.5) is necessary because of the two-electron transfer involved; simple one-electron reduction of the proton to give a hydrogen *atom* would require much more negative potentials of less than -2 V and thus completely unphysiological conditions. The inorganic components of hydrogenases serve as electron reservoirs and, presumably, as catalytic centers.

A large number of photosynthetically active bacteria and algae shows hydrogenase activity [16]; depending on the preferred direction of the reaction, unidirectional 'uptake' hydrogenases $(\text{H}_2 \rightarrow \text{H}^+)$ can be distinguished from bidirectional 'reversible' enzymes. The hydrogenases are mostly of medium size (40–100 kDa) and, in principle, reversibly functioning enzymes (9.5); however, the potentials of electron-transferring Fe/S cluster and flavin components are often such that catalysis proceeds preferentially in one direction under physiological conditions. *Evolution* of hydrogen occurs only under strictly anaerobic conditions while the oxidation of H_2 may occur aerobically as well as anaerobically. Under anaerobic conditions, hydrogenases are involved in the reduction of CO_2, NO_3^-, O_2 or SO_4^{2-} (*Desulfovibrio gigas*); in some organisms, membrane-bound hydrogenases are found in addition to those that are soluble in the cytoplasm (Figure 9.1). In these cases, the soluble form catalyzes the reduction of NAD^+ by H_2 whereas the electrons generated via the membrane-localized oxidation of dihydrogen are inserted into the respiratory chain and serve in the production of high-energy phosphates (9.6) [15].

Hydrogenases catalyze the degradation of potentially dangerous free dihydrogen, H_2, as inadvertently produced during reduction reactions; on the other hand, the hydrogenase-catalyzed 'Knallgas' reaction (9.7) protects strictly anaerobic life processes by scavenging and deactivation of O_2, e.g. during N_2 fixation (Section 11.2).

Hydrogenases are being thoroughly studied because of their potential use in the controlled microbial degradation of organic material and in the production of CH_4 or H_2 as energy carriers [13,17]. However, the enzymes have proven to be very sensitive and labile and thus difficult to characterize structurally. Sequential analyses of some Ni/Fe hydrogenases have shown substantial amino acid homology, in particular with regard to cysteine groups [16].

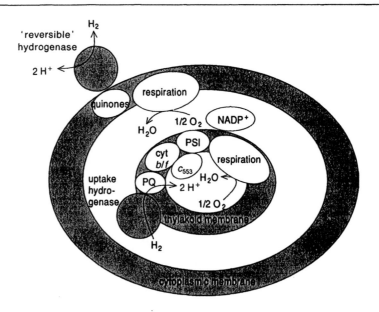

Figure 9.1
Arrangement of the protein complexes (compare Chapters 4 and 6), including both hydrogenases, in the cyanobacterium *Anacystis nidulans* (according to [13])

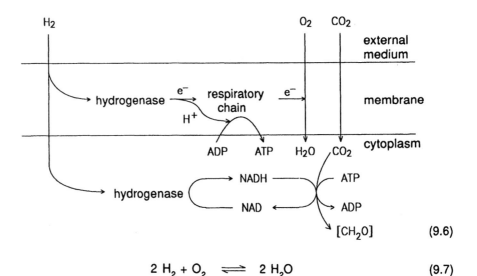

$$2\,H_2 + O_2 \;\rightleftharpoons\; 2\,H_2O \tag{9.7}$$

Although 'iron-only' hydrogenases exhibit a markedly higher activity than the
nickel-containing enzymes [15] (see Section 7.4), the latter are less sensitive and thus
better characterized. The following is known regarding the structure of the presumed
catalytically active metal center in Ni/Fe hydrogenases [18,19]: EPR signals were
obtained for the oxidized membrane-bound hydrogenase of *Methanobacterium bryan-
tii* which could only be explained assuming the presence of paramagnetic Ni^{III}
(low-spin d^7, $S = 1/2$). The presence of nickel as a paramagnetic center was confirmed

by EPR experiments with ^{61}Ni-labeled hydrogenase which showed signal splitting due to the nuclear spin of $I = 3/2$ of that isotope.

For the Ni/Fe hydrogenases of various microorganisms the geometries at the nickel center in different oxidation states of the protein have been approached using XANES and EXAFS measurements [20]. According to these data, nickel is surrounded by 2 ± 1 sulfur ligands with an average Ni—S distance of 223 pm and by 3 ± 1 light coordinating atoms such as O or N. The distance from nickel to the nearest iron center has been estimated at about 430 pm in some studies. The coordination geometry in the generally invoked oxidation states $Ni^I = d^9$, $Ni^{II} = d^8$ and $Ni^{III} = d^7$ is not unambiguously predictable; for Ni/Fe hydrogenases it is assumed that either five or six ligands coordinate in a trigonal bipyramidal or distorted octahedral arrangement.

The mechanism of catalysis by Ni/Fe hydrogenases and the nature of the participating species and oxidation states are still being controversially discussed [4,19–21]. The (dis)appearance of EPR signals suggests changes in the nickel oxidation state during the stepwise reductive activation of the catalyst and the ENDOR spectroscopic study of an active form showed ^{1}H/electron-spin interaction, probably involving cysteine β-CH$_2$ and proton(s) originating from the H$_2$ substrate [22]. However, this result is still not sufficient evidence to prove that the nickel center is the coordination site for H$_2$, especially since there are nickel-free hydrogenases (see Section 7.4).

In addition to the Ni^{III} EPR signal of the air-oxidized (inactive) form, there are other signals, attributed to Ni^{III} or Ni^I, which have been observed for different states of the enzyme. The absence of signals in EPR-silent states may be attributed to Ni^{II} with its even number of d electrons or to strong antiferromagnetic coupling between an $S = 1/2$ nickel center and a reduced [4Fe–4S] cluster.

A typical composition of inorganic components has been found in the 89 kDa hydrogenase of the sulfate-reducing bacterium *Desulfovibrio gigas*. In addition to nickel, a non-labile [3Fe–4S] cluster (see Figure 7.3) and two unusual [4Fe–4S] centers participate in (one-)electron transfer and in the two-electron reduction (9.5). In the active (reduced) form the EPR data indicate the presence of Ni^{II} and one reduced [3Fe–4S] cluster; after H$_2$-uptake the signal is first lost (Ni^{II}–H$_2$ form?) and then another EPR-active species results which can be formulated as Ni^{III}–H^{-1} or its resonance alternative Ni^I–H^{+1}. In the absence of an exactly known coordination environment it is not possible to clearly distinguish between Ni^I (d^9 [19]) and low–spin Ni^{III} (d^7 [21]) on the basis of EPR data alone [4]. Studies of the inactive Ni^{III} forms have demonstrated that the Ni^{III}/Ni^{II} potentials are surprisingly low. Such low reduction potentials can also be observed in model complexes with NiS_4, NiS_4N_2 or NiN_6 coordination environments [23].

Any mechanism for hydrogenase catalysis must take into account that these enzymes also accelerate the H/D exchange with water according to (9.8).

$$H_2 + D_2O \; \overset{\text{hydrogenase}}{\rightleftharpoons} \; HD + HDO \qquad (9.8)$$

A corresponding model (9.9) postulates the heterolytic cleavage of H$_2$ with the hydridic hydrogen atom remaining at the metal while the protic component (H$^+$) binds to a metal-coordinated sulfide center or, as the reaction proceeds, to another

$$E + H_2 \longrightarrow EH^- + H^+$$

$$EH^- + D^+ \longrightarrow HD + E \qquad\qquad \text{E: enzyme}$$

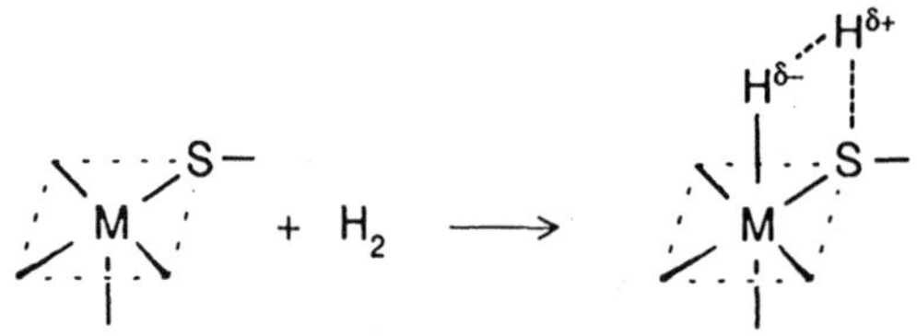

$$(9.9)$$

basic site within the protein. *Functional* model compounds with thiolate-coordinated nickel are rare [24].

The nature of the assumed bond between hydrogen and the uniquely catalytic nickel [25] is of particular interest and hydrides of nickel are well known from organometallic chemistry [26]. The possibility of side-on (η^2-)coordination of the dihydrogen molecule was discussed [27], since this binding mode has now been found in many complexes [28].

Fe/Ni hydrogenases show competitive binding of H_2 and CO; photolysis reactions of carbonyl complexes to generate catalytic intermediates as well as kinetic isotope (H/D) effects which are particularly pronounced in hydrogen complexes are being used to further elucidate the reaction mechanism.

EPR and XAS measurements on the hydrogenase from *Desulfovibrio baculatus* have indicated a coordination of selenocysteinate at the nickel atom (Ni–Se distance 244 pm); other ligands are one S center at 217 pm distance and three or four O (or N) atoms at a distance of approximately 206 pm to the metal [29]. The Ni/Fe/Se hydrogenases which also occur in some other microorganisms show a high acitivity for H_2 uptake but a markedly decreased potential for H/D exchange (9.8), suggesting a role of anionic selenium as base in the heterolytic cleavage of H_2 (9.9 [30]).

9.4 CO Dehydrogenase = CO Oxidoreductase = Acetyl–CoA Synthase

Many methanogenic and acetogenic, i.e. methane and acetic acid-producing bacteria, contain a 'CO dehydrogenase' enzyme which catalyzes the oxidation (9.10) of CO to CO_2. In biochemistry, oxidation is often equated with dehydrogenation; however, CO does not contain any hydrogen atoms and the term CO oxidoreductase instead of 'CO dehydrogenase' would thus be more appropriate [4].

$$CO + H_2O \rightleftharpoons CO_2 + 2 H^+ + 2 e^- \qquad\qquad (9.10)$$

The reaction is enzymatically reversible and may therefore serve as an alternative way of CO_2 fixation (assimilation) by photosynthetic bacteria. The other biological function of the enzyme is to catalyze the reversible formation of acetyl–coenzyme A (9.11) (see 3.15 and Table 3.1) in combination with CoA itself and a methyl source (see 9.15); corrinoid proteins with a CH_3–[Co] functionality (see Section 3.2.4), a methyl transferase and an Fe/S-containing disulfide reductase also contribute to this reaction [31,32].

$$CH_3-[Co] \; + \; CO \; + \; HS-CoA \; \rightleftharpoons \; CH_3C(O)S-CoA \; + \; H^+ \; + \; [Co]^-$$

$$\uparrow$$

$$CO_2 \qquad\qquad CoA = coenzyme \; A \; (Table \; 3.1) \qquad\qquad (9.11)$$

The resulting acetyl–CoA ('activated acetic acid') can be carboxylated to pyruvate $CH_3C(O)COO^-$ in autotrophic bacteria. In methanogenic bacteria, the further degradation of acetic acid to CO_2 and CH_4 presumably proceeds via CO as an intermediate.

It has been shown that the CO dehydrogenases of all anaerobic bacteria contain nickel while aerobic species require molybdopterin (see Section 11.1). The oxygen-sensitive CO dehydrogenase of the acetogenic bacterium *Clostridium thermoaceticum* has been most thoroughly studied. The enzyme consists of 3×2 subunits, $(\alpha\beta)_3$, with molecular masses of 82 and 73 kDa of the subunits. The metal content is not yet clear [32]; there are small amounts (1 Ni per protein hexamer?) of essential but labile nickel and, in addition, a large number (>10 per $\alpha\beta$ dimer) of iron and sulfide centers and various amounts of zinc were also found.

Structural data for the nickel center have mostly been derived from EPR, Fe-Mössbauer and EXAFS spectroscopical results [33,34]. The EPR data of an $S = 1/2$ species indicate the interaction of the unpaired electron with three or more iron atoms, one nickel and one carbon atom (^{61}Ni, ^{57}Fe and ^{13}CO hyperfine broadening in isotope-enriched material). The EXAFS data for nickel suggest a first coordination sphere with four S(thiolate) ligands at a distance of about 216 pm in what may be an approximately square planar arrangement. In accord with Mössbauer data, a Ni–Fe distance of about 325 pm points to interactions between the Ni center and one Fe/S cluster [34]. In labeling experiments, the carbonylation by CO dehydrogenases has been shown to be reversible; the mechanistic scheme (9.12) can explain this result, assuming reversibly extractable nickel as the CO coordination site [5,31].

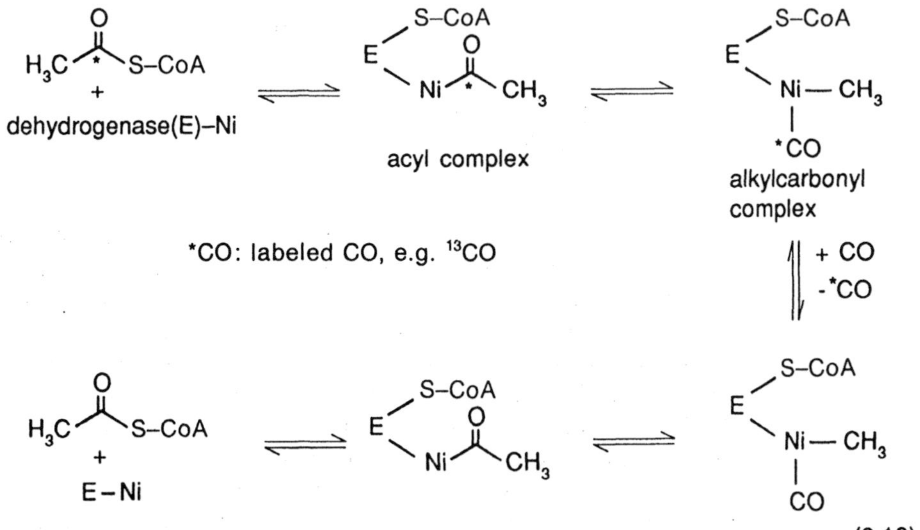

acyl complex

alkylcarbonyl complex

*CO: labeled CO, e.g. ^{13}CO

$$(9.12)$$

The insertion of CO into a metal–alkyl bond is well known from the organometallic chemistry of nickel (alkylcarbonyl → acyl rearrangement [35]). Model complexes

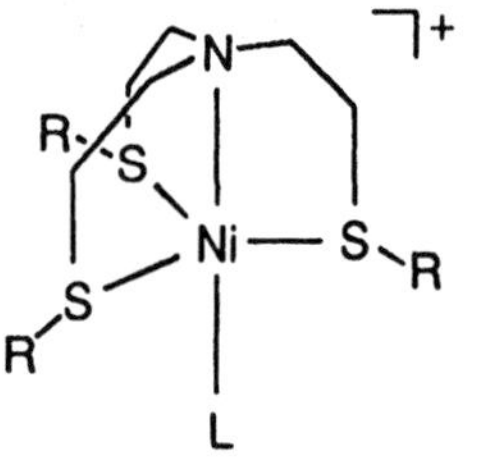

$$R = {}^iPr, {}^tBu \tag{9.13}$$

(9.13) with Ni^{II}–Me, Ni^{II}–COMe, Ni^{II}–H and Ni^{I}–CO functions were obtained with tetradentate 'tripod' ligands and the reaction sequence Ni–CH$_3$ → Ni–COCH$_3$ → CH$_3$COSR' could be established [36].

9.5 Methyl–Coenzyme M Reductase

Methyl–coenzyme M reductase serves methanogenic bacteria in the eventual formation and liberation of methane by catalyzing the reduction of methyl–coenzyme M (2-methylthioethanesulfonate). Reaction (9.14) is the last step in the energy-producing synthesis of CH_4 from CO_2 by autotrophic archaebacteria such as *Methanobacterium thermoautotrophicum* (see Figure 1.2).

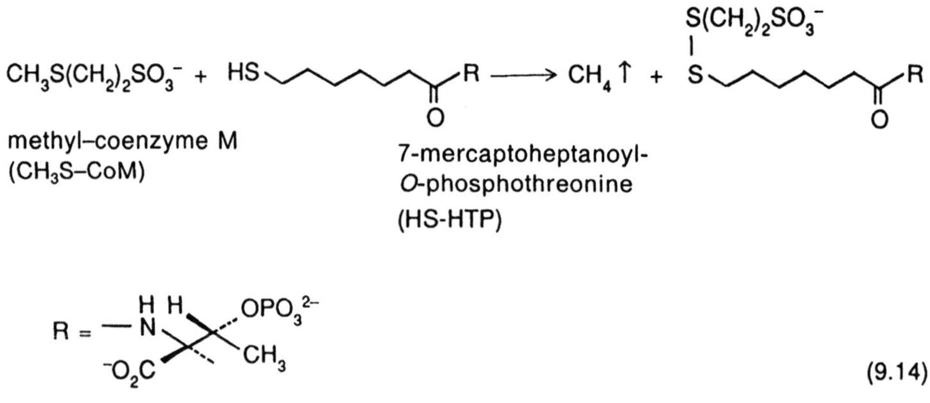

$$\tag{9.14}$$

Methanogenic bacteria participate in the anaerobic microbial degradation of the organic components of sludge in sewage plants which, by volume, is one of the largest biotechnological processes. In addition, microbial methane production takes place, e.g. in sediments (→natural gas), during the cultivation of rice and in the digestive system of ruminants; the atmospheric trace component CH_4 has received much attention lately because of its contribution to the 'greenhouse effect'. While most of the biogenic methane results from the degradation of acetate, its formation from CO_2 is more interesting from a chemical point of view.

Several coenzymes are essential for process (9.15) which requires a total of eight electrons [37,38]. Methanofuran provides for the uptake of CO_2 and for the transformation of the carboxylic function during the first $2e^-$ reduction. The resulting formyl group, CHO, is transferred to tetrahydromethanopterin which functionally resembles tetrahydrofolic acid (3.12) of eukaryotes. After two more two-electron transfer steps (formyl → hydroxymethyl → methyl) the resulting methyl group is transferred to coenzyme M which releases methane in a reaction catalyzed by methyl–coenzyme M reductase. The driving force of this reaction is the disulfide formation between coenzyme M and the additional component HS–HTP (see 9.17). The electrons required for the various reduction steps are obtained through the oxidation of molecular hydrogen which is catalyzed by various hydrogenases.

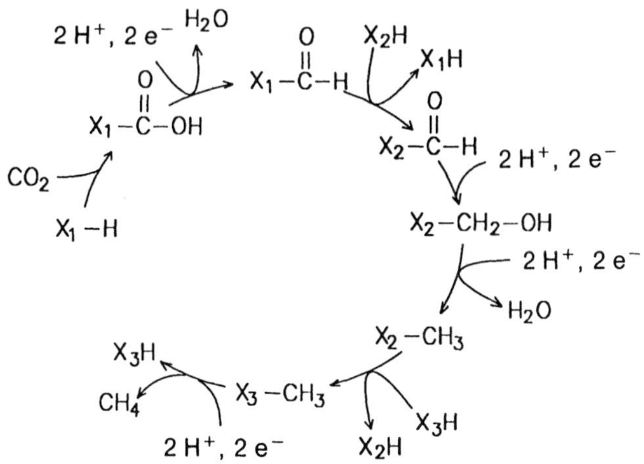

Overall process:

$$CO_2 + 8\ e^- + 8\ H^+ \rightarrow CH_4 + 2\ H_2O$$

↑ hydrogenases

'4 H_2'

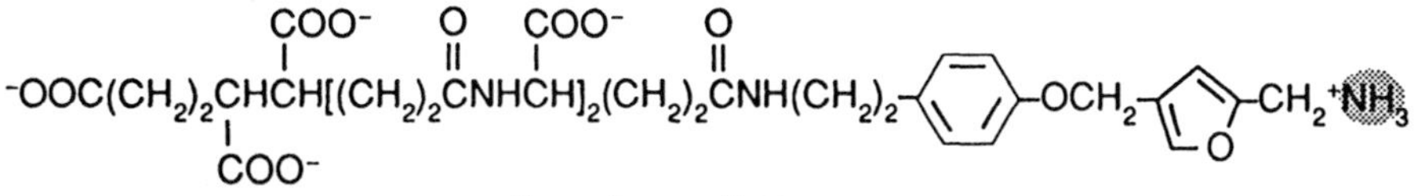

methanofuran, X_1H

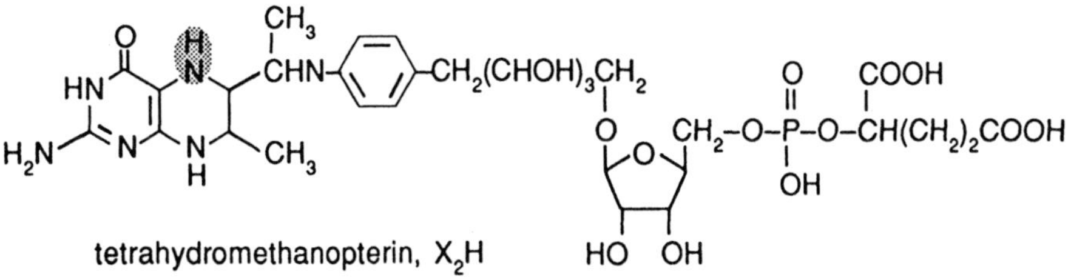

tetrahydromethanopterin, X_2H

$HS-(CH_2)_2-SO_3^-$ HS–CoM (coenzyme M), X_3H

⬤ : substitution site for C_1 fragments

(9.15)

Methyl–coenzyme M reductase is a very sensitive and complex enzyme. The dimeric protein has a molecular mass of approximately 300 kDa and is composed of three different subunits of 68, 47 and 38 kDa. From the enzyme, a yellow, low molecular weight substance was isolated which contains nickel and shows an intense absorption maximum at 430 nm: the factor F430. The structure of this first biogenic nickel–tetrapyrrole coenzyme was elucidated by extensive biosynthetic labeling and spectroscopic methods (NMR); it features a highly hydrogenated porphin system (9.16) with anellated δ-lactam and cyclohexanone rings [39]. The underlying ring structure has been named 'hydrocorphin' in order to point out the relationships both with porphyrins and with the not cyclically conjugated corrins (F430 as the 'missing link' in the evolution of tetrapyrroles [40]).

coenzyme F430   (9.16)

The partially saturated character of the macrocycle and the anellation of additional saturated rings allow a marked structural flexibility, especially with regard to a folding of the tetrapyrrole ring in the direction of an S_4-distortion (see Figure 2.9). As an important consequence, both the spin-crossover to low-spin Ni[II] ($S = 0$) and the transition to the d^9-configurated Ni[I] state is facilitated for the otherwise *not* axially activated high-spin nickel(II) center ($S = 1$). A low-spin d^8 configuration, i.e. spin pairing, only results after strong distortion of the octahedral ligand field; the degeneracy of the e_g orbitals which would disfavor spin pairing has to be effectively lifted (see 2.7 and Figure 2.10). As expected for a Ni[I] = d^9 species, the EPR spectrum of reduced F430 indicates a half-filled $d_{x^2-y^2}$ orbital with spin delocalization to the equatorial nitrogen atoms and a nucleophilic electron pair in the d_{z^2} orbital [41]. Both coordinatively unsaturated low-spin Ni[II] and Ni[I] states are thus important for the activation of methyl derivatives E—CH$_3$ in terms of an oxidative addition reaction [42]; however, they require a fairly flexible geometry of the macrocyclic ligand because of widely differing bond parameters: high-spin Ni[II]–N 210 pm, low-spin Ni[II]–N 190 pm, Ni[I]–N approximately 200 pm [43]. Like other tetrapyrrole ligands, the corphin π system itself can accommodate one additional electron, especially since there is a conjugating carbonyl group in the cyclohexanone ring. For the reduction of the complex this creates an ambivalence with regard to the alternatives of metal or ligand reduction [43]; a similar ambivalence exists for iron porphyrins (see Section 6.4). Another distinguishing aspect is the single negative charge in the inner macrocycle of the metal-binding hydrocorphin ligand; in contrast, porphyrin ligands coordinate as doubly deprotonated species.

Details of the mechanistic function of nickel in the enzyme-bound cofactor F430 are far from being complete [44,45]. In the following hypothetical scheme (9.17) nickel is assigned a role as an electron transferring center (oxidative coupling of a thiol and a thioether to a sulfuranyl radical) and as a short-lived methyl coordination site (organometallic intermediate). Direct binding of thioether or thiolate sulfur centers of the substrates are other possible functions of the nickel ion. The cleavage of alkyl-thiol(C—S) bonds by active nickel species is technically known from Raney nickel and similar desulfurization catalysts.

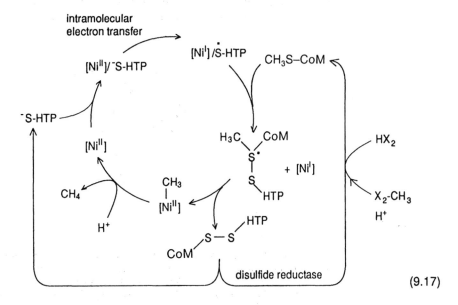

$$(9.17)$$

9.6 Model Compounds

When considering the design of model compounds for the metal centers in new and insufficiently characterized nickel proteins there is often the problem of stabilizing apparently exotic oxidation states. In organometallic compounds, nickel frequently occurs in low oxidation states like $+I$ or 0 [26] which are important for technical-catalytic hydrogenations (cf. hydrogenases), desulfurization, (cf. methyl–coenzyme M reductase) and carbonylation reactions (cf. CO dehydrogenase). On the other hand, nickel(III) has long been considered a rather unusual oxidation state of the metal. However, a large number of Ni^{III} complexes with chelating amide ligands, deprotonated peptides, oximes and thiols could be synthesized, stimulated not in the least by the detection of trivalent nickel in oxidized if not necessarily active enzymes. An important aspect in modeling Ni^{III} sites is that the coordination geometry should be restricted, e.g. by chelating ligands, since the electronic configurations d^8 (Ni^{II}) and d^7 (Ni^{III}) show markedly different geometrical preferences (see 3.3). Applying such strategies and using highly electron-rich thiolate (RS^-), sulfide (S^{2-}) or amide (R_2N^-) donor ligands, it has been possible to lower the potential for the $Ni^{II/III}$

transition to such an extent that Ni^{III} species become stable even under physiological conditions [23].

Among the stable Ni^{III} compounds are also complexes of small, C(O)NH-deprotonated peptides [46]. The nickel(III) center in (9.18) is thus located in a plane spanned by the tripeptide, while the axial positions are occupied by weakly coordinating water ligands. Such complexes with oligopeptides like Gly–Gly–Gly^{3-} (9.18) can easily be prepared by electrochemical oxidation of the corresponding Ni^{II} complexes. Similar to thiolates, the deprotonated peptide (carboxamide) nitrogen centers are strong σ and π donors, capable of stabilizing higher metal oxidation states. Whereas complexes like (9.18) still show relatively high redox potentials of about 0.9 V for the $Ni^{II/III}$ transition, the values are significantly lower for hydrogenases and other nickel enzymes, probably due to multiple cysteinate ligation [23].

(9.18)

References

1. F. H. Nielsen, Nutritional requirements for boron, silicon, vanadium, nickel, and arsenic: current knowledge and speculation, *FASEB J.*, **5**, 2661 (1991).
2. S. U. Patel, P. J. Sadler, A. Tucker and J. H. Viles, Direct determination of albumin in human blood plasma by ^{1}H NMR spectroscopy. Complexation of nickel^{2+}, *J. Am. Chem. Soc.*, **115**, 9285 (1993).
3. J. Beard, Did nickel poisoning finish off the dinosaurs?, *New Scientist*, 31, 19 May 1990.
4. R. Cammack, Nickel in metalloproteins, *Adv. Inorg. Chem. Radiochem.*, **32**, 297 (1988).
5. C. T. Walsh and W. H. Orme-Johnson, Nickel enzymes, *Biochemistry*, **26**, 4901 (1987).
6. J. R. Lancaster (ed.), *Bioinorganic Chemistry of Nickel*, VCH, Weinheim, 1988.
7. K. C. Bible, M. Buytendorp, P. D. Zierath and K. L. Rinehart, Tunichlorin: a nickel chlorin isolated from the Caribbean tunicate *Trididemnum solidum*, *Proc. Natl. Acad. Sci. USA*, **85**, 4582 (1988).
8. A. B. Costa, James Sumner and the urease controversy, *Chem. Br.*, 788 (1989).
9. N. Dixon, C. Gazzola, R. L. Blakeley and B. Zerner, Metalloenzymes. Simple biological role for nickel, *J. Am. Chem. Soc.*, **97**, 4131 (1975).
10. E. P. Day, J. Peterson, M. S. Sendova, M. J. Todd and R. P. Hausinger, Saturation magnetization of ureases from *Klebsiella aerogenes* and Jack bean: no evidence for exchange coupling between the two active site nickel ions in the native enzymes, *Inorg. Chem.*, **32**, 634 (1993).
11. R. Blakeley and B. Zerner, Jack bean urease: the first nickel enzyme, *J. Mol. Catal.*, **23**, 263 (1984).
12. R. S. Reczkowski and D. E. Ash, EPR evidence for binuclear Mn(II) centers in rat liver arginase, *J. Am. Chem. Soc.*, **114**, 10992 (1992).
13. T. Kentemich, G. Haverkamp and H. Bothe, Die Gewinnung von molekularem Wasserstoff durch Cyanobakterien, *Naturwissenschaften*, **77**, 12 (1990).
14. C. Zirngibl, W. van Dongen, B. Schwörer, R. von Bünau, M. Richter, A. Klein and R. K. Thauer, H$_2$-forming methylene tetrahydromethanopterin dehydrogenase, a novel type of

hydrogenase without iron–sulfur clusters in methanogenic archaea, *Eur. J. Biochem.*, **208**, 511 (1992).

15. M. W. W. Adams, The structure and mechanism of iron-hydrogenases, *Biochim. Biophys. Acta*, **1020**, 115 (1990).

16. T. Yagi, Hydrogenase, in *Metalloproteins; Chemical Properties and Biological Effects* (eds. S. Otsuka and T. Yamanaka), Elsevier, Amsterdam, 1988, p. 229.

17. I. Okura, Hydrogenase and its application for photoinduced hydrogen evolution, *Coord. Chem. Rev.*, **68**, 53 (1985).

18. R. P. Hausinger, Nickel utilization by microorganisms, *Microbiol. Rev.*, **51**, 22 (1987).

19. S. P. J. Albracht, Nickel in hydrogenase, *Recl. Trav. Chim. Pays-Bas*, **106**, 173 (1987).

20. C. Bagyinka, J. P. Whitehead and M. J. Maroney, An X-ray absorption spectroscopic study of nickel redox chemistry in hydrogenase, *J. Am. Chem. Soc.*, **115**, 3576 (1993).

21. M. Teixeira, I. Moura, A. V. Xavier, J. J. G. Moura, J. LeGall, D. V. DerVertanian, H. D. Peck, Jr and B.-H. Huynh, Redox intermediates of *Desulfovibrio gigas* [NiFe] hydrogenase generated under hydrogen, *J. Biol. Chem.*, **264** 16435 (1989).

22. J. P. Whitehead, R. J. Gurbiel, C. Bagyinka, B. M. Hoffman and M. J. Maroney, The hydrogen binding site in hydrogenase: 35-GHz ENDOR and XAS studies of the Ni–C active form and the Ni–L photoproduct, *J. Am. Chem. Soc.*, **115**, 5629 (1993).

23. H.-J. Krüger and R. H. Holm, Stabilization of trivalent nickel in tetragonal NiS_4N_2 and NiN_6 environments: synthesis, structures, redox potentials, and observations related to [NiFe]-hydrogenases, *J. Am. Chem. Soc.*, **112**, 2955 (1990).

24. L. L. Efros, H. H. Thorp, G. W. Brudvig and R. H. Crabtree, Toward a functional model of hydrogenase: electrocatalytic reduction of protons to dihydrogen by a nickel macrocyclic complex, *Inorg. Chem.*, **31**, 1722 (1992).

25. W. Keim, Nickel: an element with many uses in technical homogeneous catalysis, *Angew. Chem. Int. Ed. Engl.*, **29**, 235 (1990).

26. P. W. Jolly and G. Wilke, *The Organic Chemistry of Nickel*, Academic Press, New York, 1975.

27. R. H. Crabtree, Dihydrogen binding in hydrogenase and nitrogenase, *Inorg. Chim. Acta*, **125**, L7 (1986).

28. G. J. Kubas, Molecular hydrogen complexes: Coordination of a σ bond to transition metals, *Acc. Chem. Res.*, **21**, 120 (1988).

29. M. K. Eidsness, R. A. Scott, B. C. Prickril, D. V. DerVartanian, J. LeGall, I. Moura, J. J. G. Moura and H. D. Peck, Jr, Evidence for selenocysteine coordination to the active site nickel in the [NiFeSe] hydrogenases from *Desulfovibrio baculatus*, *Proc. Natl. Acad. Sci. USA*, **86**, 147 (1989).

30. O. Sorgenfrei, A. Klein and S. P. J. Albracht, Influence of illumination on the electronic interaction between ^{77}Se and nickel in active F_{420}-non-reducing hydrogenase from *Methanococcus voltae*, *FEBS Lett.*, **332**, 291 (1993).

31. D. Qiu, M. Kumar, S. W. Raysdale and T. G. Spiro, Nature's carbonylation catalyst: Spectroscopic evidence that carbon monoxide binds to iron, not nickel, in CO dehydrogenase, *Science*, **264**, 817 (1994).

32. W. Shin, M. E. Anderson and P. A. Lindahl, Heterogeneous nickel environments in carbon monoxide dehydrogenase from *Clostridium thermoaceticum*, *J. Am. Chem. Soc.*, **115**, 5522 (1993).

33. P. A. Lindahl, E. Münck and S. W. Ragsdale, CO dehydrogenase from *Clostridium thermoaceticum*, *J. Biol. Chem.*, **265**, 3873 (1990).

34. N. R. Bastian, G. Diekert, E. C. Niederhoffer, B.-K. Teo, C. T. Walsh and W. H. Orme-Johnson, Nickel and iron EXAFS of carbon monoxide dehydrogenase from *Clostridium thermoaceticum* strain DSM, *J. Am. Chem. Soc.*, **110**, 5581 (1988).

35. P. Stoppioni, P. Dapporto and L. Sacconi, Insertion reaction of carbon monoxide into metal–carbon bonds. Synthesis and structural characterization of cobalt(II) and nickel(II) acyl complexes with tri(tertiary arsines and phosphines), *Inorg. Chem.*, **17**, 718 (1978).

36. P. Stavropoulos, M. C. Muetterties, M. Carrie and R. H. Holm, Structural and reaction chemistry of nickel complexes in relation to carbon monoxide dehydrogenase: a reaction system simulating acetyl-coenzyme A synthase activity, *J. Am. Chem. Soc.*, **113**, 8485 (1991).

37. R. K. Thauer, Energy metabolism of methanogenic bacteria, *Biochim. Biophys. Acta*, **1018**, 256 (1990).

38. R. S. Wolfe, My kind of biology, *Annu. Rev. Microbiol.*, **45**, 1 (1991).
39. G. Färber, W. Keller, C. Kratky, B. Jaun, A. Pfaltz, C. Spinner, A. Kobelt and A. Eschenmoser, Coenzyme F430 from methanogenic bacteria: Complete assignment of configuration, *Helv. Chim. Acta*, **74**, 697 (1991).
40. A. Eschenmoser, Vitamin B_{12}: experimental work on the question of the origin of its molecular structure, *Angew. Chem. Int. Ed. Engl.*, **27**, 5 (1988).
41. C. Hollinger, A. J. Pierik, E. J. Reijerse and W. R. Hagen, A spectrochemical study of factor F_{430} nickel(II/I) from methanogenic bacteria in aqueous solution, *J. Am. Chem. Soc.*, **115**, 5651 (1993).
42. S.-K. Lin and B. Jaun, Coenzyme F430 from methanogenic bacteria: detection of a paramagnetic methylnickel(II) derivative of the pentamethyl ester by ^{2}H-NMR spectroscopy, *Helv. Chim. Acta*, **74**, 1725 (1991).
43. M. W. Renner, L. R. Furenlid, K. M. Barkigia, A. Forman, H.-K. Shim, D. J. Simpson, K. M. Smith and J. Fajer, Models of factor 430. Structural and spectroscopic studies of Ni(II) and Ni(I) hydroporphyrins, *J. Am. Chem. Soc.*, **113**, 6891 (1991).
44. A. Pfaltz, Control of metal-catalyzed reactions by organic ligands: from corrinoid and porphinoid metal complexes to tailor-made catalysts for asymmetric synthesis, *Chimia*, **44**, 202 (1990).
45. A. Berkessel, Methyl-coenzyme M reductase: model studies on pentadentate nickel complexes and a hypothetical mechanism, *Bioorg. Chem.*, **19**, 101 (1991).
46. A. G. Lappin, C. K. Murray and D. W. Margerum, Electron paramagnetic resonance studies of Ni(III)-oligopeptide complexes, *Inorg. Chem.*, **17**, 1630 (1978).

10 Copper-Containing Proteins: An Alternative to Biological Iron

For many iron-containing proteins there are 'parallel' copper-dependent analogues with comparable functions (Table 10.1). The correspondence between the reversibly O_2-binding proteins hemerythrin (Fe, Section 5.3) and hemocyanin (Cu, see Section 10.2) has already been mentioned; both metals are also featured in electron-transfer proteins of photosynthesis and respiration, in the metabolism of dioxygen, e.g. in oxidases or oxygenases, and in the deactivation of toxic intermediates of O_2 reduction.

Despite obvious functional similarities, however, iron and copper also show some general differences in their physiological appearance and function.

(a) Contrary to iron in heme, biological copper does not occur in tetrapyrrole coordinated compounds. It is particularly the imine-nitrogen atom in the imidazole ring of histidine that is able to form strong and, most importantly, kinetically inert bonds to copper in both relevant oxidation states, ($+$I) and ($+$II). The

Table 10.1 Correspondence of iron and copper proteins

function	Fe protein (h: heme system) (nh: non-heme system)	Cu protein
O_2 transport	hemoglobin (h) hemerythrin (nh)	hemocyanin
oxygenation	cytochrome P-450 (h) methane monooxygenase (nh) catechol dioxygenase (nh)	tyrosinase quercetinase (dioxygenase)
oxidase activity	peroxidases (h) peroxidases (nh)	amine oxidases laccase
electron transfer	cytochromes (h)	blue Cu proteins
antioxidative function	peroxidases (h) bacterial superoxide dismutases (nh)	superoxide dismutase (Cu, Zn) from erythrocytes
NO_2^- reduction	heme-containing nitrite reductase (h)	Cu-containing nitrite reductase

metal can thus be retained in the protein without having to be bound by a special macrocycle.

(b) As a general rule, the redox potentials for the $Cu^{I/II}$ transition are higher than those for $Fe^{II/III}$ pairs, both with physiological and nonphysiological ligands. Copper proteins such as ceruloplasmin are thus able to catalyze the oxidation of Fe^{II} to Fe^{III} (ferroxidase reactivity).

(c) In neutral aqueous solution and in sea water the *oxidized* form Cu^{2+} is more soluble than Cu^{I} which forms insoluble compounds with halide or sulfide ligands; in contrast, the oxidized form is less soluble in the $Fe^{II/III}$ system (5.1). In view of the biogenic O_2 production during early evolution this difference also had geochemical implications in terms of increasing iron precipitation and copper mobilization [1].

(d) Due to its later appearance and bioavailability in evolution, copper is often found in the extracellular space whereas iron occurs mainly within cells [2] (see also Section 13.1).

Since human beings do not require copper proteins for their stoichiometric oxygen transport, the total amount of about 150 mg copper in the body of an adult is relatively small. Nevertheless, there is little tolerance for deviations, mainly because of the essential role of this trace element [3] in superoxide deactivation and in the respiratory chain (see Figure 6.1). In this context, four copper-related pathological disorders should be mentioned:

(a) Wilson's disease involves a hereditary dysfunction of the primary copper storage capability of the body by the protein ceruloplasmin (Table 10.3). The metal ion is then accumulated in liver and brain, leading to dementia, liver failure and ultimately death. A therapy of this disease as well as of acute copper poisoning requires the administering of Cu-specific chelate ligands such as D-penicillamine (see 2.1). This ligand contains both S(thiolate), and N(amine) coordination sites to guarantee specificity for copper(I/II) and a hydrophilic carboxylic function which makes the resulting complex excretable.

(b) Acute copper deficiency may occur, particularly in newborn infants, since the complex metal transport and storage mechanisms involving serum albumin, ceruloplasmin and metallothionein (see Section 17.3) stabilize only several months after birth. Due to the essential role of copper in respiration (cytochrome *c* oxidase, Section 10.4) this deficiency may cause an insufficient oxygen utilization in the brain and thus permanent damage. Infants are also very sensitive to an excessive supply of copper; they usually have a high saturation concentration in the liver directly after birth.

(c) Menke's 'kinky hair' syndrome is based on a hereditary dysfunction of intracellular copper transport [4]. The resulting copper deficiency symptoms in infants include severe disturbances in the mental and physical development accompanied by the occurrence of kinky hair; an effective therapy must rely on intravenously administered copper compounds. The occurrence of sparse, kinky hair due to disorders in the copper metabolism illustrates that this element participates in the formation of connective tissue (collagen, keratin; see Section 10.3). The faulty gene responsible for Menke's syndrome was localized on the X chromosome and has been cloned, the corresponding ATPase transport protein

Table 10.2 Characteristics of 'classical' copper centers in protein

generalized coordination geometry	function, structure, characteristics

type 1

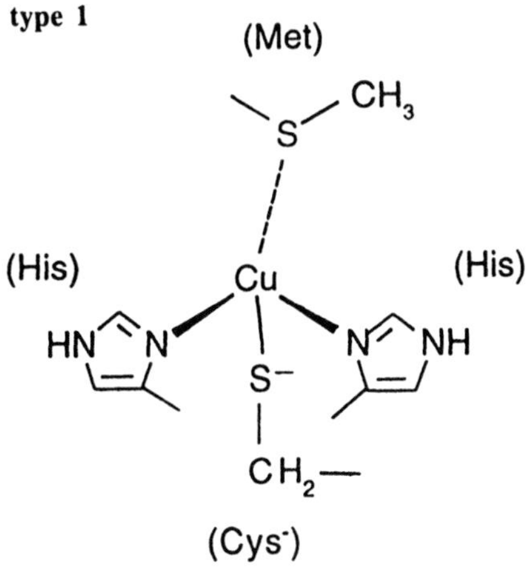

type 1: 'blue' copper centers
function: reversible electron transfer
$Cu^{II} + e^- \rightleftharpoons Cu^I$
structure: strongly distorted, (3+1) coordination
absorption of the copper(II) form at about 600 nm, molar extinction coefficient $\varepsilon > 2000$ M^{-1}cm^{-1}; LMCT transition S-(Cys-) $\rightarrow$ Cu^{II}
EPR/ENDOR of the oxidized form: small ^{63,65}Cu hyperfine coupling and g anisotropy, interaction of the electron spin with -S-CH$_2$-; Cu^{II} $\rightarrow$ S(Cys) spin delocalization

type 2

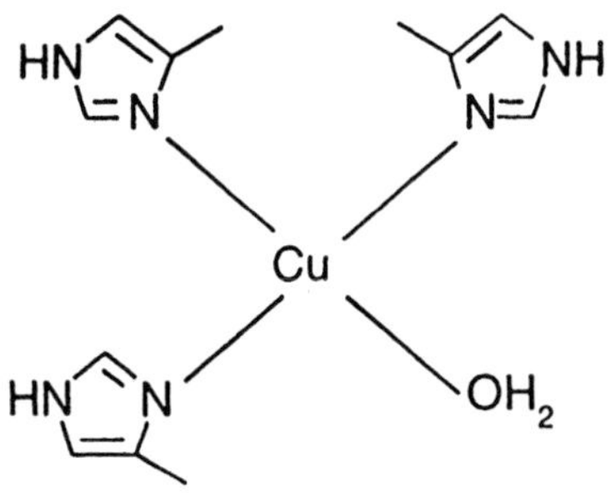

type 2: normal, 'non-blue' copper
function: O$_2$ activation from the Cu^I state in cooperation with organic coenzymes
structure: essentially planar with weak additional coordination (Jahn-Teller effect for Cu^{II})
typically weak absorptions of Cu^{II}, $\varepsilon < 1000$ M^{-1} cm^{-1}; ligand-field transitions (d $\rightarrow$ d) normal Cu^{II} EPR

type 3

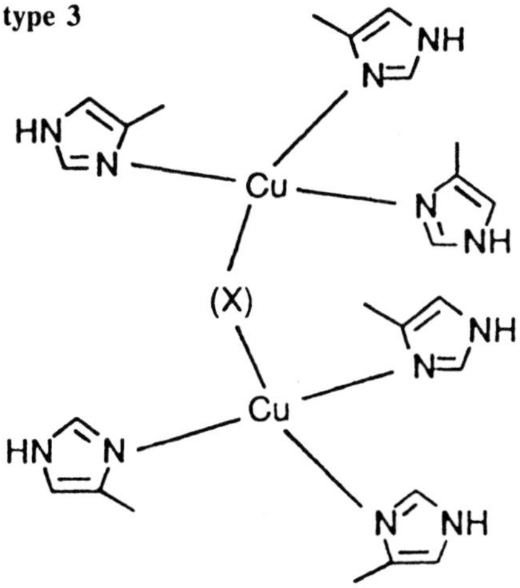

type 3: copper dimers
function: O$_2$ uptake from the Cu^I-Cu^I state
structure: (bridged) dimer, Cu–Cu distance about 360 pm after O$_2$ uptake intense absorptions around 350 and 600 nm, $\varepsilon \approx 20000$ and 1000 M^{-1}cm^{-1}; LMCT transitions O$_2^{2-}$ $\rightarrow$ Cu^{II}
EPR-inactive Cu^{II} form (antiferromagnetically coupled d^9 centers)

(see Section 13.4) features six Cys-X_2-Cys motifs as presumable copper binding sites [4].

(d) Defects (mutations) in the copper-dependent superoxide dismutase are responsible for the not uncommon neurodegenerative (paralytic) disorder known as Lou Gehrig's disease or familial amyotrophic lateral sclerosis (ALS [5]).

Copper and molybdenum (see Section 11.2) are both metals with affinity for N *and* S donor atoms and thus behave as antagonists; cattle raised on Mo-rich and Cu-poor soil may thus develop severe copper deficiencies [6]. The resulting disorders can be counteracted by supplementing copper compounds to the animals' feed. Presumably, copper is not available for the metabolism because of its tight binding to thiomolybdates, $MoO_nS_{4-n}^{2-}$, which are formed from molybdenum and sulfur-containing substances in the digestive tract of ruminants [7] (see 11.4). Similar 'secondary' copper deficiencies may occur in the presence of excess Fe, Zn or Cd.

From a structural and spectroscopic point of view, three main types of biological copper centers can be distinguished according to a generally accepted convention (Table 10.2 [8,9]); of course, these centers can occur several times in one protein as, for example, in multifunctional ceruloplasmin which *inter alia* participates in the regulation of the iron metabolism (Table 10.3). Recent studies have established special

Table 10.3 Some representative copper proteins

function and typical proteins	molecular mass (kDa)	Cu type(s)	occurrence, reactivity
electron transfer ($Cu^I \rightleftharpoons Cu^{II} + e^-$)			
plastocyanin	10.5	1 type 1 ($E = 0.3\text{-}0.4V$)	participation in plant photosynthesis (see Figure 4.9)
azurin	15	1 type 1 ($E = 0.2\text{-}0.4V$)	participation in bacterial photosynthesis
'blue' oxidases ($O_2 \rightarrow 2\ H_2O$)			
laccase	60-140	1 type 1 ($E = 0.4\text{-}0.8V$) 1 trimer	oxidation of polyphenols and polyamines in plants
ascorbate oxidase	2×75	2 type 1 ($E = 0.4V$) 2 trimers	oxidation of ascorbate to de-hydroascorbate in plants (3.12)
ceruloplasmin	130	2 type 1 ($E = 0.4V$) 1 trimer?	Cu transport and storage, Fe mobilization and oxidation, oxidase and antioxidation function in human and animal serum

Table 10.3 (*continued*)

function and typical proteins	molecular mass (kDa)	Cu type(s)	occurrence, reactivity
'non-blue' oxidases ($O_2 \rightarrow H_2O_2$)			
galactose oxidase	68	1 type 2	alcohol oxidation in fungi (10.13)
amine oxidases	>70	1 type 2	degradation of amines to carbonyl compounds (10.14), cross-linking of collagen
monooxygenases ($O_2 \rightarrow H_2O$ + substrate-O)			
dopamine β-monooxygenase	4×70	8 type 2	side chain oxidation of dopamine to norepinephrine in the adrenal cortex (10.8)
tyrosinase	42	2 type 3	ortho-hydroxylation of phenols and subsequent oxidation to *o*-quinones in skin, fruit pulp etc. (10.10)
dioxygenases ($O_2 \rightarrow$ 2 substrate-O)			
quercetinase	110	2 type 2	oxidative cleavage of quercetin in fungi
terminal oxidase ($O_2 \rightarrow$ 2 H_2O)			
cytochrome *c* oxidase	>100	Cu_A Cu_B	end point of the respiratory chain (see Figures 6.1 and 6.2)
superoxide degradation (2 $O_2^{\bullet -} \rightarrow O_2 + O_2^{2-}$)			
Cu,Zn-superoxide dismutase	2 × 16	2 type 2	$O_2^{\bullet -}$ disproportionation e.g. in erythrocytes
dioxygen transport			
hemocyanin	$n \times 50$ (molluscs) $n \times 75$ (arthropods)	2 type 3 per n	O_2 transport in hemolymph of molluscs and arthropods (n=8)
functions in the nitrogen cycle:			
nitrite reductase	3×36	3 type 1 3 type 2 (?)	NO_2^- reduction (dissimilatory)
N_2O reductase	2×70	2 Cu_A 6 type 3 (?)	reduction of N_2O to N_2 (10.15) in the nitrogen cycle

combinations (e.g. type 2/type 3 trimers [9]) as well as other copper centers like Cu_A and Cu_B of cytochrome c oxidase (see Section 10.4) which do not fit into the classical scheme. The Cu^I centers in nonenzymatic sensor, transport and storage proteins are predominantly coordinated by cysteinate residues [10] (see also the metallothioneins, Section 17.3).

10.1   Type 1: 'Blue' Copper Centers

The type 1 copper centers have received their additional attribute because of the *intensely* blue color of the corresponding Cu^{II} proteins which thus have been given appropriate names such as azurin or plastocyanin (Table 10.3; κυανος: dark blue). The color is quite distinctive because the metal centers are optically so 'diluted' in metalloproteins that only intense absorptions in the visible region resulting from symmetry allowed electronic transitions give rise to conspicuous colors. In contrast, the comparatively pale blue color of normal Cu^{2+}(aq), e.g. in crystalline copper(II) sulfate pentahydrate, is the result of 'forbidden' electronic transitions between d orbitals of different symmetry (2.7; 'ligand-field' transitions); in these cases, the molar extinction coefficients ε are less than 100 M^{-1} cm^{-1}. The copper(II) centers of 'blue' copper proteins, however, show much higher ε values of about 3000 M^{-1} cm^{-1}. In the EPR spectra of type 1 copper(II) sites, the interaction of the unpaired electron (Cu^{II} has a d^9 configuration) with the magnetically not very different isotopes ^{63}Cu and ^{65}Cu (nuclear spin $I = 3/2$ for both isotopes) leads to a markedly decreased hyperfine splitting $a_\parallel$ (Figure 10.1) as compared to that of normal Cu^{II} centers [9].

Electron Paramagnetic Resonance II (see EPR I in Section 3.2.2)

In contrast to the Co^{II} state of cobalamins with a low-spin d^7 configuration (3.7) and the unpaired electron located in the d_{z^2} orbital, normal Cu^{II} complexes with their d^9 configuration have the $d_{x^2-y^2}$ orbital occupied by the single electron. In d^9 systems, the octahedral configuration is unstable because of the ambiguity (10.1) which results from incomplete occupation of a degenerate orbital (e_g, 2.7). The system circumvents this ambiguous situation (10.1) through geometrical distortion, i.e. a lowering of the symmetry removes the orbital degeneracy—the 'first-order Jahn–Teller effect'.

The occupation of d orbitals by unpaired electrons can be deduced from the typical 'anisotropic' EPR spectra which are obtained from statistically oriented, immobilized complexes, e.g. in frozen solution or in a polycrystalline solid.

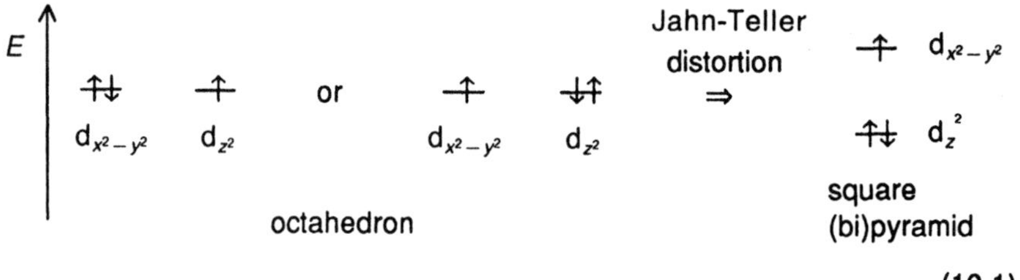

(10.1)

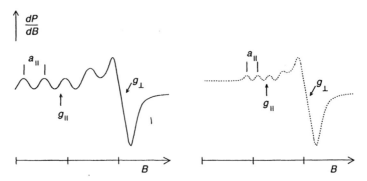

.Figure 10.1
Typical anisotropic EPR signals of a normal Cu^{II} complex (——) and of a 'blue'
copper(II) protein ($\cdots$; 1. derivative spectra, same scale of magnetic field strength B)

First of all, there is a splitting of the signal which reflects the local symmetry
of the singly occupied orbital in the homogeneous magnetic field. This 'anisotropic interaction' may be different in all three principal directions of space and
will then lead to a 'rhombic signal' with three different g factor components.
In the case of Jahn–Teller distorted Cu^{II} with square, square pyramidal or
tetragonal bipyramidal configurations, however, an 'axial' spectrum is generally
observed with two coinciding g components, $g_x = g_y = g_\perp$, perpendicular to
the axis of the magnetic field, and a separate g component, $g_z = g_\parallel$, parallel
to the magnetic field (Figure 10.1). For a d^9 configuration with $(d_{x^2-y^2})^1$ there
is $g_\parallel > g_\perp \approx 2.01$ and, furthermore, the interaction of the unpaired electron with
the nuclear spin $I = 3/2$ of the copper isotopes $^{63,65}Cu$ is markedly greater in the
axial direction than for the components perpendicular to the magnetic field:
$a_\parallel > a_\perp$ [11]. Accordingly, the low-field g component is split into four lines
(four nuclear spin orientations, $M = +3/2, +1/2, -1/2, -3/2$, relative to the
magnetic field); only transitions with $\Delta S = \pm 1$ and $\Delta I = 0$ are allowed (10.2).

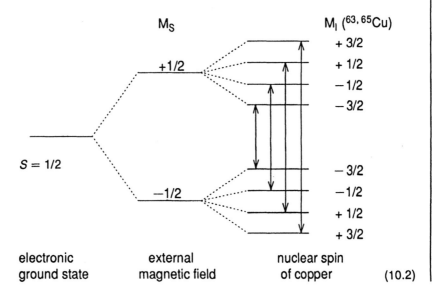

(10.2)

In blue copper proteins with their distorted metal configuration even in the Cu^{II} state and with a sizable covalent character of the Cu–thiolate bond [12] both the g anisotropy, i.e. the difference $g_{\parallel} - g_{\perp}$, as well as the $^{63,65}Cu$ hyperfine coupling are markedly reduced (Figure 10.1), reflecting a smaller contribution of the metal with its large spin–orbit coupling constant to the singly occupied molecular orbital. On the other hand, a significant interaction of the cysteinate ligand with the unpaired electron has been found in ENDOR studies [13].

Crystal structure analyses of several 'blue' copper proteins [14–17] (Figures 10.2 and 10.3) have shown that the metal centers feature a very irregular 'distorted tetrahedral' coordination shell with deviations from the ideal tetrahedral angle (about 109.5°) as large as 25°.

Two histidine residues and one hydrogen bond forming cysteinate ligand are strongly bound in an approximately trigonal planar arrangement; weakly bound (10.3) methionine or glutamine (in stellacyanin) and, in azurin, a very weakly coordinating oxygen atom from a peptide bond complete the coordination environment in the axial positions ('3 + 1' and '3 + 1 + 1' coordination, respectively). The description of these structures as trigonal pyramidal or bipyramidal is thus based on the evaluation of what still passes as a coordinative bond. However, it is the cysteinate ligand in particular that is responsible for the unusual behavior of type 1 copper

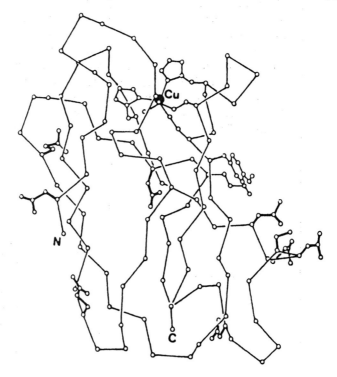

Figure 10.2
Structure of the Cu^{II} form of plastocyanin from poplar leaves (*Populus nigra*; α carbon representation of the polypeptide chain with some selected amino acid side chains; from [14]).

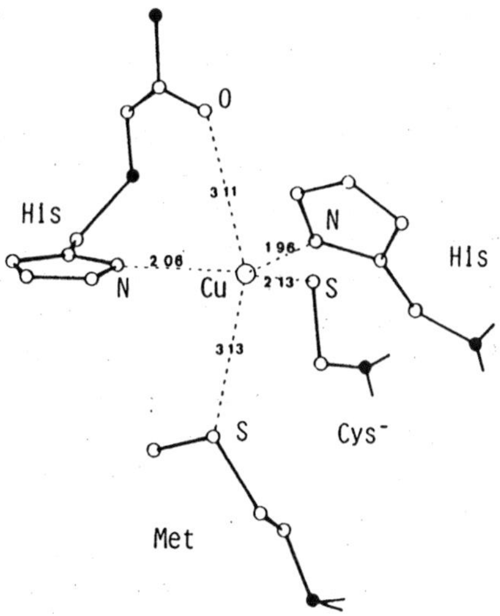

Figure 10.3
Coordination environment of the copper center in azurin from *Alcaligenes denitrificans* (α carbon centers as filled circles; from [15])

centers [9]. As in the case of oxidized rubredoxin (7.4), the intense absorption of the oxidized form is attributed to a ligand-to-metal charge transfer (LMCT) transition, i.e. electronic charge is transferred from the π and σ electron-rich thiolate ligand to the electron-poor Cu^{II} center via light excitation (10.4).

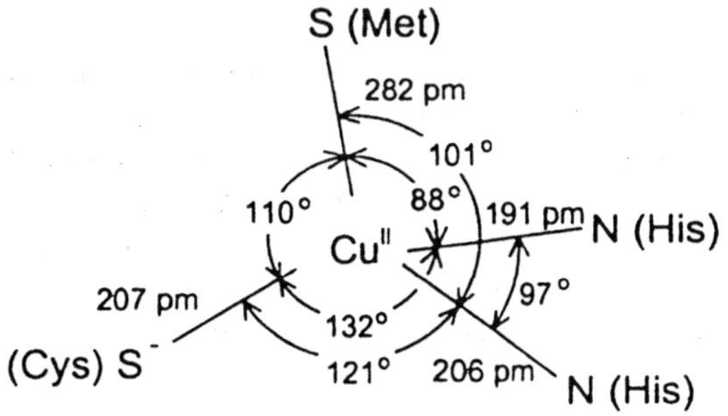

$$(10.3)$$

Even in the ground state, the cysteinate sulfur atom donates charge to the metal center which results in a delocalization of spin from the metal (smaller EPR coupling constant) to the cysteinate sulfur center (EPR/ENDOR detectable coupling with $^-S{-}CH_2{-}$ [13]).

The strong geometrical distortion is a consequence of the incorporation of the coordinating amino acid ligands in well-conserved sequences $His{-}X_k{-}Cys{-}X_n{-}His{-}X_m{-}$Met ($n$, $m = 2$–4; large k). Just like the mixture of donor centers (2 N, 2 S), this

$$(R{-}CH_2{-}S^-)Cu^{II}(His)_2(Met) \xrightarrow[\text{LMCT}]{h\nu} [(R{-}CH_2{-}S^\bullet)Cu^{I}(His)_2(Met)]^* \qquad (10.4)$$

strongly distorted arrangement represents a compromise (entatic-state situation) between $Cu^I = d^{10}$ with its preferred tetrahedral or trigonal coordination through 'soft', e.g. sulfur ligands, and $Cu^{II} = d^9$ with preferentially square planar or square pyramidal geometry and N ligand coordination. Presumably, the irregular high-energy arrangement at the metal (10.3) largely resembles the transition-state geometry between the tetrahedral and the square planar equilibrium configurations of the two involved oxidation states, resulting in a higher rate of electron transfer [17], e.g. within the photosynthetic apparatus (Figures 4.8 and 4.9). Similar 'intermediate geometries', anisotropy of the actual electron-transfer process and electron delocalization between the metal and the (porphyrin or sulfur) ligands have been found for the other two major classes of proteins with inorganic electron-transfer centers, viz. the iron/sulfur and the cytochrome proteins (Sections 6.1 and 7.2–7.4).

Good model compounds for the 'blue' type 1 copper proteins were unknown for quite some time since the typical spectroscopic properties can only be mimicked using special, multidentate chelating ligands with thiolate coordination centers (see 10.5 [18]); simple thiolates would immediately reduce Cu^{II}. The incorporation of Cu^{2+} instead of Zn^{2+} into the insulin hexamer (see Section 12.7) also results in close spectroscopic similarity with 'blue' copper centers if an external thiol is available [19]. 'Masking' of blue copper centers is possible through isomorphous metal exchange with Hg^{2+}, which prefers a similar coordination but is neither EPR nor charge-transfer active [20].

$$
\begin{array}{c}
\text{MeOOC} \quad\quad \text{H}_2\text{C} - \text{CH}_2 \quad\quad \text{COOMe} \\
\text{HC} - \text{N} \quad\quad\quad \text{N} - \text{CH} \\
\text{H}_2\text{C} - \text{S} \quad\quad \text{Cu} \quad\quad \text{S} - \text{CH}_2
\end{array}
\qquad (10.5)
$$

As for the redox pair $Fe^{II/III}$, the potentials for the $Cu^{I/II}$ transitions are generally higher in bioinorganic systems than in simple model complexes. The potential range for proteins with type 1 copper centers runs from 0.18 V (stellacyanin, with Gln instead of Met as the axial ligand) to 0.68 V (rusticyanin). The special stabilization of the lower oxidation state by ligands such as Met as well as the Cu^{II}-*destabilizing* deviation from the square planar or square pyramidal configuration are responsible for the rather high $Cu^{I/II}$ potentials.

10.2   Type 2 and Type 3 Copper Centers in O_2-Activating Proteins: Oxygen Transport and Oxygenation

The type 3 dinuclear copper centers feature an EPR-inactive oxidized state due to antiferromagnetic coupling of the Cu^{II} ions (see 4.11). These centers are always found in connection with the activation of $^3O_2$, the O_2-transporting hemocyanin (Hc) as well as several O_2-dependent oxidases (electron transfer and catalysis) and mono-oxygenases (transfer of one O from O_2) are featured in this group (see Table 10.3).

It is remarkable that triplet dioxygen is bound very rapidly by the diamagnetic Cu^I centers of the deoxy forms [21]; a diamagnetic ground state due to antiparallel spin–spin coupling of Cu^{II} centers is also observed for the oxy forms. Low-lying electronically excited (triplet) configurations such as $3d^9 4p^1$ or $3d^9 4s^1$ of the copper(I) centers are possibly important for O_2 binding by the deoxy form; such configurations have also been implicated for synthetic $3d^{10} \cdots 3d^{10}$ systems with potentially available 4s and 4p levels [22].

Preliminary structural information is available for the metal-containing center of the highly associated hemocyanin (Hc), the O_2-transferring protein of certain molluscs and arthropods (e.g. crustaceans [23]). Two coordinatively unsaturated Cu^I centers, each anchored in the protein by histidine residues, are situated at a distance of more than 350 pm in the (colorless) deoxy form (10.6) of each subunit.

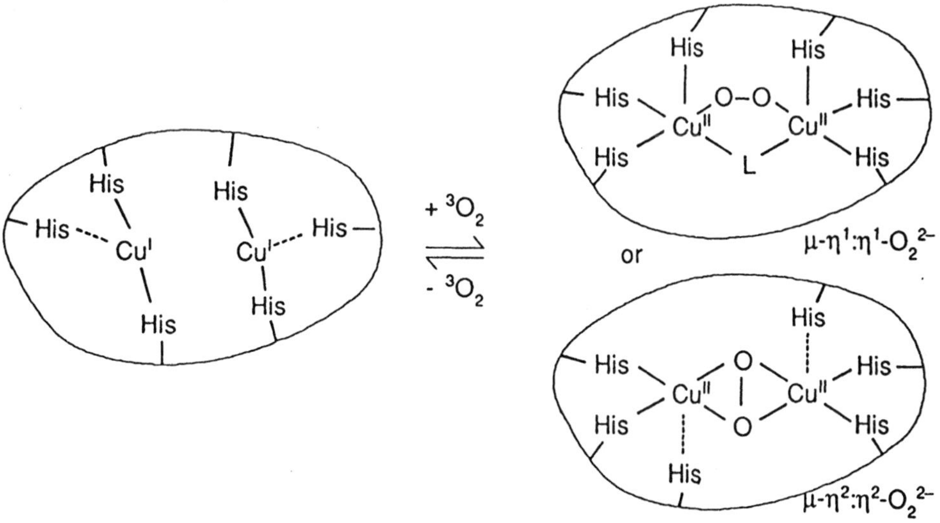

deoxyhemocyanin (deoxy-*Hc*)   oxyhemocyanin (oxy-*Hc*)

$d_{Cu-Cu} \approx 360$ pm   (10.6)

Based on numerous model compounds (see 5.6 and 10.7) and spectroscopic studies of the oxy form, it has been assumed that it contains either cis-μ-$\eta^1 : \eta^1$ [24,25] or μ-$\eta^2 : \eta^2$ coordinated 'dioxygen' [23,26,27] in the peroxide oxidation state [28]. In the first model, the Cu^{II} centers which are formed via an 'inner-sphere' two-electron transfer are additionally bridged by a hydroxo ligand, L = OH⁻, to give a five-membered ring (10.6). The alternative of bridging by doubly side-on coordinating O_2^{2-} (μ-$\eta^2 : \eta^2$) was only seriously considered after corresponding model compounds (see 5.6) had been prepared [28]; this alternative does not require an additional ligand L and is in better agreement with the strongly weakened O—O bond [27].

The assignment of the O_2 oxidation state is based on resonance Raman measurements of the O—O vibrational frequency (5.3) and on $(O_2^{2-}) \rightarrow (Cu^{II})$ ligand-to-metal charge-transfer absorptions, both in comparison with model compounds. While the reaction of O_2 with Cu^I is not totally unexpected, the reversibility of the dioxygen binding by the protein remains astonishing, given the distinct changes in oxidation states and the multiple coordination. In realistic model systems, such reversibility has been observed when protein-modeling N-polychelate ligands such as tris-pyrazolylborates (see 4.14 and 5.6) or ligands like those in (10.7) were used [26]. The high association of the Hc subunits to form protein complexes of more than 1 MDa has to be viewed in context with the (limited) degree of cooperativity achieved by this O_2 transport system [29] (see Section 5.2). Nevertheless, even the O_2 transport in giant octopus species, which may weigh up to 150 kg and may accelerate to 30 km/h, is based on the hemocyanin system.

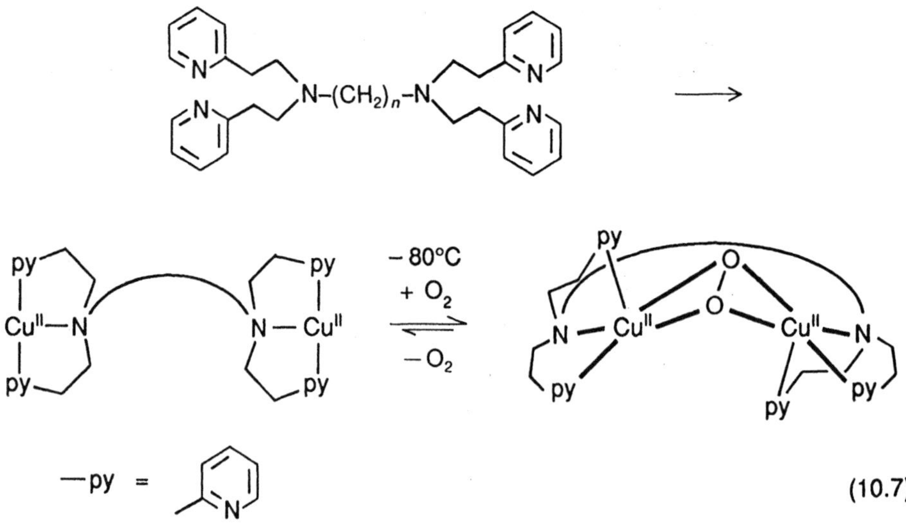

$$(10.7)$$

The group of monooxygenase (hydroxylase) enzymes does not only include iron-containing heme enzymes like cytochrome P-450 systems but also, with a more special selectivity, copper-containing enzymes such as tyrosinase or dopamine β-monooxygenase. The latter play a role in the biological oxidation of phenylalanine via the anti-Parkinson drug 'dopa' to (nor)epinephrine (10.8) and thus in the important biosynthesis of hormones and neurotransmitters. The O_2-dependent *ortho* hydroxylation of phenols by tyrosinase to give catechols with the possible subsequential oxidation to o-quinones has a quite general biochemical relevance. Numerous natural products including vitamins like ascorbic acid or hormones (see 10.8) contain unsaturated rings which are substituted with vicinal polyhydroxo or polyalkoxy groups.

The copper enzyme-catalyzed oxidative transformation of catechol derivatives to light-absorbing o-quinones is clearly visible after their polymerization to melanins, the color of which ranges from red to dark brown (10.9). The melanin pigments of skin, hair and feathers or the pigments of rotten, i.e. air-oxidized fruit (enzymatic browning), are all polymeric o-quinone derivatives [30,31]. Copper enzyme-containing banana fruit pulp tissue can thus be utilized in an electrode membrane as a dopamine-sensitive biosensor, the 'bananatrode' [32]. It might also be mentioned in this context that Cu^I centers are being discussed as binding sites for the common plant hormone ethylene, $H_2C{=}CH_2$ [33].

The oxygen atom incorporated in the substrates after copper enzyme-catalyzed monooxygenation was found to originate from O_2 as shown by isotopic labeling; a possible mechanism of monophenol oxygenation and oxidation to o-quinone is depicted in (10.10) [9,25].

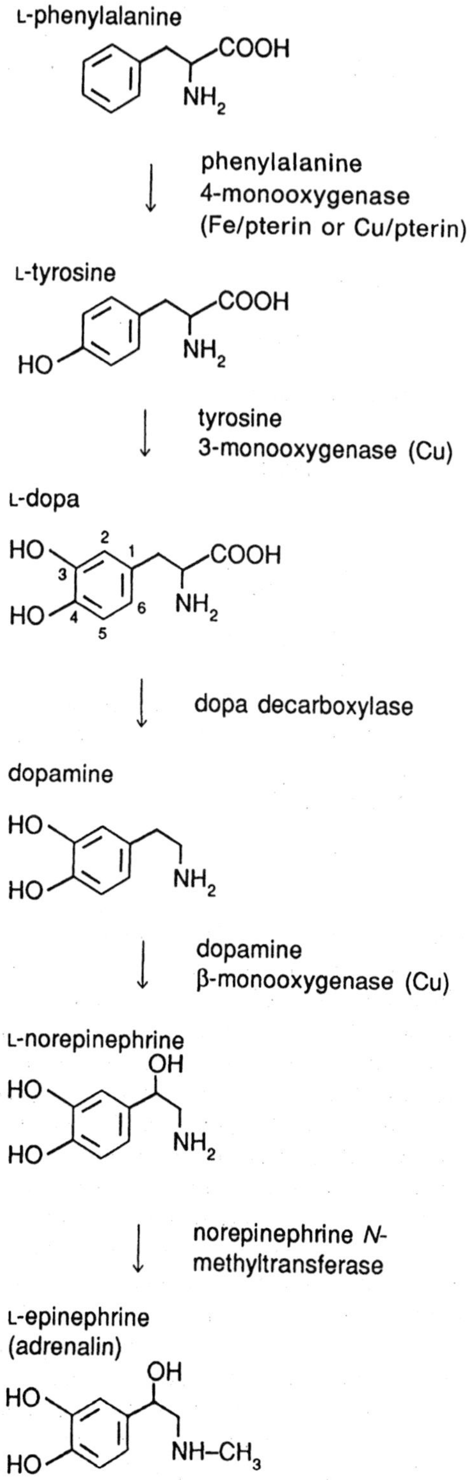

L-phenylalanine

phenylalanine
4-monooxygenase
(Fe/pterin or Cu/pterin)

L-tyrosine

tyrosine
3-monooxygenase (Cu)

L-dopa

dopa decarboxylase

dopamine

dopamine
β-monooxygenase (Cu)

L-norepinephrine

norepinephrine N-
methyltransferase

L-epinephrine
(adrenalin)

(10.8)

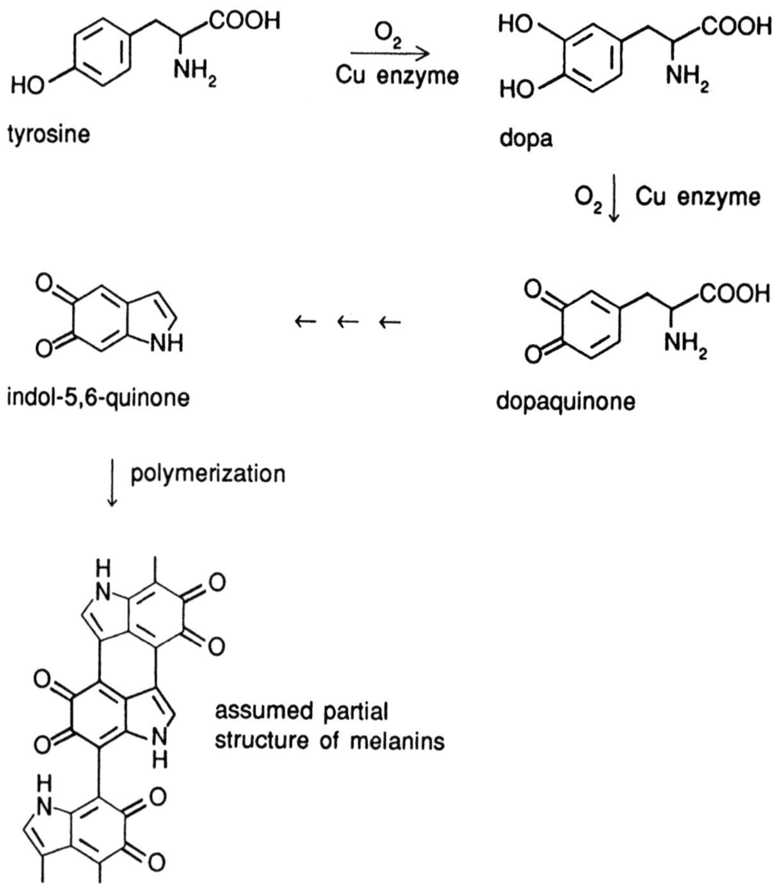

tyrosine → dopa

indol-5,6-quinone ← ← ← dopaquinone

polymerization

assumed partial
structure of melanins

(10.9)

Adduct (2) formed reversibly from the deoxy form (1) and O_2 features a cis-μ-η^1:η^1 or a μ-η^2:η^2 peroxide structure (see the alternative in 10.6) and is then able to coordinate a phenolate at the copper center via its oxygen atom (10.10: structure 3). A transition state (4) is reached through conformational changes and a chelate complex of the newly formed catecholate (10.10: structure 5) is created after electrophilic attack of one of the peroxidic oxygen atoms at the *ortho* position of the aromatic system. At this point, the O—O (single) bond is cleaved (monooxygenase activity) by shifting electron density into the antibonding σ^* orbital. Protonation of the remaining hydroxo ligand (10.10: structure 5) to give dissociable water may lead to formation of a bridging (μ-)catecholato complex (6) of the copper dimer which could then reorganize under intramolecular electron transfer to the oxidized *o*-quinone product and the catalytically active deoxy state of the enzyme (10.10).

Like catechol oxidase, tyrosinase can oxidize 1,2-dihydroxy substituted aromatic systems to *o*-quinones; a similar reactivity has been observed for aromatics with 1,2-diamino substituents. Presumably, special steric restrictions in the transition state and a positive partial charge at the peroxide oxygen atom bound to the lower coordinated copper center cause the selectivity of tyrosinase for *ortho* hydroxylation of phenols. This reaction is not only essential for the synthesis of active substances and for the formation of melanin; *o*-polyphenols can be further transformed under

monooxygenation and
oxidation by tyrosinase:

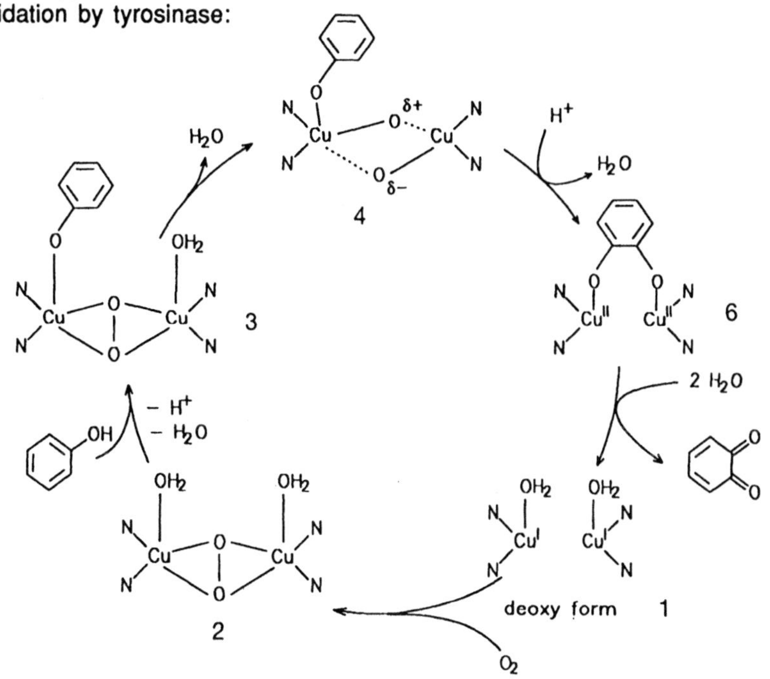

N: imine center of histidine

charges not indicated

overall reaction: $Ar(H)OH + O_2 \rightarrow Ar(O)_2 + H_2O$

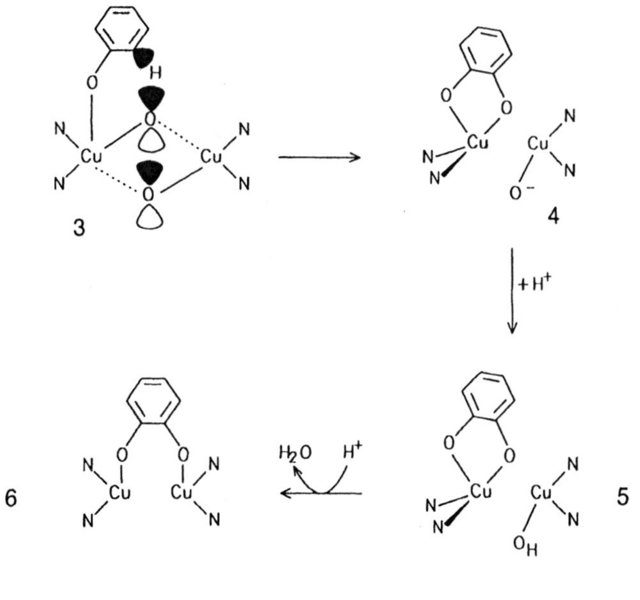

(10.10)

ring opening (see 7.16) via the help of nonheme iron-containing catechol dioxygenases (Section 7.6.4), a reaction that is important for the microbial degradation of aromatic compounds in the environment.

Dopamine β-monooxygenase catalyzes the specific monooxygenation in an aliphatic side chain (10.8). The active copper centers (two 'type 2' per protein subunit) are thus different, possibly featuring a sulfur ligand in the Cu^I state and a Cu^{II}/hydroperoxide intermediate. A similar coordination is assumed for copper-dependent phenylalanine 4-monooxygenase (10.8).

10.3   Copper Proteins as Oxidases/Reductases

In addition to Cu-dependent monooxygenases there are several copper-containing 'blue', i.e. type 1 Cu-containing, and also 'nonblue' oxidases without type 1 Cu which convert both atoms of dioxygen to H_2O (blue oxidases) or H_2O_2 (nonblue oxidases). Laccase, ascorbate oxidase and the multifunctional protein ceruloplasmin are blue oxidases while galactose oxidase and amine oxidases belong to the group of nonblue oxidase enzymes. Most of these oxidase enzymes have a complicated structure and

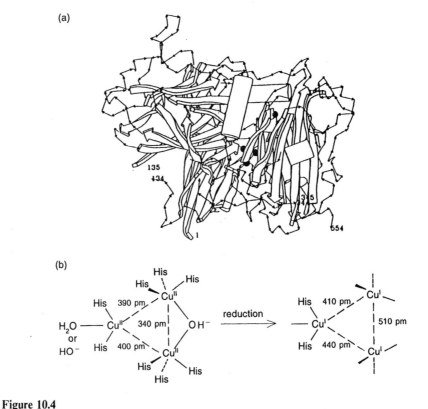

Figure 10.4
(a) Protein folding of the inactive (oxidized) form of ascorbate oxidase from zucchini with specified (●) copper centers. (b) Dimensions and ligands in the oxidized and reduced forms of the tricopper site (from [34])

often contain several types of copper centers; the large protein ceruloplasmin, for example, shows ferroxidase and antioxidation activity (see Section 10.5) even though its primary function in the cytoplasm probably lies in the transport and storage of copper as well as in the regulation and mobilization of iron. The necessity for this mutual control between Cu and Fe can be understood by considering the initially mentioned correspondence (Table 10.1); disorders of the copper storage mechanism may thus lead to anemia as a secondary iron-deficiency symptom.

Structural data for oxidized laccase and, in particular, ascorbate oxidase [34] show that type 2 and type 3 copper centers are in such close proximity to each other (Figure 10.4) that, in effect, a copper trimer with new properties may be formulated [35]. That arrangement seems to favor a four-electron reduction of O_2 to $2 H_2O$ via the following mechanism (10.11 [24]); in return, four oxidation equivalents are thus available for the polyphenolic substrates.

copper oxidation states in the laccase reaction cycle (modified from [24])

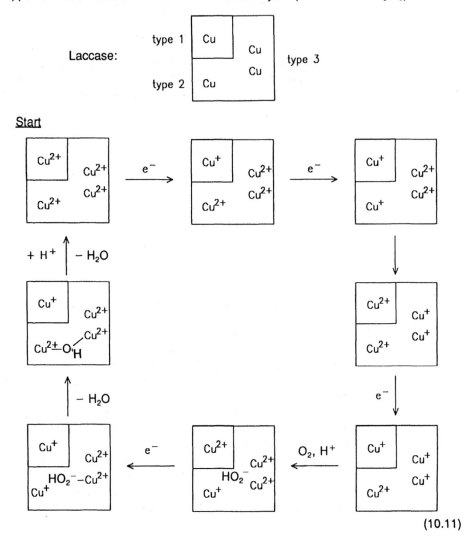

(10.11)

The crystal structure analysis of ascorbate oxidase from zucchini peels shows a copper trimer as well as a separate type 1 Cu center, more than 1200 pm away (Figure 10.4). In the trimer, two metal centers, each coordinated by three histidine residues, are bridged by dissociable hydroxide (inactive oxidized form) while the third, coordinatively unsaturated type 2 copper center is bound to two histidine ligands and one OH^- or H_2O ligand. The structural arrangement already illustrates the function of such enzymes, i.e. the transformation of the $4\,e^-$ oxidation potential of O_2 (trimer as the O_2 coordination site [34]) into separate $1\,e^-$-oxidation equivalents (see 10.11), which are then passed on vectorially to a substrate binding site through the type 1 copper center. In the fully reduced Cu^I state the bridging hydroxide ligand has dissociated, lowering the coordination number while the Cu—Cu distances experience a marked increase (Figure 10.4b). In the oxygenated form a hydroperoxide ion, HO_2^-, is end-on coordinated to one of the oxidized type 3 copper centers [34].

Detailed inspection of the amino acid sequences reveals an increasingly complex development of related copper proteins in the series plastocyanin (1 Cu), ascorbate oxidase (3 Cu) and ceruloplasmin (6 Cu).

The reactivity of nonblue copper-dependent oxidases such as the stereospecific galactose oxidase (10.12) or amine oxidases (10.13) with mainly histidine-coordinated type 2 copper is based on the interaction of the single metal center ($Cu^{I,II}$) with organic redox cofactors. The overall result is a *two-electron reactivity* which is required for the transformation $O_2 \rightarrow H_2O_2$. (Earlier suggestions involving a Cu^I/Cu^{III} transition have not been substantiated.)

$$RR'CHOH + O_2 \quad \xrightarrow{\begin{array}{c}\text{galactose}\\\text{oxidase}\end{array}} \quad RR'C{=}O + H_2O_2 \qquad (10.12)$$

$$RCH_2NH_2 + O_2 + H_2O \quad \xrightarrow{\begin{array}{c}\text{amine}\\\text{oxidase}\end{array}} \quad RCHO + H_2O_2 + NH_3 \qquad (10.13)$$

Galactose oxidase (68 kDa) as isolated from the fungus *Dactylium dendroides* [36] contains a single copper(II) center in a square pyramidal arrangement with two histidines, one special tyrosinate/tyrosyl ligand and the substrate binding site in the equatorial plane and a weakly coordinating tyrosine in the axial position. The equatorial tyrosine is substituted ('covalently linked') with the sulfur atom of a cysteine residue in the *ortho* position to the phenolic oxygen atom and, furthermore, it exhibits a π/π interaction (stacking) with an aromatic tryptophane side chain. The equatorial ligand is assumed to undergo a redox change tyrosyl radical/tyrosinate anion [36] (see Sections 4.3 and 7.6.1) which, in combination with the Cu^I/Cu^{II} transition, serves to effect catalysis of the two-electron process (10.12).

Organic redox cofactors are also present in copper-containing amine oxidases; the amino acid-derived o,p-quinonoid system 6-hydroxydopa quinone ('topaquinone', TPQ) has now been confirmed [37] (see 10.8). The role of the previously invoked pyrroloquinoline quinone (PQQ; see 3.12) is not yet clear in this context; although an established cofactor, PQQ may not be associated with copper-containing proteins. An intraenzymatic electron transfer between Cu^{II}/coenzyme catecholate and Cu^{I}/ coenzyme semiquinone forms has been suggested (10.14 [38]). Amine oxidases have a number of significant physiological functions; among other things, they are important in the metabolism and cross-linking of connective tissue such as collagen.

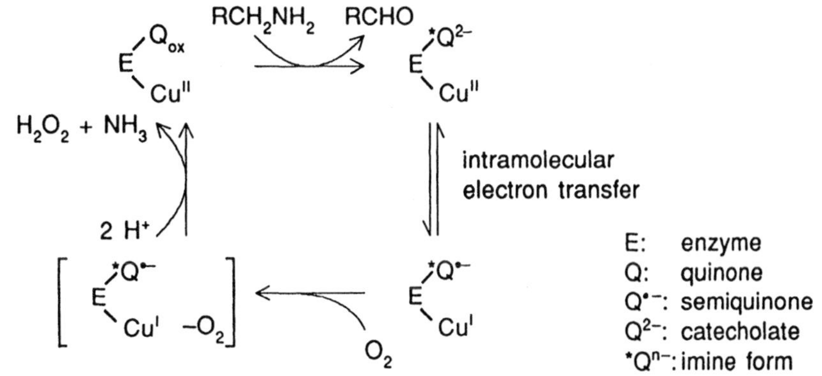

$$(10.14)$$

Copper-containing redox enzymes with quite unusual spectroscopic properties of the metal centers have been isolated from bacteria that are important for the global nitrogen cycle (see Figure 11.1), including nitrite and nitrous oxide (N_2O) reducing bacteria and microorganisms that oxidize ammonia to hydroxylamine (NH_2OH).

$$N_2O + 2\ e^- + 2\ H^+ \xrightarrow{\ \ \substack{N_2O \\ \text{reductase}}\ \ } N_2 + H_2O \qquad (10.15)$$

Some of the copper centers of N_2O reductase from *Pseudomonas stutzeri* (8 Cu, molecular mass 2×71 kDa) have cysteinate anions as ligands in addition to histidine [39]. A similar type of long wavelength-absorbing copper has been observed as component 'Cu_A' of cytochrome c oxidase, which is introduced in the following chapter. It has been suggested that Cu_A comprises a special, e.g. cysteinate-bridged, copper dimer; EPR spectra show the presence of a delocalized mixed-valent Cu^{I}/Cu^{II} species ('$Cu^{1.5}$/$Cu^{1.5}$') [40].

A nitrosyl complex structure Cu^{I}-NO was postulated as an intermediate [41,42] in the denitrification by copper-containing bacterial nitrite reductases to N_2O and N_2 (there are also polyheme-containing forms, see Section 6.5 [42,43]). Only after this suggestion have simple nitrosyl complexes of Cu^{I} been synthesized and structurally characterized [44]. The crystal structure analysis of nitrite reductase from *Achromobacter cycloclastes* shows a trimeric enzyme with 3×2 Cu centers [45], type 1 and 'pseudo type 2' copper centers being approximately 1250 pm apart. Due to their unusual tetrahedral coordination geometry the individual pseudo type 2 copper centers are assumed to be the binding and reduction sites for NO_2^- [45]; they are located at the bottom of solvent-filled channels and each such center is bound to histidine ligands of different protein subunits.

10.4 Cytochrome *c* Oxidase

Cytochrome *c* oxidase, the 'Atmungsferment' of Otto Warburg [46], is a membrane enzyme and as such the site of the last phosphorylation in the respiratory chain (see Figures 6.1 and 6.2). It serves in the transformation of O_2 and H^+ to water and, as a terminal O_2-*consuming* system, represents the counterpart to the dioxygen-*producing* manganese-containing centers in the photosynthetic membrane. The fact that both enzymes are functional and stable only within membranes results from the necessity for a controlled separation, i.e. a vectorial transmembrane transport of e^- and H^+ during redox reactions; however, in both instances the structural characterization of these very complex systems has been hampered by their membrane dependence [46–48].

Cytochrome *c* oxidase (Figure 10.5) is connected to the periplasm and to the bc_1-complex (Figures 6.1 and 6.2) via cytochrome *c*. It is one of the most important but also one of the most complex metalloproteins, containing up to thirteen different subunits (molecular mass > 100 kDa). Until 1985 a metal content of two Cu and two (heme) Fe was assumed; however, more recent refined analyses of enzymes isolated from microorganisms or from beef heart mitochondria point toward a total of three copper centers, two heme a/a_3 iron atoms, one Zn^{2+} and one Mg^{2+} per monomeric protein complex [47]. Since the enzyme is probably associated as a dimer in the

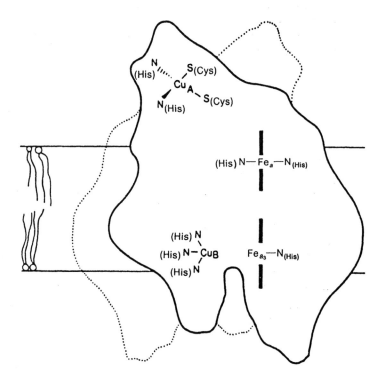

Figure 10.5
Schematic representation of redox centers in the dimeric multiprotein complex cytochrome *c* oxidase of the mitochondrial membrane (modified from [49] with mononuclear Cu_A; see also Figure 6.2)

mitochondrial membrane (Figure 10.5), the metal content of each subunit has not yet been finally defined.

The main subunit (approximately 60 kDa) of the monomeric protein contains the following inorganic components:

(a) one separate cytochrome *a* (metal–metal distances > 1.5 nm) with low-spin iron(III), two axial histidine ligands, low porphyrin symmetry and a high redox potential, and

(b) one complex composed of cytochrome a_3 (high-spin iron) and a copper center (Cu_B) which are antiferromagnetically coupled in the (Fe^{III}, Cu^{II}) oxidized state, $(S = 5/2) - (S = 1/2) \rightarrow (S = 2)$, and which do not show a conventional EPR signal. In the completely reduced form the $S = 2$ spin state remains because of the closed shell of $Cu^I = d^{10}$ and the high-spin configuration of iron(II). Due to its even number of unpaired electrons the latter would be ideally suited for an interaction with triplet dioxygen (spin balance, see Section 5.2). Additionally, the other biochemically relevant O_2-activating center, i.e. nonisolated copper in the +I oxidation state, is also present in the fully reduced 'a_3 complex' of cytochrome *c* oxidase. Unfortunately, no satisfactory structural details are known for the two antiferromagnetically coupled metal centers of the a_3 complex [48]. One histidine residue is assumed for the axial coordination of cyt a_3 with the sixth coordination site remaining empty for the binding of O_2; Cu_B is facing this free position at a distance of less than 500 pm and is probably ligated by three histidine groups (Figure 10.5).

Outside the actual membrane region, subunit II (26 kDa) contains a copper center 'Cu_A' with unusual spectroscopic characteristics in the +II state, viz. a weak long wavelength absorption at 800 nm, and a very small *g* anisotropy and $^{63,65}Cu$ hyperfine coupling in the EPR spectrum. It appears now that this Cu_A site contains a delocalized mixed-valent dimer Cu(1.5)/Cu(1.5) [50], an assumption that is supported by the $^{63,65}Cu$ EPR hyperfine splitting $a_{\|}$ into seven lines [40,51].

Cytochrome *a* and Cu_A have mainly electron-transfer functions at physiologically high potentials; like the four-electron oxidation of two H_2O molecules (4.9), the mechanism of the four-proton/four-electron reduction of O_2 can be formulated as a stepwise process [48]. The mechanism (10.16) has been proposed [48,52] for the catalysis by the cyt a_3/Cu_B complex of cytochrome *c* oxidase.

In the active, two-electron reduced form (reduction via Cu_A and cyt *a*), the a_3 complex contains high-spin iron(II). After coordination of O_2 which has reached this site through diffusion via a channel, the electronic coupling between iron and copper centers effects its rapid inner-sphere reduction to a peroxo ligand. Uptake of one proton and one electron (vectorial e^-/H^+ transport through the membrane, →phosphorylation) can lead to EPR spectroscopically detectable hydroperoxo complexes and, after proton-induced cleavage of the O—O single bond, to the known oxoferryl-heme system with an $S = 1$ state (see Section 6.3). Addition of a second electron causes oxoferryl reduction in this model; both hydroxo metal centers of this Fe^{III}/Cu^{II} resting state can then be stepwise reduced back to the active form under dissociation of water. The four-electron reduction of O_2 proceeds so rapidly that most mechanistic studies had to involve low-temperature quenching techniques or time-resolved spectroscopy [48]. It has not been conclusively determined whether the endergonic proton translocation (10.16) proceeds close to the redox center via

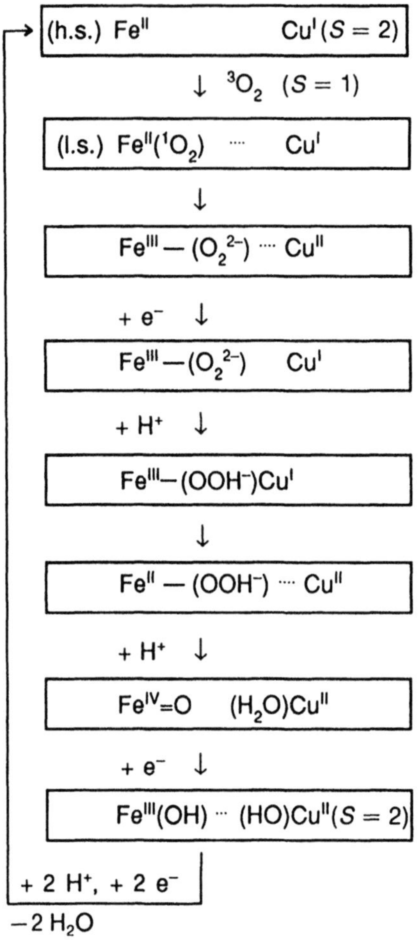

$$\boxed{\text{(h.s.) Fe}^{II} \qquad \text{Cu}^{I}(S=2)}$$

$$\downarrow \quad {}^{3}O_2 \quad (S=1)$$

$$\boxed{\text{(l.s.) Fe}^{II}({}^{1}O_2) \quad \cdots \quad \text{Cu}^{I}}$$

$$\downarrow$$

$$\boxed{\text{Fe}^{III} - (O_2{}^{2-}) \cdots \text{Cu}^{II}}$$

$$+ e^- \quad \downarrow$$

$$\boxed{\text{Fe}^{III} - (O_2{}^{2-}) \quad \text{Cu}^{I}}$$

$$+ H^+ \quad \downarrow$$

$$\boxed{\text{Fe}^{III} - (OOH^-)\text{Cu}^{I}}$$

$$\downarrow$$

$$\boxed{\text{Fe}^{II} - (OOH^-) \cdots \text{Cu}^{II}}$$

$$+ H^+ \quad \downarrow$$

$$\boxed{\text{Fe}^{IV}=O \qquad (H_2O)\text{Cu}^{II}}$$

$$+ e^- \quad \downarrow$$

$$\boxed{\text{Fe}^{III}(OH) \cdots (HO)\text{Cu}^{II}(S=2)}$$

$$+ 2 H^+, + 2 e^-$$

$$- 2 H_2O$$

overall reaction:

$$O_2 + 4 e^- + 4 H^+ \rightarrow 2 H_2O \text{ or}$$

$$O_2 + 4 \text{ cyt } c^{2+} + 8 H^+_{inside} \rightarrow 2 H_2O + 4 \text{ cyt } c^{3+} + 4 H^+_{outside}$$

$$(10.16)$$

acidic/basic ligands or at some distance, triggered by conformational change; the aqua- or hydroxo-ligated Zn^{2+} ion bound to one of the smaller protein subunits may perhaps participate in the proton transfer (see Sections 12.1 and 12.2).

In contrast to water oxidase which is an ${}^{3}O_2$-*releasing* manganese enzyme (Section 4.3), the critical states of the O_2-*consuming* enzyme cytochrome c oxidase feature an *even* number of unpaired electrons which favors the effective binding and ensuing transformation of O_2. Gaseous carbon monoxide competes with O_2 for binding by cytochrome c oxidase as in that other, nonenzymatic protein hemoglobin which transfers O_2 reversibly at the beginning of the respiratory process. Toxicologically more important, however, is the effective inhibition of the small amount of enzymatic cytochrome c oxidase by soluble cyanide which thus blocks the terminal step of respiration by irreversible coordination to the essential metal centers.

10.5 Cu,Zn and Other Superoxide Dismutases: Substrate-Specific Antioxidants

Superoxide dismutases (SODs) catalyze the disproportionation ('dismutation') of toxic $O_2^{\cdot-}$ to O_2 and H_2O_2. This reaction is also catalyzed by many nonenzymatic transition metal species, albeit in a less controlled fashion; as a rule, 'free' transition metal ions are physiologically undesirable. The hydrogen peroxide formed in this process can further disproportionate to yield O_2 and H_2O (see 6.12) in reactions catalyzed by catalases or it can be utilized via peroxidase enzymes (see Sections 6.3, 11.4 and 16.8). In addition to the Cu,Zn-containing superoxide dismutase from the cytoplasm of eucaryotes there are other forms which contain iron (bacterial and plant SODs) or manganese (SODs from mitochondria, bacteria [53,54]). Iron- and manganese-containing superoxide dismutases have been structurally characterized [55,56]; the M^{III} centers are coordinated in a trigonal bipyramidal fashion by three histidine residues, one η^1-aspartate and one solvent ligand.

Considering the metastability and reactivity of the radical anion $O_2^{\cdot-}$ there seems to be no need for a special catalyst specificity of superoxide dismutases; in essence, oxygen-affine metal cations are required which can easily change their oxidation number by one unit. In fact, O_2-activating centers such as the hemes, type 3 copper proteins, nonheme iron dimers, copper/cofactor complexes or cytochrome c oxidase contain *two* interacting redox centers (2 metals, or 1 metal + 1 π system) to circumvent just that one-electron reduction of O_2 to $O_2^{\cdot-}$; however, these measures cannot completely prevent the production of small quantities of superoxide $O_2^{\cdot-}$ or its conjugated acid $HO_2^{\cdot}$ (5.2) through 'leakage' of one-electron equivalents [53]. Under physiological circumstances the dismutation of $O_2^{\cdot-}$ must proceed very rapidly, i.e. at the diffusion limit, in order to prevent uncontrolled oxidations by this radical anion or its follow-up products in reactions with transition metal ions (see 3.11 and 4.6). A major requirement for SODs is their resistance toward the aggressive substrate, $O_2^{\cdot-}$, *and* towards the products, O_2 and O_2^{2-}.

The relatively small (2×16 kDa) Cu,Zn SOD from erythrocytes, previously also known as 'erythrocuprein', is structurally well characterized [57]; each subunit contains one copper and one zinc ion bridged by a deprotonated, resonance-stabilized imidazolate ring (10.17) of a histidine side chain. The other amino acid ligands are three His (Cu) and two His and one Asp^- (Zn), respectively. Compared to a regular tetrahedron, the geometry is more severely distorted at the copper site than at the zinc center (Figure 10.6). An additional coordination site for $O_2^{\cdot-}$ is thus created at the catalytically active copper center ($\rightarrow$square pyramidal arrangement; see Figure 10.7) which may be temporarily occupied by labile H_2O.

The detailed mechanism of the dismutation reaction, in particular the function of the Zn^{2+} ion and the imidazolate, is still being disputed; the essential part is the ability of the redox-active metal center (here Cu) to oxidize metastable superoxide in one state and to reduce it in the other oxidation state. According to (10.17) and (10.18) it is assumed that $O_2^{\cdot-}$ is oxidized to O_2 by the Cu^{II} species (1) (SODs must *not* get oxidatively attacked by O_2!); the now-reduced copper(I) center can then be replaced at the imidazole ring by a proton, resulting in a normal, geometrically relaxed zinc complex (3) with histidine. Next, the coordinatively unsaturated copper(I) which is still anchored in the protein can be oxidized by hydrogen-bonded superoxide anion (4). The thus formed basic (hydro)peroxide (5.2) is then protonated, e.g. by the imidazole ring of still Zn-coordinated histidine, and transformed into H_2O_2. The

Figure 10.6
Structure of
the dimetal center
in Cu,Zn SOD of
bovine erythrocytes
(according to [57])

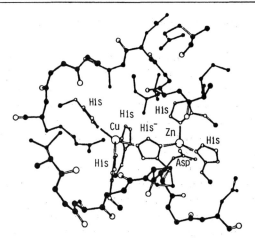

driving force for this reaction would lie in the affinity of copper(II) for the imidazole nitrogen of histidine. Removal of the Zn^{2+} ion seems to cause only a small decrease in the enzymatic activity; the role of this metal ion seems to lie in structural stabilization. On the other hand, the high concentration of Cu,Zn superoxide dismutase in erythrocytes has led to the hypothesis that it might be primarily a metal storage protein and that its SOD activity is only a secondary function.

The very rapid, almost diffusion-controlled reaction of the enzyme with $O_2^{\cdot-}$, i.e. the successful conversion with nearly every collision of the reactants, is strongly assisted by electrostatic interactions which lead the small monoanion $O_2^{\cdot-}$ into the protein through a funnel-shaped, approximately 1–2 nm deep channel. Near the catalytic site, it can be additionally positioned by the positively charged guanidinium group of an arginine residue which also offers the possibility of hydrogen bonding [58] (10.17: 2, 4 and Figure 10.7). Directed mutations can influence the field gradient inside the channel and thus even further increase the SOD activity [58]. Small halide (F^-) and pseudohalide ligands such as cyanide, CN^-, azide, N_3^-, or thiocyanate, NCS^-, can compete with $O_2^{\cdot-}$ for the binding to SOD [59].

Figure 10.7
Schematic
representation of
the cavity of $O_2^{\cdot-}$
conversion in
Cu,Zn SOD
(according to [59])

Catalysis cycle for Cu,Zn-superoxide dismutase

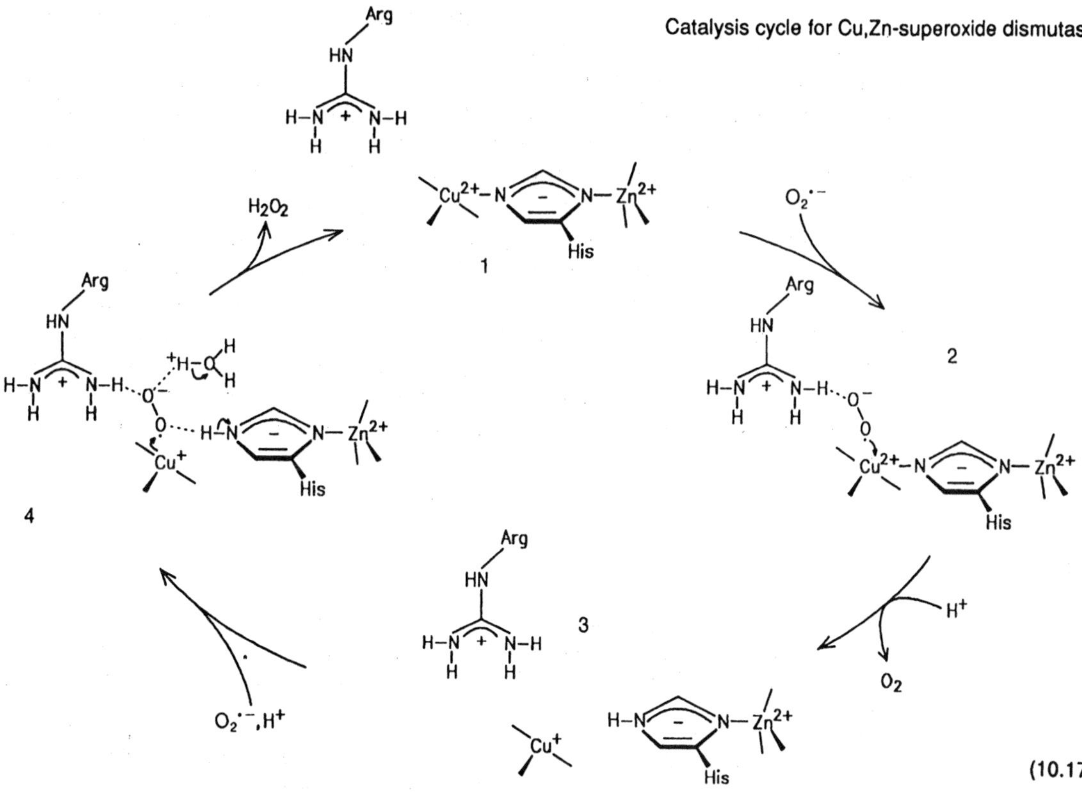

(10.17)

Following the description of H_2O_2-converting heme peroxidases in Section 6.3, the enzymes for the degradation of the other metastable and toxic reduction product of dioxygen, i.e. $O_2^{\cdot-}$, have been introduced in this chapter. A third group of biological antioxidants aimed at the detoxification of the highly reactive hydroxyl radical will be presented in Section 16.8 and Table 16.1.

hypothetical mechanism:

$$Zn\text{-}(Im^-)\text{-}Cu^{II} + O_2^{\cdot-} \rightarrow Zn\text{-}(Im^-)\text{-}Cu^{I} + O_2$$

$$Zn\text{-}(Im^-)\text{-}Cu^{I} + H^+ \rightarrow Zn\text{-}(ImH) + Cu^{I}$$

$$Cu^{I} + O_2^{\cdot-} \rightarrow Cu^{II\cdots}O_2^{2-}$$

$$Cu^{II\cdots}O_2^{2-} + H^+ + Zn\text{-}(ImH) \rightarrow Zn\text{-}(Im^-)\text{-}Cu^{II} + H_2O_2$$

ImH: imidazole ring of a histidine side chain
Zn: three-coordinate Zn^{II} center (2 His, 1 Asp$^-$)
Cu: three-coordinate copper center (3 His)

overall reaction:

$$
\begin{array}{ccccc}
(-0.5) & & & (-I) & (0) \quad \text{(oxygen oxidation states)} \\
2\,O_2^{\cdot-} + 2\,H^+ & \xrightarrow{\;SOD\;} & & H_2O_2 & + \; O_2
\end{array}
$$

(10.18)

As in the enzymatic degradation of peroxide ($\rightarrow$substrate oxidation) and in the quintessential 'environmental catastrophe' of biogenic O_2 evolution ($\rightarrow$respiration), the apparently toxic natural product $O_2^{\cdot-}$ can be utilized and even deliberately produced by organisms for special purposes. Phagocytes (neutrophils) which are essential for the immune system of higher organisms produce large amounts of superoxide and follow-up products like H_2O_2 or ClO^- with the help of a 'respiratory burst oxidase', in order to kill invading microorganisms [60]. However, this potent defense system may malfunction, giving rise to autoimmune diseases such as rheumatoid arthritis. In such cases, superoxide dismutase can be administered as an anti-inflammatory drug which is being produced in rather large quantities. A superoxide dismutase therapy is also indicated after exposure to ionizing radiation which mainly produces oxygen-containing radicals [61] (see Section 18.1); connections between SOD activity and the rate of aging [62] and the role of defect mutants for neurodegenerative diseases such as amylotrophic lateral sclerosis [5] are under investigation.

References

1. E.-I. Ochiai, Iron versus copper, *J. Chem. Educ.*, **63**, 942 (1986).
2. R. J. P. Williams, The copper and zinc triads in biology, in *The Chemistry of Copper and Zinc Triads* (eds. A. J. Welsh and S. K. Chapman), Royal Society of Chemistry, Cambridge, 1993, p. 1.
3. M. C. Linder and C. A. Goode, *Biochemistry of Copper*, Plenum Press, New York, 1991.
4. K. Davies, Cloning the Menkes disease gene, *Nature (London)*, **361**, 98 (1993).
5. H.-X. Deng *et al.*, Amyotrophic lateral sclerosis and structural defects in Cu,Zn superoxide dismutase, *Science*, **261**, 1047 (1993).
6. C. F. Mills, Trace element deficiency and excess in animals, *Chem. Br.*, **15**, 512 (1979).
7. A. Müller, E. Diemann, R. Jostes and H. Bögge, Transition metal thio anions: properties and significance for complex chemistry and bioinorganic chemistry, *Angew. Chem. Int. Ed. Engl.*, **20**, 934 (1981).
8. R. Lontie, (ed.), *Copper Proteins and Copper Enzymes*, Vols. I to III, CRC Press, Boca Raton, 1984.
9. E. I. Solomon, M. J. Baldwin and M. D. Lowery, Electronic structures of active sites in copper proteins: contributions to reactivity, *Chem. Rev.*, **92**, 521 (1992).
10. I. J. Pickering, G. N. George, C. T. Dameron, B. Kury, D. R. Winge and I. G. Dance, X-ray absorption spectroscopy of cuprous-thiolate clusters in proteins and model systems, *J. Am. Chem. Soc.*, **115**, 9498 (1993).
11. M. Symons, *Chemical and Biochemical Aspects of Electron-Spin Resonance Spectroscopy*, van Nostrand Reinhold, New York, 1978.
12. S. E. Shadle, J. E. Penner-Hahn, H. J. Schugar, B. Hedman, K. O. Hodgson and E. I. Solomon, X-ray absorption spectroscopic studies of the blue copper site: metal and ligand K-edge studies to probe the origin of the EPR hyperfine splitting in plastocyanin, *J. Am. Chem. Soc.*, **115**, 767 (1993).
13. M. M. Werst, C. E. Davoust and B. M. Hoffman, Ligand spin densities in blue copper proteins by Q-band ^1H and ^{14}N ENDOR spectroscopy, *J. Am. Chem. Soc.*, **113**, 1533 (1991).
14. J. M. Guss, H. D. Bartunik and H. C. Freeman, Accuracy and precision in protein structure analysis: restrained least-squares refinement of the structure of poplar plastocyanin at 1.33 Å resolution, *Acta Cryst.*, **B48**, 790 (1992).
15. G.E. Norris, B. F. Anderson and E. N. Baker, Blue copper proteins. The copper site in azurin from *Alcaligenes denitrificans*, *J. Am. Chem. Soc.*, **108**, 2784 (1986).

16. N. Kitajima, K. Fujisawa, M. Tanaka and Y. Moro-oka, X-ray structure of thiolato-copper(II) complexes bearing close spectroscopic similarities to blue copper proteins, *J. Am. Chem. Soc.*, **114**, 9232 (1992).

17. W. E. B. Shepard, B. F. Anderson, D. A. Lewandoski, G. E. Norris and D. N. Baker, Copper coordination geometry in azurin undergoes minimal change on reduction of copper(II) to copper (I), *J. Am. Chem. Soc.*, **112**, 7817 (1990).

18. P. K. Bharadwaj, J. A. Potenza and H. J. Schugar, Characterization of [dimethyl-*N,N'*-ethylenebis(L-cysteinato)(2-)-*S,S'*]copper(II), Cu(SCH$_2$CH(CO$_2$CH$_3$)NHCH$_2$—)$_2$, a stable Cu(II)-aliphatic dithiolate, *J. Am. Chem. Soc.*, **108**, 1351 (1986).

19. M. L. Brader and M. F. Dunn, Insulin stabilizes copper(II)-thiolate ligation that models blue copper proteins, *J. Am. Chem. Soc.*, **112**, 4585 (1990).

20. A. S. Klemens, D. R. McMillin, H. T. Tsang and J. E. Penner-Hahn, Structural characterization of mercury-substituted copper proteins. Results from X-ray absorption spectroscopy, *J. Am. Chem. Soc.*, **111**, 6398 (1989).

21. K. D. Karlin and Y. Gultneh, Binding and activation of molecular oxygen by copper complexes, *Prog. Inorg. Chem.*, **35**, 219 (1987).

22. K. M. Merz and R. Hoffmann, d^{10}–d^{10} Interactions: multinuclear copper(I) complexes, *Inorg. Chem.*, **27**, 2120 (1988).

23. K. A. Magnus, H. Ton-That and J. E. Carpenter, *Chem Rev.*, **94**, 727 (1994).

24. J.-M. Latour, Les sites actifs binucléaires des protéines à cuivre, *Bull. Soc. Chim. Fr.*, 508 (1988).

25. T. N. Sorrell, Synthetic models for binuclear copper proteins, *Tetrahedron*, **45**, 3 (1989).

26. K. D. Karlin, Z. Tyeklar, A. Farooq, M. S. Haka, P. Ghosh, R. W. Cruse, Y. Gultneh, J. C. Hayes, P. J. Toscano and J. Zubieta, Dioxygen–copper reactivity and functional modeling of hemocyanins. Reversible binding of O$_2$ and CO to dicopper(I) complexes [Cu$^I_2$(L)]$^{2+}$ (L = dinucleating ligand) and the structure of a bis(carbonyl) adduct, [Cu$^I_2$(L)(CO)$_2$]$^{2+}$, *Inorg. Chem.*, **31**, 1436 (1992).

27. M. J. Baldwin, D. E. Root, J. E. Pate, K. Fujisawa, N. Kitajima and E. I. Solomon, Spectroscopic studies of side-on peroxide-bridged binuclear copper(II) model complexes of relevance to oxyhemocyanin and oxytyrosinase, *J. Am. Chem. Soc.*, **114**, 10421 (1992).

28. N. Kitajima, Y. Moro-oka, μ–η^2; η^2-Peroxide in biological systems, *J. Chem. Soc. Dalton Trans.*, 2665 (1993).

29. C. A. Reed, Hemocyanin cooperativity: a copper coordination chemistry perspective, in *Biological and Inorganic Copper Chemistry*, (eds. K. D. Karlin and J. Zubieta), Vol. I, Adenine Press, Guilderland, 1985, p. 61.

30. M. G. Peter, Chemical modification of biopolymers with quinones and 'quinone' methides, *Angew. Chem. Int. Ed. Engl.*, **28**, 555 (1989).

31. M. G. Peter and H. Förster, The structure of 'eumelanins': identification of composition patterns by solid-state NMR spectroscopy, *Angew. Chem. Int. Ed. Engl.*, **28**, 741 (1986).

32. J. S. Sidwell and G. A. Rechnitz, 'Bananatrode'—an electrochemical biosensor for dopamine, *Biotechnol. Lett.*, **7**, 419 (1985).

33. E. W. Ainscough, A. M. Brodie and A. L. Wallace, Ethylene—an unusual plant hormone, *J. Chem. Ed.*, **69**, 315 (1992).

34. A. Messerschmidt, H. Luecke and R. Huber, X-ray structures and mechanistic implications of three functional derivatives of ascorbate oxidase from zucchini, *J. Mol. Biol.*, **230**, 997 (1993).

35. J. L. Cole, P. A. Clark and E. I. Solomon, Spectroscopic and chemical studies of the laccase trinuclear copper active site: geometric and electronic structure, *J. Am. Chem. Soc.*, **112**, 9548 (1990).

36. N. Ito, S. E. V. Phillips, C. Stevens, Z. B. Ogel, M. J. McPherson, J. N. Keen, K. D. S. Yadav and P. F. Knowles, Three dimensional structure of galactose oxidase: an enzyme with a built-in secondary cofactor, *Faraday Discuss.*, **93**, 75 (1992).

37. J. Duine, Quinoproteins: enzymes containing the quinonoid cofactor pyrroloquinoline quinone, topaquinone or tryptophan-tryptophan quinone, *Eur. J. Biochem.*, **200**, 271 (1991).

38. D. M. Dooley, M. A. McGuirl, D. E. Brown, P. N. Turowski, W. S. McIntire and P. F.

Knowles, A Cu(I)-semiquinone state in substrate-reduced amine oxidases, *Nature (London)*, **349**, 262 (1991).

39. H. Jin, H. Thomann, C. L. Coyle and W. G. Zumft, Copper coordination in nitrous oxide reductase from *Pseudomonas stutzeri*, *J. Am. Chem. Soc.*, **111**, 4262 (1989).

40. P. M. H. Kroneck, W. E. Antholine, D. H. W. Kastrau, G. Buse, G. C. M. Steffens and W. G. Zumft, Multifrequency EPR evidence for a bimetallic center at the Cu_A site in cytochrome *c* oxidase, *FEBS Lett.*, **268**, 274 (1990).

41. C. L. Hulse, B. A. Averill and J. M. Tiedje, Evidence for a copper-nitrosyl intermediate in denitrification by the copper-containing nitrite reductase of *Achromobacter cycloclastes*, *J. Am. Chem. Soc.*, **111**, 2322 (1989).

42. T. Brittain, R. Blackmore, C. Greenwood and A. J. Thomson, Bacterial nitrite-reducing enzymes, *Eur. J. Biochem.*, **209**, 793 (1992).

43. P. M. H. Kroneck, J. Beuerle and W. Schumacher, Metal-dependent conversion of inorganic nitrogen and sulfur compounds, in *Metal Ions in Biological Systems* (ed. H. Sigel), Vol. 28, Marcel Dekker, New York, 1992, p. 455.

44. S. M. Carrier, C. E. Ruggiero and W. B. Tolman, Synthesis and structural characterization of a mononuclear copper nitrosyl complex, *J. Am. Chem. Soc.*, **114**, 4407 (1992).

45. J. W. Godden, S. Turley, D. C. Teller, E. T. Adman, M. Y. Liu, W. J. Payne, J. LeGall, The 2.3 Angstrom X-ray structure of nitrite reductase from *Achromobacter cycloclastes*, *Science*, **253**, 438 (1991).

46. H. Beinert, From indophenol oxidase and atmungsferment to proton pumping cytochrome oxidase aa_3 $Cu_A Cu_B(Cu_C?)ZnMg$, *Chem. Scr.*, **28A**, 35 (1988).

47. G. C. M. Steffens, T. Soulimane, G. Wolff and G. Buse, Stoichiometry and redox behaviour of metals in cytochrome-c oxidase, *Eur. J. Biochem.*, **213** 1149 (1993).

48. G. T. Babcock and M. Wikström, Oxygen activation and the conservation of energy in cell respiration, *Nature (London)*, **356**, 301 (1992).

49. S. I. Chan and P. M. Li, Cytochrome *c* oxidase: understanding nature's design of a proton pump, *Biochemistry*, **29**, 1 (1990).

50. B. G. Malmström and R. Aasa, The nature of the Cu_A center in cytochrome *c* oxidase, *FEBS*, **325**, 49 (1993).

51. W. E. Antholine, D. H. W. Kastrau, G. C. M. Steffens, G. Buse, W. G. Zumft and P. M. H. Kroneck, A comparative EPR investigation of the multicopper proteins nitrous-oxide reductase and cytochrome *c* oxidase, *Eur. J. Biochem.*, **209**, 875 (1992).

52. C. Varotsis, Y. Zhang, E. H. Appelman and G. T. Babcock, Resolution of the reaction sequence during the reduction of O_2 by cytochrome oxidase, *Proc. Natl. Acad. Sci. USA*, **90**, 237 (1993).

53. I. Fridovich, Superoxide dismutases, *J. Biol. Chem.*, **264**, 7761 (1989).

54. A. E. G. Cass, Superoxide dismutases, in *Metalloproteins* (ed. P. Harrison), Part 1, Verlag Chemie, Weinheim, 1985, p. 121.

55. W. C. Stallings, K. A. Pattridge, R. K. Strong and M. L. Ludwig, The structure of manganese superoxide dismutase from *Thermus thermophilus* HB8 at 2.4-Å resolution, *J. Biol. Chem.*, **260**, 16424 (1985).

56. G. E. O. Borgstahl, H. E. Parge, M. J. Hickey, W. F. Beyer, Jr, R. A. Hallewell and J. A. Tainer, The structure of human mitochondrial manganese superoxide dismutase reveals a novel tetrameric interface of two 4-helix bundles, *Cell*, **71**, 107 (1992).

57. J. A. Tainer, E. D. Getzoff, J. S. Richardson and D. C. Richardson, Structure and mechanism of copper, zinc superoxide dismutase, *Nature (London)*, **306**, 284 (1983).

58. E. D. Getzoff, D. E. Cabelli, C. L. Fisher, H. E. Parge, M. S. Viezzoli, L. Banci and R. A. Hallewell, Faster superoxide dismutase mutants designed by enhancing electrostatic guidance, *Nature (London)*, **358**, 347 (1992).

59. I. Bertini, L. Banci, M. Piccioli and C. Luchinat, Spectroscopic studies on Cu_2Zn_2SOD: a continuous advancement on investigation tools, *Coord. Chem. Rev.*, **100**, 67 (1990).

60. A. W. Segal and A. Abo, The biochemical basis of the NADPH oxidase of phagocytes, *Trends Biochem. Sci.*, **18**, 43 (1993).

61. J. R. J. Sorenson, Copper complexes as 'radiation recovery' agents, *Chem. Br.*, **25**, 169 (1989).

62. R. L. Rusting, Why do we age?, *Sci. Am.*, **267**(6), 88 (1992).

11 Biological Functions of the 'Early' Transition Metals: Molybdenum, Tungsten, Vanadium and Chromium

In contrast to 'late' transition metals such as cobalt, nickel or copper, the metals from the left half of the transition element periods (see Figure 1.3) favor high oxidation states, high coordination numbers, and 'hard', in particular negatively charged oxygen donor coordination centers under physiological conditions. The resulting oxo or hydroxo complexes are thus often negatively charged, a fact that is very important with regard to physiological uptake and mobilization mechanisms. Whereas scandium and titanium at the beginning of the first (3d) row of the transition metals do not have any biochemical relevance, vanadium and chromium and the heavier homologues molybdenum and tungsten exhibit quite differentiated physiological functions. Molybdenum is the biologically most important element in this series; its chemistry and enzymatic function [1–4] in oxygen transfer and nitrogen fixation will therefore be described in more detail.

11.1 Oxygen Transfer through Tungsten- or Molybdenum-Containing Enzymes

11.1.1 Overview

The essential character of molybdenum for life processes has been known for some time from dietetics and the agricultural sciences. From the chemical point of view, however, this distinction is quite peculiar; according to current knowledge, molybdenum is the only element from the second (4d) period of the transition metals with a biological function. One explanation certainly lies in its bioavailability. Although molybdenum is quite rare in the earth's crust (Figure 2.2) like all other heavy metals from this part of the periodic table, it is quite soluble in (sea) water at pH 7 (approximately 100 mM, see Figure 2.2) in its most stable, hexavalent form as molybdate(VI), MoO_4^{2-}. This ion shows a close resemblance to the biologically important sulfur-transporting sulfate ion, SO_4^{2-}, and the two ions are also structurally very similar. In contrast to molybdate(VI), the oxometalates, MO_n^{m-}, of the heavier homologue tungsten and especially of those metals to the left in the periodic system, like niobium, tantalum, zirconium and hafnium, are nearly insoluble at pH 7 due to aggregation; on the other hand, the 4d and 5d elements to the right of Mo and W in the periodic system, i.e. technetium (see Section 18.3.2), rhenium and the platinum metals, are obviously too rare to have any biological significance.

In 1983 a formate ($HCOO^-$) dehydrogenase isolated from *Clostridium thermo-aceticum* was described which, in addition to iron, sulfur and selenium (see Section 7.4 and 16.8), contained tungsten instead of molybdenum as the active component [5] (see Table 11.1). Since then several other tungsten-incorporating microorganisms have been discovered [6], in particular thermophilic ($\approx 65\ °C$ [7]) and hyperthermophilic archaebacteria ($\approx 100\ °C$ [8]). The latter were found near the hydrothermal vents which exist on the sea floor along the middle oceanic ridges; these organisms exhibit a chemolithotrophic metabolism involving sulfide oxidation and CO_2 reduction to lead a sunlight-independent life. Analogous tungsten and molybdenum enzymes are very similar with regard to composition and oxygen-transfer function; however, in agreement with common inorganic chemical experience the tungsten-containing enzymes require higher temperatures (Mo enzymes as adaptation to the decreasing surface temperatures on earth?) and a more negative redox potential for transitions between corresponding hexavalent, pentavalent and tetravalent forms.

Table 11.1 Some molybdenum-containing hydroxylases and correspondingly catalyzed reactions

enzyme	molecular mass (kDa)	prosthetic groups	typical functions
xanthine oxidase	275 (dimer)	2 Mo, 4 Fe_2S_2, 2 FAD	oxidation of xanthine to uric acid in liver and kidney (11.12, 11.13)
nitrate reductase	228 (dimer)	2 Mo, 2 cyt b, 2 FAD	nitrate/nitrite transformation in plants and microorganisms (11.2): $$NO_3^- + 2\ H^+ + 2\ e^- \rightleftharpoons NO_2^- + H_2O$$
aldehyde oxidase	280 (dimer)	2 Mo, 4 Fe_2S_2, 2 FAD	oxidation of aldehydes, heterocycles, amines and sulfides in liver
sulfite oxidase	110 (dimer)	2 Mo, 2 cyt b	sulfite/sulfate transformation in liver (sulfite detoxification, 11.5): $$SO_3^{2-} + H_2O \rightleftharpoons SO_4^{2-} + 2\ e^- + 2\ H^+$$
arsenite oxidase	85	1 Mo, Fe_nS_n	transformation of thiolate-blocking AsO_2^- by microorganisms $$AsO_2^- + 2\ H_2O \rightleftharpoons AsO_4^{3-} + 2\ e^- + 4\ H^+$$
formate de-hydrogenase (Mo)	>100	Mo, Fe_nS_n, Se	CO_2 reduction by microorganisms $$HCOO^- \rightleftharpoons CO_2 + 2\ e^- + H^+$$
formate de-hydrogenase (W)	340	W, Fe_nS_n, Se	CO_2 reduction by microorganisms $$HCOO^- \rightleftharpoons CO_2 + 2\ e^- + H^+$$

In contrast, the chromate(VI) ion, CrO_4^{2-}, which is analogous to MoO_4^{2-} and also quite soluble at pH 7, behaves as a strong oxidant and is thus unstable under physiological conditions, in further agreement with the well-established rule that the heavier homologues in a transition metal group favor higher oxidation states. In fact, CrO_4^{2-} is mutagenic and carcinogenic, as will be pointed out in Section 17.8.

In addition to its bioavailability, a useful function has to be performed by a metal such as molybdenum or tungsten in order to make it an essential element. The molybdoenzymes [9] constitute another class of hydroxylases (oxidases) besides iron- and copper-containing proteins (see Table 10.1). Additionally, an 'FeMo cofactor' plays an essential role in the main form of the dinitrogen (N_2) fixating nitrogenase enzymes (see Section 11.2).

11.1.2 Enzymes Containing the Molybdopterin Cofactor

For oxidation states lower than $+VI$ the chemistry of molybdenum in aqueous solution is characterized by aggregation phenomena, i.e. by the formation of 'clusters'. As is shown in (11.1), dimeric and trimeric ionic systems are formed with the metals being connected by hydroxo or oxo bridges and coordinatively saturated by water ligands (from EXAFS measurements [10]).

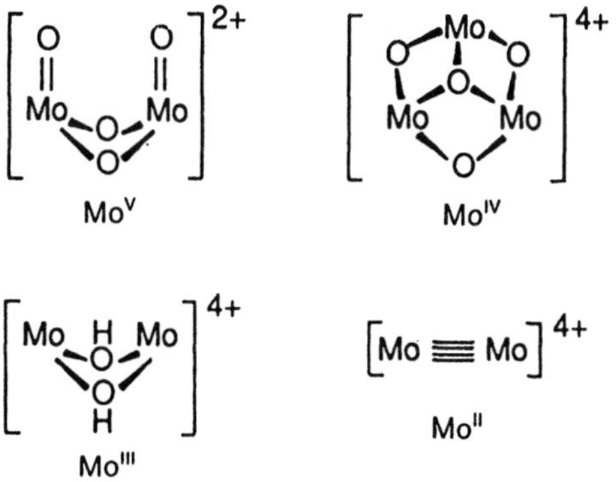

$$(11.1)$$

Such aggregation can be suppressed in the presence of a protein as a multidentate and protecting chelate system or under utilization of special cofactor ligands with the result that—as with many other metals—a rather 'unusual' biological coordination chemistry of these oxidation states results. The physiologically relevant oxidation states of molybdenum lie between $+IV$ and $+VI$ and the corresponding redox potentials of about -0.3 V are in the physiologically acceptable range, in contrast to those of homologous chromium complexes. In these oxidation states, molybdenum shows comparable affinity toward negatively charged O *and* S ligands such as oxide, sulfide, thiolates or hydroxide; nitrogen ligands are also bound quite well. One of the most important biological functions of molybdenum in enzymes is the catalysis of controlled oxygen transfer to a two-electron substrate, using spatially separated one-electron transferring compounds such as cytochromes, Fe/S centers or flavins. A coupling of electron transfer and oxide exchange leads to the direct formal transfer of an oxygen *atom* from the metal center to the substrate or vice versa (oxotransferase activity $LMo^{VI}O_2 + X \rightleftharpoons LMo^{IV}O + XO$, see 11.2 [11,12]). The transferred oxygen does not originate from O_2 (oxygenation), as is often the case with Fe- and Cu-containing hydroxylases (oxygenases) (Table 10.1); the result is a temporal and spatial separation of the oxidative electron transfer and the actual oxygen translocation. Regeneration of the reduced enzyme may proceed via O_2 as eventual oxidant, a process that yields peroxides or superoxide, as is well known, for example, from xanthine oxidase.

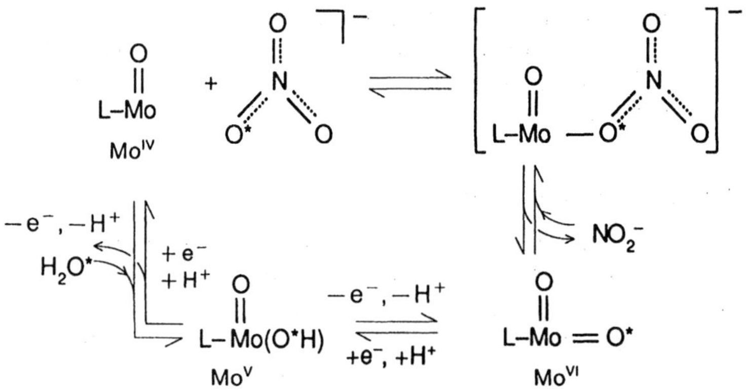

L : ligands in the coordination sphere of molybdenum
O^*: ^{18}O labelling

(11.2)

'Oxidation'

Within the biochemical terminology, oxidation either means electron loss ($\rightarrow$oxidase or oxidoreductase enzymes), elimination of hydrogen ($\rightarrow$dehydrogenase enzymes), or introduction of oxygen ($\rightarrow$oxygenase or hydroxylase enzymes). For O_2-dependent oxygenases, a distinction is made between mono- and dioxygenases (see Table 10.3) and, mechanistically, 'oxygen transfer' can proceed in sequential form: $O = O^{\cdot -} - e^-$ (P-450 systems) $= O^{2-} - 2\,e^-$ (Mo enzymes).

Several molybdenum-dependent hydroxylases are known ([9,13]) (Table 11.1), the best characterized of those being xanthine oxidase, sulfite oxidase and nitrate reductase (see 11.2). Other Mo-containing enzymes also have very important biochemical functions: aldehyde oxidases participate in the alcohol metabolism, various nitrogen heterocycles become C—H oxygenated in the α position to N, and other varieties of Mo enzymes catalyze the 'oxygen atom' transfer in amine/amine oxide, arsenite/arsenate (Section 16.4) or sulfide/sulfoxide systems. The latter reaction plays an important role in the conversion of D-biotin-5-oxide to the actual coenzymatic biotin (vitamin H) as well as in the transformation of dimethylsulfoxide (DMSO) to dimethylsulfide which is highly important for the global sulfur cycle and climate [14]. Formate and formylmethanofuran dehydrogenases containing molybdenum or tungsten are essential for the C_1 metabolism ($CO_2 \rightarrow \rightarrow \rightarrow CH_4$) of microorganisms. In general terms, molybdoenzymes catalyze the reactions depicted in (11.3).

$$\begin{array}{ccc} \diagdown\text{C--H} & \rightarrow & \diagdown\text{C--OH} \\ \diagup & & \diagup \end{array}$$

$$R_nE| \quad \rightarrow \quad R_nE{\rightarrow}O$$

$$E = N,\ S,\ As \tag{11.3}$$

Malfunctions of molybdoenzymes in higher organisms are known and are manifested, for example, in problems of the purine metabolism ($\rightarrow$gout, dysfunction of xanthine oxidase) or in neurological disorders (sulfite oxidase dysfunction [15]). The Cu/MoS antagonism known from cattle breeding has already been mentioned briefly in Chapter 10. Tetrathiomolybdates(VI) which are formed in the complex stomach of ruminants from molybdate and electron-rich sulfide do not only exhibit an intense color (low-energy LMCT transition $S^{-II} \rightarrow Mo^{+VI}$) but also act as efficient chelating ligands for positively charged albeit π electron-rich ('soft') metal ions such as Cu^+ (11.4). The thus caused secondary deficiency of copper (which is necessary for the functioning of collagenases, Section 10.3) can be responsible for a weakening of connective tissue in animals that graze on molybdenum-rich soil.

$$MoS_4^{2-} \quad : \qquad \begin{array}{c} S \diagdown \quad S^- \\ \diagup Mo \diagdown \\ S \diagup \quad S^- \end{array} \qquad\qquad Cu^+$$

soft nucleophile (sulfide centers)   soft electrophile (1+ charge)
π electron acceptor (Mo^{VI}, d^0 configuration) π electron donor (d^{10} configuration)

$$\tag{11.4}$$

In biological oxidations that are not directly dependent on O_2, the oxidation equivalents are made available in one-electron steps via electron-transfer proteins. Therefore, almost all molybdoenzymes contain electron-transfer components such as cytochromes, Fe/S centers or flavins (Table 11.1). However, the relatively large proteins have not yet been structurally characterized so that definitive mechanisms with regard to the interaction of these components cannot yet be formulated. Scheme (11.5) shows a functional catalytic cycle for the oxidation of sulfite to sulfate where d^1-configurated Mo^V is observed as EPR-detectable intermediate. Oxide ligands either become protonated after reduction due to their strongly increased basicity or they will be replaced by hydroxide from the surrounding water after complete oxygen atom transfer.

Due to the absence of unambiguous structural data, EXAFS measurements at the Mo absorption edge of approximately 20 keV and EPR studies of Mo^V intermediates have mainly been used to determine the coordination environment of the metal. In this context, the experiment of Bray and Meriwether [16] from 1966 has become famous: a Frisian cow was injected with ^{95}Mo-enriched molybdate in order to allow a better EPR detection of the coupling between the unpaired electron and the nuclear spin $I = 5/2$ of ^{95}Mo in the Mo^V form of xanthine oxidase as isolated from the cow's milk. From such experiments and from the observation of three accessible metal oxidation states the participation of Mo^{VI}, Mo^V and Mo^{IV} in the enzymatic reactions was deduced.

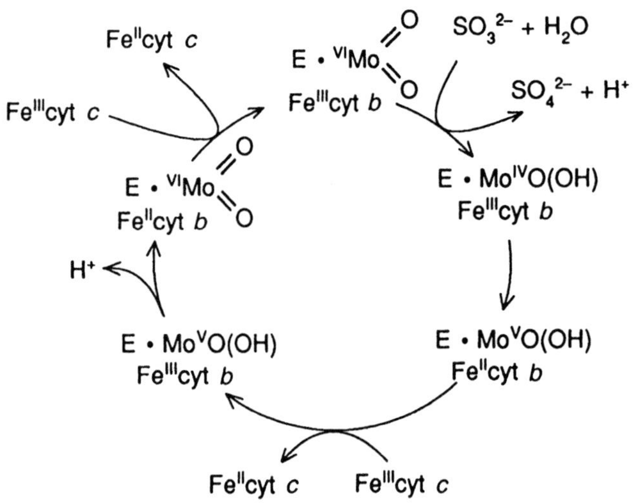

E: (apo)enzyme
Fe cyt: cytochromes with iron oxidation states   (11.5)

In the 1980s, careful reconstitution and transfer experiments have indicated that all molybdenum-containing hydroxylases contain a very labile Mo cofactor, consisting of a molybdenum oxide/sulfide fragment and a special ligand. EXAFS measurements showed that the metal center in its oxidized state always coordinates to at least one oxo group at a distance of approximately 170 pm; another constant feature is the presence of at least two (thiolate) sulfur centers at a distance of about 240 pm [3]. The nature of the organic part of the cofactor has been established via medically important degradation reactions; it is most likely a (dihydro)pterin derivative, 'molybdopterin' [15] (11.6), which contains a very characteristic heavy metal-coordinating ene-1,2-dithiolate or 'dithiolene' chelate function (11.7) in the side chain. In its metabolized form the pterin is found in human urine as urothione; nucleotide-containing 'bactopterins' with basically the same chromophor as in molybdopterin have been isolated from microorganisms [15,17].

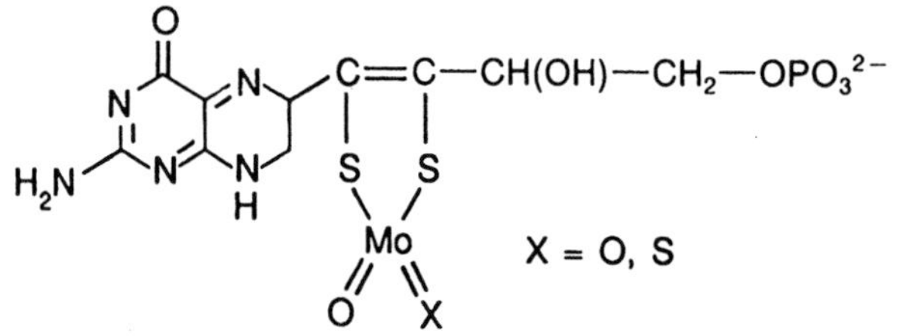

Mo cofactor ('Moco', according to [15])
(ligand: 'molybdopterin' in the quinonoid dihydro form)

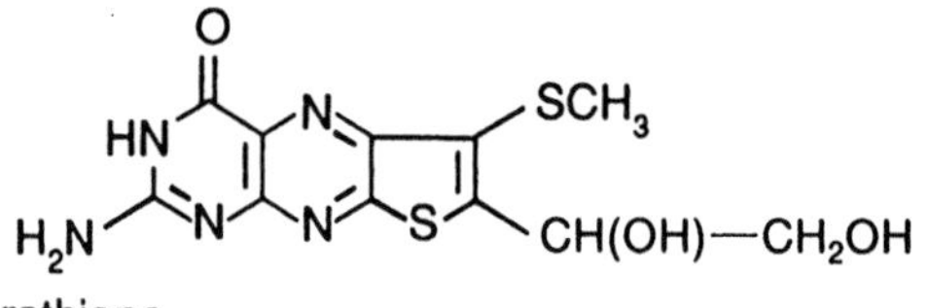

urothione   (11.6)

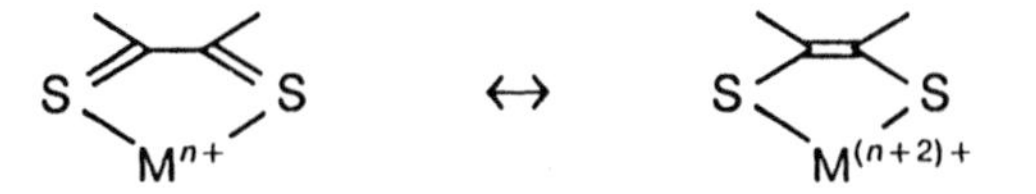

α-dithiocarbonyl complex ene-1,2-dithiolate complex (11.7)

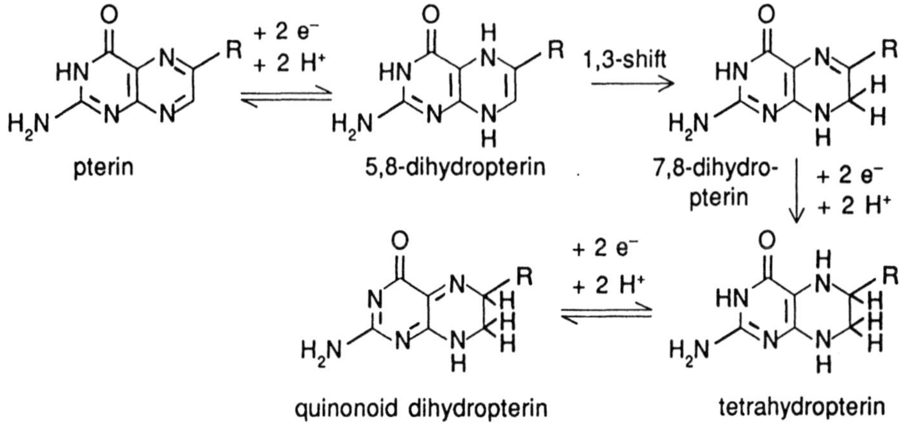

(11.8)

Both dithiolenes (11.7) [18] and pterins (11.8) [19] are potentially redox-active and metal-coordinating π systems. A hypothetical electron-transfer chain (11.9) consisting of several components (substrate, metal and other prosthetic groups) can thus be formulated.

$$\begin{array}{l}\text{Fe/S systems,}\\ \text{cytochromes,}\\ \text{or flavins}\end{array} \xrightarrow[-1\,e^-]{-1\,e^-} (\text{pterin} \longrightarrow) \text{ dithiolene} \longrightarrow \text{oxomolybdate} \xrightarrow{-2\,e^-,\,+O^{2-}} \text{substrate}$$

(11.9)

Although tetrahydropterins have been shown to reduce sulfur-coordinating molybdenum(VI) while being oxidized to the enzymatically reducable quinonoid dihydropterin [20] there has been no unambiguous evidence for that type of dihydro isomer *in vivo* or of a redox reactivity of Mo-coordinated molybdopterin in the enzymatic process. The established ability of pterins for metal coordination [19] allows for alternative coordination arrangements such as that given in (11.10) [21].

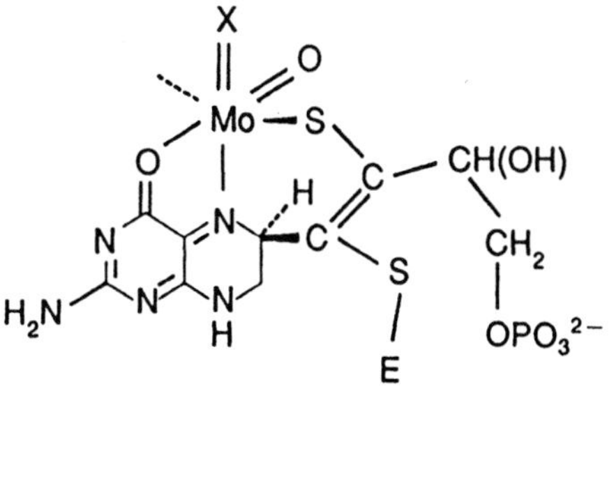

E = H, CH_3 or protein

(11.10)

In their oxidized states, sulfite oxidase and (assimilatory) nitrate reductase presumably contain a second oxo function at the metal which is converted upon reduction; either oxo transfer occurs with subsequent H_2O or OH^- replacement at the free coordination site or two successive one-electron transfer steps are followed by the protonation of one, very basic oxygen atom (11.11). In the Mo^{VI} state of xanthine oxidase a sulfide ligand is bound at a Mo—S distance of 215 pm which, after reduction at low potential, becomes protonated to give a 'sulfhydryl' ligand (11.12) as deduced from EPR spectra of Mo^V intermediates.

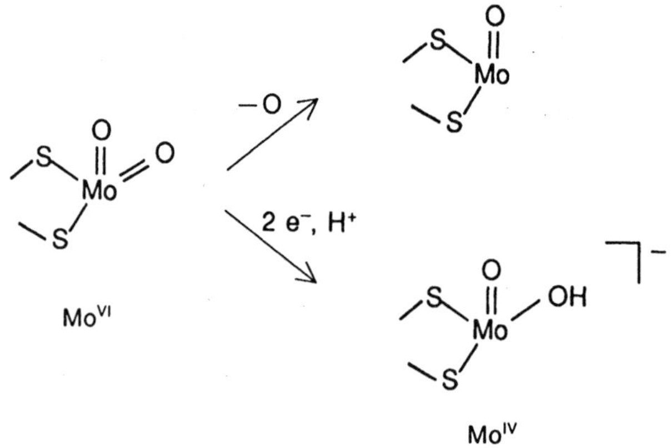

(11.11)

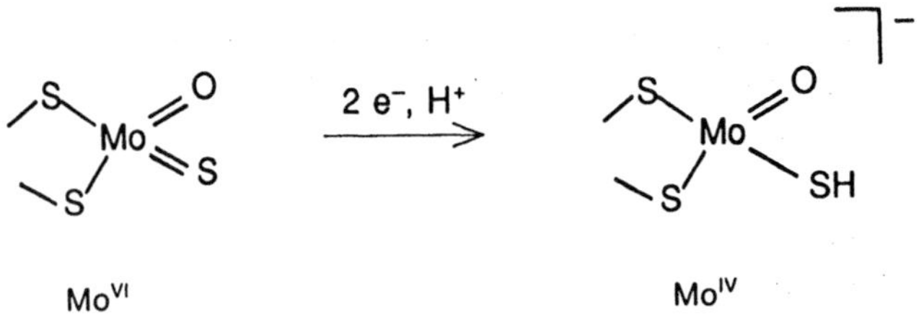

(11.12)

Recent EXAFS measurements of reduced xanthine oxidase have shown that the coordination of an additional ligand is possible which may function as substrate or as inhibitor [22]. Scheme (11.13) shows a possible reaction intermediate, various mechanisms have been discussed [9].

Mo^V species were frequently observed by EPR spectroscopy during the stepwise reoxidation of Mo^{IV} after completed oxygen transfer to the substrate [23]. The

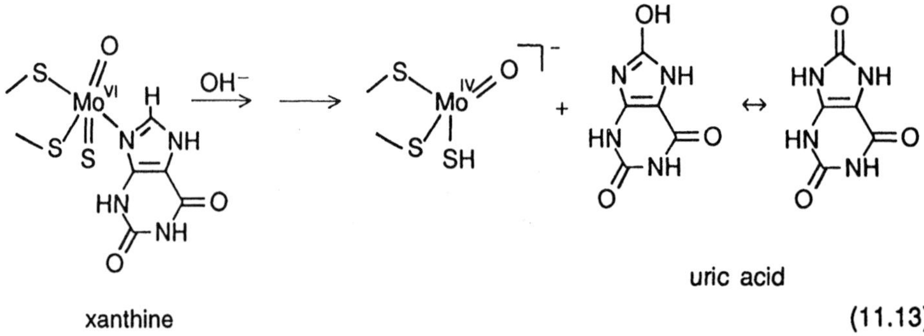

uric acid

xanthine

(11.13)

occurrence of several EPR signals with a partially detectable proton 'superhyperfine structure' can be rationalized according to scheme (11.14).

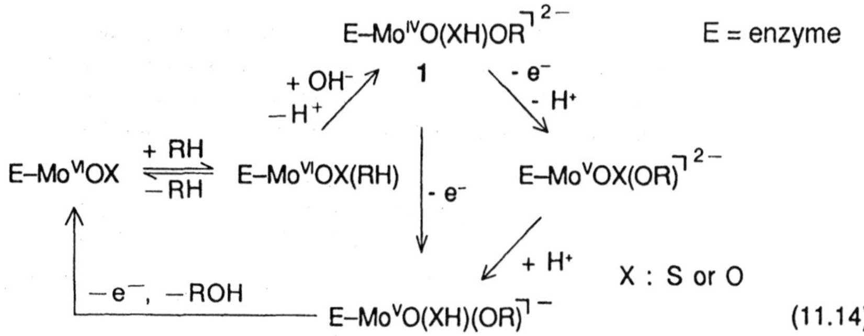

$$[E\text{-}Mo^{IV}O(XH)OR]^{2-}$$

E = enzyme

$$+ OH^- \nearrow \quad \mathbf{1} \quad \searrow \begin{array}{l} \cdot e^- \\ - H^+ \end{array}$$
$$- H^+$$

$$E\text{-}Mo^{VI}OX \underset{-RH}{\overset{+RH}{\rightleftharpoons}} E\text{-}Mo^{VI}OX(RH) \quad \downarrow \cdot e^- \qquad [E\text{-}Mo^VOX(OR)]^{2-}$$

$$\begin{array}{l} \nwarrow \\ | \\ -e^-, -ROH \end{array} \qquad \swarrow + H^+ \qquad X : S \text{ or } O$$

$$E\text{-}Mo^VO(XH)(OR)]^{1-} \tag{11.14}$$

The activated complex (1) of tetravalent molybdenum with one oxo, one hydroxyl or sulfhydryl, and one indirectly coordinated substrate ligand, R, can be one-electron oxidized under optional deprotonation. Loss of the second (and last) d electron causes release of the oxidized substrate and the resting state of the enzyme is obtained which can again react with substrate and base to complex (1) in an intramolecular electron-transfer process. The mechanistic hypothesis (11.14) is supported by studies of model complexes such as (11.15) [24,25].

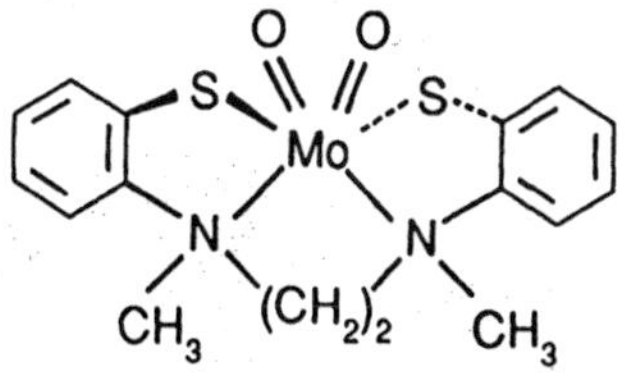

$$\tag{11.15}$$

Similar model complexes were used in the stepwise simulation of an enzyme-analogous cycle for the energetically much favored oxygen transfer from dimethylsulf-oxide to the physiologically not relevant triphenylphosphine [12] (11.16).

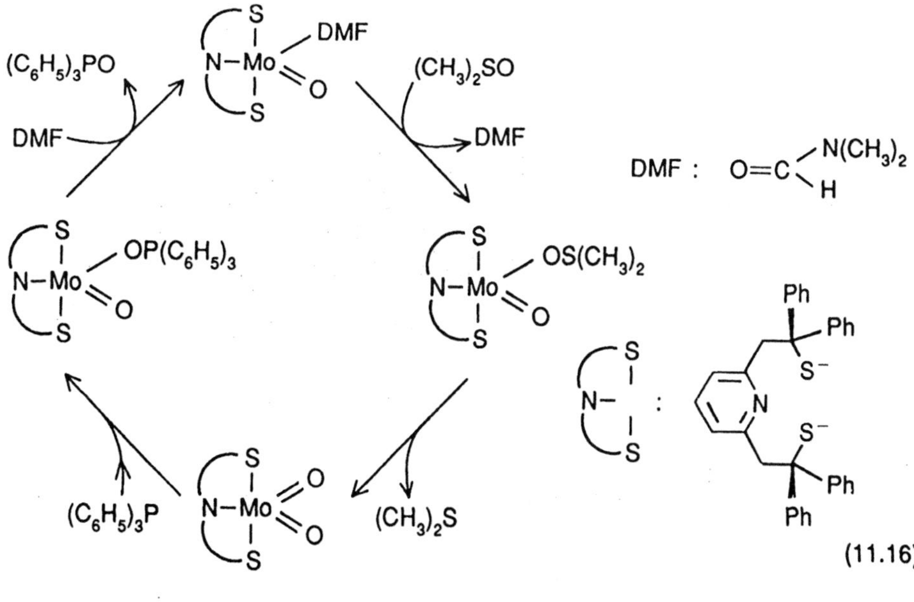

$$DMF : O{=}C\underset{H}{\overset{N(CH_3)_2}{\diagdown}}$$

$$\tag{11.16}$$

11.2   Metalloenzymes in the Biological Nitrogen Cycle: Molybdenum-Dependent Nitrogen Fixation

The importance of the inorganic-biological nitrogen cycle (Figure 11.1) for life on earth can hardly be overestimated. For instance, only through the technical nitrogen fixation as realized in the ammonia synthesis according to F. Haber and C. Bosch could the often growth-limiting nitrogen content of the soil be supplemented to

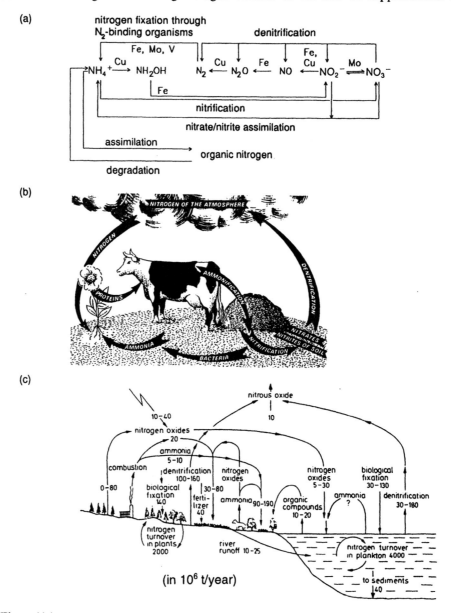

Figure 11.1
Chemical (a), biological (b), and ecological representation (c) of the nitrogen cycle (according to [3,26])

guarantee a food production adequate for the growing population on earth. The importance of nitrogen compounds as fertilizers in agriculture may be taken from the fact that ammonia continues to be one of the leading products of the chemical industry; other large-scale chemicals such as ammonium nitrate, urea and nitric acid are follow-up products of the technical 'fixation' of nitrogen as obtained from air. Accordingly, the overall turnover of products from synthetic nitrogen fixation may be estimated at about 40% or more of the biological process [26], a third source being physical atmospherical reactions which transform N_2, e.g. during thunderstorm discharges.

The drawbacks of excessive fertilizer use lie in the pollution of soil or ground and surface water with ammonium or nitrate; in addition, the toxic gases NO and NO_2 (NO_x) as well as the greenhouse gas N_2O are formed in increasing concentrations through agricultural activity or in combustion processes (catalyzed ozone formation, 'summer smog').

Most of the biological systems which participate in the global nitrogen cycle contain metal-requiring enzymes (Figure 11.1a). Three main processes can be distinguished: nitrogen fixation (11.17), nitrification (11.18) and denitrification (11.19).

$$\text{nitrogen fixation (technical):}$$

$$N_2 + 3\,H_2 \quad \xrightarrow[\text{metal oxide catalyst}]{>400\,°C,\ >100\ \text{bar}} \quad 2\,NH_3 \quad \text{(gas phase)}$$

$$\text{nitrogen fixation (biological):}$$

$$N_2 + 10\,H^+ + 8\,e^- \quad \xrightarrow{\text{nitrogenases}} \quad 2\,NH_4^+ + H_2 \tag{11.17}$$

$$NH_4^+ + 2\,O_2 \quad \xrightarrow{\text{nitrification}} \quad NO_3^- + H_2O + 2\,H^+ \tag{11.18}$$

$$2\,NO_3^- + \underbrace{12\,H^+ + 10\,e^-}_{\substack{\text{(from 'biomass',}\\ \text{i.e. reduced carbon}\\ \text{compounds)}}} \xrightarrow{\text{denitrification}} N_2 + 6\,H_2O \tag{11.19}$$

The oxidative process of nitrification yields nitrate, i.e. the oxidation state in which nitrogen can be assimilated by most higher plants. The crucial problem of this reaction as performed by *Nitrosomas* bacteria is to avoid the formation of the stable dinitrogen molecule, N_2, the zero-valent state. Aerobic conditions and sufficient buffer capacity for the resulting protons in soil or mineral material (atmospheric weathering) are indispensable. Some metalloenzymes for certain stages of *denitrifica-tion*, a process which is gaining importance for ecological reasons [27], have already been introduced: the molybdenum-containing nitrate(N^{+V}) reductase ($\rightarrow$nitrite), the copper or heme-iron containing nitrite(N^{+III}) reductases (see Section 6.5 and 10.3), and the copper-containing dinitrogen monoxide(N^{+I}) reductase (10.15). Denitrifica-tion requires rather anaerobic conditions and organic substances as reductants. The potential physiological importance of nitrogen monoxide as a free radical, $NO^{\cdot}$, or as a metal-coordinated NO^+ intermediate (Sections 6.5 and 10.3) is only just beginning to be realized. The final product of denitrification is the extremely stable, inert and volatile dinitrogen molecule. Its recycling, i.e. its reuse in the biological cycle, proceeds via the energetically and mechanistically challenging process of nitrogen fixation as based on nitrogenase (or better 'dinitrogenase') enzymes.

This process which is comparable only to photosynthesis in its biological importance is exclusively confined to procaryotic 'diazotrophic' organisms, e.g. free bacteria of the *Azotobacter* strains or *Rhizobium* bacteria which exist in symbiosis with the root system of leguminous plants. The restriction of the ability for dinitrogen fixation to relatively few specialized organisms and their symbionts becomes obvious through the appearance of certain 'pioneer plants' in the (re)population of nutrient-poor soil, e.g. as left behind by receding glaciers. After nonvolatile ('fixated') inorganic nitrogen compounds have been made available via (11.17), the element is incorporated as amino group into organic carrier molecules such as glutaric acid to be used further in the biosynthesis of, for example, proteins and nucleobases.

Due to the thermodynamic stability of the dinitrogen molecule, its reduction requires a large amount of energy in the form of several ATP equivalents (with Mg^{2+} as hydrolysis catalyst, Section 14.1) and (at least) six electrons per N_2 at a physiologically quite negative potential of less than -0.3 V. In addition, the extremely low reactivity of the dinitrogen molecule requires efficient catalysts, viz. the dinitrogenase enzymes [28–31]. Because of both thermodynamic and kinetic difficulties, the enzymatic nitrogen fixation proceeds rather slowly with turnover numbers on the order of 1/s; (11.17) is an eight-electron process coupled with ATP hydrolysis.

Although a large number of stable synthetic complexes of N_2 is now known [32], the first of these substances, a complex (11.20) of the 4d transition metal ruthenium, has been reported only in 1965. In contrast, metal carbonyl complexes of the carbon monoxide ligand which is isoelectronic with N_2 have been known for more than 100 years. Coordination of centrosymmetrical N_2 (mostly end-on) requires a twofold attack: in addition to the normal coordinative σ bond involving the free electron pair at one nitrogen center and the electrophilic metal ion, the transition metal center should provide electron density for the relatively low-lying unoccupied molecular orbitals of the triple-bond system in $N\equiv N$ via π interactions (11.20; π back bonding, push–pull mechanism; compare also 5.7). Thus, N_2 complexation and fixation requires d_π electron-rich metal centers as found, for example, in many organometallic compounds [32,33].

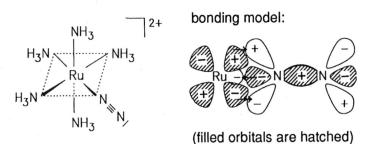

bonding model:

(filled orbitals are hatched)   (11.20)

The low redox potentials and the necessarily high reactivity of nitrogenase enzymes further require the absence of competing, i.e. related but better coordinating molecules

Table 11.2 Reduction (hydrogenation) reactions catalyzed by nitrogenases

substrate	products	number of electrons required
$\vert N{\equiv}N \vert$	$2\ NH_3 + H_2$	$8\ e^-$
$H-C{\equiv}C-H$	C_2H_4 or $Z\text{-}C_2H_2D_2$ (from C_2D_2)^a	$2\ e^-$
$H-C{\equiv}N\vert$	$CH_4 + NH_3\ (CH_3NH_2)$	$6\ e^-\ (4\ e^-)$
$CH_3-\overset{+}{N}{\equiv}C\vert^{\ -}$	$CH_3NH_2 + CH_4$	$6\ e^-$
$^-\langle N{=}\overset{+}{N}{=}N \rangle^{\ -}$	$N_2H_4 + NH_3\ (N_2 + NH_3)$	$6\ e^-\ (2\ e^-)$
$^-\langle N{=}\overset{+}{N}{=}O \rangle$	$N_2 + H_2O$	$2\ e^-$
$\begin{smallmatrix} & CH_2 \\ / & \backslash \\ HC & = CH \end{smallmatrix}$	$1/3\ \begin{smallmatrix} & CH_2 \\ / & \backslash \\ H_2C & - CH_2 \end{smallmatrix} + 2/3\ CH_3-CH{=}CH_2$	$2\ e^-$
$2\ H^+$	H_2	$2\ e^-$

^{a}Partially four-electron reduction to C_2H_6 in vanadium-dependent nitrogenase

such as dioxygen, O_2. Therefore, N_2-assimilating microorganisms are either obligatory anaerobes or they possess complex protection mechanisms for the exclusion of O_2 from the active site of nitrogenase enzymes. These mechanisms include iron-containing proteins which serve as O_2 sensors. Nitrogenase activity is also inhibited by the isoelectronic substrates carbon monoxide, $^-C{\equiv}O^+$, and NO^+ ($\rightarrow$NO); characteristic reaction products are obtained with several other small molecules containing multiple bonds (Table 11.2).

In the case of conventional, i.e. molybdenum-containing nitrogenase, both the exclusive Z(cis)-hydrogenation of acetylene only to the ethylene state and the cleavage of the isocyanide triple bond are remarkable. Furthermore, nitrogenases possess an intrinsic hydrogenase activity which leads to the obligatory production of dihydrogen, H_2, during the biological N_2 fixation reaction (11.21); no H_2 evolution was observed during the reduction of the unphysiological substrates listed in Table 11.2.

$$N_2 + 8\ H^+ + 8\ e^- \longrightarrow H_2 + 2\ NH_3 \underset{pK_a = 9.2}{\overset{2\ H^+}{\rightleftarrows}} 2\ NH_4^+ \qquad (11.21)$$

Even N_2 pressures of up to 50 bar did not preclude the formation of 25% H_2 per reduction equivalent according to (11.21); therefore, a simple displacement equilibrium cannot explain this H_2 evolution. On the other hand, dihydrogen (in equilibrium) is inhibiting the N_2 fixation process. It is now assumed (11.22) that the reduction of the enzyme, E, proceeds via one-electron/one-proton addition steps and involves binding of N_2 only after addition of the third reduction equivalent [31]. In the hypothetical sequence (11.22) the immediately bound and reduced hydrogen, formulated as side-on coordinated H_2, is replaced by N_2.

$$E \xrightarrow{e^-/H^+} E\text{-}H \xrightarrow{e^-/H^+} E\text{-}\overset{H}{\underset{H}{|}} \xrightarrow{e^-/H^+} E\text{-}\overset{H}{\underset{H}{|}}\overset{H}{\underset{}{|}} \underset{N_2 \quad\quad H_2}{\overset{N_2 \quad\quad H_2}{\rightleftharpoons}} \overset{H}{\underset{}{|}}E(N_2) \xrightarrow{5e^-, 5H^+} 2\,NH_3 + E$$

$$(11.22)$$

In most cases, the dihydrogen thus obtained is immediately reoxidized by hydrogenases (Section 9.3) to yield protons and energy.

Nitrogenase activity is also inhibited by an excess of the product, NH_4^+, by the presence of the assumed intermediate hydrazine, N_2H_4 (see 11.25), and in the absence of essential inorganic components. Required are the already mentioned Mg^{2+} for ATP hydrolysis, sulfur in the form of sulfide or transformable sulfate, iron and, at least for the 'classical' nitrogenase, molybdenum; the heavier homologue tungsten cannot serve as a substitute.

The most extensively studied molybdenum-dependent form of nitrogenase (see Figure 11.4) has an $(\alpha_2\beta_2)(\gamma_2)$ two-protein composition, the special dimeric iron protein (γ_2), the 'dinitrogenase reductase' (about 60 kDa), being essential for the function of the enzyme. Bound *between* the two γ subunits of this protein is a single [4Fe–4S] cluster which can be reduced to a paramagnetic form at the physiologically very negative potential of -0.35 V. This 'Fe protein' of nitrogenase contains two Mg^{2+}/ATP receptors because two ATP molecules have to be hydrolyzed for each transferred electron. Binding of ATP or ADP further lowers the potential by about 0.1 V.

The second component, the actual 'dinitrogenase' or 'FeMo protein', is an $\alpha_2\beta_2$ tetramer (220 kDa; see Figure 11.4) which contains two very special [8Fe–8S] systems, the P clusters (cysteinate-bridged [4Fe–4S] clusters; 7.11) and two 'FeMo cofactors' (FeMo-co or M clusters), each with the inorganic composition of $MoFe_7S_8$. The breakthrough structural elucidation of a nitrogenase from *Azotobacter vinelandii* with a resolution of 0.27 nm in 1992 [34–36] has confirmed some of the assumptions made previously on the basis of physical measurements and chemical models. However, the actual structure of the $MoFe_7S_8$ cluster (11.23) had not been deduced by the model studies. It has long been known from Fe- and Mo-EXAFS measurements of the protein and the anionic FeMo-cofactor (1.5 kDa, extractable with N-methylformamide, $HN(CH_3)C(O)H$) that a molybdenum-modified cuboidal Fe/S cluster is present and that the tetravalent molybdenum (reduced form) is bound to some O or N donor atoms. In fact, the six-coordinate molybdenum of the M cluster is located at the corner of a metal sulfide heterocubane structure and is coordinated to the outside by a histidine and a tetraanionic chelating homocitrate ligand (11.24: homocitric acid = (R)-2-hydroxy-1,2,4-butanetricarboxylic acid). A definite function of the homocitrate ligand has not yet been established although its synthesis is encoded by a separate gene.

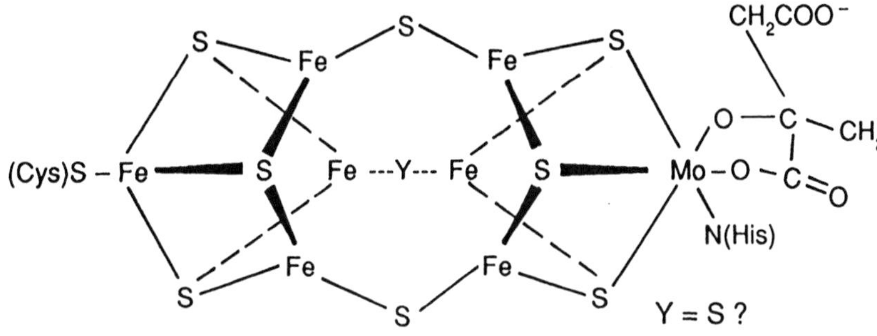

$$Y = S ? \tag{11.23}$$

homocitrate:

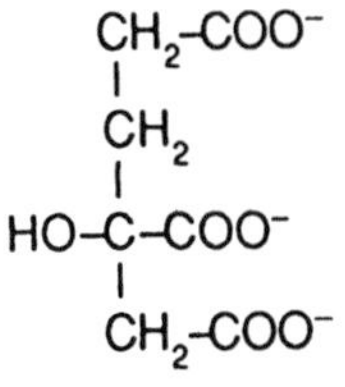

$$
\begin{array}{c}
CH_2\text{-}COO^- \\
| \\
CH_2 \\
| \\
HO\text{-}C\text{-}COO^- \\
| \\
CH_2\text{-}COO^-
\end{array} \tag{11.24}
$$

Asymmetry and high complexity of the M clusters can also be deduced from heteroatom ENDOR measurements [37]. In the reduced state, the cofactor contains tetravalent molybdenum; however, the total electron spin of $S = 3/2$ is mostly located at the various different sulfur-coordinated iron centers which, according to Mössbauer data, show some degree of electron delocalization. In the C_2 symmetrical (Figure 11.4) dinitrogenase protein the M clusters are separated by about 7 nm whereas the closest distance between M and P clusters is about 2 nm.

In the reduced Fe protein (dinitrogenase reductase) the function of the unusual [4Fe–4S] cluster is to provide an electron flow at low potential using simultaneous ATP hydrolysis as the driving force (2 ATP per electron). Inside the FeMo protein the polynuclear P clusters may be reduced to the all-Fe^{II} form and presumably regulate the low-potential transfer of electrons to the FeMo cofactors.

Concerning the actual mechanism of the metalloenzyme-catalyzed transformation of N_2 to NH_3, only hypotheses are available that take into account the results obtained for nonenzymatic complexes, e.g. of zerovalent molybdenum. A multistep reaction sequence is certainly necessary for this energetically and mechanistically demanding process (11.21); according to (11.25) and in analogy to the heterogeneously catalyzed technical synthesis of ammonia the catalyst is required to prevent the formation of 'free', high-energy intermediates (see Figure 2.7).

To begin with, both the actual coordination site and the coordination mode of N_2 in the enzyme are still a matter of debate. The coordinative saturation of the Mo center (11.23) and the existence of heteroatom-free nitrogenases (see below) seem to suggest an iron rather than molybdenum coordination of N_2; as with O_2 (see Table 5.1) the alternatives [38] for the coordination mode can range from end-on binding (η^1) to side-on bridging, e.g. in the central position 'Y' of (11.23). In synthetic chemistry, the μ-η^2:η^2-coordination of N_2 is rarely observed [39], end-on bridging (η-μ^1:η^1) being much more common [32]. The cluster asymmetry caused by the molybdenum center presumably facilitates electrophilic attack during the stepwise reduction of dinitrogen.

Stabilization of high-energy intermediates in the conversion $N_2 \rightarrow NH_3$
through binding to a metal center, M:

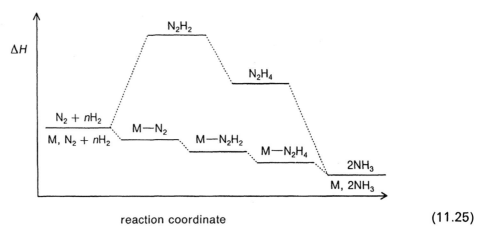

reaction coordinate   (11.25)

A catalysis mechanism (11.26) based on model reactions of complexes with non-bridging, end-on coordinated N_2 can then be formulated (see [31]).

Possible intermediate states in the metal-catalyzed reduction of N_2 with
reference to the different nitrogen ligands (LM = metal complex with
one free coordination site):

$$
\begin{array}{ccccccc}
 & +\,H^+,\,e^- & & +\,H^+,\,e^- & & +\,H^+ & \\
\text{LM–N} \equiv \text{N} & \longrightarrow & \text{LM=N–NH} & \longrightarrow & \text{LM} \equiv \text{N–NH}_2 & \longrightarrow & \text{LM} \equiv \text{N–}^+\text{NH}_3 \\
\text{dinitro-} & & \text{diazenido(1-)} & & \text{hydrazido(2-)} & & \\
\text{gen} & & & & & & \\
\end{array}
$$

$+\,N_2$ (upward) ... $+\,H^+,\,e^-$ (downward)

$$
\begin{array}{ccccccc}
 & +\,H^+ & & +2\,H^+,\,2\,e^- & & +\,H^+,\,e^- & \\
\text{NH}_4^+ + \text{LM} & \longleftarrow & \text{LM–NH}_3 & \longleftarrow & \text{LM=NH} & \longleftarrow & \text{LM} \equiv \text{N} + \text{NH}_4^+ \\
 & & \text{ammine} & & \text{imido} & & \text{nitrido} \\
\end{array}
$$

(11.26)

Multiple stepwise addition of e^-/H^+ before the actual dinitrogen fixation results in the displacement of H_2 (11.22) as caused by the coordination of N_2, followed by the combination of N_2 and e^-/H^+ to form a diazenido(1-) ligand, HNN^- (11.26). This ligand can react with another e^-/H^+ equivalent to yield a terminal hydrazido(2-) ligand, $H_2N\text{—}N^{2-}$, which could also be formulated as a neutral aminonitrene ligand, $H_2N\text{—}N$, after different distribution of oxidation states. In fact, hydrazine can be detected as an incompletely reduced form of dinitrogen after rapid quenching of the nitrogenase reaction sequence. Nonreductive protonation of the diazenido ligand would yield the N^{-1} species diazene, $HN{=}NH$, which is unstable in free form; however, metal complexes may stabilize that molecule through multiple coordination and hydrogen bond interactions, $N\text{—}H\cdots S$ (Figure 11.2 [40]).

Continuing with the hydrazido(2-) complex, an $e^-/2\,H^+$ addition (11.26) would result in N—N bond cleavage and the dissociation of a first ammonium ion, leaving behind a metal-nitrido function, $M{\equiv}N$ [41] with a high formal oxidation state of

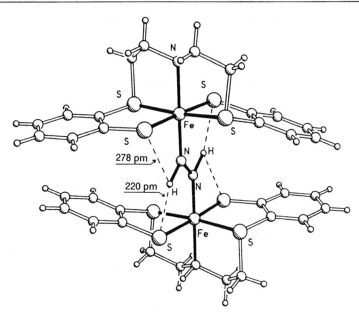

Figure 11.2
Molecular structure of a dinuclear iron complex of diazene, HN = NH (from [40])

the metal. Successive addition of further e^-/H^+ equivalents leads to the ammine complex via imido (HN^{2-}) and amido (H_2N^-) complexes and, after final protonation, the second ammonium ion is obtained.

Many of the stages in sequence (11.26) have been verified in model complexes with metals in rather *low* oxidation states; however, there are some organometallic complexes such as $[(C_5Me_5)Me_3Mo]_2(\mu\text{-}N_2)$ containing high oxidation state Mo and partially reduced dinitrogen ligands [42,43]. With regard to a full reaction cycle and hydrazine or ammonia production only model complexes of low-valent molybdenum and tungsten have been successful; immobilized systems (11.27 [44]) or electrochemical studies (11.28 [45]) need to be mentioned in this context.

$$\text{(11.27)}$$

In addition to the structural modeling of the nitrogenase metal cluster as based on EXAFS data [31] and the experiments directed at a simulation of at least parts of the biocatalytic reactivity, the genetic variations of corresponding microorganisms and the effect on proteins synthesized by the resulting mutants have contributed to catalyzed electrochemical reduction of N_2 to NH_3:

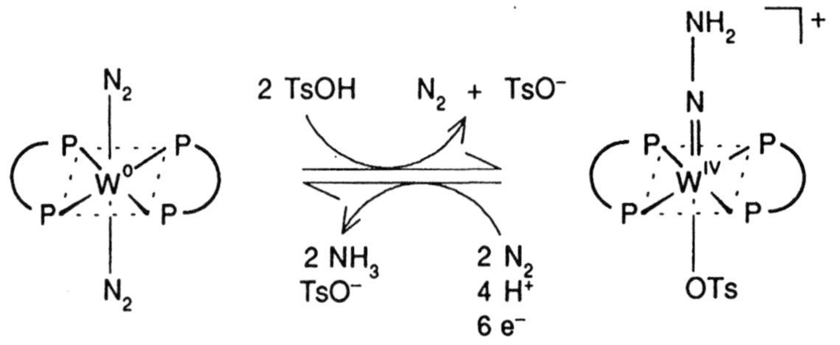

Overall reaction:

$$2\ TsOH\ +\ 2\ N_2\ +\ 4\ H^+\ +\ 6\ e^-\ \longrightarrow\ N_2\ +\ 2\ TsO^-\ +\ 2\ NH_3$$

$$N_2\ +\ 6\ H^+\ +\ 6\ e^-\ \longrightarrow\ 2\ NH_3$$

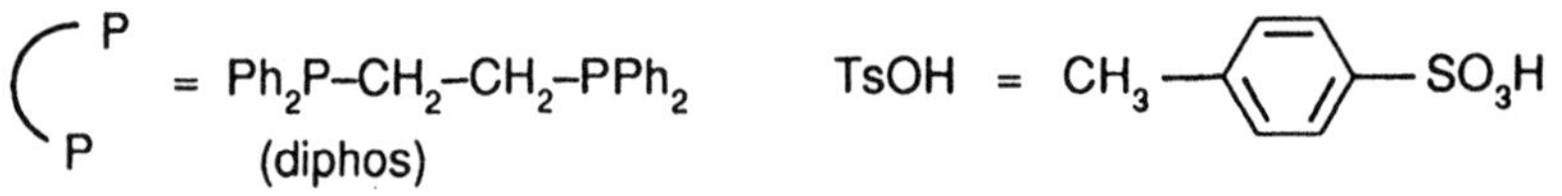

$$\overset{P}{\underset{P}{\Big(}}\ =\ Ph_2P{-}CH_2{-}CH_2{-}PPh_2 \qquad TsOH\ =\ CH_3{-}\!\!\left\langle\bigcirc\right\rangle\!\!{-}SO_3H$$
(diphos)

(11.28)

the current understanding of biological nitrogen fixation. One main objective of such
rational protein design [31] is to eventually transfer the set of at least seventeen *nif*
(*nitrogen fixation*) genes from microorganisms to important crops. The knowledge
gained for the classical, i.e. molybdenum-dependent nitrogenase, has also turned out
to be very valuable for the characterization of two more recently established variants,
the vanadium-dependent and the heteroatom-independent nitrogenase enzymes.

11.3 Alternative Nitrogenases

Contrary to a proposition by H. Bortels from the 1930s with respect to the possible
role of vanadium, molybdenum was for a long time considered to be indispensable
for nitrogen fixation. Only the more recent advances in ultra-trace elemental analysis
and in genetic analysis of N_2-fixating organisms have provided unambiguous evidence for the existence of *alternative* nitrogenases which are produced in the
absence of molybdenum or FeMo cofactor synthesizing genes [31,46]. In the presence
of vanadium a V-dependent nitrogenase is formed; if this element is also not available,
a third, 'iron-only' form can be synthesized by some organisms. In the case of a
sufficient supply of molybdenum, the formation of such 'alternative' nitrogenases is
suppressed so that these, under normal conditions less effective enzymes may be
tentatively regarded as 'back-up' systems.

There are numerous parallels between the three types of nitrogenases but also
characteristic differences [30,31]. First of all, it should come as no surprise [47] that
the 'second best' form of nitrogenase contains vanadium since V and Mo are
connected by a diagonal relationship in the periodic table of the elements; other
(bio)inorganic examples such as Mn/Ru with regard to O_2 production (see Section

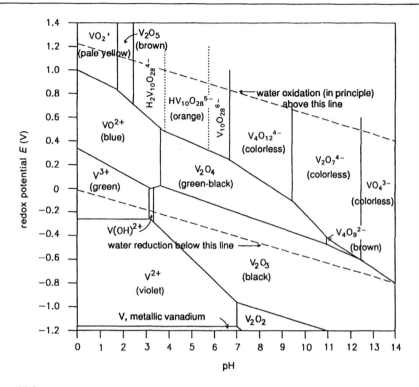

Figure 11.3
Stability or 'Pourbaix' diagram of vanadium in water: regions of various hydrated species and solid forms as depending on redox potentials and pH values (according to [48])

4.3) or Fe/Rh with regard to H^+/H_2 conversion (Sections 7.4 and 9.3) may be mentioned here. The correspondence between V and Mo manifests itself, for example, by a similar chemistry of both elements in aqueous solution. The fairly complicated stability diagram in Figure 11.3 shows that vanadium, like molybdenum (see 11.1), has a strong tendency to form oxo/hydroxo-bridged aggregates.

Vanadium is more abundant than molybdenum in the earth's crust but not in sea water; nevertheless, it is the most abundant first-row (3d) transition metal in that medium (Figure 2.1). The V-dependent nitrogenase has a slightly lower activity regarding N_2 reduction and exhibits small amounts of N_2H_4 among its reduction products; however, its main disadvantage is that about 50% of the reduction equivalents are being used in the formation of H_2 as compared to only 25% in the case of the Mo-enzyme (see 11.21). Typically, the V-dependent nitrogenase shows rather little activity in the reduction of acetylene, yielding not only ethylene, C_2H_4, but also some ethane, C_2H_6, as four-electron hydrogenation products (Figure 11.4). Remarkably, however, the V-dependent nitrogenase is more efficient than the molybdenum system at temperatures around 5 °C, a fact that might have favored the conservation of this form during evolution [31]. The most sensitive, heterometal-independent (Fe-only) nitrogenases have not yet been well characterized; their activity seems to be lower than that of the vanadium- or molybdenum-containing enzymes.

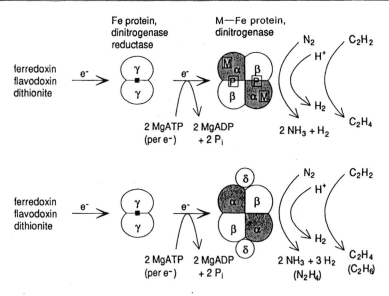

Figure 11.4
Schematic composition of nitrogenase enzymes: Mo-dependent (top) and V-dependent form (bottom)

As the molybdenum-dependent system, both alternative nitrogenases are composed of two proteins. The V-dependent enzyme in particular exhibits many common features when compared to the Mo analogue (Figure 11.4). A dimeric Fe protein with Mg^{2+}/ATP binding sites faces a larger protein which, in the case of the V system, is an $\alpha_2\beta_2\delta_2$ hexamer with two small δ subunits. Not only the basic enzyme structure but also certain invariant amino acids in the protein sequence and the resulting inorganic composition of the cluster centers are comparable for vanadium- and molybdenum-dependent nitrogenases [31]. In both cases, EXAFS measurements have indicated the existence of MFe_3S_4 heterocubane structural motifs; the heteroatom in vanadium-dependent nitrogenase is surrounded by 3 ± 1 sulfide centers at a distance of about 232 pm, 3 ± 1 iron atoms at a distance of 274 pm, and approximately three light donor atoms (O, N) at a distance of about 214 pm [49]. Iron-EXAFS studies and results obtained for model complexes like (11.29) have supported this description [50].

X = solvent molecules,
      e.g. DMF

(11.29)

Characteristically, the geometrical changes at the metal cluster site are small during redox reactions of both Mo and V nitrogenases whereas the Mo centers of oxo-transferases show a marked structural difference between the $+VI$ and $+IV$ oxidation states. The heteroatom cofactor of vanadium-dependent nitrogenase is also extractable with N-methylformamide; its approximate 'inorganic' composition has been determined as $VFe_{5-6}S_{4-5}$. Another analogy with the molybdenum system is the $(S = 3/2)$ EPR signal of reduced V-dependent nitrogenase and the d^2 configuration of the heterometal (here V^{III}). Like molybdenum, vanadium is known to form low molecular weight complexes in which the low-valent ($\leq +II$) metal can bind N_2 [51]. The apparent requirement of certain heteroatoms for an efficient N_2 fixation may be rationalized considering the pronounced ability of polythiometal centers for multielectron 'buffering' in the negative potential range. Müller et al. [47] have therefore suggested that rhenium (which is also connected to molybdenum via a diagonal relationship) could similarly function as a heteroelement in nitrogenases; however, its extremely low natural abundance restricts its bioavailability.

11.4 Biological Vanadium Outside Nitrogenases

While vanadium-dependent nitrogenases were established only in about 1985, the occurrence of vanadium, often in a highly enriched form, in marine organisms such as tunicates and brown algae (seaweeds) or in lichen and mushrooms (toadstools) has been known for some time [52–54]. With regard to higher organisms, the inhibiting effect of vanadate(V) on phosphate-depending enzymes (see Sections 13.4 and 14.1) is well recognized, as is the possible therapeutic function of vanadyl(IV) complexes as insulin 'mimics' in the glucose metabolism [55]. Considerable amounts of vanadium, in particular in the form of tetrapyrrole complexes, can be found in some kinds of biogenic crude oil (Table 2.7).

As in the molybdate/sulfate pair, the *ortho*-vanadate ion, VO_4^{3-}, resembles quite well—but not completely [56]—the (*ortho*)phosphate ion, PO_4^{3-}, e.g. in terms of structural characteristics (tetrahedral ion with a certain tendency for the trigonal bipyramidal coordination) and a shared preference for aggregation (see, for example, ATP, 14.2). As a transition metal, however, vanadium has empty, low-lying 3d orbitals in its maximal oxidation state of $+V$; it can thus be reduced fairly easily (see Figure 11.3) to lower oxidation states such as $+IV$ or $+III$ with biological reductants such as glutathione (16.4). While V^V and V^{III} compounds can be examined by heteroatom NMR spectroscopy [53] (^{51}V: $I = 7/2$, 99.75 natural abundance; see also the boxed insert in Section 13.1), the vanadyl(IV) ion, VO^{2+}, and its complexes are particularly well suited for EPR spectroscopical detection because of the d^1 configuration.

For humans, the essentiality and biochemical function of the ultra-trace element vanadium (uptake 10–60 μg/day) has not yet been well defined [57]. Experiments on the vanadate-supported stimulation of cardiovascular activity in insulin-treated diabetic rats were able to demonstrate that the glucose metabolism itself can be favorably influenced by the oral administering of certain vanadium compounds (see also Section 11.5). Among effective insulin mimics are peroxovanadium(V) and vanadyl(IV) complexes, $L_nV^{IV}O$, which have the ability to cross membrane barriers if L is a lipophilic ligand [55]. On the other hand, vanadium and its aggregates [58] inhibit enzymes like the Na^+/K^+-ATPase (see Section 13.4) and related phosphate transferases (kinases, cyclases, phosphatases, ribonucleases; see Section 14.1), even in rather low concentrations. Also, the biosynthesis of cysteine which is an essential amino acid for metal coordination and protein conformation is inhibited by vanadate. The iron-transport protein apotransferrin (Section 8.4.1) binds vanadium in the oxidation states III–V.

It has been known since about 1910 that certain sessile marine organisms, sea squirts (*Ascidiae*) of the tunicate group, can accumulate vanadium in their 'blood cells' by a factor of more than 10^7 as compared to the surrounding sea water. Earlier assumptions which attributed an O_2 transport function to these large amounts of reduced vanadium($+$III, partly $+$IV) have turned out to be untenable; instead they may possibly serve as a primitive immune system, in the template synthesis of tunichrome peptides (11.30), or as components in the anaerobic metabolism [59]. At which pH value (neutral or strongly acidic) and in which coordination arrangement vanadium occurs in these sulfate-rich cells has been strongly debated [59–61]. An interaction is assumed between dissolved vanadium and the labile tunichrome dipeptides (11.30) which contain potentially metal-coordinating and redox-active *o*-polyphenol groups; they are also very characteristic natural products of such animals but their function has not been well defined.

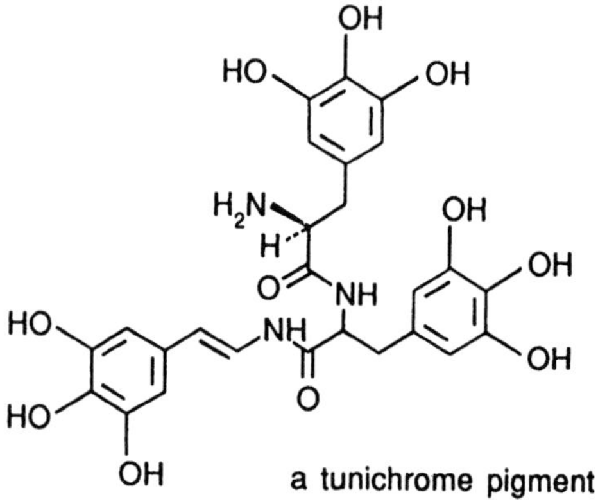

a tunichrome pigment   (11.30)

Another vanadium-containing natural product, 'amavadin', can be isolated from mushrooms of the toadstool kind (genus *Amanita*), its exact identity having been disputed for a long time [62,63]. The eventually successful crystal structure analysis shows eight-coordinate vanadium (V^V or V^{IV} [63]) with two 2,2'-(oxyimino)dipropionate ligands (11.31). This complex is extremely stable (compare the related hydroxamate function in iron siderophore complexes, Section 8.2) which explains the remarkable vanadium accumulation of up to 200 ppm from the surrounding soil. The function of such a vanadium accumulation is still unclear; mushrooms and fungi are generally known to synthesize amazingly efficient low molecular weight chelate ligands for metal ions (see Sections 17.3 and 18.2).

Certain red and brown algae ('seaweeds' such as *Ascophyllum nodosum*) and also some species of lichen and fungi contain vanadium-dependent haloperoxidases [64,65]. These enzymes, which also occur in a more 'conventional' form as hemoproteins (see Section 6.3), catalyze the halogenation of organic substrates with the help

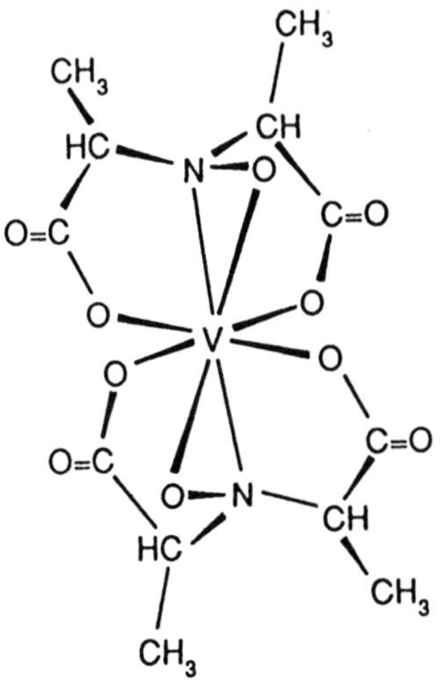

amavadin from toadstool (Amanita),
charge 1- (V^V) or 2- (V^{IV})

of hydrogen peroxide according to (11.32). Like many of their synthetic analogues, the 'natural' halogenated hydrocarbons which are formed in quite large amounts on a global scale may serve as biocides in the defense system of organisms.

$$R{-}H \ + \ H_2O_2 \ + \ Hal^- \ + \ H^+ \ \xrightarrow{\text{halo-peroxidase}} \ R{-}Hal \ + \ 2\ H_2O$$

R: organic alkyl or aryl group
Hal: Cl, Br, I   (11.32)

While land-living organisms like fungi or lichen feature mainly chlorinating and iodinating enzymes (cf. thyreoperoxidases, Sections 6.3 and 16.7), the bromination to yield, for example, bromoform, $CHBr_3$, plays an important role for marine organisms. Hypobromite, ^-OBr, is generally assumed to be the final brominating species. The exact function of vanadium in the rather acidic bromoperoxidase enzymes (about 110 kDa) is still unclear. Coordination and activation of (hydro)peroxide by the metal is assumed to be essential for the controlled formation of reactive bromooxygen compounds, e.g. ^-OBr; oxidation of the far more abundant chloride requires higher potentials and proceeds more slowly. However, some substrates are also efficiently chlorinated by vanadium-containing 'bromo'peroxidases [65]. In agreement with the peroxidase function only the + V oxidation state of vanadium is assumed to be relevant under physiological conditions; the metal is probably coordinated by six O or N donor atoms including one terminal oxo group (EXAFS measurements [66]). Because of their high thermal and chemical stability, the V-dependent peroxidases are being considered as components for medical diagnostics, biotechnology, or even of household detergents.

11.5   Chromium(III) in the Metabolism?

While chromium(VI) as chromate, CrO_4^{2-}, has been recognized as a mutagenic and even carcinogenic species (see Section 17.8), the metal has been proposed as an essential trace element in its $+III$ oxidation state, which is the most stable state in aqueous solution. Like iron(III) (see Chapter 8) or aluminum(III) (see Section 17.6), trivalent chromium can only be resorbed very slowly due to the insolubility of the hydroxide at pH 7 and because of very slow substitution reactions. Responsible for this extraordinary sluggishness, even with regard to the exchange of coordinated water, is the stable d^3 configuration with half-filled t_{2g} orbitals in octahedral symmetry (see 2.8); any structural deviation caused by associative or dissociative processes requires a very high activation energy.

Apparently the best-established function of Cr^{III} concerns the participation of a chromium-containing glucose tolerance factor (GTF) in optimizing the effect of insulin [67]. Chromium-accumulating plants like shepherd's purse have thus long been used in the natural remedy treatment of diabetes mellitus, type II [68], while, more recently, chromium(III) therapy was reported to help reduce the insulin requirement in hypoglycaemic diabetes patients. The GTF is presumably a typical, i.e. octahedrally configured and substitutionally inert Cr^{III} complex, usually formulated with two nicotinic acid ligands in the *trans* position; the remaining four coordination sites could at least partially be occupied by sulfur ligands, e.g. from the peptide glutathione (11.33).

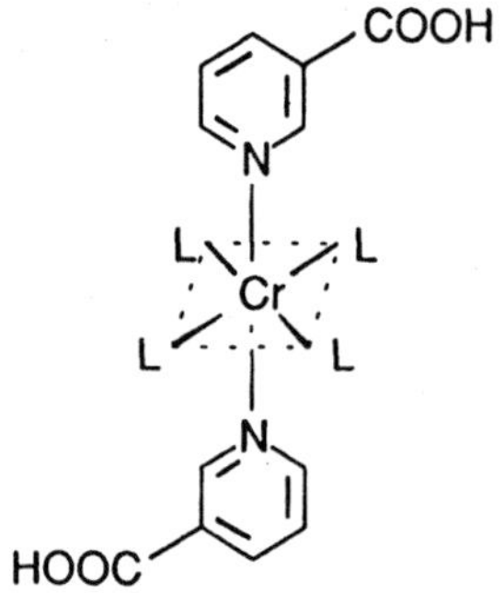

L = glycine, cysteine?   (11.33)

It has been speculated that this complex interacts with the membrane receptor for insulin (see Section 12.7). However, in microorganisms such as yeast cells the glucose tolerance factor does not necessarily contain chromium which, in any case, is an ubiquitous element (occurring in stainless steel) and thus hard to exactly specify in ultra-trace quantities in biological systems (see also Section 17.6).

References

1. T. G. Spiro (ed.), *Molybdenum Enzymes*, Wiley, New York, 1985.
2. R. C. Bray, The inorganic biochemistry of molybdoenzymes, *Rev. Biophys.*, **21**, 299 (1988).
3. S. J. N. Burgmayer and E. I. Stiefel, Molybdenum enzymes, cofactors, and model systems, *J. Chem. Educ.*, **62**, 943 (1985).

4. E. I. Stiefel, D. Coucouvanis and W. E. Newton (eds.), *Molybdenum Enzymes, Cofactors, and Model Systems*, ACS Symposium Series 535, 1993.

5. I. Yamamoto, T. Saiki, S. M. Liu and L. G. Ljungdahl, Purification and properties of NADP-dependent formate dehydrogenase from *Clostridium thermoaceticum*, a tungsten–selenium–iron protein, *J. Biol. Chem.*, **258**, 1826 (1983).

6. R. Wagner and J. R. Andreesen, Accumulation and incorporation of [185]W-tungsten into proteins of *Clostridium acidiurici* and *Clostridium cylindrosporum*, *Arch. Microbiol.*, **147**, 295 (1987).

7. R. A. Schmitz, S. P. J. Albracht and R. K. Thauer, Properties of the tungsten-substituted molybdenum formylmethanofuran dehydrogenase from *Methanobacterium wolfei*, *FEBS Lett.*, **309**, 78 (1992).

8. G. N. George, R. C. Prince, S. Mukund and M. W. W. Adams, Aldehyde ferredoxin oxidoreductase from the hyperthermophilic archaebacterium *Pyrococcus furiosus* contains a tungsten oxo-thiolate center, *J. Am. Chem. Soc.*, **114**, 3521 (1992).

9. R. S. Pilato and E. I. Stiefel, Catalysis by molybdenum-cofactor enzymes, in *Bioinorganic Catalysis* (ed. J. Reedijk), Marcel Dekker, New York, 1993, p. 131.

10. S. P. Cramer, P. K. Eidem, M. T. Paffett, J. R. Winkler, Z. Dori and H. B. Gray, X-ray absorption edge and EXAFS spectroscopic studies of molybdenum ions in aqueous solution, *J. Am. Chem. Soc.*, **105**, 799 (1983).

11. R. H. Holm, Metal-centered oxygen atom transfer reactions, *Chem. Rev.*, **87**, 1401 (1987).

12. R. H. Holm, The biologically relevant oxygen atom transfer chemistry of molybdenum: from synthetic analogue systems to enzymes, *Coord. Chem. Rev.*, **100**, 183 (1990).

13. J. C. Wootton, R. E. Nicolson, J. M. Cock, D. E. Walters, J. F. Burke, W. A. Doyle and R. C. Bray, Enzymes depending on the pterin molybdenum cofactor: sequence families, spectroscopic properties of molybdenum and possible cofactor-binding domains, *Biochim. Biophys. Acta*, **1057**, 157 (1991).

14. J. H. Weiner, R. A. Rothery, D. Sambasivarao and C. A. Trieber, Molecular analysis of dimethylsulfoxide reductase: a complex iron–sulfur molybdoenzyme of *Escherichia coli*, *Biochim. Biophys. Acta*, **1102**, 1 (1992).

15. K. V. Rajagopalan and J. L. Johnson, The pterin molybdenum cofactors, *J. Biol. Chem.*, **267**, 10199 (1992).

16. R. C. Bray and L. S. Meriwether, Electron spin resonance of xanthine oxidase substituted with molybdenum-95, *Nature (London)*, **212**, 467 (1966).

17. B. Krüger and O. Meyer, Structural elements of bactopterin from *Pseudomonas carboxydoflava* carbon monoxide dehydrogenase, *Biochim. Biophys. Acta*, **912**, 357 (1987).

18. R. P. Burns and C. A. McAuliffe, 1,2-Dithiolene complexes of transition metals, *Adv. Inorg. Chem. Radiochem.*, **22**, 303 (1979).

19. A. Abelleira, R. D. Galang and M. J. Clarke, Synthesis and electrochemistry of pterins coordinated to tetraammineruthenium(II), *Inorg. Chem.*, **29**, 633 (1990).

20. S. J. N. Burgmayer, A. Baruch, K. Kerr and K. Yoon, A model reaction for Mo(VI) reduction by molybdopterin, *J. Am. Chem. Soc.*, **111**, 4982 (1989).

21. B. Fischer, J. Strähle and M. Viscontini, Synthese und Kristallstruktur des ersten chinoiden Dihydropterinmolybdän(IV)-Komplexes, *Helv. Chim. Acta*, **74**, 1544 (1991).

22. R. Hille, G. N. George, M. K. Eidsness and S. P. Cramer, EXAFS analysis of xanthine oxidase complexes with alloxanthine, violapterin, and 6-pteridylaldehyde, *Inorg. Chem.*, **28**, 4018 (1989).

23. R. J. Greenwood, G. L. Wilson, J. R. Pilbrow and A. G. Wedd, Molybdenum(V) sites in xanthine oxidase and relevant analog complexes: comparison of oxygen-17 hyperfine coupling, *J. Am. Chem. Soc.*, **115**, 5385 (1993).

24. D. Dowerah, J. T. Spence, R. Singh, A. G. Wedd, G. L. Wilson, F. Farchione, J. H. Enemark, J. Kristofzski and M. Bruck, Electrochemical and electron paramagnetic resonance models for molybdenum(VI/V) centers of the molybdenum hydroxylases and related enzymes, *J. Am. Chem. Soc.*, **109**, 5655 (1987).

25. Z. Xiao, C. G. Young, J. H. Enemark and A. G. Wedd, A single model displaying all the important centers and processes involved in catalysis by molybdoenzymes containing $[Mo^{VI}O_2]^{2+}$ active sites, *J. Am. Chem. Soc.*, **114**, 9194 (1992).

26. R. Söderlund and T. Rosswall, The nitrogen cycles, in *Biochemistry of the Elements* (ed. E. Frieden), Vol. 1, Part B, Plenum Press, New York, 1982, p. 61.

27. S. J. Ferguson, Denitrification: a question of the control and organization of electron and ion transport, *Trends Biochem. Sci.*, **12**, 354 (1987).
28. A. Müller and W. E. Newton (eds.), *Nitrogen Fixation: The Chemical–Biochemical–Genetic Interface*, Plenum Press, New York, 1983.
29. D. J. Lowe, R. N. F. Thorneley and B. E. Smith, Nitrogenase, in *Metalloproteins* (ed. P. Harrison), Part 1, Verlag Chemie, Weinheim, 1985, p. 207.
30. E. I. Stiefel, H. Thomann, H. Jin, R. E. Bare, T. V. Meorgan, S. J. N. Burgmayer, C. L. Coyle, Nitrogenase, in *Metal Clusters in Proteins* (ed. L. Que, Jr), ACS Symposium Series 372, 1988, p. 372.
31. B. E. Smith and R. R. Eady, Metalloclusters of the nitrogenases, *Eur. J. Biochem.*, **205**, 1 (1992).
32. R. A. Henderson, G. J. Leigh and C. J. Pickett, The chemistry of nitrogen fixation and models for the reactions of nitrogenase, *Adv. Inorg. Chem. Radiochem.*, **27**, 198 (1983).
33. P. Pelikan and R. Boca, Geometric and electronic factors of dinitrogen activation on transition metal complexes, *Coord. Chem. Rev.*, **55**, 55 (1984).
34. M. M. Georgiadis, H. Komiya, P. Chakrabarti, D. Woo, J. J. Kornuc and D. C. Rees, Crystallographic structure of the nitrogenase iron protein from *Azotobacter vinelandii*, *Science*, **257**, 1653 (1992).
35. J. Kim and D. C. Rees, Structural models for the metal centers in the nitrogenase molybdenum–iron protein, *Science*, **257**, 1677 (1992).
36. J. Kim and D. C. Rees, Crystallographic structure and functional implications of the nitrogenase molybdenum–iron protein from *Azotobacter vinelandii*, *Nature (London)*, **360**, 553 (1992).
37. A. E. True, M. J. Nelson, R. A. Venters, W. H. Orme-Johnson and B. M. Hoffman, ^{57}Fe hyperfine coupling tensors of the FeMo cluster in *Azotobacter vinelandii* MoFe protein: determination by polycrystalline ENDOR spectroscopy, *J. Am. Chem. Soc.*, **110**, 1935 (1988).
38. H. Deng and R. Hoffmann, How N_2 might be activated by the FeMo cofactor of nitrogenase, *Angew. Chem. Int. Ed. Engl.*, **32**, 1062 (1993).
39. J. Ho, R. J. Drake and D. W. Stephan, $[Cp_2Zr(\mu\text{-PPh})]_2[((THF)_3Li)_2(\mu\text{-N}_2)]$: a remarkable salt of a zirconene phosphinidene dianion and lithium dication containing side-bound dinitrogen, *J. Am. Chem. Soc.*, **115**, 3792 (1993).
40. D. Sellmann, W. Soglowek, F. Knoch and M. Moll, Transition metal complexes with sulfur ligands. 48. Nitrogenase model compounds: $[\mu\text{-N}_2\text{H}_2\{Fe('NHS}_4')\}_2]$, the prototype for the coordination of diazene on iron–sulfur center and its stabilization via strong N—H···S hydrogen bonds, *Angew. Chem. Int. Ed. Engl.*, **28**, 1271 (1989).
41. K. Dehnicke and J. Strähle, The transition-metal–nitrogen multiple bond, *Angew. Chem. Int. Ed. Engl.*, **20**, 413 (1981).
42. R. R. Schrock, R. M. Kolodziej, A. H. Liu, W. M. Davis and M. G. Vale, Preparation and characterization of two high oxidation state molybdenum dinitrogen complexes: $[MoCp*Me_3]_2(\mu\text{-N}_2)$ and $[MoCp*Me_3](\mu\text{-N}_2)[WCp'Me_3]$, *J. Am. Chem. Soc.*, **112**, 4338 (1990).
43. R. R. Schrock, T. E. Glassman and M. G. Vale, Cleavage of the N—N bond in a high-oxidation-state tungsten or molybdenum hydrazine complex and the catalytic reduction of hydrazine, *J. Am. Chem. Soc.*, **113**, 725 (1991).
44. B. B. Kaul, R. K. Hayes and T. A. George, Reactions of a resin-bound dinitrogen complex of molybdenum, *J. Am. Chem. Soc.*, **112**, 2002 (1990).
45. C. J. Pickett and J. Talarmin, Electrosynthesis of ammonia, *Nature (London)*, **317**, 652 (1985).
46. J. R. Chisnell, R. Premakumar and P. E. Bishop, Purification of a second alternative nitrogenase from a nifHDK deletion strain of *Azotobacter vinelandii*, *J. Bacteriol.*, **170**, 27 (1988).
47. A. Müller, R. Jostes, E. Krickemeyer and H. Bögge, Zur Rolle des Heterometall-Atoms im N_2-reduzierten Protein der Nitrogenase, *Naturwissenschaften*, **74**, 388 (1987).
48. R. M. Garrels and C. L. Christ, *Solutions, Minerals, and Equilibria*, Freeman & Cooper, San Francisco, 1965.
49. C. D. Garner, J. M. Arber, I. Harvey, S. S. Hasnain, R. R. Eady, B. E. Smith, E. de Boer

and R. Wever, Characterization of molybdenum and vanadium centers in enzymes by x-ray absorption spectroscopy, *Polyhedron*, **8**, 1649 (1989).

50. S. Ciurli and R. H. Holm, Insertion of $[VFe_3S_4]^{2+}$ and $[MoFe_3S_4]^{3+}$ cores into a semirigid trithiolate cavitand ligand: regiospecific reactions at a vanadium site similar to that in nitrogenase, *Inorg. Chem.*, **28**, 1685 (1989).

51. D. Rehder, C. Woitha, W. Priebsch and H. Gailus, trans-$[Na(thf)][V(N_2)_2$-$(Ph_2PCH_2CH_2PPh_2)_2]$: structural characterization of a dinitrogenvanadium complex, a functional model for vanadium nitrogenase, *J. Chem. Soc., Chem. Commun.*, 364 (1992).

52. R. Wever and K. Kustin, Vanadium: a biologically relevant element, *Adv. Inorg. Chem.*, **53**, 81 (1990).

53. D. Rehder, Bioinorganic chemistry of vanadium, *Angew. Chem. Int. Ed. Engl.*, **30**, 148 (1991).

54. A. Butler and C. J. Carrano, Coordination chemistry of vanadium in biological systems, *Coord. Chem. Rev.*, **109**, 61 (1991).

55. Y. Shechter, Insulin-mimetic effects of vanadate: possible implications for future treatment of diabetes, *Diabetes*, **39**, 1 (1990).

56. M. Krauss and H. Basch, Is the vanadate anion an analogue of the transition state of RNAse A?, *J. Am. Chem. Soc.*, **114**, 3630 (1992).

57. F. H. Nielsen, Nutritional requirements for boron, silicon, vanadium, nickel, and arsenic; current knowledge and speculation, *FASEB J.*, **5**, 2661 (1991).

58. D. C. Crans, E. M. Willging and S. R. Butler, Vanadate tetramer as the inhibiting species in enzyme reaction *in vitro* and *in vivo*, *J. Am. Chem. Soc.*, **112**, 427 (1990).

59. M. J. Smith, D. Kim, B. Horenstein, K. Nakanishi and K. Kustin, Unraveling the chemistry of tunichrome, *Acc. Chem. Res.*, **24**, 117 (1991).

60. E. Bayer, G. Schiefer, D. Waidelich, S. Scippa and M. de Vincentiis, Structure of the tunichrome of tunicates and their role in vanadium enrichment, *Angew. Chem. Int. Ed. Engl.*, **31**, 52 (1992).

61. J. P. Michael and G. Pattenden, Marine metabolites and the complexation of metal ions: facts and hypotheses, *Angew. Chem. Int. Ed. Engl.*, **32**, 1 (1993).

62. E. Bayer, E. Koch and G. Anderegg, Amavadin, an example of selective vanadium binding in nature. Complexation studies and a new structure proposal, *Angew. Chem. Int. Ed. Engl.*, **26**, 545 (1987).

63. E. M. Armstrong, R. L. Beddoes, L. J. Calviou, J. M. Charnock, D. Collison, N. Ertok, J. H. Naismith and C. D. Garner, The chemical nature of amavadin, *J. Am. Chem. Soc.*, **115**, 807 (1993).

64. J. W. P. M. van Schijndel, E. G. M. Vollenbroek and R. Wever, The chloroperoxidase from the fungus *Curvularia inaequalis*; a novel vanadium enzyme, *Biochim. Biophys. Acta*, **1161**, 249 (1993).

65. A. Butler and J. V. Walker, Marine haloperoxidases, *Chem. Rev.*, **93**, 1937 (1993).

66. J. Arber, E. De Boer, C. D. Garner, S. S. Hasnain and R. Wever, Vanadium K-edge x-ray absorption spectroscopy of bromoperoxidase from *Ascophyllum nodosum*, *Biochemistry*, **28**, 7968 (1989).

67. J. Barrett, P. O'Brien and J. P. de Jesus, Chromium(III) and the glucose tolerance factor, *Polyhedron*, **4**, 1 (1985).

68. A. Müller, E. Diemann and P. Sassenberg, Chromium content of medicinal plants used against diabetes mellitus type II, *Naturwissenschaften*, **75**, 155 (1988).

12 Zinc: Structural and Gene-Regulatory Functions and the Enzymatic Catalysis of Hydrolysis or Condensation Reactions

12.1 Overview

With approximately 2 g per 70 kg body weight, zinc is the second most abundant transition element in the human organism, following iron; the metal also plays an important role in many other living beings [1–5]. Under physiological conditions this element occurs only in the dicationic state; the Zn^{2+} ion is diamagnetic and colorless in its complexes, due to its closed d shell (d^{10} configuration). While the metal ion itself is thus not easily excited electronically there are also no naturally occuring zinc(II) complexes with (colored) tetrapyrrole ligands, although synthetic complexes of this kind are very stable. As a consequence of this, the ubiquituous zinc-containing proteins could only be characterized as such after more advanced analytical methods had become available; today, more than 200 different zinc proteins are known (see Table 12.1). These include numerous essential enzymes which catalyze the metabolic conversion (synthases, polymerases, ligases, transferases) or degradation (hydrolases) of proteins, nucleic acids, lipids, porphyrin precursors and other important bioorganic compounds. Other functions lie in the structural fixation of special rate- and/or stereoselectivity-determining protein conformations in oxido-reductases and in the structural stabilization of insulin, of hormone/receptor complexes or of transcription-regulating factors for the transfer of genetic information. It is thus not surprising that zinc deficiency may lead to severe pathological effects [6] and that the heavier homologues, cadmium and mercury, are toxic not in the least because they may replace zinc in its enzymes (see Sections 17.3 and 17.5).

The wound-healing effect of zinc-containing ointments was already known in the ancient world ($\rightarrow Zn^{2+}$-containing collagenase) and, during the last decades, zinc has increasingly been used as a remedy for growth disorders due to malnutrition ($\rightarrow$interactions between Zn^{2+} and growth hormones [7]). Zinc deficiency may also cause a variety of other symptoms such as a lack of appetite, a reduced sense of taste, an enhanced disposition for inflammations and an impairment of the immune system (AIDS-like symptoms [6]). Particularly high zinc contents have been found in tissues of the fetus and of infants as well as in the reproductive organs, especially in seminal fluid—a further indication of the essential catalytic function of zinc for metabolic processes. The daily zinc requirement which is not always met in contemporary diets and which increases with alcohol consumption has been estimated between 3 mg

Table 12.1 Some zinc-containing proteins

zinc protein	molecular mass (kDa)		ligands	function
carboanhydrase (CA)	30		3 His 1 H_2O	hydrolysis (12.6)
carboxypeptidase (CPA)	34		2 His 1 η^2-Glu 1 H_2O	hydrolysis (12.2, 12.11)
thermolysin	35		2 His 1 η^1-Glu 1 H_2O	hydrolysis (12.2)
alkaline phosphatase	2 × 47	2x $\{$	2 His 1 η^2-Asp H_2O	phosphate ester hydrolysis (14.1)
		2x $\{$	1 His 2 η^1-Asp H_2O?	2 Mg^{2+} centers: 2x $\{$ 1 η^1-Asp / 1 η^1-Glu / 1 Thr / 3 H_2O
5-aminolevulinic acid dehydratase	8 × 35	8x $\{$	3 S 1 N/O	condensation (12.18)
alcohol dehydrogenase (ADH)	2 × 40	2x $\{$	2 Cys 1 His 1 H_2O	oxidation of 1° or 2° alcohols via NAD^+ (12.19)
		2x	4 Cys	
glyoxalase	2 × 23	2x $\{$	2 His 2 Glu? 2 H_2O	reduction of α-dicarbonyl compounds by glutathione (12.21)
superoxide dismutase (SOD)	2 × 16	2x $\{$	2 His 1 μ-His$^-$ 1 Asp	disproportionation of $O_2^{\bullet -}$ (10.18)
transcription factors	TFIIIA: 40	n x $\{$	2 His 2 Cys	structural function: formation of specifically folded domains
	GAL4: 17		2x 4 Cys	
insulin hexamer	6 × 6	2 x $\{$	3 His n L	structural function: stabilization of oligomeric storage forms
metallothionein	6		≤ 7x 4 Cys	transport and storage protein (?)

(small children) and 25 mg (pregnant women). A relatively large tolerance exists for higher doses before symptoms of zinc poisoning become manifest.

From a chemical point of view, the most obvious biologically effective function of divalent zinc is its Lewis acidity, i.e. its ability to catalyze condensation and (in reverse) hydrolysis reactions at *physiological* pH by polarizing the substrates, including H_2O (12.1). Such reactions include the polymerization of RNA (condensation) or the cleavage of peptides, proteins or esters (hydrolysis) (12.2).

$$Zn^{2+} \quad \longleftarrow \quad \overset{\delta^-}{} substrate \overset{\delta^+}{} \qquad (12.1)$$

$$R-XH + HO-A \quad \underset{\text{hydrolysis}}{\overset{\text{condensation}}{\rightleftharpoons}} \quad R-X-A + H_2O$$

e.g. $X = NH$, $A = -\underset{O}{\overset{\|}{C}}-R'$ peptidases, lactamases, collagenases; dehydratases

$X = O$, $A = -\underset{O}{\overset{\|}{C}}-R'$ esterases

$X = O$, $A = PO_3^{2-}$ phosphatases, nucleases   (12.2)

In chemical synthesis such reactions are generally catalyzed by strong acids or bases; however, with the exception of gastric fluid (see Figure 13.3), such extreme pH conditions are physiologically not feasible. An alternative is thus the use of an electrophilic polarizing agent, viz. a Lewis acidic metal cation with a rather high effective charge [8].

Whereas a direct attack by the Lewis *acidic* metal can occur at nucleophilic substrates, the Zn^{2+} ion can also undergo an 'Umpolung' to a Lewis *basic* species $[Zn-OH]^+$ after reaction with H_2O, a process that is very important for enzymatic hydrolysis. The underlying deprotonation of an aqua complex (12.3) is a well-known reaction that precedes every metal hydroxide precipitation. However, the polymerization of deprotonated aqua complexes (see 8.20) is not possible for an immobilized monofunctional system which is buried inside a protein (see also 5.12). The pK_a value, which is still about 10 for free $[Zn(OH_2)_6]^{2+}$, can drop to less than 7 in enzymatic systems (12.3) while the ability of metal-bound hydroxide to attack electrophilic centers in hydrolyzable substrates is largely retained.

$$\underset{\diagup}{\overset{\diagdown}{}}Zn-OH_2^{2+} \quad \overset{K_a}{\rightleftharpoons} \quad \underset{\diagup}{\overset{\diagdown}{}}Zn-OH^{+} + H^+$$

(water activation!)   (12.3)

The function of substrate activation requires first of all a firm anchoring of the metal ion in the enzyme; like Cu^{2+}, Zn^{2+} forms kinetically inert bonds in particular to histidine (see Section 2.3.1). Zinc(II) ions thus differ from other, in some regards quite similar divalent ions such as Mg^{2+}, high-spin Mn^{2+}, Fe^{2+} or Co^{2+} which also have a lower Lewis acidity [8]. In contrast to Cu^{2+} and Ni^{2+}, however, Zn^{2+}

is not redox active which excludes unwanted electron-transfer processes; due to the d^{10} configuration (no ligand field effects) and the position in the periodic table, zinc(II) centers prefer rather low coordination numbers and exhibit isotropic, i.e. spatially not directed polarization effects. The apoenzyme can thus well determine the metal coordination geometry, and, during enzymatic catalysis, rather large substrates may coordinate to the metal in such a way that a distorted geometry resembling the transition state of the reaction becomes sterically possible. In fact, the entatic state concept (Section 2.3.1) had been developed in view of the first structures of zinc-containing enzymes with their typically unsaturated metal coordination regarding amino acid ligands [4,5]. In contrast to the kinetically inert bonds between the metal and some protein side-chains, the activation of water ($\rightarrow$hydrolysis function) requires a very labile binding of this molecule during catalysis; in fact, Zn^{2+} belongs to those metal ions that exchange water very rapidly.

The ligand field stabilization phenomenon (see 2.9) which often contributes to the adoption of an octahedral configuration is not effective in ions with a filled (Zn^{2+}), half-filled (high-spin Mn^{2+}) or empty d shell (Mg^{2+}). Physicochemical measurements have strongly benefited from the possibility of an isomorphous metal exchange in zinc enzymes between Zn^{2+} and high-spin Co^{2+} (d^7); the latter does not show a marked preference for octahedral (O_h) versus tetrahedral (T_d) coordination [9] due to a rather favorable situation with completely filled bonding e orbitals (T_d) versus not yet filled bonding t_{2g} orbitals in the O_h situation (12.4). The open d shell of the Co^{2+} ion, however, allows several electronic transitions in the visible region of the electromagnetic spectrum. Therefore, the presumed isomorphous substitution of Zn^{2+} by Co^{2+} has often been used e.g. in the determination of pK_a values of ionizable groups in the coordination sphere of the metal ion.

d electron configuration for high-spin Co^{2+} (d^7):

octahedral symmetry, O_h tetrahedral symmetry, T_d

$$(12.4)$$

Starting from the typical coordination number four for Zn^{II}, the additional coordination of a substrate (metal catalysis as a reaction of coordinated ligands) leads to structures with five-coordinate metal. Such systems do not unambiguously prefer one of the prototypical geometries, the square pyramid or the trigonal bipyramid (12.5), and thus do not impose a metal-based restriction on the substrate specificity of the enzyme (protein $\rightarrow$ selectivity; metal $\rightarrow$ activity, see Section 2.3.1).

In addition to its polarization function in hydrolyzing or condensation-catalyzing enzymes, zinc may also have a structural, conformation-determining role. Special examples include the zinc-containing superoxide dismutase (Section 10.5) where Zn^{II} binds and activates a crucial histidin(at)e ligand, or alcohol dehydrogenase (Section 12.5) where one of the two different Zn^{II} centers positions and electronically activates the substrate for the redox reaction with an organic coenzyme. Protein-bound zinc is not only coordinated to histidine but also, in part or even exclusively, to negatively trigonal bipyramid   square pyramid   (12.5)

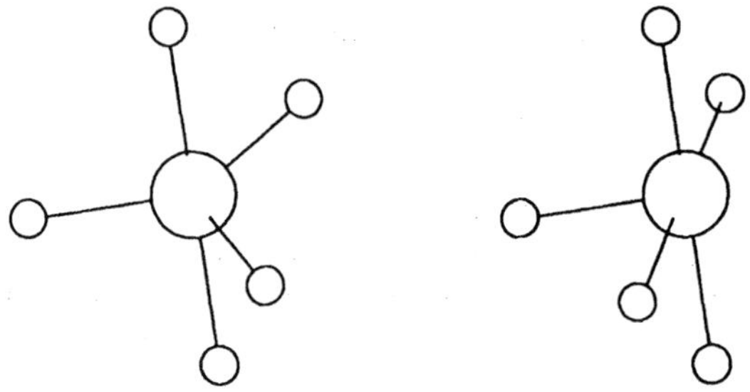

charged sulfur (cysteinate) or oxygen (e.g. glutamate) donor ligands. In addition to a rather variable coordination number, Zn^{2+} thus shows a useful flexibility with regard to the kind and charge of coordinated amino acid side chains. Unlike copper, most of the biological zinc is found *inside* cells. As a first, well-documented example of zinc-containing enzymes the carboanhydrase is introduced in the following section.

12.2   Carboanhydrase (CA)

Carboanhydrases catalyze the hydrolysis equilibrium (12.6) for CO_2.

$$H_2O \; + \; CO_2 \; \rightleftharpoons \; HCO_3^- \; + \; H^+ \qquad\qquad (12.6)$$

This reaction, which normally proceeds quite slowly (see 12.7), can be enzymatically accelerated by a factor of 10^7; some forms of carboanhydrase have thus been considered as 'perfectly evolved enzymes' with maximal turnover and diffusion-controlled reactivity. Therefore, it is not surprising that the details of the enzymatic catalysis of this apparently simple inorganic reaction (12.6) have been studied with great interest [10], including quantum chemical approaches [11,12]. Carboanhydrases are biologically very important enzymes which play an essential role in processes like photosynthesis (efficient CO_2 uptake), respiration (rapid CO_2 disposal), (de)calcification (formation and degradation of carbonate-containing skeletons, Section 15.3.2), or in pH control (buffering). In human erythrocytes, for example, one isozymic form of CA is the most abundant protein component after hemoglobin. Carboanhydrase is an important component for the necessary 'bio'-recycling of CO_2, following the additional, anthropogenically caused CO_2 introduction into the atmosphere with the presumable result of an enhanced 'greenhouse effect'.

Due to the quite varied sites of utilization in higher organisms there are often several, structurally very similar but differently effective and pH-dependent variants (isozymes) of the enzyme carboanhydrase. Figure 12.1 shows a structural representation of human CA, form II(c). The protein is of medium size ($4 \times 4 \times 5.5$ nm), composed of 259 amino acids with a molecular mass of approximately 30 kDa; the dipositive zinc ion is coordinated to three neutral histidine residues and located at the bottom of a 1.6 nm deep conical cleft which contains hydrophilic and lipophilic regions.

The structural analysis of substrate-free enzyme shows that the fourth coordination site is occupied by a water molecule which is connected to other amino acid side

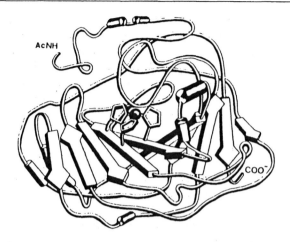

Figure 12.1
Structural representation of human carboanhydrase II, showing the protein folding and the Zn^{2+} coordination to three five-membered imidazole rings of histidine side chains (from [13])

chains and water molecules via hydrogen bonds (see 12.10a). The coordination geometry of zinc is a distorted tetrahedron; the metal can be removed by chelators such as 2,2′-bipyridine, leaving behind an inactive apoprotein. As in many other proteins, the water molecules retained by hydrogen bonds constitute an integral component for structure *and* function, inside and at the surface of the enzyme (Figure 12.2). In the case of carboanhydrase, the ordered network of water molecules (Figure 12.3) is of special importance because CA is a very efficient *hydrolysis enzyme* [15].

Figure 12.2
Protein backbone structure of human carboanhydrase I, supplemented by 503 calculated water molecules (two zinc-coordinated, 16 internal and 485 surface-bound H_2O; from [14])

In addition to quantum-mechanical calculations, many detailed experimental studies involving, for example, ^{18}O-labeling and measurements of the effects of metal substitution, H/D exchange, inhibitors or pH on reaction rates and equilibrium position, have suggested that the rate determining step in the CO_2 hydrolysis by CA is not the CO_2/HCO_3^- conversion but rather a proton shuttling under participation of amino acid side chains and the H_2O network [11,14,15]. Before current mechanistic hypotheses can be discussed, the problems of the inorganic reaction (12.6) should be pointed out.

As can be seen from the equilibrium (12.6), the hydrolysis of CO_2 is a strongly pH-dependent process. In addition to 'physically' dissolved (hydrated) carbon dioxide, the carbonic acid molecule, H_2CO_3, which is present in small amounts (12.7) has to be considered. The astonishingly slow hydration of CO_2 with a half-life of about 20 s at 25 °C [2] cannot be influenced significantly by conventional acid or base catalysis. Due to the symmetrical character of linear O=C=O, i.e. a molecule without a permanent dipole moment, the activation of this polarizable molecule (12.8) requires a combination of (Lewis) acid attack at the oxygen *and* (Lewis) base attack at the carbon atom ('push-pull effect', bifunctional catalysis [16]).

$$H_2O + CO_2 \underset{}{\overset{K_1}{\rightleftharpoons}} H^+ + HCO_3^-$$

$$K_3 \diagdown \quad \diagup K_2$$

$$H_2CO_3$$

equilibrium constants:
$K_1 = 4.44 \times 10^{-7}$
$K_2 = 1.72 \times 10^4 \ M^{-1}$
$K_3 = 2.58 \times 10^3 \ M$

$$HO^- + CO_2 \underset{k_{-4}}{\overset{k_4}{\rightleftharpoons}} HCO_3^-$$

rate constants
(uncatalyzed):
$k_4 \ = {\sim}8500 \ M^{-1}s^{-1}$
$k_{-4} = {\sim}2 \times 10^{-4} s^{-1}$ (12.7)

$$\overset{\delta-}{}\ \overset{\delta+}{}\ \overset{\delta-}{}$$
$$\langle O = C = O \rangle \longrightarrow \text{(Lewis) acid}$$
$$\uparrow$$
$$\text{(Lewis) base}$$ (12.8)

Non-enzymatic catalysis of this reaction can thus be achieved to some extent by using ambivalent 'amphoteric' systems like arsenic, sulfurous or hypobromous acids or monohydroxo metal complexes [2]. These compounds feature both basic free electron pairs as well as an electron pair acceptor (Lewis acid) function; however, the efficiency of such small, unspecific systems and of model complexes is much smaller than that of carboanhydrase.

In CA, a zinc-bound *hydroxide* group which is formed via the indirect deprotonation of the originally coordinated water ligand caused by histidine-64 serves as the center for nucleophilic attack at the carbon atom of CO_2. Extensive model studies have shown that the Zn^{II}—OH entity which does not polymerize due to shielding by the protein features a nucleophilicity that is sufficient for an attack at CO_2. While the metal center itself continues to attract, orient and polarize the CO_2 molecule in a nearly linear arrangement $Zn{\cdots}O=C=O$ [12], the actual productive attack at the oxygen centers of the substrate occurs from acidic *protons* inside the mentioned

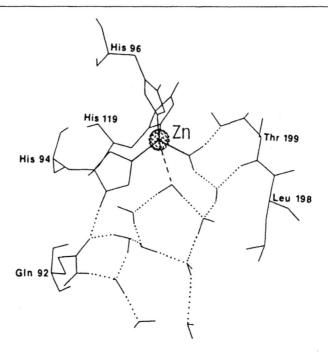

Figure 12.3
Schematic representation of the water network and proton relay system ($\cdots$: hydrogen bonds)
surrounding the zinc ion in carboanhydrase I (from [14])

hydrogen bond network; the very rapid rate of hydrolysis at physiological pH is
favored by the poised 'proton-relay system' depicted in Figure 12.3. Proton shifts in
hydrogen-bonded systems may be extremely rapid, in particular when there is a fairly
rigid structure. In effect, the productive attacks on CO_2 occur from the components
of water, H^+ and OH^-, and the autoprotolysis of water has thus been proposed
as the rate-determining step for catalysis by CA [15]. The most simple reaction
sequence, in accord with the experiments but without structural details, is depicted
in the following (12.9), where E is the apoprotein.

$$[E-Zn-OH]^+ \xrightarrow{CO_2} [E-Zn(OH)CO_2]^+ \rightarrow [E-Zn-HCO_3]^+ \xrightarrow{H_2O} [E-Zn-H_2O]^{2+} + HCO_3^-$$

$$-H^+$$

$$(12.9)$$

Several alternatives have been discussed for the binding of the resulting hydrogen
carbonate in the transition state; an unsymmetrical (4 + 1) coordination (3 N, 2 O;
compare 12.10b) at the metal seems to be the most plausible arrangement [14]. Metal
complexes with hydrogen carbonate ligands are generally very soluble (cf. the water
hardness equilibrium, 15.2) and labile, i.e. they tend to transform into the much more
stable carbonate complexes. However, if the proton in the HCO_3^- ligand is
replaced by an organic residue R, model complexes of the type $L^3Zn(RCO_3)$ can be
isolated (L^3: a tris(pyrazolyl)borate ligand; see 4.14 [17,18]). Studies on correspond-

ing nitrate complexes suggest that the monodentate η^1 coordination of the potential bidentate chelate ligands XO_2^- (X = HCO or NO) is characteristic for L^3Zn^{II} complexes, contributing to facile dissociation and rapid catalysis. Substrates related to hydrogen carbonate such as formate, hydrogen sulfite or sulfonamides inhibit CA reactivity because they bind more strongly to the Zn^{II} center and thus stabilize reaction intermediates.

One of the more detailed hypothetical mechanisms for CA catalysis is depicted in scheme (12.10b) which illustrates intermediate states and includes histidine-64, which is connected to the zinc ion via the water network and features a pH-dependent conformation [11].

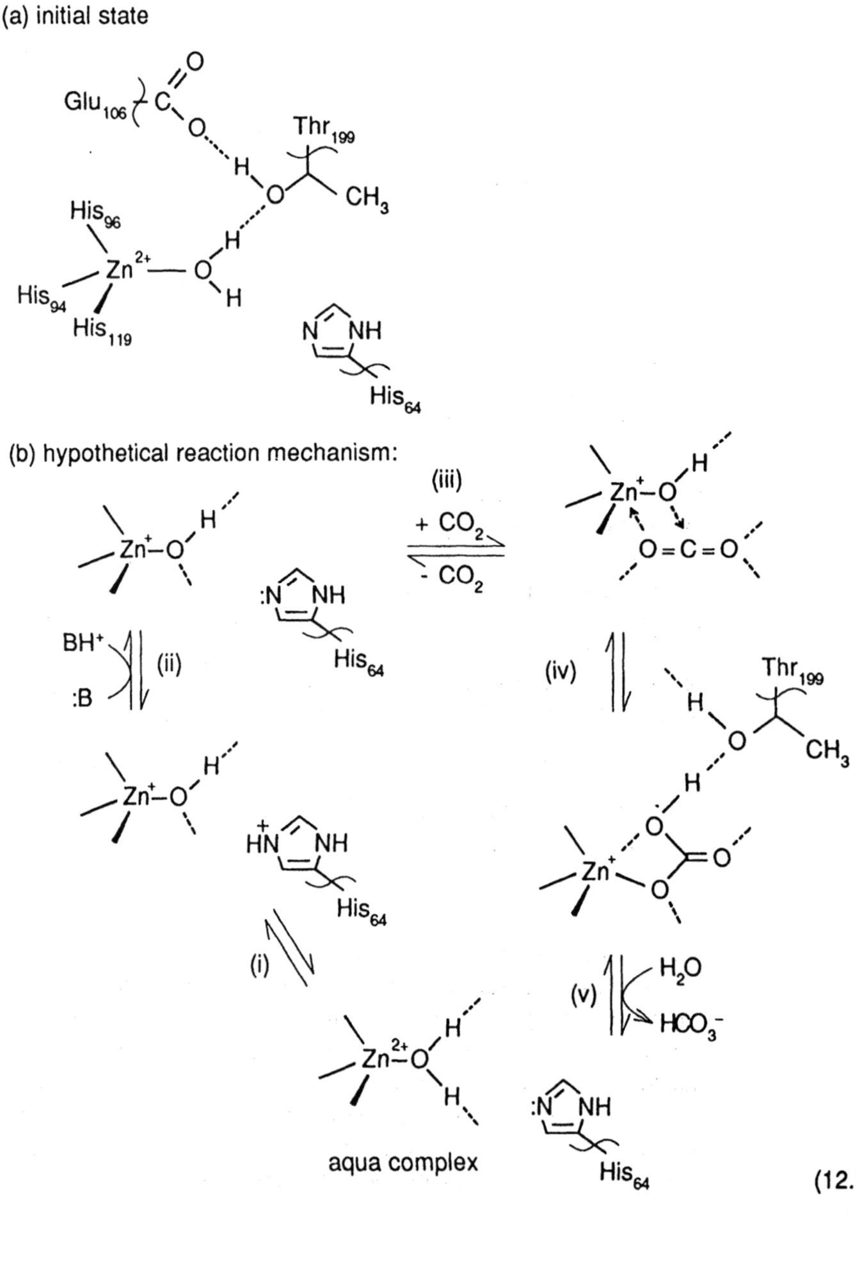

(a) initial state

(b) hypothetical reaction mechanism:

aqua complex

(12.10)

Starting from the aqua complex, a proton is transferred indirectly (i) to histidine-64 via interposed molecules of the water network; one proton as a stoichiometric product of the forward reaction (12.6) is then delivered to buffer molecules situated farther away (ii). The hydroxide ligand remaining at the zinc center very rapidly reacts and forms the transition state (iii) through hydrogen bond interactions with already partly activated CO_2. Following this transition state, the system yields the product complex (iv) and, in a final step (v), the labile HCO_3^- ligand is very rapidly replaced by water.

12.3   Carboxypeptidase A (CPA) and Other Hydrolases

One of the most thoroughly studied peptide-hydrolyzing enzymes is carboxypeptidase A which is a digestive enzyme and typically isolated from the bovine pancreas. According to its specificity it may be more precisely referred to as peptidyl-L-amino acid hydrolase; the enzymatic binding sites for the substrate favor a C-terminal cleavage of L-amino acids with large hydrophobic, preferably aromatic, side chains such as phenylalanine (12.11). Nevertheless, CPA can also catalyze the hydrolysis of certain esters.

$$
\underset{\substack{\text{R–C–NH–CH–COO}^-}}{\overset{\substack{O\quad\quad CH_2Ph\\ \parallel\quad\quad\ \ |}}{}} \ \xrightarrow{H_2O}\ \text{R–C}\!\!\overset{\nearrow OH}{\underset{\searrow O}{}}\ +\ \underset{\substack{|}}{\overset{\substack{CH_2Ph}}{\text{H}_2\text{N–CH–COO}^-}} \tag{12.11}
$$

Carboxypeptidase A has long been investigated with regard to its structure (crystallographic resolution 154 pm), its selectivity, and the underlying reaction mechanism [19]. Although the enzyme itself may not be necessarily essential for the mammalian organism since no deficiency symptoms are known, it can function as a model for metal-containing or even metal-free (serine or thiol) proteases which certainly play a central role in the metabolism.

Crystal structure analyses are not only available for the substrate-free CPA enzyme with 'natural' zinc or some other metal ions [20] but also for various enzyme/inhibitor complexes. It is assumed that the various phases of the multistage enzymatic catalysis can thus be examined in a 'frozen' state. Such studies, in particular on zinc-containing proteases, have provided the experimental basis for a computer-aided 'molecular modeling' of enzyme-substrate and in particular of protease enzyme-inhibitor interactions. Due to the many ways in which small peptides can influence the action of hormones, the search for inhibitors of special Zn-containing metallo-peptidases such as the 'angiotensin converting enzyme' (ACE) has meanwhile led to a very successful development of tailor-made pharmaceuticals, e.g. for the treatment and control of high blood pressure ('drug design' [21]).

CPA is very similar in size to carboanhydrase with about 300 amino acids and a molecular mass of approximately 34 kDa; the catalytically essential zinc center is again located at the bottom of a cavity in the otherwise rather globular protein (Figure 12.4). The metal is coordinated to two histidine ligands and a chelating (η^2-)glutamate anion as well as to one H_2O molecule, this water molecule being linked via hydrogen bonds to serine-197 and glutamate-270. As for carbonic anhydrase, several acidic and basic amino acid residues in the vicinity of the active center are

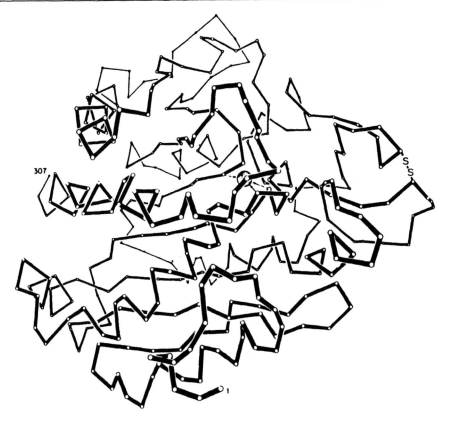

Figure 12.4
Polypeptide chain of carboxypeptidase A (α carbon representation) with Zn^{2+} shown as a sphere (from [22])

important for the activity of the enzyme (see 12.14). In addition to the acidic phenol group of tyrosine-248 these include the basic carboxylate of glutamate-270 as well as the arginine groups Arg-127 and Arg-145, each in their protonated forms $C(\alpha)—CH_2—CH_2—CH_2—NH—C(=^{+}NH_2)NH_2$.

Peptidase and esterase activity differ in some kinetic aspects, e.g. in the rate-determining step, and also in the effect of metal substitution. The peptidase function is restricted mainly to the 'natural' zinc and the Co^{2+} substituted enzyme (see 12.4) while esterase activity can be found with a variety of introduced divalent ions, the activity decreasing in the series $Mn^{2+} > Cd^{2+} > Zn^{2+} > Co^{2+} > Hg^{2+} > Pb^{2+} > Ni^{2+}$. It can thus be assumed that the reaction mechanisms for ester and peptide hydrolysis differ in some details. Also, significant structural differences have been detected after metal substitution, even though the coordination environment remained the same (2 His, η^2-Glu$^-$, H_2O [20]). The deviation from the trigonal bipyramidal configuration (12.5) of the five-coordinate metal was smallest in the 'natural' zinc system followed by that of the Co^{2+} and Mn^{2+} complexes. With Cd^{2+}, Hg^{2+} and particularly Ni^{2+}, an approximately square pyramidal structure (12.5) results which can also be described as an octahedral arrangement with one open coodination site. A direct correlation of these structural variants with the corresponding different reactivities has not yet been established; however, the trigonal bipyramidal arrangement obviously seems to have a greater structural flexibility.

Rapid and selective hydrolysis of peptides and proteins under physiological conditions requires sophisticated catalysis; the half-life for noncatalyzed hydrolysis of the carboxamide bond at pH 7 is about 7 years! However, both the ester and the carboxamide groups are polarizable systems with a carbonyl (acceptor) and an alkoxy or amino (donor) component (12.12). As in the activation of CO_2, a bifunctional attack by electrophilic *and* nucleophilic reagents is thus necessary and promising; due to the hydrolysis character of the reaction this requires the participation of hydrogen bond forming acidic or basic components such as polar amino acid residues or coordinated water. On the other hand, the multitude of hydrolyzable systems (autoproteolysis of the enzyme itself is also a possibility) requires a rather high substrate specificity; therefore, several other enzyme/substrate adhesion or recognition sites have to be present (cooperativity, 'induced fit'), in addition to the enzymatic centers poised for multiple attack at the actual functional group to be cleaved in the substrate. These numerous stringent requirements for a rather common but also selective type of enzyme-catalyzed reactions have made CPA one of the most widely studied enzymes.

carboxylic ester:

$$\overset{\delta-}{O}$$
$$\|$$
$$\overset{-}{O}-\underset{\delta+}{C}-$$

carboxamide:

$$\overset{\delta-}{O}$$
$$\|$$
$$\overset{-}{\underset{|}{N}}-\underset{\delta+}{C}-$$

(12.12)

Two mechanistical hypotheses for CPA-catalyzed hydrolysis are presented in the following. Alternative (12.13) features a direct coordination of the electrophilic metal at the carbonyl oxygen atom of the peptide or ester substrate; as a consequence, the activated carbonyl carbon center is attacked by nucleophilic glutamate-270 to form an intermediate mixed anhydride function, Glu—C(O)—O—C(O)—R [23]. Hydrolysis of that anhydride, e.g. by zinc-coordinated water or hydroxide, leads to the formation of the products.

In fact, there is direct evidence for the occurrence of anhydrides during the hydrolysis of certain substrates by CPA [24]. In contrast, Christiansen and Libscomb [19] suggested a mechanism (12.14) which shows more parallels to the carboanhydrase mechanism with regard to the metal function.

In this scheme, the water ligand bound to five-coordinate zinc is converted to a nucleophilic, metal-modified hydroxide (12.3) through deprotonation by glutamate-270 and can then attack the carbonyl carbon center of the peptide or ester (12.14b).

The electrophilic attack at the carbonyl oxygen atom occurs from the conjugated acid form of an arginine by way of hydrogen bond interactions; hydrogen bonds also participate in the selective binding and further activation of the substrate (12.14).

$$
\begin{array}{c}
\text{Zn}^+ \quad \text{Arg}_{145} \\
\text{H}_2\text{O} \quad {}^+\text{NH}_2 \\
\text{Glu}_{270}\!-\!\text{C}\!-\!\text{O}^- \quad \text{Ph} \; \text{CH}_2 \\
\text{R}\!-\!\text{C}\!-\!\text{O}\!-\!\text{CH}\!-\!\text{COO}^-
\end{array}
\rightleftharpoons
\begin{array}{c}
\text{Zn}^+ \quad \text{Arg}_{145} \\
\text{H}_2\text{O} \; \text{O}^- \quad {}^+\text{NH}_2 \\
\text{Glu}_{270}\!-\!\text{C}\!-\!\text{O}\!-\!\text{C} \; \text{Ph} \; \text{CH}_2 \\
\text{R} \quad \text{CH}\!-\!\text{COO}^-
\end{array}
$$

$$\downarrow\ +\text{H}^+$$

$$
\begin{array}{c}
\text{Zn} \quad \text{Arg}_{145} \\
\text{HO} \quad \text{O} \quad {}^+\text{NH}_2 \\
\text{Glu}_{270}\!-\!\text{C}\!-\!\text{O}\!-\!\text{C}\!-\!\text{R} \quad \text{Ph} \; \text{CH}_2 \\
\text{HOCH}\!-\!\text{COO}^-
\end{array}
\begin{array}{c} +\text{H}^+ \\ \rightleftharpoons \\ -\text{H}^+ \end{array}
\begin{array}{c}
\text{Zn}^+ \quad \text{Arg}_{145} \\
\text{H}_2\text{O} \quad {}^+\text{NH}_2 \\
\text{Glu}_{270}\!-\!\text{C}\!-\!\text{O}\!-\!\text{C}\!-\!\text{R} \quad \text{Ph} \; \text{CH}_2 \\
\text{HOCH}\!-\!\text{COO}^-
\end{array}
$$

$$\downarrow$$

$$
\begin{array}{c}
\text{Zn} \quad \text{Arg}_{145} \\
\text{OH} \; \text{O} \; \text{O} \quad {}^+\text{NH}_2 \\
\text{Glu}_{270}\!-\!\text{C}\!=\!\text{O} \quad \text{C} \quad \text{Ph} \; \text{CH}_2 \\
\text{R} \quad \text{HOCH}\!-\!\text{COO}^-
\end{array}
\begin{array}{c} +\text{OH}^- \\ \longrightarrow \end{array}
\begin{array}{c}
\text{Zn}^+ \quad \text{Arg}_{145} \\
\text{H}_2\text{O} \quad {}^+\text{NH}_2 \\
\text{Glu}_{270}\!-\!\text{C}\!-\!\text{O}^- \quad \text{Ph} \; \text{CH}_2 \\
\text{R}\!-\!\text{C}\!-\!\text{O}^- \quad \text{HO}\!-\!\text{CH}\!-\!\text{COO}^-
\end{array}
$$

$$(12.13)$$

Additional direct polarization of the carbonyl function by the zinc ion in combination with an attack of metal and glutamate-activated water or hydroxide at the carbonyl carbon (12.13, 12.14) ultimately brings about hydrolysis. In any case, the metal site with its single positive net charge serves to electrostatically stabilize and compensate the negative partial charges occurring during enzymatic catalysis, in particular in the transition state.

The structural elucidation of an *amino*peptidase which attacks at the N-terminus of a polypeptide chain [25] (12.15) has shown the unexpected existence of two neighboring zinc centers at a distance of 288 pm; the ligands are glutamate and aspartate groups coordinating through carbonyl or carboxylate oxygen atoms [26]. One of the two zinc centers is coordinately unsaturated with regard to amino acid ligands, as might be expected from the above.

The proteinase (endopeptidase, 12.15) thermolysin is another structurally well-documented zinc-containing protease. This enzyme from *Bacillus thermoproteolyticus* features four-coordinated Ca^{2+} ions (see Figure 2.7 and Section 14.2) and is thus stabilized both thermally and chemically, i.e. against autoproteolysis, the degradation of its own polypeptide backbone. In contrast to peptidases and aminopeptidases, proteinases such as thermolysin, the digestive enzyme astacin [27] or matrix metallo-proteinases (MMPs [28]) do not hydrolyze at one of the termini but within the protein. In collagenases and astacin the zinc ion features three histidine ligands [27]. Zinc-dependent tissue-dissolving MMPs such as collagenases, gelatinases or strome-lysin are essential, for example, for the development of the embryo, for wound healing, tumor metabolism, arthritic processes or the degradation of amyloid proteins

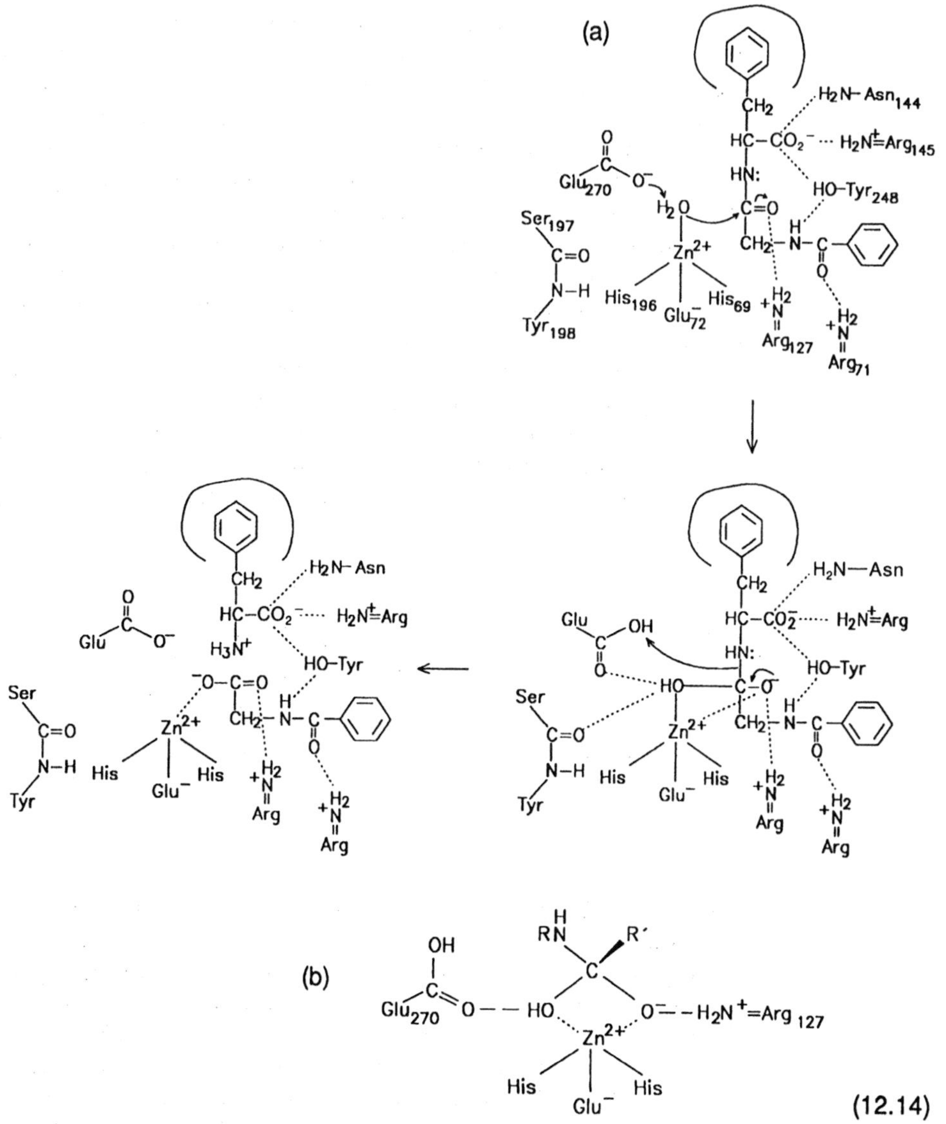

(a)

$$(12.14)$$

$$(12.15)$$

($\rightarrow$Alzheimer syndrome [29]). Characteristically, MMPs are first produced as inactive 'zymogen' forms which can then be activated through the removal of a blocking cysteinate ligand from the catalytic Zn^{2+} center [28]. Essential structural characteristics of zinc-containing proteases which are highly conserved [27,30] include the coordination environment at the metal as well as the relative positions of base (glutamate), zinc, and acid (arginine for CPA, histidine in the case of thermolysin) at the catalytically active site.

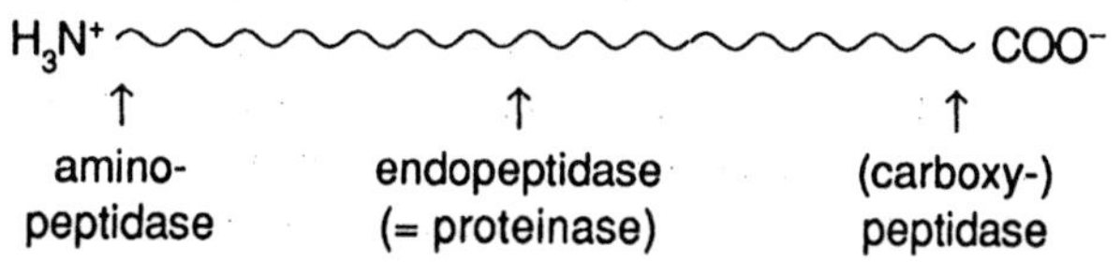

An interesting aspect relating to the problem of autoproteolysis is the occurrence of zinc-containing proteases as essential constituents in the toxins of poisonous snakes which very effectively dissolve connective tissue and inhibit blood clotting. For example, five such aggressive metalloproteases have been isolated from the toxin of the western diamond rattlesnake, *Crotalus atrox*; after removal of the metal by complexation with EDTA they completely lose their activity [31]. The even more effective (neuro)toxins of the tetanus or botulinum type have also been recognized as zinc-dependent proteases which specifically degrade synaptic membrane proteins [32]. On the other hand, nonmetalloproteases such as the aspartyl protease of the human immunodeficiency virus, HIV, may be inhibited by Zn^{2+} through binding of the metal ion close to the active center [33].

Using a model complex (12.16), Groves and Olson [34] were able to demonstrate that a single water ligand coordinated to zinc(II) can in fact be quite acidic ($pK_a \approx 7$; compare 12.3). Furthermore, the hydrolysis of a carboxamide function by metal hydroxide attack at the (unoccupied) π^* orbital of the carbonyl group is accelerated by several orders of magnitude, even in a small model complex (12.16), provided the groups are properly oriented.

$$-H^+ \quad\rightleftharpoons\quad +H^+ \qquad pK_a \approx 7$$

optimal position for
nucleophilic attack

(12.16)

Among the frequently zinc-containing hydrolytic enzymes are not only proteases, peptidases, lactamases and ester-cleaving phospholipases but also phosphate ester-cleaving phosphatases and nucleases [35]. Depending on their pH optimum, phosphatases may be classified as acidic (see Section 7.6.3) or alkaline [36]. Although phosphoryl transfer is the domain of Mg^{2+} catalysis (see Section 14.1), the alkaline phosphatases for phosphate monoester hydrolysis seem to require zinc as well. In its most active form, the enzyme isolated from *E. coli* features three neighboring metal centers, two directly catalytic Zn^{II} centers at 0.4 nm distance and one more remote (0.5–0.7 nm) Mg^{2+} ion (see Table 12.1 [36]). The special requirements for phosphate hydrolysis with a five-coordinate phosphorus atom in the transition state are discussed in more detail in Section 14.1.

12.4 Catalysis of Condensation Reactions by Zinc-Containing Enzymes

Whereas hydrolases (with or without Zn) serve in the cleavage of peptide or (phosphate-)ester bonds there are also zinc-containing enzymes that catalyze the reverse reaction, i.e. the linking of smaller molecules through condensation reactions (12.2). These enzymes have not been as well characterized as the proteases; the better-known representatives include, for example, metalloaldolases which reversibly catalyze metabolically essential aldol condensation reactions (12.17), 5-aminolevulinate dehydratase (12.18), and DNA or RNA polymerases and synthetases (ligases).

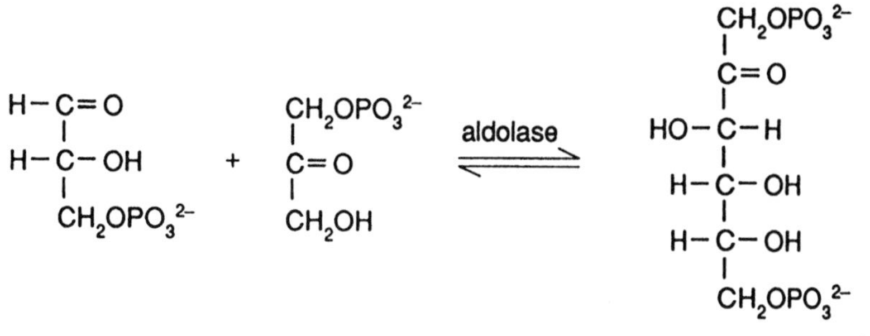

$$
\begin{array}{c}
\text{H–C=O} \\
\text{H–C–OH} \\
\text{CH}_2\text{OPO}_3^{2-}
\end{array}
\quad + \quad
\begin{array}{c}
\text{CH}_2\text{OPO}_3^{2-} \\
\text{C=O} \\
\text{CH}_2\text{OH}
\end{array}
\quad \xrightleftharpoons{\text{aldolase}} \quad
\begin{array}{c}
\text{CH}_2\text{OPO}_3^{2-} \\
\text{C=O} \\
\text{HO–C–H} \\
\text{H–C–OH} \\
\text{H–C–OH} \\
\text{CH}_2\text{OPO}_3^{2-}
\end{array}
$$

(12.17)

The zinc content of the biologically important RNA polymerases (380 kDa) has been established at two metal centers per mole enzyme [37]; however, the details of the coordination environment and of the catalytic role of the metals have yet to be confirmed for those very complex enzymes.

In view of the much debated toxicity of lead (see Section 17.2) it is remarkable that the heavy metal ion Pb^{2+} inhibits a zinc-requiring enzyme which serves in the formation of an essential precursor for the biosynthesis of tetrapyrroles. This enzyme, 5-aminolevulinate dehydratase, catalyzes the condensation of two molecules of 5-aminolevulinic acid to give the functionalized pyrrole 'porphobilinogen' (12.18 [38]).

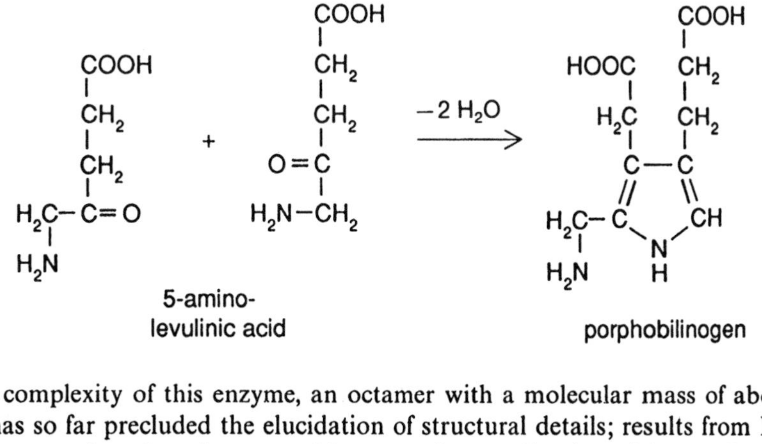

$$
\begin{array}{c}
\text{COOH} \\
\text{CH}_2 \\
\text{CH}_2 \\
\text{H}_2\text{C–C=O} \\
\text{H}_2\text{N}
\end{array}
\quad + \quad
\begin{array}{c}
\text{COOH} \\
\text{CH}_2 \\
\text{CH}_2 \\
\text{O=C} \\
\text{H}_2\text{N–CH}_2
\end{array}
\quad \xrightarrow{-2\,\text{H}_2\text{O}} \quad
\begin{array}{c}
\text{porphobilinogen}
\end{array}
$$

5-amino-
levulinic acid   porphobilinogen   (12.18)

The complexity of this enzyme, an octamer with a molecular mass of about 280 kDa, has so far precluded the elucidation of structural details; results from EXAFS measurements have been interpreted in terms of a coordination of three sulfur ligands and one lighter atom (N or O) per metal center [39].

12.5 Alcohol Dehydrogenase (ADH) and Related Enzymes

Primary alcohols, in particular ethyl alcohol (ethanol), are metabolized in two steps: zinc-containing alcohol dehydrogenases (ADHs) produce aldehyde intermediates which are then oxidized by aldehyde dehydrogenases (ALDHs; see Table 11.1) to

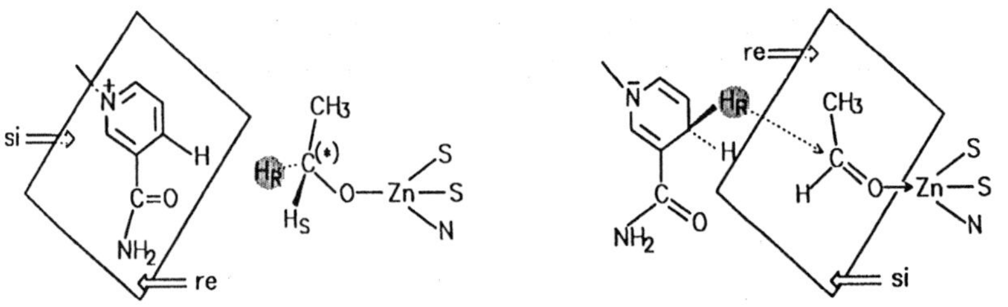

Figure 12.5
Stereospecificity of the ADH-catalyzed oxidation/reduction, using the ethanolate/acetaldehyde substrate pair as an example. Due to the fixed orientation and restricted mobility of the substrate and coenzyme in the metalloenzyme, only the hydrogen atom H_R which, for example, could be labeled as ^2H (D) can be transferred as 'hydride' (formally: $H^- = H^+ + 2 e^-$). The potential chirality (*) at the alcohol is thus transferred to the C(4) center of the (dihydro)pyridine ring (enantioselectivity). The labels R/S or re/si with regard to chiral centers or planes correspond to organic chemical convention (according to [40]).

give carboxylates, e.g. acetate from ethanol. Enzymes which produce or metabolize alcohols are particularly important for grazing animals or fermenting microorganisms such as yeast; furthermore, the stereospecific enantioselective catalysis (see Figure 12.5) of the reverse process of carbonyl *reduction* by ADH has rendered these enzymes extremely interesting for organic synthetic chemistry [41].

The ADH enzyme which has been most thoroughly studied is dimeric liver alcohol dehydrogenase (LADH), isolated from horses. It is a metalloenzyme which features two very different zinc ions per subunit and a molecular mass of about 2×40 kDa. The catalytically active metal center is anchored in the protein by one histidine residue and two negatively charged cysteinate groups; in the resting state it also exhibits a labile water ligand at the fourth coordination site [42,43]. Coordination of two thiolate ligands results in a *neutral* overall charge at the metal site (cf. the charged Zn^{II} centers in CA or CPA) and in the restriction to the coordination number four, even in the transition state, due to the steric demand and the mutual repulsion of the large sulfur donor atoms. The second zinc ion in each subunit is coordinated by four cysteinate ligands and does not directly participate in the enzymatic catalysis; just as the similarly coordinated zinc centers of aspartate transcarbamoylase or of hormone receptor proteins (Section 12.6), it probably has mainly a structural function. Yeast ADH (YADH) and glycerol(1,2)-diol dehydrogenase contain only one zinc center per subunit [44]. Certain bacterial alcohol dehydrogenases may contain presumably six-coordinate high-spin iron(II) with three or four histidine and two or three oxygen ligands [45] instead of Zn^{II} with its typical distorted tetrahedral configuration. Since zinc itself is not redox active, the ADH enzyme requires a dehydrogenase coenzyme, the NAD^+/NADH system (3.12). For a primary alcohol, the chemical equation of the enzyme-catalyzed (reversible) reaction thus has to be written as follows:

$$R-CH_2-OH + NAD^+ \xrightleftharpoons{ADH} R-CHO + NADH + H^+ \qquad (12.19)$$

As suggested by Figure 12.5, the function of the metal is to bind the oxygen atom of the substrate, thus effecting a spatial fixation and, possibly, an additional electronic activation for this strictly stereospecific reaction.

Widely varying reaction rates have been observed with different substrates which include mainly short-chain primary and secondary aliphatic alcohols (aldehydes or ketones as products). From these results one can deduce structures for the arrangement of the substrate in the transition state, involving, for example, a spatial differentiation between large and small substituents at the O-bound carbon atom (12.20).

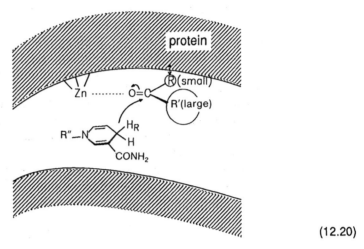

(12.20)

For the most important natural substrates such as simple primary alcohols there exists a specificity which is dependent on the type of ADH enzyme. While methanol is only slowly oxidized to formaldehyde (neurotoxicity of methanol), the higher primary alcohols n-propanol or n-butanol are oxidized quite rapidly to the respective aldehydes; however, further oxidation of these higher aldehydes is relatively sluggish which leads to unpleasant physiological effects such as the 'hangover' syndrome. Slow follow-up oxidation of acetaldehyde by aldehyde oxidases (see Table 11.1) may be genetically caused and occurs quite frequently among people of East Asian origin [46]. High aldehyde intermediate concentrations causing very unpleasant symptoms may also result during alcohol abuse therapy after administering SH-modifying drugs like tetraethylthiuram disulfide, $Et_2N—C(S)—S—S—C(S)—NEt_2$ (disulfiram, antabuse). Other known variations in the tolerance for ethanol, a potentially teratogenic 'social' drug, are based on different ADH availability; for instance, women seem to have generally less ADH enzyme in the gastric system than men so that more nonmetabolized alcohol can pass through the blood stream into the central nervous system. Liver cirrhoses are generally connected with disorders in the zinc metabolism [47].

Although LADH has been very thoroughly studied for decades [42], this enzyme has remained a rather intricate system because of the many possible interactions between protein, metal, coenzyme, substrate and the aqueous medium. The mode of action of the catalytic zinc center in LADH can be described as follows [43]. After coenzyme binding the dimeric enzyme changes its conformation in terms of a rotation from an open to a closed form. The hydrophilic nucleotide section and the actually redox-active nonpolar 1,4-dihydropyridine ring of NADH are then surrounded by different regions of the enzyme subunits. The active site is located about 2 nm deep within the protein at the bottom of rather hydrophobic channels which enable the access of substrate. Coenzyme binding and conformational change lead to the displacement of water molecules from the active center; this and the lack of free charges at the metal site are favoring this catalysis in view of the (formal) transfer of a hydride ($H^- \triangleq H^+ + 2\,e^-$).

While zinc is not necessary for coenzyme binding and the conformational change, it obviously plays an important role in the binding, orientation and activation of the substrate. Binding occurs via the substrate oxygen atom after displacement of the water ligand, a metal alkoxide complex, $[(His)(Cys^-)Zn^{2+}(^-OR)]^-$, is presumably formed as intermediate. The highly specific orientation allows two electrons and one proton to be transferred directly and enantiospecifically from the coenzyme to the substrate (Figure 12.5). As experiments with model complexes between zinc and macrocyclic polyamine ligands have shown [48], the activating function of the metal consists in the electrostatic stabilization of negative partial charges in the transition state by the Lewis-acidic metal center. Not only the transfer of hydroxide, OH^-, by CA and CPA but also the transfer of electrons or a hydride, H^-, require such a stabilization.

Zinc-containing glyoxalase I which participates in the reductive degradation of potentially toxic α-dicarbonyl compounds (12.21 [49]) contains a different organic redox coenzyme, glutathione, G–SH (see Section 16.8), instead of NADH.

$$G\text{–}SH \;+\; H\text{–}C(=O)\text{–}C(=O)\text{–}R \;\rightarrow\; G\text{–}S\text{–}C(=O)\text{–}C(H)(OH)\text{–}R$$

G–SH: glutathione, γ-Glu-Cys-Gly   (12.21)

The metal ion is apparently not coordinated by sulfur ligands and a higher coordination number of up to six is thus possible.

12.6   The 'Zinc Finger' and Other Gene Regulatory Zinc Proteins

The empirically well-established importance of zinc for the growth of organisms and, in particular, its high concentration in the reproductive organs indicate that this metal does not only participate in the catalysis of essential metabolic (anabolic, catabolic) reactions but also in the reliable transfer of genetic information (transcription, replication). However, only since the early 1980s have special proteins become known which recognize DNA base sequences and thus serve in the selective activation and regulatory control of genetic transcription; the prototypical forms contain several protein domains about 30 amino acids long which exhibit invariable zinc-coordinating residues [50]. Because of their originally assumed peptide conformation (Figure 12.6) these modular units have been called 'zinc fingers (zif)' [51].

The zinc content of these factors which regulate gene transcription was discovered by the fact that their stability and function strongly decreased when metal-complexing agents like EDTA were added to the buffer solution. The first transcription factor (TF) IIIA formulated as zinc finger protein was isolated from the ovaries of immature South African clawed frogs (*Xenopus laevis*) and was found to contain approximately nine domains with the amino acid sequence depicted in (12.22); these domains are typically arranged in strings, each domain being conformationally stabilized through

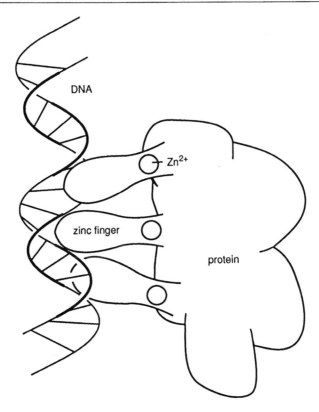

Figure 12.6
Schematic representation of the interaction between DNA and a zinc finger protein

one Zn^{2+} center. In the meantime, such zinc finger motifs with variable 'chain' lengths and DNA binding modes were detected in many other organisms; for instance, about 1% of the human DNA possibly encodes for various zinc finger proteins [51].

$$\underline{Cys}\text{-}X_{2,4}\text{-}\underline{Cys}\text{-}X_3\text{-}Phe\text{-}X_5\text{-}Leu\text{-}X_2\text{-}\underline{His}\text{-}X_{3,4}\text{-}\underline{His} \qquad (12.22)$$

Each one of these domains which could also be synthesized as single units shows a high affinity for the binding of Zn^{2+} and, to a smaller extent, of Co^{2+}, Ni^{2+} or Fe^{2+} [52]; the coordination of two deprotonated cysteine and two neutral histidine side chains to the metal results in a typical, compact folding (structural function of Zn^{2+}). The thus formed protrusions, the zinc 'fingers', feature a characteristic β sheet structure on the Cys^- side and an α helical arrangement on the His side (12.23). According to structure analyses [53], the zinc finger proteins wrap closely around the double-stranded DNA with the actual *zif* regions making multiple specific protein/base pair contacts and thus allowing mutual recognition (Figure 12.6).

The advantages of a modular design using zinc finger domains lie in the variability and the stability. Specifically folded protein loops could also be held together, for example, by disulfide bridges which, however, can be easily cleaved via reduction in contrast to bonds between Zn^{2+} and His or Cys^-.

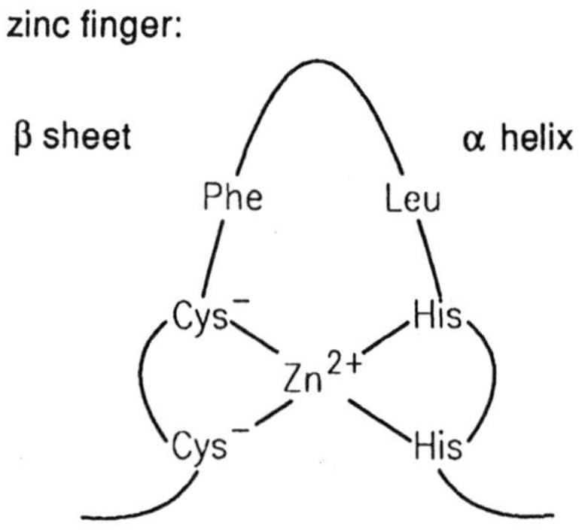

zinc finger:

zinc finger modular unit from TF IIIA   (12.23)

The transcription-activating protein GAL4 from yeast also requires Zn^{2+}; in contrast to the classical zinc finger with its neutral mononuclear $Zn^{II}(His)_2(Cys^-)_2$ entity the GAL4 factor exhibits two neighboring metal centers which are coordinated to a total of six cysteinate residues (12.24 [54]).

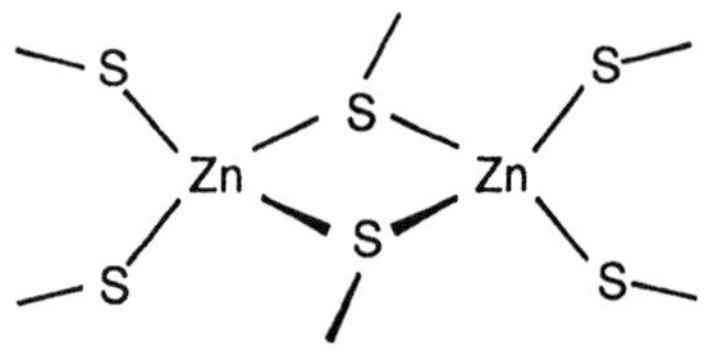

zinc coordination by 6 Cys⁻ in the yeast transcription factor GAL4   (12.24)

A third, structurally related group of zinc-dependent transcription and recognition factors is formed by those hormone receptor proteins that are able to activate genes only after binding of thyroid, glucocorticoid or other steroid hormones [7,51]. In these cases, one zinc atom per domain was found to be surrounded by four cysteinate centers. Zinc binding through cysteinate residues was also found in proteins with repair [55] or developmentally regulatory functions as well as in nucleic acid binding proteins of the HIV retrovirus [50,56].

While the usefulness of coordinatively inert metal ions in maintaining specific protein structures is quite obvious, the high affinity of many gene regulatory proteins [57] for zinc is probably due to the typical tolerance of Zn^{2+} with regard to severe distortions of the tetrahedral coordination. There is no ligand field stabilization energy (compare 12.4) for the zinc(II) ion with its fully occupied d shell.

12.7 Insulin, hGH, Metallothionein and DNA Repair Systems as Zinc-Containing Proteins

A structural function—albeit with a different objective as in the examples mentioned above—can be attributed to Zn^{2+} in its complexes with peptide hormones where it serves to connect two or more single hormone units in oligomeric 'storage forms'. For instance, the dimerization of human growth hormone (hGH) is induced by

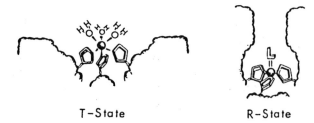

T–State   R–State

Figure 12.7
Arrangement of the imidazole rings of histidine ligands from three different insulin peptide chains around one structurally coordinating Zn^{2+} center in T- and R-states of 2-Zn/insulin hexamer (e.g. L = Cl; according to [58])

coordination (2 His, Glu^-) with two zinc centers [7]. A similar metal-induced oligomerization has been observed for the pancreatic hormone insulin which consists of two relatively short peptide chains. Different modifications of the allosteric insulin hexamers with two or four zinc ions (besides Ca^{2+} ions) have been established as storage forms for insulin in the pancreas and were crystallographically characterized [58]. The six-coordinate metal in the T-state of the 2-Zn form is connected to three histidine ligands of *different* insulin dimers (Figure 12.7); the three remaining coordination sites are occupied by water which should allow a facile and reversible removal of Zn^{2+}, e.g. through chelating agents, in order to mobilize individual insulin dimers. Another, less activated form of the allosteric insulin hexamer is the R-state which features a more closed arrangement and a lower coordination number at the metal [58].

A possible transport and storage system for zinc exists in form of the small, cysteine-rich metallothionein proteins which shall be discussed in more detail in Section 17.3. Their preference for soft metal dications with the favored coordination number four results in particularly efficient binding of Cd^{2+}, the heavier homologue of Zn^{2+}; a heavy metal detoxification function has thus been discussed as another possible major role of these fairly ubiquitous proteins. The large concentration of partially noncoordinated cysteine residues suggests still a third function for metallothioneins, viz. the efficient trapping of damaging oxidizing radicals such as $OH^\cdot$.

$$Cys^{42}-S \diagdown \\ M \cdots S-Cys^{38} \\ Cys^{72}-S \diagup \diagdown S-Cys^{69}$$

$$H_3C \\ | \\ O \\ | \\ P \\ RO \diagup \|\ \diagdown OR \\ O$$

(12.25)

More than a structural role is played by fourfold cysteinate-coordinated zinc(II) in the *Ada* DNA repair protein of *E. coli*. This protein recognizes mutagenic methylated forms like phosphate triesters or 6-*O*-methylated guanine bases in DNA and removes the methyl group by transfer to a special, previously zinc-bound cysteine residue (12.25), which in turn triggers the activation of methylation-resistance genes [55]. Both the activation of the active cysteine and the conformational change ('switching') following methyl transfer may directly involve that particular metal center.

References

1. I. Bertini, C. Luchinat, W. Maret and M. Zeppezauer (eds.), *Zinc Enzymes*, Birkhäuser, Boston, 1986.
2. R. H. Prince, Some aspects of the bioinorganic chemistry of zinc, *Adv. Inorg. Chem. Radiochem.*, **22**, 349 (1979).
3. R. J. P. Williams, The biochemistry of zinc, *Polyhedron*, **6**, 61 (1987).
4. B. L. Vallee and D. S. Auld, Activated-site zinc ligands and activated H_2O of zinc enzymes, *Proc. Natl. Acad. Sci. USA*, **87**, 220 (1990).
5. B. L. Vallee and D. S. Auld, Zinc: biological functions and coordination motifs, *Acc. Chem. Res.*, **26**, 543 (1993).
6. D. Bryce-Smith, Zinc deficiency–the neglected factor, *Chem. Br.*, **25**, 783 (1989).
7. B. C. Cunningham, M. G. Mulkerrin and J. A. Wells, Dimerization of human growth hormone by zinc, *Science*, **253**, 545 (1991).
8. E.-I. Ochiai, Uniqueness of zinc as a bioelement, *J. Chem. Educ.*, **65**, 943 (1988).
9. J. M. Berg and D. L. Merkle, On the metal ion specificity of 'zinc finger' proteins, *J. Am. Chem. Soc.*, **111**, 3759 (1989).
10. F. Botrè, G. Gros and B. T. Storey (eds.), *Carbonic Anhydrase*, VCH, Weinheim, 1991.
11. Y.-J. Zheng and K. M. Merz, Jr, Mechanism of the human carbonic anhydrase II catalyzed hydration of carbon dioxide, *J. Am. Chem. Soc.*, **114**, 10498 (1992).
12. J. Y. Liang and W. N. Lipscomb, Binding of substrate CO_2 to the active site of human carbonic anhydrase II: a molecular dynamics study, *Proc. Natl. Acad. Sci. USA*, **87**, 3675 (1990).
13. A. Liljas, K. K. Kannan, P. C. Bergsten, I. Waasa, K. Fridborg, B. Strandberg, U. Carlbom, L. Järup, S. Lövgren and M. Petef, Crystal structure of human carbonic anhydrase *c*, *Nature (London)*, **235**, 131 (1972).
14. A. Vedani, D. W. Huhta and S. P. Jacober, Metal coordination, H-bond network formation, and protein-solvent interactions in native and complexed human carbonic anhydrase I: a molecular mechanics study, *J. Am. Chem. Soc.*, **111**, 4075 (1989).
15. D. N. Silverman and S. Lindskog, The catalytic mechanism of carbonic anhydrase: implications of a rate-limiting protolysis of water, *Acc. Chem. Res.*, **21**, 30 (1988).
16. R. Breslow, D. Berger and D.-L. Huang, Bifunctional zinc-imidazole and zinc-thiophenol catalysts, *J. Am. Chem. Soc.*, **112**, 3686 (1990).
17. A. Looney, G. Parkin, R. Alsfasser, M. Ruf and H. Vahrenkamp, Pyrazolylboratozinc complexes with reference to carboanhydrase biological function, *Angew. Chem. Int. Ed. Engl.*, **31**, 92 (1992).
18. A. Looney, R. Han, K. McNeill and G. Parkin, Tris(pyrazolyl)hydroboratozinc hydroxide complexes as functional models for carbonic anhydrase: on the nature of the bicarbonate intermediate, *J. Am. Chem. Soc.*, **115**, 4690 (1993).
19. D. W. Christianson and W. N. Lipscomb, Carboxypeptidase A, *Acc. Chem. Res.*, **22**, 62 (1989).
20. D. C. Rees, J. B. Howard, P. Chakrabarti, T. Yeates, B. T. Hsu, K. D. Hardman and W. N. Lipscomb, Crystal structures of metallosubstituted carboxypeptidase A, in *Zinc Enzymes* (eds. I. Bertini, C. Luchinat, W. Maret and M. Zeppezauer), Birkhäuser, Boston, 1986, p. 155.
21. E. W. Petrillo and M. A. Ondetti, Angiotensin-converting enzyme inhibitors: medicinal chemistry and biological actions, *Med. Res. Rev.*, **2**, 1 (1982).

22. W. Lipscomb, Three-dimensional structures and chemical mechanisms of enzymes, *Chem. Soc. Rev.*, **1**, 319 (1972).

23. M. W. Makinen, Assignment of the structural basis of catalytic action of carboxypeptidase A, in *Zinc Enzymes* (eds. I. Bertini, C. Luchinat, W. Maret and M. Zeppezauer), Birkhäuser, Boston, 1986, p. 215.

24. B. M. Britt, W. L. Peticolas, Raman spectral evidence for an anhydride intermediate in the catalysis of ester hydrolysis by carboxypeptidase A, *J. Am. Chem. Soc.*, **114**, 5295 (1992).

25. A. Taylor, Aminopeptidases: structure and function, *FASEB J.*, **7**, 290 (1993).

26. S. K. Burley, P. R. David, A. Taylor and W. N. Lipscomb, Molecular structure of leucine aminopeptidase at 2.7-Å resolution, *Proc. Natl. Acad. Sci. USA*, **87**, 6878 (1990).

27. W. Bode, F. X. Gomis-Rüth, R. Huber, R. Zwilling and W. Stöcker, Structure of astacin and implications for activation of astacins and zinc-ligation of collagenases, *Nature (London)*, **358**, 164 (1992).

28. J. F. Woessner, Jr, Matrix metalloproteinases and their inhibitors in connective tissue remodeling, *FASEB J.* **5**, 2145 (1991).

29. K. Miyazaki, M. Hasegawa, K. Funahashi and M. Umeda, A metalloproteinase inhibitor domain in Alzheimer amyloid protein precursor, *Nature (London)*, **362**, 839 (1993).

30. B. W. Matthews, Structural basis of the action of thermolysin and related zinc peptidases, *Acc. Chem. Res.*, **21**, 333 (1988).

31. J. B. Bjarnason and A. T. Tu, Hemorrhagic toxins from western diamondback rattlesnake (*Crotalus atrox*) venom: isolation and characterization of five hemorrhagic toxins, *Biochemistry*, **17**, 3395 (1978).

32. G. Schiavo, F. Benfenati, B. Poulain, O. Rossetto, P. Polverino de Laureto, B. R. DasGupta and C. Montecucco, Tetanus and botulinum-B neurotoxins block neurotransmitter release by proteolytic cleavage of synaptobrevin, *Nature (London)*, **359**, 832 (1992).

33. Z.-Y. Zhang, I. M. Reardon, J. O. Hui, K. L. O'Connell, R. A. Poorman, A. G. Tomasselli and R. L. Heinrickson, Zinc inhibition of renin and the protease from human immunodeficiency virus type I, *Biochemistry*, **30**, 8720 (1991).

34. J. T. Groves and J. R. Olson, Models of zinc-containing proteases. Rapid amide hydrolysis by an unusually acidic Zn^{2+}–OH_2 complex, *Inorg. Chem.*, **24**, 2715 (1985).

35. P. S. Freemont, J. M. Friedman, L. S. Beese, M. R. Sanderson and T. A. Steitz, Cocrystal structure of an editing complex of Klenow fragment with DNA, *Proc. Natl. Acad. Sci. USA*, **85**, 8924 (1988).

36. J. B. Vincent, M. W. Crowder and B. A. Averill, Hydrolysis of phosphate monoesters: a biological problem with multiple chemical solutions, *Trends Biol. Sci.*, **17**, 105 (1992).

37. F. Y. H. Wu and C. W. Wu, The role of zinc in DNA and RNA polymerases, in *Zinc Enzymes* (eds. I. Bertini, C. Luchinat, W. Maret and M. Zeppezauer), Birkhäuser, Boston, 1986, p. 157.

38. D. Beyersmann, Zinc and lead in mammalian 5-aminolevulinate dehydratase, in *Zinc Enzymes* (eds. I. Bertini, C. Luchinat, W. Maret and M. Zeppezauer), Birkhäuser, Boston, 1986, p. 525.

39. S. Hasnain, E. M. Wardell, C. D. Garner, M. Schlösser and D. Beyersmann, Extended X-ray-absorption fine structure investigations of zinc in 5-aminolevulinate dehydratase, *Biochem. J.*, **230**, 625 (1985).

40. E. S. Cedergren-Zeppezauer, Coenzyme binding to three conformational states of horse liver alcohol dehydrogenase, in *Zinc Enzymes* (eds. I. Bertini, C. Luchinat, W. Maret and M. Zeppezauer), Birkhäuser, Boston, 1986, p. 393.

41. J. D. Wuest (ed.), Formal transfers of hydride from carbon–hydrogen bonds, Tetrahedron Symposia-in-Print Number 25, *Tetrahedron*, **42**, 941–1046 (1986).

42. M. Zeppezauer, The metal environment of alcohol dehydrogenase: aspects of chemical speciation and catalytic efficiency in a biological catalyst, in *Zinc Enzymes* (eds. I. Bertini, C. Luchinat, W. Maret and M. Zeppezauer), Birkhäuser, Boston, 1986, p. 417.

43. H. Eklund, A. Jones and G. Schneider, Active site in alcohol dehydrogenase, in *Zinc Enzymes* (eds. I. Bertini, C. Luchinat, W. Maret and M. Zeppezauer), Birkhäuser, Boston, 1986, p. 377.

44. G. Pfleiderer, M. Scharschmidt and A. Ganzhorn, Glycerol dehydrogenase—an additional Zn-dehydrogenase, *Zinc Enzymes* (eds. I. Bertini, C. Luchinat, W. Maret and M. Zeppezauer), Birkhäuser, Boston, 1986, p. 507.

45. P. Tse, R. K. Scopes and A. G. Wedd, Iron-activated alcohol dehydrogenase from *Zymomonas mobilis*: isolation of apoenzyme and metal dissociation constants, *J. Am. Chem. Soc.*, 111, 8793 (1989).

46. H. W. Goedde and D. P. Agarwal, Polymorphism of aldehyde dehydrogenase and alcohol sensitivity, *Enzyme*, 37, 29 (1987).

47. B. L. Vallee, A synopsis of zinc biology and pathology, in *Zinc Enzymes* (eds. I. Bertini, C. Luchinat, W. Maret and M. Zeppezauer), Birkhäuser, Boston, 1986, p. 1.

48. E. Kimura, M. Shionoya, A. Hoshino, T. Ikeda and Y. Yamada, A model for catalytically active zinc(II) ion in liver alcohol dehydrogenase: a novel 'hydride transfer' reaction catalyzed by zinc(II)-macrocyclic polyamine complexes, *J. Am. Chem. Soc.*, 114, 10134 (1992).

49. B. Mannervik, S. Sellin and L. E. G. Eriksson, Glyoxalase I —an enzyme containing hexacoordinate Zn^{2+} in its active site, in *Zinc Enzymes* (eds. I. Bertini, C. Luchinat, W. Maret and M. Zeppezauer), Birkhäuser, Boston, 1986, p. 518.

50. J. M. Berg, Metal-binding domains in nucleic acid-binding and gene-regulatory proteins, *Prog. Inorg. Chem.*, 37, 143 (1988).

51. D. Rhodes and A. Klug, Zinc fingers, *Sci. Am.*, 268(2), 56 (1993).

52. B. A. Krizek and J. M. Berg, Complexes of zinc finger peptides with Ni^{2+} and Fe^{2+}, *Inorg. Chem.*, 31, 2984 (1992).

53. N. P. Pavletich and C. O. Pabo, Crystal structure of a five-finger GLI-DNA complex: new perspectives on zinc fingers, *Science*, 261, 1701 (1993).

54. P. J. Kraulis, A. R. C. Raine, P. L. Gadhavi and E. D. Laue, Structure of the DNA-binding domain of Zinc GAL4, *Nature (London)*, 356, 448 (1992).

55. L. C. Myers, M. P. Terranova, A. E. Ferentz, G. Wagner and G. L. Verdine, Repair of DNA methylphosphotriesters through a metalloactivated cysteine nucleophile, *Science*, 261, 1164 (1993).

56. A. J. van Wijnen, K. L. Wright, J. B. Lian, J. L. Stein and G. S. Stein, Human H4 histone gene transcription requires the proliferation-specific nuclear factor HiNF-D, *J. Biol. Chem.*, 264, 15034 (1989).

57. T. O'Halloran, Transition metals in control of gene expression, *Science*, 261, 715 (1993).

58. M. L. Brader and M. F. Dunn, Insulin hexamers: new conformations and applications, *Trends Biochem. Sci.*, 16, 341 (1991).

13 Unequally Distributed Electrolytes: Function and Transport of Alkali and Alkaline Earth Metal Cations

13.1 Characterization of K^+, Na^+, Ca^{2+} and Mg^{2+}

All metals discussed so far can be cataloged as 'trace metals' if a daily requirement of less than 25 mg for an adult human being is used as criterion (see Table 2.3). The ions of magnesium and calcium (divalent) and of sodium and potassium (monovalent [1]) certainly do not belong in this category, as can also be inferred from their share in the elemental composition of the human organism (Table 2.1). Because of their high abundance in the earth's crust as well as in sea water these cations are predestined for noncatalytic functions (Figure 2.2); in combination with anions such as chloride they are often referred to as 'electrolytes', 'mass elements' or 'macro nutrients'. For the three metal ions K^+, Na^+ and Ca^{2+}, the (co-) catalytic role comes only second after those functions that require large amounts; the two most important such functions are described in the following:

(a) The construction of supporting and confining *structures* definitely requires large amounts of material. In addition to silicates, the calcium-containing biominerals play a particularly important role as components of endo- and exoskeletons, teeth or (egg) shells (see Chapter 15). Less obvious is the participation of alkali and alkaline earth metal cations in the stabilization of cell membranes and enzyme or polynucleotide (DNA^{n-}, RNA^{n-}) conformations via electrostatic interactions and osmotic effects; many biomolecules are thus denatured in deionized water.

(b) The binding of alkali and alkaline earth metal cations to ligands is generally weak and thus often requires elaborate molecular constructions. On the other hand, this ionic mobility can be used for *information transfer* through the free diffusion of these electrically charged particles which can occur extremely rapidly along a particularly created concentration gradient. While a direct charge *separation* would lead to high potential differences which could be used for chemical synthesis (see Chapter 4), the smaller membrane potential differences used in signal generation result from *concentration differences of individually different ions*. The required specificity is given mainly through the radius-to-charge and surface area-to-charge ratios which characterize simple atomic, i.e. spherical, ions; the four cations discussed in this chapter differ significantly with respect to just these aspects, as shown in Table 13.1. These cations are also significantly different from protons which are ubiquitous due to the autodissociation of water; proton gradients play a very important role in biological energy transfer (chemiosmotic effect; see the ATP synthesis, Section 14.1). Obviously, the maximal possible rate of a chemical reaction under 'diffusion control' limit is desirable for many processes requiring information transfer (compare Figure 13.1).

Figure 13.1
The necessity of rapid information processing (from G. Larson, Copyright 1988 by Universal Press Syndicate, reprinted by permission of Editors Press Service, Inc.)

The main requirement for the utilization of ionic diffusion as one of the most rapid 'chemical' processes is a concentration gradient which has to be maintained in a continuously energy-consuming 'entatic state' (see Section 2.3.1). This situation can be illustrated by a pump storage model as shown in the diagram in Figure 13.2. Here the ions are *actively* 'pumped' through the biological membrane *against* the concentration gradient until a certain stationary nonequilibrium state is reached; the diffusion-controlled concentration equilibration can then occur *passively* via ion channels with differently (chemically, electrically) regulated gate functions.

Some characteristic features of the four cations are summarized in Table 13.1.

From the simple radius-to-charge (r/q) and, in particular, surface area-to-charge ratios, O/q, (Table 13.1: differences corresponding to powers of 2), it is evident that the four cations can have distinctly individual functions in cooperation with size- and

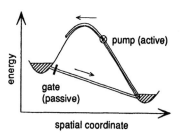

Figure 13.2
Pump storage model for maintaining a local nonequilibrium ion concentration (see text)

Table 13.1 Characteristics of the four biologically relevant alkali and alkaline earth metal cations

property	K$^+$	Na$^+$	Ca^{2+}	Mg^{2+}
ionic radius r (pm)	138	102	100	72
r/q^a	138	102	50	36
relative value r/q (Mg^{2+} = 1)	3.83	2.83	1.39	1.00
surface area O=$4\pi r^2$ of the spherical ion (in pm^2)	239300	130700	125700	65100
relative value O/q (Mg^{2+} = 1)	7.35$\approx 2^3$	4.02$\approx 2^2$	1.93$\approx 2^1$	1.00=2^0
preferred coordination number	6-8	6	6-8	6
preferred coordination centers	O	O	O	O,N,O=P(O$-$)$_3$
preferred type of ligand	multidentate chelate ligands, particularly macrocycles		bidentate ligands, e.g. polycarboxylates	
distribution:^{b}				
in human erythrocytes (intracellular)	92	11	0.1	2.5
in human blood plasma (extracellular)	5	152	2.5	1.5
squid nerve (inside)	300	10	0.0005	7
squid nerve (outside)	22	440	10	55

^{a}Ratio ionic radius/charge [pm/elemental charge] for six-coordinate metal ions (ionic radii from [2])
^{b}In mmol/kg (from [3])

charge-specific ligands. The larger cations in each group of the periodic system tend to prefer even higher coordination numbers than six with a lower coordination symmetry; they are thus well suited for structural, conformation-specific functions in enzymes (see Section 14.2). Numerous enzymes are activated by the coordination of K$^+$ (e.g. pyruvate kinase, dialkylglycine decarboxylase [4]) or Ca^{2+} (see Figure 14.5 [4]). Except for the most polarizing Mg^{2+} ion with its tendency to form strong bonds with N-ligands, e.g. in chlorophyll (see Chapter 4), and with phosphates (see Section 14.1), the other 'hard' cations from Table 13.1 almost exclusively prefer ligands with oxygen donor centers. Negatively charged ligands which contain, for example, η^2-carboxylate groups are usually sufficient to tightly bind alkaline earth metal ions inside a protein due to the strong electrostatic interactions with M^{2+} ions. Alkali metal monocations, on the other hand, can only be retained by multidentate chelate ligands, preferably macrocycles or quasi-macrocycles.

The markedly different concentrations of electrolyte ions in intra- and extracellular compartments are of essential biochemical significance, e.g. for the transfer of information (Table 13.1). While such unequal spatial distributions have been examined in very great detail in the case of the four cations K^+, Na^+, Ca^{2+} and Mg^{2+}, they are, however, also found for anionic electrolytes (Figure 13.3) and for many trace elements: copper, for instance, is mainly found in extracellular, zinc mainly in intracellular regions [7]. For a further discussion it should be kept in mind that the ions of potassium and magnesium as well as hydrogen phosphates are more abundant within cells while Na^+, Ca^{2+} and Cl^- are dominant in the extracellular space. By

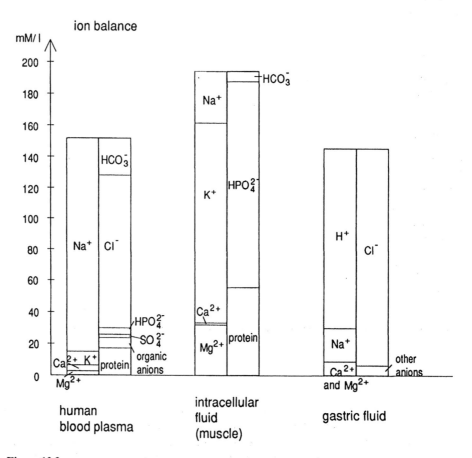

Figure 13.3
Distribution of cationic and anionic electrolytes in three typical fluids of the human body (according to [6])

far the largest transmembrane gradient is that of Ca^{2+}, its intracellular concentration being lower by a factor of 10^4 relative to that outside cells.

The great importance of the four cations discussed here is also evident from the fact that disorders in the metabolism of these 'electrolytes' can severely affect the state of health (remember the flow equilibrium, Figure 2.1). Thus, an excessive intake of sodium ions in combination with high chloride concentrations (e.g. as heavily salted food) is discussed as a major factor contributing to the development of high blood pressure. On the other hand, it seems to be difficult for the aged organism to prevent excretion of the very labile K$^+$ ion because of a disturbed membrane permeability. Magnesium and calcium deficiencies have increasingly become recurring topics of discussions in a health-conscious public and in advertised supplements, e.g. isotonic beverages. A deficiency of magnesium ions can be responsible for diminished mental and physical ability due to its importance in the ATP and general phosphate metabolism [8,9]; severe calcium deficiency, e.g. through insufficient absorption and utilization due to hormonal disorders, can lead to a variety of symptoms, including skeletal diseases (see Sections 14.2 and 15.1).

Detailed knowledge of biochemical reactions involving the cations from Table 13.1 has been obtained only quite recently since their analytical detection is rather difficult. Because of their closed electron shells (electronic configuration of the noble gases), the alkali and alkaline earth metal cations are colorless and diamagnetic; furthermore, they are soluble and very mobile due to the labile bonds formed with normal ligands. For this reason, some spectroscopically better-suited 'Ersatz' ions with characteristics as similar as possible have been used for physical studies.

For sodium, the nuclear magnetic resonance (NMR) spectroscopy of the isotope ^{23}Na (100% natural abundance) with its nuclear spin of $I = 3/2$ has become a widely used method to obtain information on the environment of this ion in biological probes [10]. Unfortunately, the poor time-resolution of this 'slow' spectroscopy (time-scale of the order of seconds) gives only statistically averaged information. For the K$^+$ ion the NMR method is less applicable due to the very small magnetic moment of ^{39}K (93.1%, $I = 3/2$); suitable substitute ions with a comparable radius-to-charge ratio are Ag$^+$ (115 pm), the NMR active ^{205}Tl$^+$ (150 pm; 70.5% natural abundance, $I = 1/2$), and as radioactive isotopes the nuclei ^{42}K (half-life 12.4 h), ^{43}K (22 h), ^{81}Rb (4.6 h) or ^{86}Rb (18.7 d; see also Table 18.1).

Heteroatom Nuclear Magnetic Resonance (NMR)

A prerequisite for nuclear magnetic resonance (NMR) of an isotope is a nuclear spin quantum number $I \neq 0$. In an external magnetic field, such a nucleus can assume $m_I = I, I - 1, \ldots, (-I + 1), -I$ orientations, each with different energies (13.1: $I = 1/2$).

In NMR spectroscopy, transitions are induced between these nuclear spin states which show a slightly different occupation according to the Boltzmann distribution; at magnetic field strengths of a few tesla the resonances are found in the region of the radio frequencies. The resonance energy does not only depend on the nature of the nucleus and its characteristic value γ but also on its electronic and thus chemical environment ($\rightarrow$'chemical shift'). Due to the often very high spectral resolution, even small chemical shift effects of a few

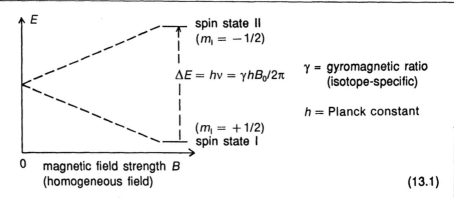

$$\Delta E = h\nu = \gamma h B_0/2\pi$$

γ = gyromagnetic ratio (isotope-specific)

h = Planck constant

magnetic field strength B (homogeneous field)

(13.1)

ppm or less can be detected and interpreted. Interactions between nuclear spins and between nuclear and electron spins (in paramagnetic species) can provide further details of the electronic and geometric structure.

In heteroatom NMR, i.e. NMR spectroscopy of nuclei other than ^1H, some isotopes can be studied only with great difficulties due to small values of γ or because of a large quadrupole moment, when $I > 1/2$. Another problem is the often low natural abundance of spectroscopically interesting nuclei; in view of the generally low sensitivity of this method, relatively high concentrations of isotopically enriched materials have then to be used in order to obtain spectra of sufficient quality.

From a biological point of view, ions such as $^{23}Na^+$, $^{39}K^+$, $^{43}Ca^{2+}$ and $^{25}Mg^{2+}$ are particularly interesting. $^{23}Na^+$ and $^{39}K^+$ NMR spectroscopy can be used to determine the concentrations of these ions in the intra- or extra-cellular regions. For instance, the addition of specific 'paramagnetic shift reagents' in the extracellular region shifts the signal of these ions relative to the unchanged signal of the ions within the cell [11].

Magnesium ions in natural abundance contain only 10% of the isotope $^{25}Mg^{2+}$ with $I = 5/2$ and a small nuclear magnetic moment. However, Mg^{2+} can often be substituted by the paramagnetic $(S = 5/2)$, easily EPR-detectable Mn^{2+} which features a half-filled 3d shell.

For the study of calcium ions there are several alternatives. The NMR-active isotope ^{43}Ca with a nuclear spin of $I = 7/2$ and only 0.13% natural abundance is suitable for biochemical studies only after isotopic enrichment (see Section 14.2). The europium(II) ion Eu^{2+} with a half-filled 4f shell as a substitute with a slightly larger ionic radius (117 pm) than that of Ca^{2+} can be studied via EPR or Mössbauer spectroscopy. For Ca^{2+} it is particularly important to monitor rapid concentration changes on a micromolar scale; Ca^{2+}-specific chelating agents with short response times are used for this purpose as color or fluorescence indicators (see 14.11).

In the following we shall now describe solutions to the complexation and transport problem [5] for alkaline earth and, in particular, alkali metal cations by using the examples of natural and artificial macrocyclic complexing agents [12,13].

13.2 Complexes of Alkali and Alkaline Earth Metal Ions with Macrocycles

Ions such as Na^+, K^+, Mg^{2+} or Ca^{2+} exist in aqueous solution as very labile hydrated 'aqua' complexes $[M(H_2O)_n]^{m+}$ which undergo ligand exchange with water molecules from the surrounding solution within nanoseconds or less. The formation of stable complexes with multidentate chelate ligands L from aqueous solutions is thus always a substitution reaction involving the water ligands, at least from the first coordination sphere (13.2); H_2O molecules from more remote coordination spheres also play an important role, especially with regard to the energy balance.

$$[M(H_2O)_n]^{m+} + L \; \rightleftharpoons \; [ML]^{m+} + nH_2O \qquad (13.2)$$

The generally observed kinetic and thermodynamic stabilization through the formation of chelate complexes can be attributed to several factors. First of all, an increase in the number of free particles typically occurs during reaction (13.2), resulting in increased entropy; the strongly charge-dependent long-range hydration involving the outer coordination spheres must also be considered here. Furthermore, an intelligent architecture of the chelate complex can lead to a number of conformationally and electrostatically favorable interactions between donor atoms and the metal cation in a fixed chelate ring structure and thus to a contribution from the side of enthalpy change. Finally, a 'statistical', i.e. kinetic stabilization results because of the low probability for 'simultaneous' cleavage of *all* metal-donor bonds which would be necessary for the dissociation of a chelate complex. As long as there is only a partial breaking of bonds, the 'virtual' concentration of donor atoms D around the metal center is very high because of the spatial proximity of uncoordinated, but not really free donor centers (13.3 [14]). Therefore, both probability *and* equilibrium effects favor recombination and thus nondissociation of the chelate complex.

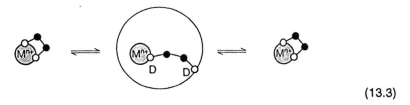

$$(13.3)$$

For the dipositively charged alkaline earth metal ions there is a number of long-established multidentate open-chain chelate ligands such as EDTA available (see 2.1). On the other hand, efficient synthetic complex ligands for the alkali metal monocations have become available only since about 1970 with the design of multidentate macrocyclic ligands (13.4) such as crown ethers [15], cryptands [16] and related components for 'supramolecular chemistry' and 'molecular recognition' (Nobel prize 1987 for Pedersen, Lehn and Cram [15,17,18]).

In synthetic systems such as (13.4) and in the numerous naturally occurring 'ionophores' which could be isolated from fungi or lichen [13,19,20] the metal ion coordination occurs via several strategically placed heteroatom donor centers. These include ether, alcohol or carbonyl oxygen centers in carboxylate, carboxylic ester or carboxamide groups; furthermore, the sulfur analogues of these oxygen donor groups

[18]crown-6   dibenzo[30]crown-10

cryptand[2.2.2]   cryptate (complex)

$$(13.4)$$

or nitrogen components like amine (NR_3), pyridine or imine ($RN\!=\!C$) functions are also suitable. A particular feature of *macrocyclic* complexing ligands is that the ring size can be tailored to fit the metal's ionic radius (size selectivity [21]). In contrast to the essentially planar, tetradentate and also size-selective tetrapyrrole macrocycles (Table 2.7), the efficient ligands for alkaline metal monocations feature a three-dimensional encapsulation of the ion in the complex (see Figures. 13.4 and 2.11).

For this reason, the polycyclic cryptands (13.4) with their largely preformed cavity are superior to the monocyclic crown ethers, *cyclo*-$(OCH_2CH_2)_n$, with regard to complex stability and ion size selectivity [18]. Once a suitable molecular architecture has been designed, the number of individually weak coordinative interactions can lead to a collectively strong and inert coordinative binding of the metal ion by the macrocycle. The sometimes considerable conformational change illustrates the structuring 'template' effect of even weakly coordinating alkali metal cations on suitable polyfunctional substrates which can, for example, wrap around such a template ion (Figure 13.4).

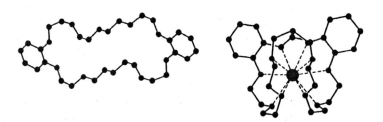

Figure 13.4
Molecular structures of dibenzo[30]crown-10 (left, see 13.4) and its K^+ complex (right), each from single crystal studies (from [16])

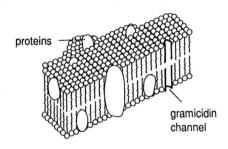

Figure 13.5
Simplified representation of a biological membrane (according to [12])

The biological and physiological [22] as well as organic-synthetic significance of such complexes is related to the fact that the polar heteroatom donor centers of the macrocycle face inward to the metal cation while the outside features alkyl or aryl groups and is thus rather lipophilic. Using such ligands it is therefore possible to dissolve ionic compounds like $KMnO_4$ at least partly in nonpolar organic solvents; on the other hand, this situation allows for a mediated transport of (polar) metal cations through biological membranes (Figure 13.5) with their hydrophobic, approximately 5–6 nm spanning phospholipid double layer. The complexation by macrocycles is thus one of the possibilities (compare Figure 13.8) to effect the passive transport of hydrophilic metal cations through membranes.

Natural ionophores like those shown in (2.10), (13.5) and (13.6) have meanwhile been isolated in large numbers as potential sensor components and as pharmacologically active natural products from fungi, lichen or marine organisms [19,20,23]. Many of these molecules act as antibiotics because they can perturb the stationary ionic nonequilibrium (Section 13.1) and thus the membrane function of bacteria without affecting the more complex ion-transport mechanisms of their higher (host) organisms: the defense function of antibiotics. The two best-known examples, valinomycin (2.10) and nonactin (13.5), illustrate that many of these ionophores are macrocyclic oligopeptides or -esters with a number of chiral centers. This latter feature is of importance regarding recognition and a possible receptor selectivity; furthermore, of the numerous possible conformations there is frequently only one optimal for metal complexation [24]. The K^+/nonactin system with its high coordination number of

nonactin   (13.5)

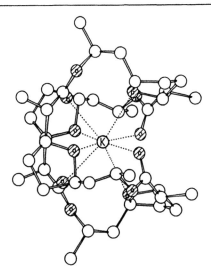

Figure 13.6
Molecular structure of the K^+/nonactin complex in the crystal (oxygen centers hatched)
(according to [25])

eight for the metal center and a macrocycle folded around this ion is quite typical
(Figure 13.6).

In the case of the dodecadepsipeptide valinomycin (2.19) astonishingly high
selectivities of greater than 10^3 are observed for the discrimination between K^+ and
Na^+; however, this differentiation may strongly depend on the solvent [18] (remember (13.2)). Some important representatives such as the acyclic, Na^+-specific
polyether monensin A (13.6 [22]) show a cyclization to give a quasi-macrocycle only
after metal coordination, i.e. hydrogen bonds are then formed between both ends of
the open-chain ligand (Figure 13.7).

monensin A

(⊙ : coordination centers)   (13.6)

The discovery of many more natural and physiologically active ionophores, often
labeled only by a number, and the development of synthetic analogues are being
pursued because of the potential pharmacological activity of such compounds. Both
the selectivity toward the inside, i.e. regarding charge and size of the metal ion, and
toward the outside, e.g. with regard to a membrane bound receptor, have to be taken
into account. The example of the small synthetic cyclic peptide (13.7) illustrates the

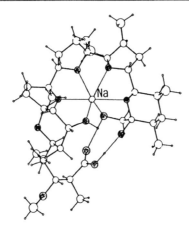

Figure 13.7
Molecular structure of the Na^+/monensin complex in the crystal (oxygen centers as filled circles; according to [26])

possible design of synthetic ionophores; the critical high coordination number for effective K^+ complexation is attained only after linking two peptide rings via a disulfide bridge [27].

$$(13.7)$$

13.3 Ion Channels

The passive cation transport along a concentration gradient via a carrier mechanism (Figure 13.8) proceeds relatively slowly because of the necessary steps of complexation, migration and decomplexation. A more efficient albeit biosynthetically more complex realization of controlled cation diffusion consists in the integration of ion channels of various complexity into the fluid double layer of biological phospholipid membranes (see Figures 13.5 and 13.10).

Ion channels can be formed from integral or quasi-integral membrane proteins; a particularly simple *model* for such proteins is the pentadecapeptide gramicidin A (13.8) which has been used as an antibiotic since 1940. Due to an antiparallel helical aggregation of two molecules, gramicidin A from *Bacillus brevis* forms a channel structure approximately 3 nm long (Figure 13.9) with an inner diameter of 385–547 pm [28,29].

Gramicidin A:

$$HC(O)NH-\text{(L)-Val}-Gly-\text{(L)-Ala}-\text{(D)-Leu}-\text{(L)-Ala}-\text{(D)-Val}-\text{(L)-Val}-\text{(D)-Val}-\text{(L)-Trp}-$$
$$\text{(D)-Leu}-\text{(L)-Trp}-\text{(D)-Leu}-\text{(L)-Trp}-\text{(D)-Leu}-\text{(L)-Trp}-C(O)NH-CH_2-CH_2-OH$$

$$(13.8)$$

Since the thickness of biological phospholipid double-layer membranes is about 5–6 nm, two gramicidin dimer channels have to be arranged in sequence in the fluid membrane in order to permit a very rapid transmembrane cation flow. Ionic transport via such membrane 'pores' (Figure 13.8) is much more efficient than the ionophore-mediated process. While some alkali metal ions (Na^+) and other monopositive ions of suitable size can pass consecutively ('single file') through the channel in rudimentary solvated form, e.g. as monoaqua complexes, the dipositively charged Ca^{2+} ion blocks the gramicidin channel.

As integral membrane proteins the actual ion channels have a much more complex structure. The general construction principle of several groups of ion channels seems to be an arrangement with a four- or fivefold symmetry axis at the center of the pore. Four or five homologous proteins consisting of several transmembrane α helices each form a bundle which confines the pore from the membrane side (Figure 13.10). Not unexpectedly, those parts of the helices which function as immediate lining of the pore contain fixed charges in the form of polar amino acid residues like serine [30–32].

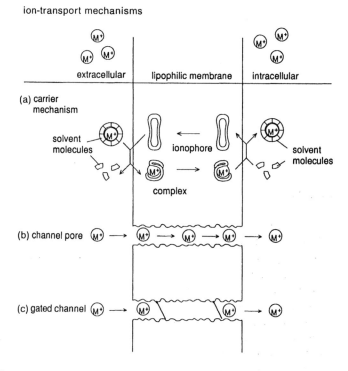

Figure 13.8
Mechanistic alternatives for passive ion transport through membranes (according to [12]; charge compensation not considered)

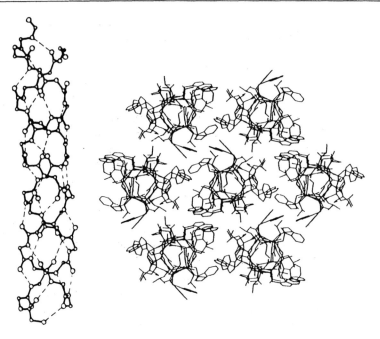

Figure 13.9
Structure of the polypeptide chains in crystalline gramicidin A dimer (left) and a view along hexagonally packed channels in the crystal (right) (from [28], Copyright 1986 by the AAAS)

Of special importance is the entrance region of the channel where negatively charged amino acids promote cation diffusion and thus may contribute to the selectivity and to the gate mechanisms that control the ion flux.

Due to the essential physiological importance of ion channels, the external control of the ion-selective gates through development of suitable inhibitors ('blockers') or stimulating agents has become one of the most active fields of pharmaceutical and medical research (cardiology, oncology, neurology [32]). The 1991 Nobel prize for medicine was thus awarded to E. Neher and B. Sakman for the development of the

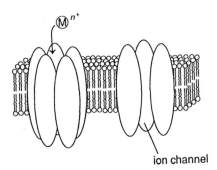

Figure 13.10
Simple model of an ion channel consisting of five homologous transmembrane proteins arranged in an axial symmetric fashion

'patch-clamp' method [33,34] which allows preparation of membrane segments containing a single ion channel; studies with regard to the electrical conductivity, opening time, and reaction of ion channels toward external stimulants have since been greatly facilitated.

In channel proteins the gates (Figure 13.8) are normally closed to guarantee the maintenance of the concentration gradient. The opening of these gates can be influenced by exogeneous or endogeneous low molecular weight compounds ('ligands', e.g. neurotransmitters, nucleotides), by released Ca^{2+} (see Section 14.2), by other proteins, or by a change in the electrical potential difference (voltage) across the membrane. Voltage-controlled channels are thus biological switching elements which serve in the transformation of electrical into chemical signals. The development of receptor-specific organic compounds (compare 14.10) for the blocking of ion channels is one of the main targets of 'molecular modeling'; on the other hand, the unspecific blocking, e.g. of K^+ channels by $^+N(C_2H_5)_4$, Cs^+ or Ba^{2+}, can be easily understood on the basis of size and charge effects. The blocking of K^+ channels in the taste receptors by H^+ is probably responsible for the sensing of 'sour'.

Another example, the blocking of channels in the 'disc'-containing rod cells of the retina (Figure 13.11) which are permeable for Na^+ in their 'resting' state, is an essential step in the transformation of a light stimulus into (electrical) nerve impulses [35,36]. The very sensitive rod cells for black/white vision contain a membrane pigment, 'rhodopsin', formed from the polyene retinal and the protein opsin. The polyene which is connected to the opsin via a protonated azomethine function $C{=}NH^+$ undergoes a single photon-induced isomerization of a double bond (Z-11, 12 → E-11,12) and, via several consecutive steps, this charge shift results in a degradation of cyclic guanosine monophosphate (cGMP). Only in the presence of cGMP, however, is the continuous, energy-consuming flow of Na^+ through the inner membrane of the rod cells maintained ('dark current': entatic state). The cGMP degradation leads to a blocking of the Na^+ channels (Figure 13.11) and thus to a marked ionic 'hyperpolarization' (→amplification) which finally creates an electric signal from this sensory cell.

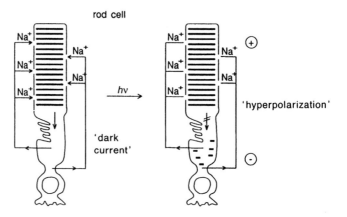

Figure 13.11
Rod cell of the retina with Na^+ channel-containing discs

13.4 Ion Pumps

Passive ion diffusion along transmembrane concentration gradients is possible only because the concentration differences in the stationary 'resting states' are maintained through active, energy-consuming ion-transport mechanisms against the tendency for rediffusion and thus against increasing entropy. The complex protein systems required for this task, the ion pumps, have to operate against controlled (Section 13.3) and uncontrolled charge compensation processes ('leakage'); due to their high energy consumption in the form of hydrolyzable ATP equivalents they belong to the group of ATPase enzymes (14.2 [37]). A large part of the basic, continuous energy requirement of the nonactive cell (flow equilibrium, see Figure 2.1) is used for maintaining the nonequilibrium situation with the help of ion pumps. Even when resting, the daily turnover of continuously recycled ATP in an adult human being corresponds to about half of the body mass. The ion pumps are large, complex systems and thus very sensitive entities; their activity, for example, is strongly temperature dependent. There are usually several kinds of ion pumps for one individual ion; two alternatives can be realized due to the always necessary charge compensation. One alternative is the 'symport' process where cations and anions are simultaneously transported in the *same direction*; in the 'antiport' process, on the other hand, ions of the same charge sign are exchanged by movement in *opposite directions*. When balancing the charges, the protons formed in the hydrolysis of ATP also have to be taken into account (see 13.9) which allows for the existence of some 'uniport' processes involving only one inorganic ion and H^+.

The best-known ion pump is the Na^+/K^+-ATPase, a major component of the sodium/potassium pump system which is essential in creating membrane potentials [38]. Incidentally, the comparable problem of a proton transport across membranes has already been mentioned in the context of electron transfer (H^+/e^- symport) in photosynthesis and respiration (Sections 4.1 and 10.4).

The integral membrane protein Na^+/K^+-ATPase is very similarly structured in all eucaryotic organisms; it consists of two subunit peptide pairs (heterodimer, $\alpha_2\beta_2$) with an overall molecular mass of $2 \times 112(\alpha) + 2 \times 35(\beta) \approx 294$ kDa. The function of this protein oligomer which has not yet been defined in molecular detail is an exchange of sodium and potassium ions in an antiport fashion against the respective concentration gradients (compare Figure 14.3). This process requires Mg^{2+}-catalyzed hydrolysis of ATP and takes places until a certain high-energy state has been reached (stationary nonequilibrium, Figure 13.2). The relevant overall equation reads as in (13.9).

$$3\ Na^+(ic) + 2\ K^+(ec) + ATP^{4-} + H_2O \xrightarrow{\ Mg^{2+}\ }$$
$$3Na^+(ec) + 2\ K^+(ic) + ADP^{3-} + HPO_4^{2-} + H^+$$

ic: intracellular, ec: extracellular space   (13.9)

It is believed that the protein can assume at least two markedly different conformations, E_1 and E_2, in which the binding of the metal ions has to be very different [39]. Figure 13.12 illustrates this hypothesis. Because of the alternatives Na^+/K^+,

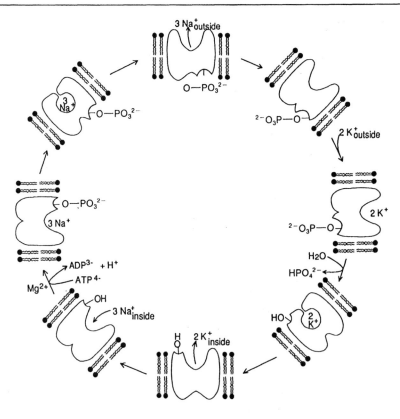

Figure 13.12
Schematic mechanism of the functional cycle of Na$^+$/K$^+$-ATPase (according to [40]; see also Figure 14.3)

ATP^{4-}/ADP^{3-}, and the conformations E$_1$/E$_2$, there have to be at least $2^3 = 8$ essentially different states of this protein system. Functional requirements are the possibility for a translocation of the ions, i.e. their transport between intracellular and extracellular regions, and the (intracellular) energetic coupling with ATP hydrolysis; the dimeric structure of the protein is characteristic and points to a 'flip-flop' mechanism. From an alternative point of view, Na$^+$ co-catalyzes (together with Mg^{2+}, Figure 14.3) the phosphorylation while K$^+$ activates dephosphorylation or at least does not inhibit it (Na$^+$ pump function of Na$^+$/K$^+$-ATPase [41]).

All details known so far, e.g. pertaining to binding sites, have been localized at the larger α protein; the role of the β glycoprotein is not yet well understood [38]. An inhibition of Na$^+$/K$^+$-ATPase is possible with low molecular weight compounds. For example, extracellularly binding steroids like ouabain or the active substance digitoxigenin from *Digitalis* species lead to an increase of the Na$^+$ concentration inside heart muscle cells due to an inhibition of the enzyme function. As a consequence, the Na$^+$/Ca^{2+} exchange through a corresponding antiport system is slowed down [42], leading to increased contractility (cardiotonic activity) due to a rising intracellular Ca^{2+} concentration. Trace amounts of intracellular vanadate(V) also strongly inhibit the Na$^+$/K$^+$-ATPase; due to its larger size and the resulting preference for a coordination number five, the group 5 transition metal vanadium blocks the necessary ATP hydrolysis via stabilization of the enzymatic transition state with a five-coordinated pentavalent element (see 14.7 and Figure 14.3).

Other transport processes are frequently linked to the transport of ions, in particular to that of Na^+; examples include the transmembrane transport of carbohydrates and amino acids, or the modification of proton gradients via the bioenergetically important Na^+/H^+ antiport system [22]. On the other hand, hormones such as steroids, small peptides or the thyroid hormones (16.2) can effectively stimulate the function of Na^+-dependent ATPases. This capacity confers a high pharmaceutical potency to these low molecular weight compounds due to the great importance of the Na^+ transport for the biological energy balance [43], for electrical membrane potentials, and for processes mediated by Ca^{2+}. Another extremely efficient cation pump system is the H^+/K^+-ATPase which, through an antiport process coupled to a K^+/Cl^- symport, is responsible for the extraordinary, more than 10^6 fold enrichment of H^+ in the stomach (pH $\approx$ 1).

While few molecular details are known about anion-specific ion pumps (see [44] and Section 16.4), a passive antiport system HCO_3^-/Cl^- important for respiration (CO_2 disposal; Figure 13.13) was identified in erythrocytes.

Figure 13.13
Function of the HCO_3^-/Cl^- antiport system in erythrocytes (Hb: hemoglobin)

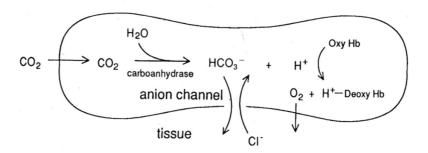

The relatively common hereditary disease cystic fibrosis is known to result from a genetically caused misregulation of chloride channels.

Efforts are being made to understand and reconstitute (Figure 13.14) the extraordinarily effective Ca^{2+}-specific pumps which are monomeric and vanadate-inhibited ATPases with a molecular mass of 134 kDa [45] and which are concentrated

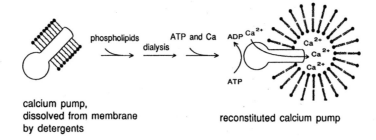

calcium pump,
dissolved from membrane
by detergents

reconstituted calcium pump

Figure 13.14
Stabilization and reconstitution of the Ca^{2+} pump (monomeric protein) using vesicle-forming detergents

in the sarcoplasmic reticulum of muscle cells (Section 14.2). To generate the huge concentration gradient between the inside (approximately 10^{-7} M) and the outside of cells (about 10^{-3} M) these particular ion pumps have to be present in high concentrations and function very efficiently. The ATP/ADP conversion coupled to the Ca^{2+} transport is reversible so that an ATP synthesis is possible via formation of Ca^{2+} concentration gradients, e.g. by Ca^{2+} complexation.

References

1. C. A. Pasternak (ed.), *Monovalent Cations in Biological Systems*, CRC Press, Boca Raton, 1990.
2. W. L. Jolly, *Modern Inorganic Chemistry*, McGraw-Hill, New York, 1984.
3. M. N. Hughes, *The Inorganic Chemistry of Biological Processes*, 2nd edn, Wiley, Chichester, 1981.
4. M. D. Toney, E. Hohenester, S. W. Cowan and J. N. Jansonius, Dialkylglycine decarboxylase structure: bifunctional active site and alkali metal sites, *Science*, **261**, 756 (1993).
5. P. B. Chock and E. O. Titus, Alkali metal ion transport and biochemical activity, *Prog. Inorg. Chem.*, **18**, 287 (1973).
6. A. L. Lehninger, *Biochemistry*, 2nd edn, Worth, New York, 1975.
7. R. J. P. Williams, Inorganic elements in biological space and time, *Pure Appl. Chem.*, **55**, 1089 (1983).
8. N. Brautbar, A. T. Roy, P. Horn and D. B. N. Lee, Hypomagnesemia and hypermagnesemia, and other papers in this volume, in *Metal Ions in Biological Systems* (ed. H. Sigel), Vol. 26, Marcel Dekker, New York, 1990, p. 285.
9. H. Schmidbaur, H. G. Classen and J. Helbig, Asparagine and glutamine as ligands for alkali and alkaline earth metals: structural chemistry contribution of the complex to magnesium therapy, *Angew. Chem. Int. Ed. Engl.*, **29**, 1090 (1990).
10. P. Laszlo, Nuclear resonance spectroscopy with sodium-23, *Angew. Chem. Int. Ed. Engl.*, **17**, 254 (1978).
11. T. Ogino, G. I. Shulman, M. J. Avison, S. R. Gullans, J. A. den Hollander and R. G. Shulman, ^{23}Na and ^{39}K NMR studies of ion transport in human erythrocytes, *Proc. Natl. Acad. Sci. USA*, **82**, 1099 (1985).
12. F. Vögtle, E. Weber and U. Elben, Neutrale organische Komplexliganden und ihre Alkalikomplexe III–Biologische Wirkungen synthetischer und natürlicher Ionophore, *Kontakte (Darmstadt)*, **3**, 32 (1978) and **1**, 3 (1979).
13. B. Dietrich, Coordination chemistry of alkali and alkaline-earth cations with macrocyclic ligands, *J. Chem. Educ.*, **62**, 954 (1985).
14. D. H. Busch and N. A. Stephenson, Molecular organization, portal to supramolecular chemistry, *Coord. Chem. Rev.*, **100**, 119 (1990).
15. C. J. Pedersen and H. K. Frensdorff, Macrocyclic polyethers and their complexes, *Angew. Chem. Int. Ed. Engl.*, **11**, 16 (1972).
16. B. Dietrich, J. M. Lehn and J. P. Sauvage, Kryptate–makrocyclische Metallkomplexe, *Chem. Unserer Zeit*, **7**, 120 (1973).
17. D. J. Cram, Molecular hosts and guests, and their complexes (Nobel address), *Angew. Chem. Int. Ed. Engl.*, **27**, 1009 (1988).
18. J. M. Lehn, Cryptates: the chemistry of macropolycyclic inclusion complexes, *Acc. Chem. Res.*, **11**, 49 (1978).
19. V. Prelog, Role of certain microbial metabolites as specific complexing agents, *Pure Appl. Chem.*, **25**, 197 (1971).
20. B. C. Pressman, The discovery of ionophores: a historical account, in *Metal Ions in Biological Systems* (ed. H. Sigel), Vol. 19, Marcel Dekker, New York, 1985, p. 1.
21. R. D. Hancock, Chelate ring size and metal ion selection, *J. Chem. Ed.*, **69**, 615 (1992).
22. H. H. Mollenhauer, D. J. Morre and L. D. Rowe, Alteration of intracellular traffic by

monensin; mechanism, specificity and relationship to toxicity, *Biochim. Biophys. Acta*, **1031**, 225 (1990).

23. J. P. Michael and G. Pattenden, Marine metabolites and the complexation of metal ions: facts and hypotheses, *Angew. Chem. Int. Ed. Engl.*, **32**, 1 (1993).
24. F. Vögtle, *Supramolecular Chemistry*, Wiley, New York, 1993.
25. B. T. Kilbourn, J. D. Dunitz, L. A. R. Pioda and W. Simon, Structure of the K⁺ complex with nonactin, a macrotetrolide antibiotic possessing highly specific K⁺ transport properties, *J. Mol. Biol.*, **30**, 559 (1967).
26. D. L. Ward, K. T. Wei, J. G. Hoogerheide and A. L. Popov, The crystal and molecular structure of the sodium bromide complex of monensin, $C_{36}H_{62}O_{11} \cdot Na^+Br^-$, *Acta Cryst.*, **B34**, 110 (1978).
27. R. Schwyzer, A. Tun-Kyi, M. Caviezel and P. Moser, S,S'-Bis-cyclo-glycyl-L-hemicystyl-glycyl-glycyl-L-prolyl, ein künstliches bicyclisches Peptid mit Kationenspezifität, *Helv. Chim. Acta*, **53**, 15 (1970).
28. D. A. Langs, Three-dimensional structure at 0.86 Å of the uncomplexed form of the transmembrane ion channel peptide gramicidin A, *Science*, **241**, 188 (1988).
29. B. A. Wallace and K. Ravikumar, The gramicidin pore: crystal structure of a cesium complex, *Science*, **241**, 182 (1988).
30. H. Betz, Homology and analogy in transmembrane channel design: lessons from synaptic membrane proteins, *Biochemistry*, **29**, 3591 (1990).
31. K. S. Åkerfeldt, J. D. Lear, Z. R. Wasserman, L. A. Chung and W. F. DeGrado, Synthetic peptides as models for ion channel proteins, *Acc. Chem. Res.*, **26**, 191 (1993).
32. J.-P. Changeux, Chemical signaling in the brain, *Sci. Am.*, **269**(5), 30 (1993).
33. E. Neher, Ion channels for communication between and within cells (Nobel address), *Angew. Chem. Int. Ed. Engl.*, **31**, 837 (1992).
34. B. Sakmann, Elementary ion flow and synaptic transition (Nobel address), *Angew. Chem. Int. Ed. Engl.*, **31**, 830 (1992).
35. L. Stryer, The molecules of visual excitation, *Sci. Am.*, **257**(1), 32 (1987).
36. J. L. Schnapf and D. A. Baylor, How photoreceptor cells respond to light, *Sci. Am.*, **256**(4), 32 (1987).
37. P. L. Pedersen and E. Carafoli, Ion motive ATPases. I. Ubiquity, properties, and significance to cell function, *Trends Biochem. Sci.*, **12**, 146, 186 (1987).
38. B. C. Rossier, K. Geering and J. P. Kraehenbuhl, Regulation of the sodium pump: how and why?, *Trends Biochem. Sci.*, **12**, 483 (1987).
39. J. D. Robinson and P. R. Pratap, Indicators of conformational changes in the Na⁺/K⁺-ATPase and their interpretation, *Biochim. Biophys. Acta*, **1154**, 83 (1993).
40. P. Karlson, *Kurzes Lehrbuch der Biochemie für Mediziner und Naturwissenschaftler*, 13th edn, Thieme, Stuttgart, 1988.
41. C. L. Bashford and C. A. Pasternak, Plasma membrane potential of some animal cells is generated by ion pumping, not by ion gradients, *Trends Biochem. Sci.*, **11**, 113 (1986).
42. H. Reuter, Ins and outs of Ca^{2+} transport, *Nature (London)*, **349**, 567 (1991).
43. P. A. Dibrov, The role of sodium ion transport in *Escherichia coli* energetics, *Biochim. Biophys. Acta*, **1056**, 209 (1991).
44. M. Ikeda, R. Schmid and D. Oesterhelt, A Cl⁻-translocating adenosinetriphosphatase in *Acetabularia acetabulum*, *Biochemistry*, **29**, 2057 (1990).
45. E. Carafoli, The plasma membrane calcium pump. Structure, function, regulation, *Biochim. Biophys. Acta*, **1101**, 266 (1992).

14 Catalysis and Regulation of Bioenergetic Processes by the Alkaline Earth Metal Ions Mg^{2+} and Ca^{2+}

14.1 Magnesium: Catalysis of Phosphate Transfer by Divalent Ions

Among the bioessential metal cations without redox functions Mg^{2+} is distinguished by its small ionic radius (see Table 13.1) [1,2]. Due to its rather low radius-to-charge ratio and the resulting Lewis acidity, this ion prefers multiply negatively charged ligands, especially polyphosphates. In contrast to the related Zn^{2+} ion with its sometimes similar catalytic activity, Mg^{2+} is definitely a 'hard' electrophile (see Figure 2.6) which does not form inert bonds to simple N and S donor ligands such as histidine or deprotonated cysteine. Furthermore, Mg^{2+} strongly prefers the coordination number six with close to octahedral configuration while other ions with comparable biological functions prefer either lower (Zn^{2+}) or higher (Ca^{2+}) coordination numbers. However, the example of Mg^{2+}-dependent enolase (see 14.9) shows that deviations from this rule are possible due to the 'entatic strain' imposed in an enzyme.

The magnesium ion as a constituent of the chlorophylls has already been introduced in Section 4.2. Further important roles are its carbamate-stabilizing function in the photosynthetic CO_2 fixation by the most abundant enzyme on earth, ribulose-1,5-bisphosphate carboxylase ('rubisco' [3]), and its Ca^{2+}-analogous function in exo- and endoskeletons (see Chapter 15) and in the stabilization of cell membranes. While a long-term magnesium deficiency thus inhibits growth, a temporary lack of Mg^{2+} causes a relative overabundance of Ca^{2+} within cells due to the antagonistic relation between Mg^{2+} and Ca^{2+}. Higher intracellular Ca^{2+} concentrations induce increased muscle excitability (cramps); other effects of Mg^{2+} deficiency are reduced mental and physical performance due to insufficient energy production via phosphate transfer (see below) and due to the inhibition of protein metabolism. Therefore, a hormonal control of Mg^{2+} transport exists, for example, in heart muscle cells, and serious Mg^{2+} deficiency requires a 'magnesium therapy' [2].

With regard to enzymatic activity, Mg^{2+} is an essential factor for the biochemical transfer of phosphates and for many nonoxidative cleavage reactions of nucleic acids through nucleases or ribozymes [4]. Mono-, di- and triphosphate groups are not only parts of the nucleotide components of RNA and DNA but also essential constituents of intermediate energy carrier molecules in organisms that can be converted by 'simple' hydrolysis, i.e. through PO_3^- transfer between a substrate and water (14.1 [5]).

$$X-O-PO_3{}^{n-} + H_2O \xrightarrow{\;M^{2+}\;} X-O^{(n-1)-} +$$

$$H_3PO_4 \qquad pK_a:$$

$$+H^+ \uparrow\downarrow -H^+ \qquad 1.96$$

$$H_2PO_4{}^-$$

$$+H^+ \uparrow\downarrow -H^+ \qquad 7.21$$

$$HPO_4{}^{2-}$$

$$+H^+ \uparrow\downarrow -H^+ \qquad 12.32$$

$$PO_4{}^{3-} \qquad \text{(in pure } H_2O)$$

$$\text{(14.1)}$$

In addition to the well-known adenosine triphosphate (ATP^{4-}*, 14.2), creatine phosphate (14.3) must also be mentioned in this context because it is important with regard to short-term *anaerobic* hydrolysis and can be detected through *in vivo* ^{31}P NMR spectroscopy of muscle tissue.

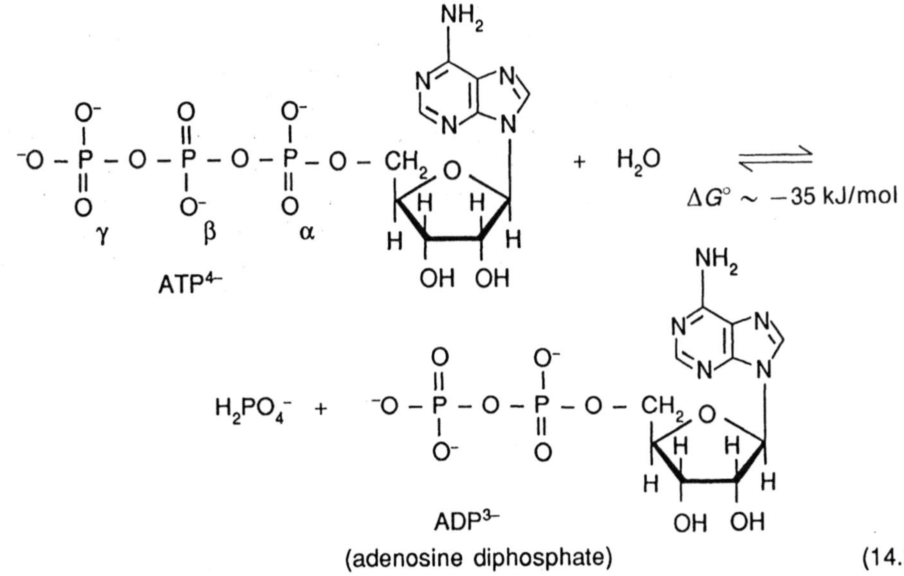

(14.2)

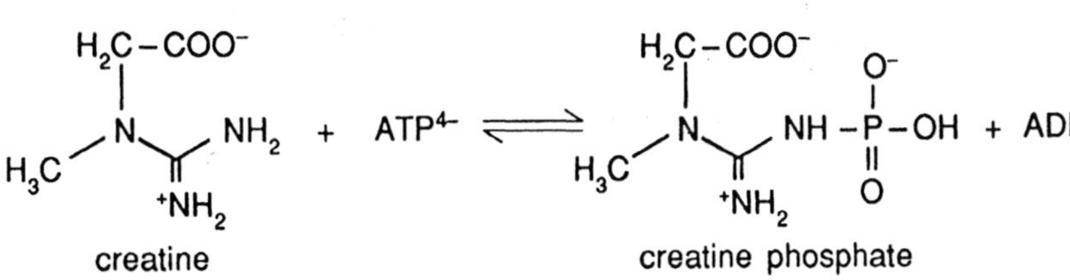

(14.3)

* Following biochemical practice we do not regularly use the (4−) superscript although it should be kept in mind when balancing charges as in equation (14.2).

On the average, a normally active human adult synthesizes and uses an amount of ATP per day that corresponds to his or her own body weight. Overall, the components of equation (14.2) participate in more chemical reactions than any other compound on the surface of the earth.

All biological phosphate transfer reactions such as phosphorylations by kinases or dephosphorylations by phosphatases [6] require the presence of catalyzing dipositively charged metal ions. In addition to Mg^{2+} (ionic radius 72 pm for a coordination number six), Zn^{2+} (74 pm) in alkaline phosphatase (Section 12.3), high-spin Fe^{2+} (78 pm) in purple acid phosphatases (Section 7.6.3), and the comparatively large ions high-spin Mn^{2+} (83 pm) [7,8] and Ca^{2+} (100 pm) can perform this task *in vivo*. In principle, Cd^{2+} (95 pm) and Pb^{2+} (119 pm) could also be suited; however, due to their rather soft character they tend to form strong bonds with sulfur ligands (see Sections 17.2 and 17.3).

Several aspects have to be considered for a catalysis by metal ions. First, the function of dipositive metal catalysts in phosphate transfer, including hydrolysis, can be attributed to an effective compensation of the high negative charge which is a consequence of the ionization of mono- and polyphosphates at physiological pH. The charge compensation by M^{2+} ions concerns *both* sides of the reaction, thus contributing to a reduction in activation energy [9]. Trivalent metal ions, M^{3+}, can compensate negative charges even better; however, they do no longer catalyze the reaction efficiently due to the unproductive stabilization of reaction intermediates (see Section 17.6). Furthermore, the metallic electrophiles M^{2+} activate weak Lewis bases such as water and thus create nucleophiles $(M^{2+})-OH$ via 'Umpolung' (see 12.3) under physiological conditions. It is obvious that a strongly polarizing dication can coordinate polyphosphates in a chelating manner by binding to the oxygen centers of *several* phosphate moieties; the effect is a spatial fixation including an activating ring strain (see 14.5). Finally, metal ions can generally lower the transition state of an associative reaction through intermediate coordination of *both* reactants (14.4; see 14.7).

simple model of a M^{2+}-catalyzed phosphate hydrolysis:

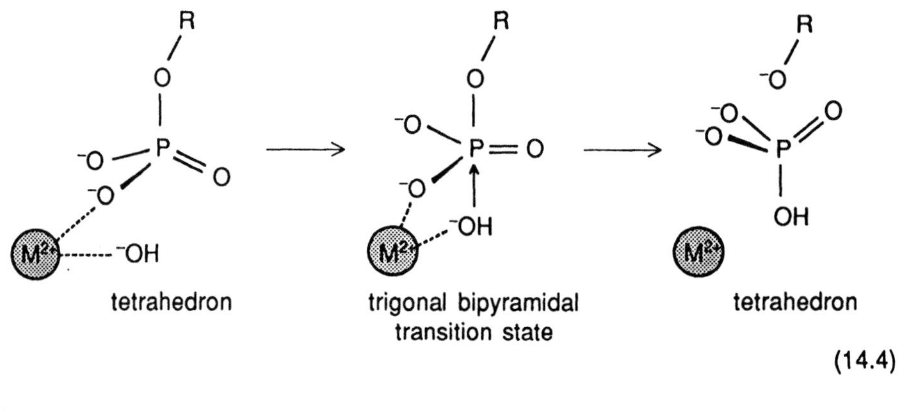

tetrahedron   trigonal bipyramidal   tetrahedron
                                  transition state

(14.4)

The following conclusions regarding the general reaction mechanism for the hydrolysis of ATP and other nucleoside triphosphates have been drawn from numerous model studies which were summarized by Sigel [10].

A (partially hydrated) metal dication can typically coordinate to one oxygen center of each of the α-, β- and γ-phosphate groups [11,12] and—in a free nucleotide—also to the imine nitrogen center of the purine heterocycle (macrochelate structures; 14.5).

proposed hydrolysis-productive $(ATP^{4-})(M^{2+})$ structures [10]:

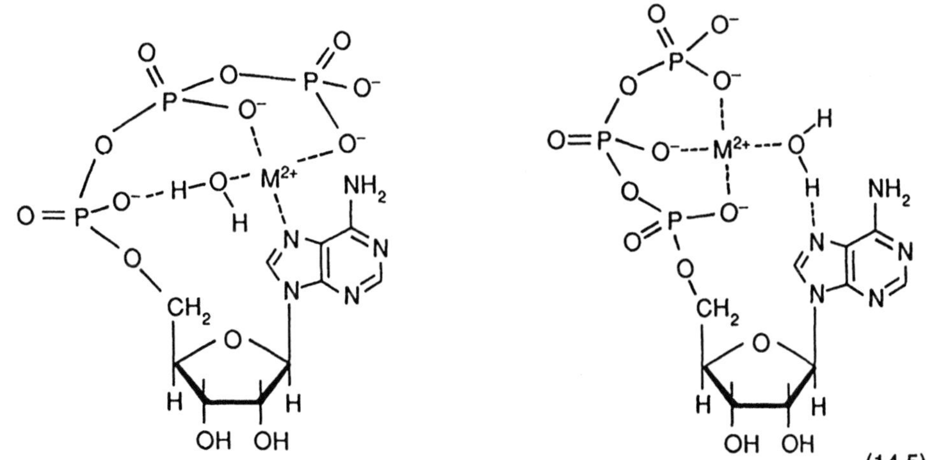

$$(14.5)$$

In the enzyme, this coordinative variability may be reduced [12] (Figure 14.1); however, an additional activation is conceivable after dimerization of complexes as in (14.5) which involves stacking of the heterocyclic bases [10,11]. The reactive species may even require two metal ions (Figure 14.1), one of which attacks at the more basic γ phosphate group and provides a bound hydroxide ion, i.e. a deprotonated water ligand (14.4).

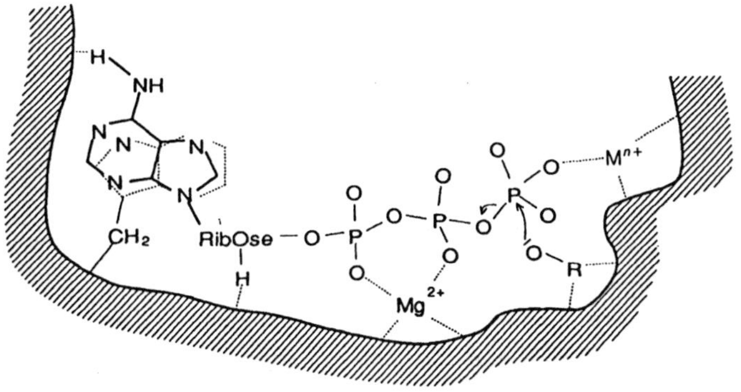

Figure 14.1
Hypothetical arrangement of a reactive $Mg(ATP)^{2-}$ complex in the enzyme. Partially enzyme-bound metal ions activate the triphosphate chain for a nucleophilic attack, in this case the attack by an alkoxide or ester at the terminal phosphate. Binding of the adenine heterocycle possibly involves π/π interactions with a tryptophane. Eventual dissociation of $Mg(ADP)^-$ can be visualized as the consequence of a stronger bond between Mg^{2+} and the (then terminal) β phosphate and thus weakened Mg^{2+}/enzyme binding (according to [10])

$$X-PO_3^{2-} + Y \rightarrow Y-PO_3^{2-} + X$$

X,Y: carboxyl functions, phosphates, guanidines, alcohols, water   (14.6)

Since the overall phosphate transfer reaction (14.6) is a nucleophilic substitution, it can mechanistically proceed as a dissociative (S_N1) or as an associative process (S_N2). A dissociative process would involve a reduction of the intermediate coordination number at the phosphorus center to three while the associative pathway (which implicates the simultaneous binding of both reactants) leads to an increase of the coordination number at phosphorus from four to five in the transition state (14.7). In the latter, biochemically relevant case, the reaction is susceptible to stereochemical control. A compelling inorganic chemical indication for a coordination number five in the transition state is the inhibition of ATPases by trace amounts of vanadate(V); the larger transition metal vanadium from group 5 in the periodic table has a higher tolerance of coordination number five than the smaller phosphorus atom in phosphates and can thus stabilize five-coordinate intermediate or transition states up to the point of inhibiting the catalytic cycle [13]. The aggregated oligovanadates which are present in the equilibrium at pH 7 (see Figure 11.3) are also known inhibitors of phosphate transferring enzymes [14].

mechanistic alternatives for the substitution at a tetrahedral center:

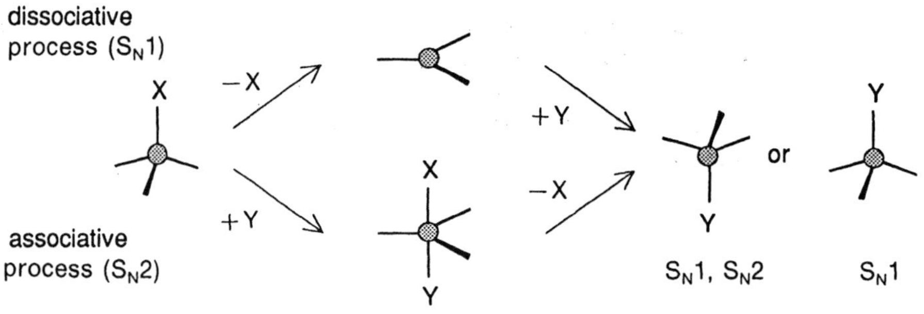

(14.7)

Within metabolic cycles the 'kinase' enzymes catalyze the transfer (14.8) of phosphoryl groups from ATP to other substrates X such as carbohydrates (e.g. glucose), carboxylates (e.g. pyruvate, $CH_3-C(=O)-COO^-$; see 14.9) or guanidines (e.g. creatine, 14.3). For the elucidation of the extensive regulatory functions of protein kinases and phosphatases the 1992 Nobel prize for medicine was awarded to E. Krebs and E. Fischer [15,16]. Available crystal structure determinations of kinases show ATP binding to Mg^{2+} via one oxygen atom from each of the three phosphate groups and completion of the hexacoordination through water and amino acid side chains [12]. During actual catalysis, the Mg^{2+} ion may migrate between $\alpha\beta$ and $\beta\gamma$ phosphate groups.

$$ATP^{4-} + X-H \xrightarrow{\text{kinase}} ADP^{3-} + X-PO_3^{2-} + H^+ \qquad (14.8)$$

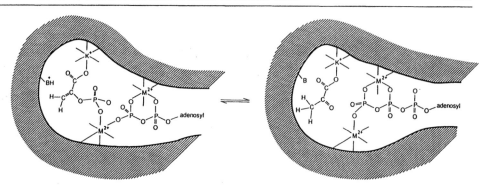

Figure 14.2
Assumed reaction mechanism for metal-dependent pyruvate kinase

In kinases and related enzymes the Mg^{2+} ion has often been substituted by Mn^{2+} in order to obtain information about the coordination environment, either directly through the EPR signal of this high-spin d^5 ion, or via its influence on other nuclei [17]. A classical example for this type of approach is pyruvate kinase (Figure 14.2) which requires the coordination of a large monocation, M^+, in particular K^+, in addition to two divalent metal cations. Using nuclear magnetic resonance, conformational changes were detected in the enzyme after coordination of the monovalent metal ion Tl^+ via Mn^{2+}-induced linewidth effects in ^{205}Tl NMR [18]—an example of double metal substitution (K^+, $Mg^{2+} \rightarrow Tl^+$, Mn^{2+}) for spectroscopic reasons.

Mechanistic models for the catalytic role of Mg^{2+} have also been put forward for the previously discussed Na^+/K^+-ATPase (Section 13.4 [19]). As the sequence in Figure 14.3 illustrates, the role of the Mg^{2+} ion presumably consists in a chelate-type coordination to the triphosphate oxygen centers (see Figure 14.1) with a resulting activation of the terminal phosphate for esterification by an amino acid side chain, e.g. Glu^-, of the protein. In the transition state, the $P(O_5)$ system is assumed to be five-coordinate in a trigonal-bipyramidal arrangement; its isomerization through a 'pseudorotation' could be linked to the Na^+-translocating conformational change in the protein which in turn gives rise to the hydrolysis to ADP (see Figure 13.12). By pseudorotation back to the original configuration at the phosphorus center the monophosphate magnesium complex still anchored in the protein could then trigger the reverse conformational change and thus a translocation of K^+. In this process, the monophosphate is dissociated through hydrolysis of the phosphate ester bond and the starting point is reached again.

Magnesium can also occur as an essential component of non-phosphate-transferring enzymes, e.g. of carbohydrate isomerases, DNA-activating topoisomerases and of enolases. In an elimination reaction (dehydration), the latter catalyze the synthesis of reactive phosphoenolpyruvate which, together with ADP, forms the energy storage molecule ATP and pyruvate in a pyruvate kinase-catalyzed reaction at the end of glycolysis (14.9).

According to structural data [20], yeast enolase requires a dipositive metal center (Mg^{2+} as natural cofactor, lower activity with Zn^{2+}) which is coordinated in a trigonal bipyramidal fashion by two water molecules and a glutamate residue in the trigonal plane and by two aspartate groups in the axial positions. This coordination

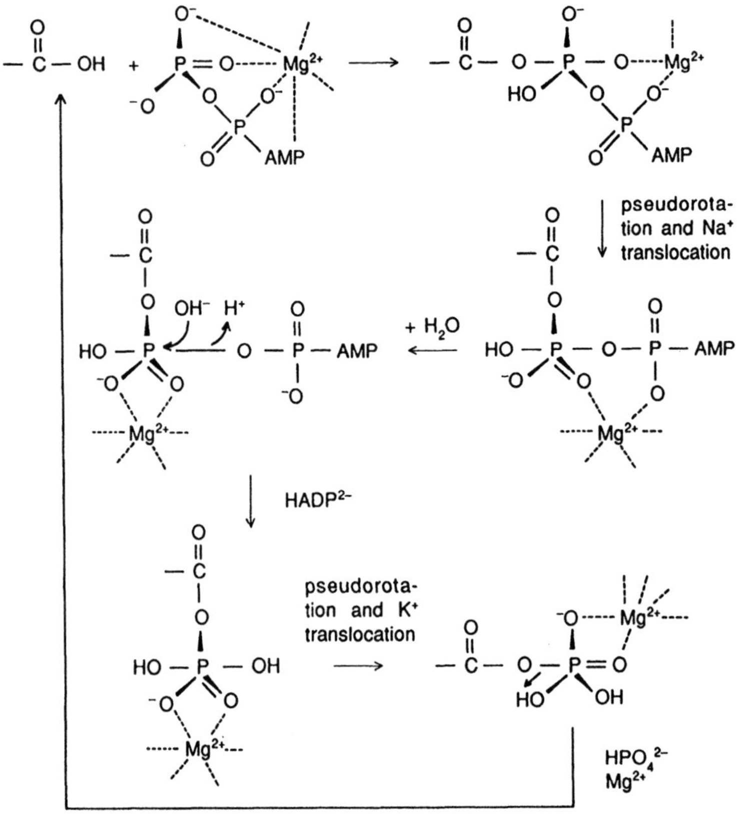

Figure 14.3
Mechanistic hypothesis regarding the role of Mg^{2+} in Na^+/K^+-ATPase (modified from [19]); AMP: adenosine monophosphate

geometry is very unusual for Mg^{2+} and has apparently not yet been observed outside this enzyme; it is achieved through very strong hydrogen bonds which widen the angle $O(Glu)-M^{2+}-OH_2(1)$ in the trigonal plane to the required 120°. During catalysis, the second, obviously more labile water molecule $OH_2(2)$ is presumably substituted by the hydroxyl group from the substrate 2-phosphoglycerate (14.9). The reason for this unusual and certainly high-energy coordination arrangement is an

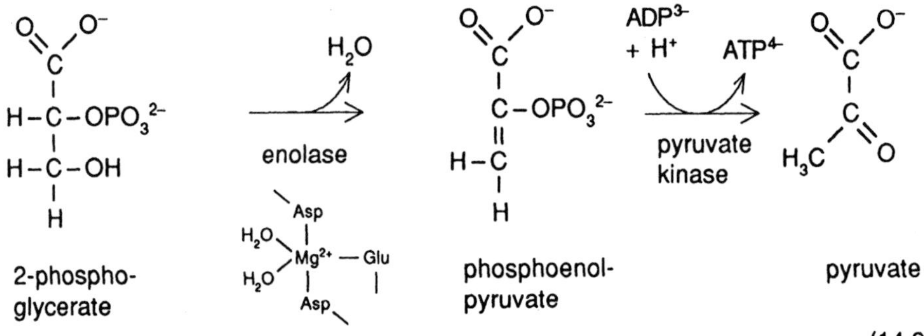

$$(14.9)$$

acceleration of the substitution ($\rightarrow$entatic state). With Ca^{2+} instead of Mg^{2+} there is stronger binding to the apoenzyme and an increase of the coordination number to six (size effect); however, the enzyme then becomes inactive [20]. An unusual metal coordination for a dehydrating catalytic center was already observed in the case of aconitase (Section 7.4).

14.2 The Ubiquitous Regulatory Role of Ca^{2+}

'Calcium probably fulfills a greater variety of biological functions than any other cation.'
F. L. Siegel, Calcium-binding proteins, *Struct. Bonding (Berlin)*, 17, 221 (1973).

'Without any doubt calcium is the chemical element that was the most researched ... in biology during the last decennium.'
L. J. Anghileri, *The Role of Calcium in Biological Systems*, Vol. IV, CRC Press, Boca Raton, 1987.

Besides iron and probably even surpassing it, calcium (in ionic form as Ca^{2+}) is the most important and most versatile 'bioinorganic' element. Its wide distribution in bound form in the earth's crust as well as in dissolved form in sea water (Figure 2.2) has certainly facilitated its numerous uses in biology ($\rightarrow$bioavailability). There are various inorganic calcium compounds with often strongly pH-dependent solubilities (see Tables 15.1 and 15.2). Their importance for biological solid-state materials, e.g. for exo- and endoskeletons, is presented separately in Chapter 15.

Compared to the large amounts of Ca^{2+} stored in the skeleton (about 1.2 kg in an adult human, turnover up to 0.7 g/day), the approximately 10 g of calcium which are *not* constituents of solid-state material seem rather unpretentious. However, calcium ions play a central role in many fundamental physiological processes, starting from cell division via hormonal secretion (e.g. the provision of insulin), blood clotting (the 'coagulation cascade'), antibody reactions, photosynthesis (Table 4.1), sensory functions and energy generation (ATP dephosphorylation, degradation of glycogen) to muscle contraction [21–23]. With Ca^{2+}, as with the alkali metal ions, the very specific ligands have received much more attention than the relatively inert metal center itself; only a brief overview regarding the biochemical importance of calcium can be given here from an 'inorganic' point of view.

In general, Ca^{2+} ions can be regarded as information mediators, i.e. as 'second messengers' as opposed to, for example, hormonal 'first messengers', or as triggering, regulating and signal amplifying species [24,25]. On the other hand, the uptake, storage and release processes of calcium are regulated through hormonally influenced control circuits via complex feedback mechanisms [26].

Disorders of these complex regulatory mechanisms are of great importance in medicine and pharmacology. At this point we mention the following:

(a) the necessary activation of Ca^{2+} resorption via specific calcium-binding proteins in the intestinal tissue [27] through 1,25-dihydroxy-colecalciferol, the physiologically active metabolite of vitamin D formed by P-450 catalyzed oxidation (see Section 6.2),

(b) the unwelcome deposition of calcium salts, e.g. oxalates, phosphates or steroids in blood vessels or in excretory organs (formation of 'stones', calculi) due to malfunctioning control mechanisms, or

(c) the excessive excitation of heart muscle tissue through Ca^{2+} ions permeating too easily into the cells; Ca^{2+} channel-blocking 'calcium antagonists' of the 1,4-dihydropyridine type are therefore used on a large scale in the therapy of cardiovascular diseases (14.10 [28,29]).

nifedipine   (14.10)

Several neuronal diseases are also believed to be caused by an endogenously disturbed calcium metabolism or by corresponding effects of exogenous toxic substances.

The control of the Ca^{2+} concentration in body fluids is crucial because this ion can show extremely high concentration differences of up to four orders of magnitude across cellular and other membranes (Table 13.1). Within the cell (in the cytosol), the concentration of 'free' calcium is normally very low (about 10^{-7} M) while the extracellular value is approximately 10^{-3} M. Ca^{2+} concentration gradients do not only exist between the inside and outside of cells but also between compartments of more complex cells such as the mitochondria or the cell nucleus [30]. The pH-dependent anion concentrations of phosphates and carbonates have to be considered here since the solubility product may otherwise be exceeded with the result of undesired precipitation (see Table 15.1). Only the extremely low intracellular Ca^{2+} concentration as maintained by the continuously operating Ca^{2+} pumps allows the diverse control functions and, in particular, the *amplification* of protein activity (see Figure 14.5).

The quantitative assay of calcium ions, in particular during rapidly proceeding Ca^{2+} exchange processes, has been enormously facilitated through the development of Ca^{2+}-specific ligands which show a rapidly, i.e. within milliseconds decaying coordination-dependent fluorescence. These compounds permit microscopic studies with high temporal and spatial resolution in concentrations of 10^{-1}–10^{-5} M Ca^{2+} [31,32]. Synthetic reagents such as 'Quin 2AM' (14.11) or the protein aequorin from bioluminescent organisms (jellyfish [33]) are being used.

Of course, there are also nonluminescing Ca^{2+}-specific ionophores such as calcimycin and similar substances from *Streptomyces* strains [34], or the synthetic 1,2-bis-(o-aminophenoxy)ethane-N,N,N',N'-tetraacetate ('BAPTA', 14.11). Other means of quantitative calcium determination such as precipitation with oxalate or the detection by ion-sensitive microelectrodes have become less popular; ^{43}Ca NMR spectroscopy is dependent on the availability of isotope-enriched material due to the low natural abundance of 0.13% [35].

In order to maintain the large concentration difference various Ca^{2+} pumps are required which, for example, are the main constituents of the sarcoplasmic reticulum of muscle cells. Among the well-known systems are the Ca^{2+}-dependent ATPase, a monomeric protein with a molecular mass of more than 100 kDa (see Figure 13.4)

R = CH₂OC(O)CH₃   'Quin 2AM'

calcimycin   'BAPTA'   (14.11)

which is inhibited by vanadates, and the already mentioned sodium/calcium antiport system (Section 13.4).

Bioavailability and the obviously possible control notwithstanding, why is Ca^{2+} so well suited for effecting transfer, conversion and amplification of information? Ca^{2+} is a divalent ion without redox function which typically exhibits high coordination numbers and often irregular coordination geometry in its complexes due to the ionic radius of about 100–120 pm [24,27,36,37]. The Cd^{2+} (95 pm) and Pb^{2+} (119 pm) ions are similar to Ca^{2+} but biologically harmful due to their strong coordination with thiolates (Cys⁻); Mn^{2+} (83 pm) and the heavier homologue Sr^{2+} (118 pm) are less toxic as calcium 'substitutes', the possible biological importance of Sr^{2+} perhaps being obscured by the much more abundant Ca^{2+} (see Table 2.1). Coordination numbers of seven or eight are quite common for Ca^{2+} in proteins in contrast to the strong preference by Mg^{2+} for octahedral configuration [37], the latter being less specific as it is very stable toward external influences. Typically realized geometries for the Ca^{2+} ion with a coordination number of seven are the pentagonal bipyramid [36,38], as in α-lactalbumin of milk, a trigonal prismatic arrangement with a capped

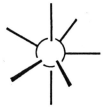

pentagonal   trigonal prism with
bipyramide   capped rectangular face   (14.12)

rectangular face [36] (14.12), or a distorted octahedron with an additional coordination site through (η^2-)carboxylate chelate coordination at one corner [27,39].

High coordination numbers are attained rather easily since the large Ca^{2+} ion likes to coordinate small water molecules, the carbonyl oxygen atoms of peptide bonds [40], the hydroxyl groups of chelating carbohydrates [41], and the potentially chelating (2.3) carboxylate groups which are abundant in acidic proteins. A well-documented example is parvalbumin (Figure 14.4) which is present in smooth muscle and binds Ca^{2+} but also tolerates Mg^{2+}. In contrast to the unspecific, approximately octahedral configuration of the magnesium center, the calcium analogue features an irregular, i.e. specific and thus protein-determined coordination geometry; at the same time, the larger ionic radius of Ca^{2+} guarantees a higher rate of (de)complexation and thus more rapid information transfer. Yet another emerging feature of Ca^{2+} with its propensity for large coordination numbers is the ability to specifically mediate protein–protein or protein–carbohydrate interactions [42].

Several types of Ca^{2+}-containing proteins are now rather well understood with regard to their functions. Calcium-releasing regions close to membranes contain very acidic Ca^{2+} storage proteins, the calsequestrins (approximately 40 kDa), each of which can bind up to 50 calcium ions [43]. The large amounts of Ca^{2+} ions required for information transmission and amplification and the triggering of muscle contractions are released from such proteins. The activation of stored calcium proceeds by mechanisms which are not yet understood in full detail; nucleotides whose formation can be influenced by Ca^{2+} ($\rightarrow$feedback) presumably serve as anionic 'second messengers'. Membrane depolarization through an electric nerve impulse as well as local hormone/receptor interactions can also cause the release of Ca^{2+}.

In addition to a protein-structure stabilizing function as found, for example, in thermolysin (see Figure 2.7) or proteinase K [44], Ca^{2+} ions can also show hydrolysis-catalyzing activity. One of the best documented examples is the phosphodiester-cleaving nuclease from staphylococcal strains [9,45] which features a 'Mg^{2+}-like' coordination sphere of the catalytic metal center (2 η^1-Asp, 1 Thr, 2 H_2O, 1 substrate-O).

Amino acid sequences and structures have been determined for another group of ubiquituous, very stable and, from an evolution point of view, very old Ca^{2+}-specific

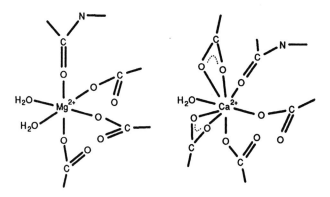

Figure 14.4
Metal coordination in Mg^{2+}- and Ca^{2+}-containing parvalbumin (according to [24])

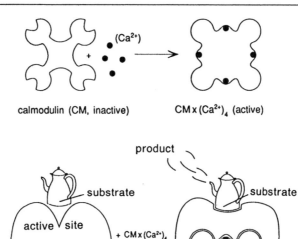

Figure 14.5
Model for the activation of enzymes by Ca^{2+}-containing proteins of the calmodulin type (according to [51])

proteins, the 'calmodulins' [46]. These are rather small proteins (molecular mass approximately 17 kDa) with a number of calcium-binding acidic regions, i.e. carboxylate-containing side chains: glutamate, aspartate. The function of such Ca^{2+} receptor proteins which in turn serve to activate many 'calcium-dependent' enzymes [47–49] is to cooperatively bind 2–4 Ca^{2+} ions and thus change the conformation so that the recognition [50] and activation of an enzyme can take place through specific calmodulin–protein interactions (Figure 14.5).

Among the enzymes that are activated by calmodulin/Ca^{2+} complexes are:

(a) adenylate and guanylate cyclases for the formation of cAMP and cGMP,
(b) NO synthase (see Section 6.5),
(c) Ca^{2+}-ATPase (feedback),
(d) NAD kinase for the synthesis of NADP (3.12) and
(e) phosphorylase kinase which contributes in the degradation of the energy storage molecule glycogen.

Parvalbumins which are present in smooth muscle and presumably assist in muscle relaxation (see Figure 14.4), the troponins of skeletal (striped) muscle (Figure 14.7), and the S100 proteins [52] which are found in the nervous system all belong to the extended calmodulin family with its typical 'EF-hand' protein structure (Figure 14.6). About 170 proteins are now known in which often several neighboring Ca^{2+}-selective 'EF-hand' binding sites exist [49] (see Figure 14.7). Another different but also conformationally flexible class of Ca^{2+}-dependent phospholipid- and membrane-binding proteins is referred to as 'annexins' [53]; they are important for the

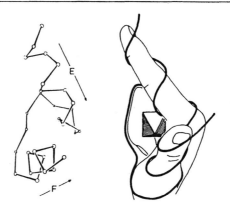

Figure 14.6
'EF-hand' structure. Ca^{2+} is bound to the protein in a distorted octahedral configuration at
the intersection of E-α-helix and F-α-helix

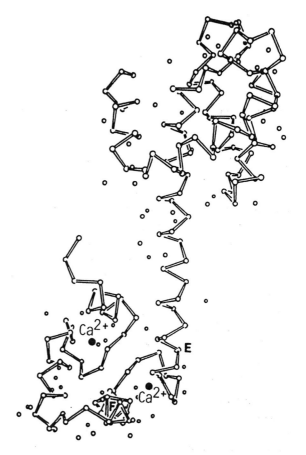

Figure 14.7
Structure of crystallized troponin C from chicken skeletal muscle: α-C backbone representation
with identification of the helices E and F; water molecules are depicted as open circles, while
both seven-coordinate Ca^{2+} ions in the lower half of the protein appear as filled circles (from
[39])

regulation of cell growth and coagulation and exhibit a pentagonal-bipyramidal coordination geometry for the cross-linking Ca^{2+} ion [54].

In this context, the muscle contraction is presented in a very basic form as a well-researched example for the messenger function of calcium ions. Ca^{2+}-triggered release of a neurotransmitter from a nerve cell leads to an opening of K^{+} channels, which in turn causes the depolarization of the normally polarized biological membrane. In an as yet little understood step, the release of Ca^{2+} from the storage proteins in the sarcoplasmic reticulum is effected via activation of voltage-controlled Na^{+} channels. Incidentally, calcium-specific channels can be voltage-controlled (opening times approximately 1 ms) or controlled by nucleotides such as cGMP or inositol-1,4,5-triphosphate (IP$_3$ [55]). Blocking of these channels is not only caused by low molecular weight 'antagonist' molecules (see 14.10) but also by other metal cations such as Co^{2+} or the lanthanoid ions La^{3+}–Lu^{3+} (ionic radii 103–86 pm) which are quite similar to Ca^{2+} (100 pm) in size [56].

The large increase in the concentration of Ca^{2+} ions after their release from the sarcoplasmic reticulum leads to the binding of these ions by troponin C [39]. This protein with a molecular mass of approximately 18 kDa and 'EF-hand' binding sites

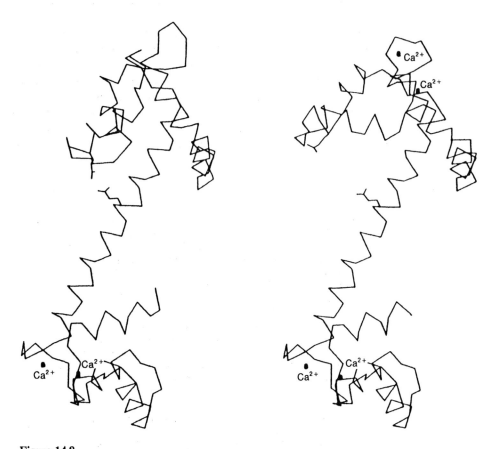

Figure 14.8
Structural change of the dicalcium/troponin C complex (left) after additional Ca^{2+}-binding (right, α-C backbone representation of the protein; from [57])

(Figure 14.7) is similar to calmodulin; it can adopt various conformations with different Ca^{2+} affinities (Figure 14.8). Through deblocking of the initial entatic state situation, Ca^{2+}-activated troponin C causes an interaction of the thin, also M^{2+}-binding actin fiber with the thick myosin fiber in the muscle cell combined with a spatial displacement. The simultaneous displacement of the fibers triggered by Ca^{2+} binding actin fiber with the thick myosin fiber in the muscle cell, combined with a spatial displacement. The simultaneous displacement of the fibers triggered by Ca^{2+} is thus transformed into mechanical energy through an electric impulse (membrane depolarization) using Ca^{2+} for rapid amplification and transformation into a specific chemical signal. Only after Mg^{2+}-dependent binding and hydrolysis of ATP is Ca^{2+} released and pumped back to the storage proteins; this energy-consuming process leads to a separation of myosin and actin fibers, i.e. to a lifting of the rigor state and thus to a restoration of the entatically blocked starting situation.

For longer-lasting muscle contraction a continuous generation of ATP for the myosin ATPase and the rapid pumping of Ca^{2+} by the membrane pumps are necessary. A constant cyclic Ca^{2+} flow has been established as prerequisite for continuous contraction in the smooth muscle system [25].

The activation of phosphate-transferring kinases (14.8) through Ca^{2+} plays an important role in combination with muscle contraction and other Ca^{2+}-controlled processes. As triggers of reaction 'cascades', calcium ions and their calmodulin complexes are important for the generation of ATP through the degradation of glycogen, for blood coagulation and other secretory processes. In some cases, e.g. in the visual process (Section 13.4), where calcium ions had been regarded as actual activators, most recent studies have rather indicated the activation of specific nucleotides through dephosphorylation ($\rightarrow$anionic second/third/fourth ... messengers). However, in almost all cases there are complex feedback mechanisms between calcium ions and nucleotides which are very sensitive to the ratios of reaction rates.

References

1. R. B. Martin, Bioinorganic chemistry of magnesium, and following contributions in *Metal Ions in Biological Systems* (ed. H. Sigel), Vol. 26, Marcel Dekker, New York, 1990, p. 1.
2. H. Schmidbaur, H. G. Classen and J. Helbig, Asparagine and glutamine as ligands for alkali and alkaline earth metals: structural chemistry contribution of the complex to magnesium therapy, *Angew. Chem. Int. Ed. Engl.*, **29**, 1090 (1990).
3. I. Andersson, S. Knight, G. Schneider, Y. Lindqvist, T. Lundqvist, C.-I. Bränden and G. H. Lorimer, Crystal structure of the active site of ribulose-bisphosphate carboxylase, *Nature (London)*, **337**, 229 (1989).
4. A. M. Pyle, Ribozymes: a distinct class of metalloenzymes, *Science*, **261**, 709 (1993).
5. F. H. Westheimer, Why nature chose phosphates, *Science*, **235**, 1173 (1987).
6. J. B. Vincent, M. W. Crowder and B. A. Averill, Hydrolysis of phosphate monoesters: a biological problem with multiple chemical solutions, *Trends Biochem. Sci.*, **17**, 105 (1992).
7. J. F. Davies, II, Z. Hostomska, Z. Hostomsky, S. R. Jordan and D. A. Matthews, Crystal structure of the ribonuclease H domain of HIV-1 reverse transcriptase, *Science*, **252**, 88 (1991).
8. L. Beese and T. A. Steitz, Structural basis for the 3'-5' exonuclease activity of *Escherichia coli* DNA polymerase I: a two metal ion mechanism, *EMBO J.*, **10**, 25 (1991).
9. J. Aqvist and A. Warshel, Free energy relationships in metalloenzyme-catalyzed reactions. Calculations of the effects of metal ion substitutions in staphylococcal nuclease, *J. Am. Chem. Soc.*, **112**, 2860 (1990).

10. H. Sigel, Mechanistic aspects of the metal ion promoted hydrolysis of nucleoside 5'-triphosphates, *Coord. Chem. Rev.*, **100**, 453 (1990).

11. R. Cini, X-ray structural studies of adenosine 5'-triphosphate metal compounds, *Comments Inorg. Chem.*, **13**, 1 (1992).

12. H. L. De Bondt, J. Rosenblatt, J. Jancarik, H. D. Jones, D. O. Morgan and S.-H. Kim, Crystal structure of cyclin-dependent kinase 2, *Nature (London)*, **363**, 595 (1993).

13. A. S. Tracey, J. S. Jaswal, M. J. Gresser and D. Rehder, Condensation reactions of aqueous vanadate with the common nucleosides, *Inorg. Chem.*, **29**, 4283 (1990).

14. D. Crans, C. D. Rithner and L. A. Theisen, Application of time-resolved ^{51}V 2D NMR for quantitation of kinetic exchange pathways between vanadate monomer, dimer, tetramer, and pentamer, *J. Am. Chem. Soc.*, **112**, 2901 (1990).

15. E. G. Krebs, Protein phosphorylation and cell regulation I (Nobel address), *Angew. Chem. Int. Ed. Engl.*, **105**, 1173 (1993).

16. E. H. Fischer, Protein phosphorylation and cell regulation II (Nobel address), *Angew. Chem. Int. Ed. Engl.*, **105**, 1181 (1993).

17. D. T. Lodato and G. H. Reed, Structure of the oxalate-ATP complex with pyruvate kinase: ATP as a bridging ligand for the two divalent cations, *Biochemistry*, **26**, 2243 (1987).

18. F. J. Kayne and J. Reuben, Thallium-205 nuclear magnetic resonance as a probe for studying metal ion binding to biological macromolecules. Estimate of the distance between the monovalent and divalent activators of pyruvate kinase, *J. Am. Chem. Soc.*, **92**, 220 (1970).

19. K. R. H. Repke and R. Schön, Chemistry and energetics of transphosphorylations on the mechanism of Na^+/K^+-transporting ATPase: an attempt at a unifying model, *Biochim. Biophys. Acta*, **1154**, 1 (1992).

20. L. Lebioda and B. Stec, Crystal structure of holoenolase refined at 1.9 Å resolution: trigonal-bipyramidal geometry of the cation binding site, *J. Am. Chem. Soc.*, **111**, 8511 (1989).

21. L. J. Anghileri (ed.), *The Role of Calcium in Biological Systems*, Vol. IV, CRC Press, Boca Raton, 1987.

22. C. Gerday, L. Bolis and R. Gilles (eds.), *Calcium and Calcium Binding Proteins*, Springer-Verlag, Berlin, 1988.

23. D. Pietrobon, F. Di Virgilio and T. Pozzan, Structural and functional aspects of calcium homeostasis in eukaryotic cells, *Eur. J. Biochem.*, **193**, 599 (1990).

24. E. Carafoli and J. T. Penniston, The calcium signal, *Sci. Am.*, **254**(1), 76 (1986).

25. H. Rasmussen, The cycling of calcium as an intracellular messenger, *Sci. Am.*, **261**(4), 66 (1989).

26. E. Carafoli, Intracellular calcium homeostasis, *Ann. Rev. Biochem.*, **56**, 395 (1987).

27. D. M. E. Szebenyi and K. Moffat, The refined structure of vitamin D-dependent calcium-binding protein from bovine intestine, *J. Biol. Chem.*, **261**, 8761 (1986).

28. S. Goldmann and J. Stoltefuss, 1,4-Dihydropyridines: effect of chirality and conformation on the calcium-antagonistic and -agonistic effects, *Angew. Chem. Int. Ed. Engl.*, **30**, 1559 (1991).

29. R. Fossheim, K. Svarteng, A. Mostad, C. Romming, E. Shefter and D. J. Triggle, Crystal structures and pharmacological activity of calcium channel antagonists, *J. Med. Chem.*, **25**, 126 (1982).

30. O. Bachs, N. Agell and E. Carafoli, Calcium and calmodulin function in the cell nucleus, *Biochim. Biophys. Acta*, **1113**, 259 (1992).

31. G. Grynkiewicz, M. Poenie and R. Y. Tsien, A new generation of Ca^{2+} indicators with greatly improved fluorescence properties, *J. Biol. Chem.*, **260**, 3440 (1985).

32. S. R. Adams, J. P. Y. Kao, G. Grynkiewicz, A. Minta and R. Y. Tsien, Biologically useful chelators that release Ca^{2+} upon illumination, *J. Am. Chem. Soc.*, **110**, 3212 (1988).

33. A. K. Campbell, Living light: chemiluminescence in the research and clinical laboratory, *Trends Biochem. Sci.*, **11**, 104 (1986).

34. A. M. Albrecht-Gary, S. Blanc-Parasote, D. W. Boyd, G. Dauphin, G. Jeminet, J. Juillard, M. Prudhomme and C. Tissier, X-14885A: an ionophore closely related to calcimycin (A-23187). NMR, thermodynamic, and kinetic studies of cation selectivity, *J. Am. Chem. Soc.*, **111**, 8598 (1989).

35. Y. Ogoma, T. Shimizu, M. Hatano, T. Fujii, A. Hachimori and Y. Kondo, ^{43}Ca nuclear magnetic resonance spectra of Ca^{2+}-S100 protein solutions, *Inorg. Chem.*, **27**, 1853 (1988).
36. A. L. Swain and E. L. Amma, The coordination polyhedron of Ca^{2+}, Cd^{2+} in parvalbumin, *Inorg. Chim. Acta*, **163**, 5 (1989).
37. O. Carugo, K. Djinovic and M. Rizzi, Comparison of the co-ordinative behaviour of calcium(II) and magnesium(II) from crystallographic data, *J. Chem. Soc. Dalton Trans.*, 2127 (1993).
38. N. K. Vyas, M. N. Vyas and F. A. Quiocho, A novel calcium binding site in the galactose-binding protein of bacterial transport and chemotaxis, *Nature (London)*, **327**, 635 (1987).
39. K. A. Satyshur, S. T. Rao, D. Pyzalska, W. Drendel, M. Greaser and M. Sundaralingam, Refined structure of chicken skeletal muscle troponin C in the two-calcium state at 2-Å resolution, *J. Biol. Chem.*, **263**, 1628 (1988).
40. P. Chakrabarti, Systematics in the interaction of metal ions with the main-chain carbonyl group in protein structures, *Biochemistry*, **29**, 651 (1990).
41. W. I. Weis, K. Drickamer and W. A. Hendrickson, Structure of a C-type mannose-binding protein complexed with an oligosaccharide, *Nature (London)*, **360**, 127 (1992).
42. P. J. McLaughlin, J. T. Gooch, H. G. Mannherz and A. G. Weeds, Structure of gelsolin segment 1-actin complex and the mechanism of filament severing, *Nature (London)*, **364**, 685 (1993).
43. M. Ohnishi and R. A. F. Reithmeier, Fragmentation of rabbit skeletal muscle calsequestrin: spectral and ion binding properties of the carboxyl-terminal region, *Biochemistry*, **26**, 7458 (1987).
44. J. Bajorath, S. Raghunathan, W. Hinrichs and W. Saenger, Long-range structural changes in proteinase K triggered by calcium ion removal, *Nature (London)*, **337**, 481 (1989).
45. F. A. Cotton, E. E. Hazen and M. J. Legg, Staphylococcal nuclease: proposed mechanism of action based on structure of enzyme-thymidine 3',5'-bisphosphate-calcium ion complex at 1.5-Å resolution, *Proc. Natl. Acad. Sci. USA*, **76**, 2551 (1979).
46. A. S. Babu, J. S. Sack, T. J. Greenhough, C. E. Bugg, A. R. Means and W. J. Cook, Three-dimensional structure of calmodulin, *Nature (London)*, **315**, 37 (1985).
47. W. Y. Cheung, Many of the bioregulatory functions of Ca^{2+} are mediated through calmodulin, *Rec. Trav. Chim. Pays-Bas*, **106**, 262 (1987).
48. P. Cohen and C. B. Klee (eds.), *Calmodulin*, Elsevier, Amsterdam, 1988.
49. S. Forsen and J. Kördel, The molecular anatomy of a calcium-binding protein, *Acc. Chem. Res.*, **26**, 7 (1993).
50. W. E. Meador, A. R. Means and F. A. Quiocho, Target enzyme recognition by calmodulin: 2.4 Å Structure of a calmodulin-peptide complex, *Science*, **257**, 1251 (1992).
51. S. Klumpp and J. E. Schultz, Calcium und Calmodulin, *Pharm. Unserer Zeit*, **14**, 19 (1983).
52. D. Kligman and D. C. Hilt, The S100 protein family, *Trends Biochem. Sci.*, **13**, 437 (1988).
53. C. B. Klee, Ca^{2+}-dependent phospholipid- (and membrane-) binding proteins, *Perspectives in Biochemistry*, 37 (1989).
54. N. O. Concha, J. F. Head, M. A. Kaetzel, J. R. Dedman and B. A. Seaton, Rat annexin V crystal structure Ca^{2+}-induced conformational changes, *Science*, **261**, 1321 (1993).
55. E. I. Ochiai, Why calcium?, *J. Chem. Educ.*, **68**, 10 (1991).
56. C. H. Evans, Interesting and useful biochemical properties of lanthanides, *Trends Biochem. Sci.*, 445 (1983).
57. K. Fujimori, M. Sorenson, O. Herzberg, J. Moult and F. C. Reinach, Probing the calcium-induced conformational transition of troponin C with site-directed mutants, *Nature (London)*, **345**, 182 (1990).

15 Biomineralization: The Controlled Assembly of 'Advanced Materials' in Biology

15.1 Overview

Much more obviously than the metalloproteins or ionic electrolytes do the chemically and morphologically diverse inorganic biominerals belie the impression of life as being monopolized by organic chemistry. Even our knowledge about earlier forms of life is largely based on biominerals, i.e. fossils, some of which have accumulated to an enormous, 'geological' extent. Many mountain ranges, islands and coral reefs consist of biogenic material such as limestone. This immense bioinorganic production over hundreds of millions of years has significantly changed the conditions for life itself, for instance, CO_2 has been bound in the form of carbonates, thus diminishing the early greenhouse effect at the earth's surface. In addition to the well-known calcium-containing shells, teeth and skeletons, a variety of other materials and objects can be classified as biominerals. This includes the aragonite pearls produced by molluscs, the hulls and spicules of diatoms, radiolaria and certain plants, the Ca-, Ba- and Fe-containing crystallites of gravity and magnetic field sensors as well as some of the pathological 'stones' (calculi) formed in kidney or urinary tracts. The iron storage protein ferritin introduced in Chapter 8.4.2 can also be regarded as a biomineral based on its structure and content of inorganic material.

The relatively new and highly interdisciplinary research field of biomineralization [1–6] encompasses such diverse areas as geology, classical (descriptive) biology and the modern 'biomimetic' material sciences [2,3,7]; under more chemical aspects, it is concerned with the molecular control mechanisms which are operating in biological systems to achieve the formation of well-defined inorganic solid-state materials. The production of morphologically complex minerals according to a genetically determined blueprint mainly results from the necessity for robust support and defense structures. In principle, there is no natural preference for inorganic or organic support materials in endo- or exoskeletons; for instance, the relatively rapidly assembled chitin (polysaccharide) structures of the invertebrates and the skeletons of sharks consist largely of organic-chemical material. In fact, most biomineral constructions feature an organic/inorganic composite texture; the bones of vertebrates thus consist of the calcium 'mineral' hydroxyapatite and an organic matrix. The advantage of the inorganic component is its hardness and pressure-resistance, which allows for the existence of larger land-living creatures; the organic matrix consisting of collagen fibers, glycoproteins and mucopolysaccharides guarantees elasticity as well as tensile, bending and breaking strength. Because of such advantages much of the modern material sciences is focused on the development of composite materials, in particular

Table 15.1 The most important biominerals

chemical composition	mineral form (phase)	solubility exponent pK_{sp}^{*a} at pH 7	occurrence and function (examples)
calcium carbonate			
$CaCO_3^b$	calcite	8.42	exoskeletons (e.g. egg
	aragonite	8.22	shells, corals, mollusc
	vaterite	7.6	shells), spicules,
	amorphous	7.4	gravity sensor
calcium phosphates	(see Table 15.2)		
$Ca_{10}(OH)_2(PO_4)_6$	hydroxyapatite	$\approx 13^a$	endoskeletons (vertebrate bones and teeth)
$Ca_{10}F_2(PO_4)_6$	fluoroapatite	$\approx 14^a$	
calcium oxalate			
$CaC_2O_4(\cdot\, n\; H_2O)$ $n = 1,2$	whewellite, weddelite	8.6	calcium storage and passive defense of plants, calculi of excretory tracts
metal sulfates			
$CaSO_4 \cdot 2\; H_2O$	gypsum	4.2	gravity sensors
$SrSO_4$	celestite	6.5	exoskeletons (*Acantharia*)
$BaSO_4$	baryte	10.0	gravity sensors
amorphous silica			
$SiO_2 \cdot n\; H_2O =$ $SiO_n(OH)_{4-2n}$	amorphous	solubility < 100 mg/l	valves of diatoms and radiolarians, defense functions in plants
iron oxides			
Fe_3O_4	magnetite		magnetic sensors, teeth of chitons
$\alpha,\gamma\text{-}Fe(O)OH$ '$5Fe_2O_3 \cdot 9\; H_2O$'	goethite, lepidocrocite ferrihydrite (see Figure 8.6)		teeth of chitons teeth of chitons, iron storage (Section 8.4.2)

^{a} Solubility products K_{sp}^* in pure water have been reduced to the unit M^{-2} for sake of comparison.
^{b} Often occurring with variable small amounts of $MgCO_3$ ($pK_{sp}^* = 5.2$).

fiber-enhanced materials, and microstructural physical methods of analysis such as high-resolution electron microscopy are now being used in both areas of natural and synthetic composite systems.

The most important biominerals, some in different polymorphic forms, are summarized in Table 15.1 with respect to occurrence and main functions. An example for the morphological complexity of biominerals is depicted in Figure 15.1.

An essential requirement for a biomineral is its low solubility under normal physiological conditions. For comparison, Table 15.1 lists several solubility product constants K_{sp} (15.1) which, for better comparison, are reduced to the common unit M^{-2}.

The biominerals can occur as pure or mixed phases, in amorphous or (micro)-crystalline form, or as composites with polymeric organic 'matrix' materials such as proteins, lipids or polysaccharides. They can be formed intracellularly, at the cell

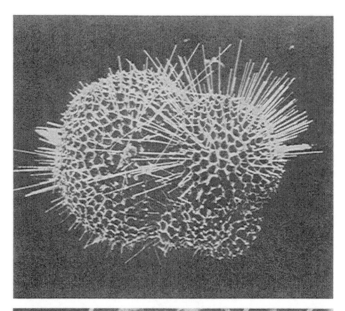

enlargement:

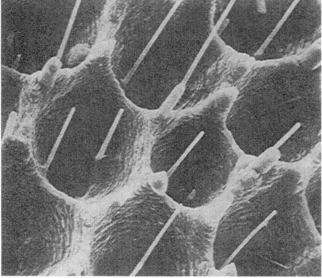

Figure 15.1
Foraminifer with spine-armored, porous hull made from calcite (*Globigerinoides sacculifer*, from [8])

cation + anion $\rightleftharpoons$ solid + soluble cation/anion aggregate

$K_{sp} = [\text{cation}] \cdot [\text{anion}];\ pK_{sp} = -\log K_{sp}$

[ ]: molar concentrations

Equation (15.1) is valid only for heterogeneous equilibria with unhindered exchange between solid and solution phases.

(15.1)

surface (epicellularly) or in the extracellular space. Table 15.1 reflects only partly the possible variety of inorganic components; for example, the more exotic fluoride and sulfide biominerals have not been included. Overall, the following biominerals are predominant:

(a) the calcium carbonate phases aragonite and calcite which often contain some Mg^{2+},
(b) the calcium phosphates, most frequently as hydroxyapatite,
(c) amorphous silica, and
(d) the iron oxides/hydroxides ferrihydrite and magnetite.

All other biominerals occur either as minor components or in only a few species.

In addition to the already mentioned *mechanical support function* the biominerals also assume a *storage function* because of the large amounts usually present. On the other hand, the formation of biominerals can have a *deposition* character (placer gold [9]) or even a *detoxification function* (e.g. CdS, see Section 17.3), particularly in microorganisms. This implies the existence of active regulatory and transport systems (see Chapter 8 and Section 14.2) which control (de)mineralization and regeneration.

Magnetite, for instance, is a widely distributed mineral in the earth's crust which is formed geologically at high temperatures and pressures. In contrast, 'magneto-tactic' bacteria are able to synthesize this mineral in biologically useful form under physiological conditions (see Section 15.3.4).

In another example, some marine unicellular organisms utilize celestite, $SrSO_4$, for their exoskeletons [10]. Sea water is undersaturated with respect to this mineral so that only *active* accumulation mechanisms within the organism can guarantee the existence of these solids. After the death of the organism these celestite structures dissolve rapidly.

Further important functions of the biominerals consist in:

(a) their use as essential constituents of mechanically robust *instruments* and *weapons*, e.g. of teeth for killing and processing of food,
(b) the formation of *sensor components*, e.g. of specifically heavy crystallites in gravity-sensitive organs or of magnetic microcrystallites in magnetotactic bacteria, and
(c) the passive *mechanical protection* of animals (e.g. the shells of molluscs) and plants (e.g. silica-containing spikes) against predators or climatic effects.

The last example clearly shows that both chemical composition ($\rightarrow$hardness) and morphology contribute to the full functionality of the biomineral.

Due to the usually large amounts of required material, the chemical composition of biominerals is determined mainly by the availability of the components (which is also true for the 'geological' mineral formation). The difference, however, is that the material properties in the biological realm are determined by the chemical composition *and* by the carefully controlled ('enforced') morphology. The organic components in particular can have a 'matrix' or 'template' function with regard to vectorial, i.e. directed, phase-independent crystal growth via specific catalysis and control of nucleation. Biological calcite, for example, usually does not form rhomboid crystals, as is the case for the unrestrained system, but is synthesized, for example, in the functionally much more useful form of Figure 15.1. In contrast to the 'geological' minerals, biominerals also have to be formed and redissolved ('demineralized') in a much shorter, biologically acceptable time period; therefore, relatively small crystalline domains or single crystals with large surface areas such as spicules are often formed because of their better accessibility by the solvent. Nevertheless, the half-life of the calcium exchange in an adult human being amounts to several years because of the relatively slow turnover in the solid-state skeleton. Pathological effects due to deviations from the normal metabolic rates for biominerals are quite common; they include calcium-containing deposits in blood vessels and the formation of calculi in the excretory organs (CaC_2O_4, apatite, $MgNH_4PO_4$) as well as insufficient skeletal mineralization in children (rickets) or unwanted demineralization processes such as dental caries or bone resorption (osteoporosis) in the aging organism.

Different degrees of biomineralization can be distinguished, depending on the type and complexity of the already mentioned control mechanisms [4]. The most primitive type is the biologically induced mineralization which occurs mainly in bacteria and algae. In these cases the biominerals are formed by spontaneous crystallization, via supersaturation through the action of ion pumps (see Section 13.4); polycrystalline aggregates with random orientations are then formed in the extracellular space. Gases formed in biological processes, e.g. by bacteria, frequently react with metal ions from the external medium to form such biomineral deposits.

As an example, biomineralization results from the decreasing CO_2 content in water as caused by photosynthetically active algae (15.2).

$$Ca^{2+} + 2\ HCO_3^- \rightleftharpoons CaCO_3\ (s) + CO_2\ (g) + H_2O$$

$$\phantom{Ca^{2+} + 2\ HCO_3^- \rightleftharpoons CaCO_3\ (s) + }\text{photosynthetic}$$
$$\phantom{Ca^{2+} + 2\ HCO_3^- \rightleftharpoons CaCO_3\ (s) + }\text{assimilation} \qquad (15.2)$$

Equilibrium (15.2) is shifted to the right side by photosynthetic CO_2 assimilation and $CaCO_3$ precipitates. As in the process of global iron(II) oxidation through the action of biogenic O_2, the photosynthetic activity has thus caused geologically and climatically important chemical changes on the earth's surface; carbonate biominerals may be viewed as a long-term CO_2 sink. The reversible process (15.2) is commonly known as the 'water hardness' equilibrium; there is no well-controlled $CaCO_3$ crystal growth in simple organisms.

Of greater interest with regard to material sciences are the biologically better controlled processes. The resulting 'advanced' bioinorganic solid materials are mostly formed as defined composites from inorganic and organic matter. The organic phase may consist of fibrous proteins, lipids or polysaccharides, its properties being relevant for the resulting morphology and the structural integrity of the composite material. Four types of biocomposites can be distinguished according to the degree of participation of the organic phase [1]:

(a) Type I (example: iron oxide-containing teeth of chiton molluscs) consists of randomly arranged crystallites, the structure of which is determined by the physicochemical properties of the mineralization zone. The organic matrix confers only mechanical stability.

(b) Type II (e.g. avian egg shells) shows matrix-supported crystal formation at predetermined sites but little control over the actual crystal growth.

(c) Type III (silica deposits in plants or valves of diatoms) features an amorphous mineral phase, the organic matrix directing nucleation *and* vectorial growth of the inorganic phase.

(d) Type IV (bones, teeth, shells of molluscs) involves a high degree of control through the organization of the matrix with regard to nucleation as well as oriented, e.g. epitaxial crystal growth.

15.2 Nucleation and Crystal Growth

Nucleation and crystal growth are processes that occur in supersaturated media and that have to be carefully controlled in a directed mineralization process. These requirements can be met in an organism through highly regulated active transport mechanisms and through specific modulation of the surface reactivity. The transport mechanisms may include transmembrane ion flux (see Chapter 13), ion (de)complexation, enzymatically catalyzed gas exchange (CO_2, O_2 or H_2S), local changes in redox potential ($Fe^{II/III}$) or pH, and variations in the ionic strength of the medium. All these factors can create and sustain a supersaturated solution in a biological compartment. Nucleation, on the other hand, is linked to the kinetics of (heterogeneous) surface reactions; processes such as cluster formation, anisotropic crystal growth and phase transformations are also determined by the properties of the surface. In the biological realm there are also a number of surface structures that specifically prevent undesired nucleation; for instance, fishes in polar waters can thus protect themselves from ice formation in their bodily fluids at temperatures slightly below 0 °C [11].

The growth of a crystal or of an amorphous solid from the nucleus which is formed in an activation step (crystal nucleation energy) can proceed directly from the surrounding solution or by the continuous supplementation of the necessary ions or molecules (Figure 15.2). On the other hand, the diffusion of particles can be drastically

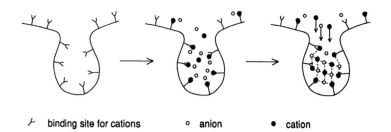

↲   binding site for cations   ○   anion   ●   cation

Figure 15.2
Model representation for nucleation and growth limitation of a microcrystallite (according to [1])

altered through significant changes in the viscosity of the medium, e.g. through gel formation of a biological matrix—a mechanism which is presumably responsible for the deposition of amorphous silica in plants [12].

The controlled growth of biominerals can also proceed stepwise with small activation energies for the individual steps (see Figure 2.8) through phase transformations of one or more solid-state precursors. This mechanism is particularly interesting if the minerals also serve as a main storage form of the involved components; in general, the precursors have a higher total energy and can be more easily mobilized. In most instances, the phase transitions are chemically induced; for example, the redox potential seems to play an important role in the transformation of ferrihydrite to magnetite in magnetotactic bacteria or in the teeth of chiton molluscs. A comparable function can be attributed to the pH in many condensation processes.

Two limiting cases can be visualized for the formation of actual microstructures: purely epitactical crystal growth on (organic) matrices or the linking of preformed inorganic crystallites by organic 'mortar' material [13]. The importance of the matrix thus lies in the control of nucleation, in the orientation and limitation of crystal growth, and in the immobilization of the crystallites.

One of the most important and most fascinating aspects of biominerals is the fact that their shapes do not have to coincide at all with the regular crystallographic forms of the inorganic material. Far more significant for the final morphology are the spatial limitations (see Figure 15.2) through biopolymers, membranes or vesicles during the formation of biominerals, even if the completed structures are finally assembled outside such confinements. Of course, modifications of crystal growth may also be induced nonbiologically; in fact, rather simple chemicals in the growth medium can influence the shape of the mineral formed. For example, the spindle shape of calcite crystals in gravity sensors (see below) can be reproduced by crystallization in the presence of 5 mM malonic acid, and monomolecular layers of stearic acid induce the crystallization of $CaCO_3$ as disc-shaped vaterite crystals rather than as rhombic calcite [14].

15.3 Examples of Biominerals

15.3.1 Calcium Phosphate in the Bones of Vertebrates

The dry support structure of vertebrate long bones consists of elastic fibrous proteins (approximately 30%, mainly collagen) and the inorganic components which are imbedded in 'cementing' glycoproteins: calcium phosphate, microcrystallized mainly as hydroxyapatite (approximately 55%), small amounts of calcium carbonate, silica, magnesium carbonate, other metal ions and citrate (remaining 15%). Collagens are fibrous proteins with a molecular mass of about 300 kDa; three polypeptide chains are wound together in the fibrilles as a superhelix (dimensions approximately 1.3×300 nm [15]). It has been assumed for a long time that the inorganic phase would largely consist of amorphous calcium phosphate which, in a process of aging, rearranges to microcrystalline hydroxyapatite. Recent results from solid-state ^{31}P NMR spectroscopy have shown that the amorphous form is never present in large amounts during the development of the bone; instead, acidic phosphate groups were detected with this method. These phosphate functions belong to proteins with

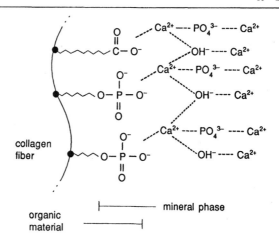

Figure 15.3
Schematic representation of the connection between collagen and hydroxyapatite through carboxylate and phosphate groups (molecular complementarity of the organic/inorganic interface; according to [16])

O-phosphoserine and O-phosphothreonine groups which are probably used to connect the inorganic mineral component and the organic matrix (Figure 15.3). The phosphoproteins are arranged at the collagen fibers in such a way that Ca^{2+} can be bound in regular intervals corresponding to the inorganic crystal structure, thus providing a condition for the crystallinity of the inorganic phase [16]. The main building blocks of the inorganic component are small crystallites (approximately 5×50 nm) of hydroxyapatite.

Apatite is also well known as a nonbiological mineral; it has assumed great importance as fertilizer and as the basic material for phosphorus chemistry. The inorganic material of bones, the bone meal, is also being used as fertilizer while the main organic component can be converted to collagen glue. The complex crystal structure [17] of hydroxyapatite and other apatites is illustrated in Figure 15.4.

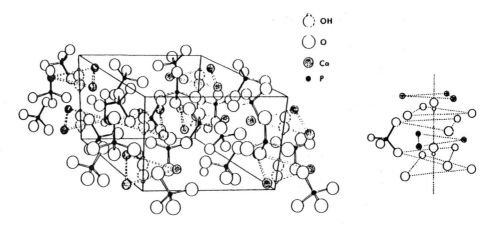

Figure 15.4
Unit cell of hexagonal hydroxyapatite with half-occupied OH^- positions (left) and view along a threefold axis with the anion channel (right; from [17])

The hydroxyl ions are located on threefold axes (hexagonal structure, space group $P6_3/m$). Phosphate oxygen atoms are arranged around these axes in such a way that the packing leaves holes for the Ca^{2+} ions. Each hydroxyl ion is then surrounded by a triangular arrangement of three calcium ions; however, the OH^- groups are randomly distributed 30 pm above or below the plane spanned by the Ca^{2+} ions. This hexagonal structure is particularly susceptible to substitution and defect formation (ion exchange behavior, H^+ diffusion) which is very advantageous for a high metabolic rate, i.e. the (de)mineralization of the 'living' endoskeleton. Hydroxide can be substituted by fluoride or chloride, phosphate by carbonate or sulfate, and Ca^{2+} by other divalent ions such as Sr^{2+}. In human tooth enamel (see below) F^- is thus present in amounts between 30 and 3000 ppm and there is an established but poorly understood dependence of the resistance against dental caries on the fluoride ion content (see Section 16.6). Chloride is present in even larger amounts (0.1–0.5%) than F^-; however, there is no established function of the chloride in dental apatite [17].

Crystallization of the complex and little soluble hydroxyapatite structures proceeds favorably through kinetically controlled formation of metastable intermediates (Table 15.2) within the concept of the 'Ostwald rule'. The *in vitro* transformation of initially precipitated amorphous calcium phosphate to hydroxyapatite (HAP) occurs via octacalcium phosphate (OCP) at higher pH values; at lower pH values dicalcium phosphate dihydrate (DCPD) may be an intermediate.

The bones as supporting scaffold of the vertebrate body can feature different types of integration of organic and inorganic material which results in a considerable variability of mechanical properties [18]. The ratio of both components reflects the compromise between hardness (high inorganic content) and elasticity or breaking strength (low inorganic content). The hitherto only partially successful attempts to synthesize suitable, i.e. physiologically tolerated ('biocompatible') and long-term stable bone substitutes for medical purposes ('bioceramics' [7,19]) have substantiated the superiority and complexity of the natural structure. In addition to the microstructural composition, the macroscopic architecture (lightweight construction) determines the mechanical properties of bones; for instance, the human femur can tolerate loads of up to 1650 kg.

The continuous formation of bone tissue takes place in a peripheral zone which consists of an outer and inner layer of connective tissue containing osteoblast cells. These osteoblasts are rich in phosphatases and excrete a gelatinous substance, the

Table 15.2 Biologically relevant calcium phosphates (compare Table 15.1)

mineral	formula	$pK_{sp}*$ [a]
dicalcium phosphate dihydrate (DCPD)	$Ca_2(HPO_4)_2 \cdot 2\,H_2O$	6.7
dicalcium phosphate (DCPA)	$Ca_2(HPO_4)_2$	6.0
octacalcium phosphate (OCP)	$Ca_8(HPO_4)_2(PO_4)_4 \cdot H_2O$	≈12
ß-tricalcium phosphate (TCP)	$Ca_3(PO_4)_2$	11.6
hydroxyapatite (HAP)	$Ca_{10}(PO_4)_6(OH)_2$	≈13
defect apatites	$Ca_{10-x}(HPO_4)_x(PO_4)_{6-x}(OH)_{2-x}$ $0 \geq x \geq 2$	

[a] Solubility products $K_{sp}*$ in pure water have been reduced to the unit M^{-2} for the sake of comparison.

osteoid; through gradual deposition of inorganic material the osteoid hardens and the thus walled-in osteoblasts turn into actual bone cells (osteocytes). For the purpose of transformation and to prevent excessive growth of the bone there are degradation processes occurring simultaneously with the bone formation. Multinucleate giant cells, the osteoclasts, catabolize bones, perhaps using citrate as the chelating agent; the control of osteoclastic activity occurs via the parathyroid hormone which promotes demineralization, and via its antagonist, thyreocalcitonin (see Section 14.2).

The Ca^{2+} deposited and thus stored in the skeleton continuously exchanges with dissolved calcium ions. For bone growth to occur, a relative excess of Ca^{2+} and corresponding anions such phosphate and carbonate has to be actively created in the bone matrix. This is guaranteed through the action of efficient ATP-consuming ion pumps such as the Ca^{2+} ATPases for active calcium transport. Physiologically, carbonate and phosphate exist as hydrogen anions HCO_3^-, HPO_4^{2-} and $H_2PO_4^-$; when these are incorporated into the bone, protons will be released which are mobile within the bone tissue (hydrogen ion conductivity) and can thus be readily removed from the area of nucleation and mineralization.

Favored by the continuous metabolism of the bone substance and by the substitution-prone structure of hydroxyapatite (Figure 15.4), a chronic heavy metal poisoning (Sections 17.2 and 17.3) may lead to the substitution of Ca^{2+} by ions such as Cd^{2+} or Pb^{2+} in the bone structure. The results are altered mechanical properties (brittleness) or even very painful bone deformations.

The permanent teeth of higher vertebrates feature as the outer layer the tooth enamel which, in an adult organism, does *no longer* contain any living cells. Up to 90% of this enamel may consist of inorganic material, mainly hydroxyapatite [20]. It is the enamel that experiences the most extensive changes during tooth development. Initially, it is deposited with a mineral content of only 10–20%, the remaining 80–90% being special matrix proteins and fluids; in later developmental stages the organic components of enamel are nearly completely substituted by the biomineral. The special features of tooth enamel as compared to bone material are the significantly larger crystalline domains in the form of long, highly oriented enamel prisms made from hydroxyapatite [20,21]. This 'dead' solid-state material is unsurpassed in the biological realm with regard to hardness and durability; however, as is generally (and often painfully) known, a regeneration is then no longer possible.

The role of the trace amounts of fluoroapatite, $Ca_{10}F_2(PO_4)_6$, for the prevention of microbially enhanced decay of tooth enamel, i.e. dental caries, is still being disputed. The mechanisms discussed include surface enamel hardening (anticorrosion effect), enhanced ionic remineralization, and the deactivation of acid-forming enzymes through the fluoride deposited on the tooth surface [22] (see also Section 16.6).

15.3.2 Calcium Carbonate

In egg and molluscan shells, the $CaCO_3$ crystals grow in a preformed arrangement of proteins and polysaccharides. The pure organic matrix can be obtained from these shells after dissolution of $CaCO_3$ through chelating agents such as EDTA [13]. The matrix is composed of a water-soluble protein and oligosaccharide component and of water-insoluble hydrophobic proteins, also with a significant polysaccharide content. The soluble proteins and sulfate-containing oligosaccharides are strongly acidic and therefore well suited to bind Ca^{2+}. Significant amounts of carboanhydrase are also present (see Section 12.2); they are obviously necessary to produce HCO_3^--supersaturated solutions, thus allowing for a rapid growth of shells.

The mineralization process starts at the insoluble matrix in the egg shell membrane which binds Ca^{2+} and HCO_3^-, the latter through scavenging of the released protons with ammonia. Ammonium ions, NH_4^+, can then be bound by the sulfate-containing soluble matrix [13]. During the actual growth of the shells the $CaCO_3$ mineral and the soluble protein are deposited; growth is stopped when NH_3 production by mucosa cells is halted.

Marine organisms such as algae, sponges, corals or molluscs form $MgCO_3$-containing calcium carbonate in large amounts as a consequence of photosynthetic activity (15.2). When CO_2 dissolved in water is photosynthetically assimilated, the pH value of the surrounding medium rises according to (12.7), the concentration of CO_3^{2-} increases and the equilibrium is shifted towards the precipitation of $CaCO_3$ (15.2, 15.3).

$$HCO_3^- \rightleftharpoons H^+ + CO_3^{2-} \qquad \begin{aligned} pK_a &= 10.33 \text{ (fresh water)} \\ &= 10.89 \text{ (sea water)} \end{aligned} \qquad (15.3)$$

The limestone deposits in corals reefs which display a highly species-specific architecture are thus a direct consequence of the symbiosis of polyps with photosynthetically active algae. CO_2 uptake and $CaCO_3$ precipitation by marine organisms may depend on temperature, salinity, buffer capacity of the medium, and on the original pH value. $CaCO_3$ deposition is particularly favored in compartmentalized cells with small volumes since diffusion effects are negligible in such cases. Because of the slightly higher solubility of $MgCO_3$ (see Table 15.1) its share is often so small that the typical $CaCO_3$ structures such as calcite or aragonite dominate. However, even very small amounts of 'fiber-enhancing' proteins can be sufficient to transform the brittle calcite into, for example, the morphologically and mechanically far more functional spikes of the sea-urchin [23].

Gravity- or inertia-sensitive sensors, e.g. in the human inner ear, often contain spindle-shaped mineral deposits ('statoconia', 'otoconia' from calcite or larger 'statoliths', 'otoliths' from aragonite) associated with membrane-linked sensory cells. Functionally, the specifically heavy ($\rho \approx 2.9$ g/cm^3) inorganic minerals lend mass to the membrane so that accelerations can be sensed very accurately. The movement of the statoconia in relation to the sensory cells gives information on the direction and intensity of the acceleration. The structural similarity of the organic matrix of the statoconia with that of other $CaCO_3$-containing biominerals could be demonstrated.

15.3.3 Amorphous Silica

While silicon, in form of the silicates, comes second relative to all other elements in the earth's crust with regard to quantity (Figure 2.2), it plays only a marginal role in the biosphere (Table 2.1 and Section 16.3). This can be partially attributed to the low solubility of 'silicic acid', H_4SiO_4, and its oligomeric condensation product 'amorphous silica', $SiO_n(OH)_{4-2n}$ (see Figure 15.5); in water between pH 1 and 9 this solubility is about 100–140 ppm. In the presence of cations such as calcium, aluminum or iron, the solubility decreases markedly, and in sea water, the solubility

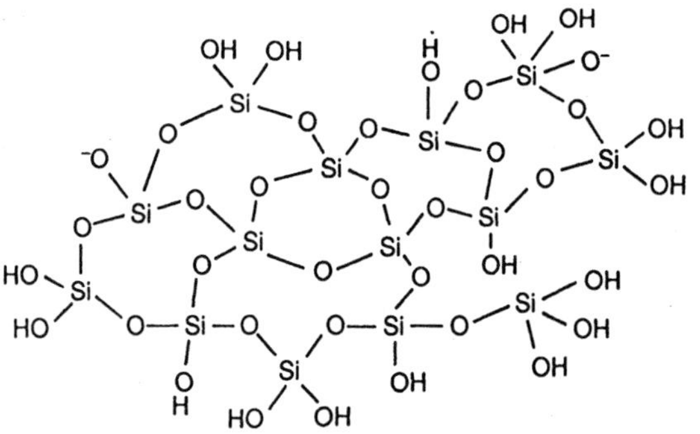

Figure 15.5
Representative linkage of atoms in amorphous silica

is only about 5 ppm. In the biosphere, dissolved amorphous silica is resorbed by organisms and then polymerized or connected to other solid structures [24].

Amorphous silica as biomineral is featured mainly in unicellular organisms [25], in siliceous ('glass') sponges and in several plants [12], where it occurs in the cell membranes of grains, grass or horsetail in the form of 'phytolithes' with a passive deterrent function. The brittle tips of the stinging hairs of various nettle plants are also made of amorphous silica.

Accumulated in deposits of diatomaceous earth, the siliceous remains of diatoms illustrate that these are a rather old group of diverse unicellular organisms which form two silica-containing box-and-cover structures, the valves, as exoskeletons. After each cell division, two new valves have to be formed from amorphous, polymeric $SiO_n(OH)_{4-2n}$ (Figure 15.5) to protect the daughter cells. The morphology of biogenic silica is determined by membrane proteins where nucleation can take place through condensation (H_2O elimination) between silicic acid or its derivatives and the hydroxyl groups of the organic matrix.

Brittleness and the chemical surface properties of polycondensated silica can be dangerous to human beings beyond its function in nettle plants. The frequent occurrence of oesophageal cancer in certain areas may be connected with dust from silica-containing grains [12] where the fibrous microstructure (passive defense function) of the silica particles may resemble that of the carcinogenic asbestos mineral fibers. The chemical basis of the membrane-dissolving and carcinogenic effect of only slowly degraded mineral or biogenic silica as fibers or dust particles is still unclear; acid–base effects, irreversible condensation processes and even surface-based oxygen radicals are being discussed [26].

15.3.4 Iron Biominerals

The biomineralization of iron oxides is a relatively well documented field of bioinorganic chemistry (compare Chapter 8), not in the least because of the available methods, in particular Mössbauer spectroscopy. Phosphate- and silica-containing biominerals, on the other hand, have become spectroscopically well accessible only recently through solid-state NMR methods (^{29}Si, ^{31}P). In addition to the iron(III) hydroxide condensation products mentioned in Section 8.4.2, biogenic iron oxide also exists in the form of magnetite (Fe_3O_4). This $Fe^{III,II}$-containing mineral was detected in the fairly ubiquitous magnetotactic bacteria [27,28], in chitons, molluscs, pidgeons, bees, fishes and even in humans [29]. The magnetotactic bacteria can exhibit an orientation along the earth's magnetic field based on these iron minerals. Under an electron microscope, dark, membrane-coated iron-containing particles, the magnetosomes, are observed in such bacteria. The particles consist mainly of magnetite, Fe_3O_4, and in some instances even of greigite (Fe_3S_4 [30]); their size of 40–120 nm corresponds to that of the single magnetic domains of Fe_3O_4. The magnetosomes are normally arranged as chains along the direction of bacterial movement so that this string of particles may function as a biomagnetic compass. To form the magnetosomes, iron is taken up as chelated Fe^{3+} and then made available through reduction to Fe^{2+} (compare Chapter 8). After controlled oxidation a water-containing iron(III) oxide precipitates (8.20); dehydration first leads to ferrihydrite, $Fe^{III}_{10}O_6(OH)_{18}$, and then, under partial reduction, to magnetite, $Fe^{II}Fe^{III}_2O_4$.

Marine molluscs of the chiton family are found in tidal zones where they feed on algae which grow on rock. For that purpose they use a tongue-shaped organ, the radula, on which mineralized 'teeth' are found, consisting of iron oxides with organic inclusions. Since the worn-out teeth in the front of the radula are continuously replaced by fresh material formed further back, it became possible to study the developmental phases of the formation of these teeth simply by looking at the spatial distribution of material on the radula. Young teeth consist of purely organic substances without inclusion of an inorganic phase. After the onset of mineralization, the inorganic elements Fe, Zn, S, Ca, Cl, P and K can be detected in various amounts by X-ray emission spectroscopy. In the first phases of mineralization, the iron content increases strongly only to remain constant later at about 10%. Again, ferrihydrite was found as the first unambiguously detected mineral (Figure 8.6) which is then replaced with goethite or lepidocrocite, $Fe(O)OH$, and finally magnetite. The different minerals occur simultaneously but at spatially different sites of the radula. Adult chiton teeth contain magnetite as the main iron-containing mineral; however, calcium and phosphate are also found, presumably for anchoring purposes, in a structure similar to that of apatite.

15.3.5 Strontium and Barium Sulfates

The unicellular plankton organism *Acantharia* which belongs to the radiolarians features exoskeletons consisting of exactly 20 strontium sulfate single crystals which can assume very complex shapes (Figure 15.6). The origin of these forms is connected to the morphology of the cell since nonbiogenic strontium sulfate (celestite) crystallizes as flat rhombs when left undisturbed. Vesicles in which strontium sulfate crystals grow are formed along radial filaments starting from the center of the cell. Some species are known which show a nearly perfect D_{4h} symmetry of the 20-spicule system [10]. While the general structure is determined by the position of the vesicles, the inherent crystal structure of $SrSO_4$ is responsible for the angles. The high symmetry and simplicity of this special mineral have made the *Acantharia* a preferred object for studies of biomineralization.

Certain unicellular algae (*Desmidiacea*) contain vesicles with barium sulfate in their crystalline form as baryte ($BaSO_4$, $\rho = 4.5$ g/cm^3). The cells of *Closterium*, for

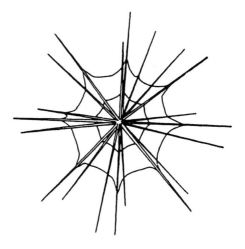

Figure 15.6
Twenty-spicule *Acantharia* species with inner cell membranes (total diameter approximately 1 mm; from [10])

example, are bent in a crescent-like shape with two baryte crystal-containing vesicles at opposing ends of the cell which suggests a function as inertia sensors. The crystals show the normal rhombic form of baryte, i.e. the morphology of the cell does not influence the crystal shape in contrast to the *Acantharia*.

References

1. S. Mann, J. Webb and R. J. P. Williams (eds.), *Biomineralization*, VCH, Weinheim, 1990.
2. S. Mann, Biomineralization: the hard part of bioinorganic chemistry, *J. Chem. Soc., Dalton Trans.*, 1 (1993).
3. S. Mann, D. D. Archibald, J. M. Didymus, T. Douglas, B. R. Heywood, F. C. Meldrum and N. J. Reeves, Crystallization at inorganic-organic interfaces: biominerals and biomimetic synthesis, *Science*, **261**, 1286 (1993).
4. H. A. Lowenstam, Minerals formed by organisms, *Science*, **211**, 1126 (1981).
5. L. Addadi and S. Weiner, Control and design principles in biomineralization, *Angew. Chem. Int. Ed. Engl.*, **31**, 153 (1992).
6. G. Krampitz and W. Witt, Biochemical aspects of biomineralization, *Top. Curr. Chem.*, 57 (1978).
7. A. H. Heuer *et al.*, Innovative materials processing strategies: a biomimetic approach, *Science*, **255**, 1098 (1992).
8. *Phil. Trans. Roy. Soc. London* **B304**, 425 (1984).
9. S. Mann, Bacteria and the Midas touch, *Nature (London)*, **357**, 358 (1992).
10. C. C. Perry, J. R. Wilcock and R. J. P. Williams, A physicochemical approach to morphogenesis: the roles of inorganic ions and crystals, *Experientia*, **44**, 638 (1988).
11. A. L. De Vries, Role of glycopeptides and peptides in inhibition of crystallization of water in polar fishes, *Phil. Trans. Roy. Soc. London*, **B304**, 575 (1984).
12. D. W. Parry, M. J. Hodson and A. G. Sangster, Some recent advances in studies of silicon in higher plants, *Phil. Trans. Roy. Soc. London*, **B304**, 537 (1984).
13. G. Krampitz and G. Graser, Molecular mechanisms of biomineralization in the formation of calcareous shells, *Angew. Chem. Int. Ed. Engl.*, **27**, 1145 (1988).
14. S. Mann, B. R. Heywood, S. Rajam and J. D. Birchall, Controlled crystallization of $CaCO_3$ under stearic acid monolayers, *Nature (London)*, **334**, 692 (1988).

15. A. Miller, Collagen: the organic matrix of bone, *Phil. Trans. Roy. Soc. London*, **B304**, 455 (1984).

16. M. J. Glimcher, Recent studies of the mineral phase in bone and its possible linkage to the organic matrix by protein-bound phosphate bonds, *Phil. Trans, Roy. Soc. London*, **B304**, 479 (1984).

17. K. Sudarsanan and R. A. Young, Structural interactions of F, Cl and OH in apatites, *Acta Cryst.*, **B34**, 1401 (1978).

18. J. D. Currey, Effects of differences in mineralization on the mechanical properties of bone, *Phil. Trans. Roy. Soc. London*, **B304**, 509 (1984).

19. G. Heimke, Bioactive ceramics, *Adv. Mater.*, **3**, 320 (1991).

20. W. von Koenigswald, Biomechanische Anpassungen im Zahnschmelz von Säugetieren, *Biol. Unserer Zeit*, **20**, 110 (1990).

21. C. Robinson, J. A. Weatherell and H. J. Höhling, Formation and mineralization of dental enamel, *Trends Biochem. Sci.*, 24 (1983).

22. C. Dawes, J. M. ten Cate (eds.), International symposium on fluorides: mechanism of action and recommendations for use, *J. Dent. Res. (Special Issue)*, **69**, 505–831 (1990).

23. A. Berman, L. Addadi, A. Krick, L. Leiserowitz, M. Nelson and S. Weiner, Intercalation of sea urchin proteins in calcite: study of a crystalline composite material, *Science*, **250**, 664 (1990).

24. J. D. Birchall, Silicon in the biosphere, in *New Trends in Bio-inorganic Chemistry* (eds. R. J. P. Williams and J. R. R. Frausto da Silva), Academic Press, London, 1978.

25. D. H. Robinson and C. W. Sullivan, How do diatoms make silicon biominerals?, *Trends Biochem. Sci.*, **12**, 151 (1987).

26. B. Fubini, E. Giamello and M. Volante, The possible role of surface oxygen species in quartz pathogenicity, *Inorg. Chim. Acta*, **162**, 187 (1989).

27. R. B. Frankel and R. P. Blakemore, Precipitation of Fe_3O_4 in magnetotactic bacteria, *Phil. Trans. Roy. Soc. London*, **B304**, 567 (1984).

28. R. P. Blakemore and R. B. Frankel, Magnetic navigation in bacteria, *Sci. Am.*, **245**(6), 42 (1981).

29. J. L. Kirschvink, A. Kobayashi-Kirschvink and B. J. Woodford, Magnetite biomineralization in the human brain, *Proc. Natl. Acad. Sci. USA*, **89**, 7683 (1992).

30. B. R. Heywood, D. A. Bazylinski, A. Garratt-Reed, S. Mann and R. B. Frankel, Controlled biosynthesis of greigite (Fe_3S_4) in magnetotactic bacteria, *Naturwissenschaften*, **77**, 536 (1990).

16 Biological Functions of the Nonmetallic Inorganic Elements

16.1 Overview

As members of the group of nonmetals in the periodic table (Figure 1.3) the elements carbon, hydrogen, nitrogen, oxygen, sulfur, phosphorus and chlorine [1] traditionally belong to 'conventional' biochemistry. Among the other less metallic elements, the noble gases and the rare, nonradioactive elements germanium, antimony, bismuth and tellurium have not been found to possess any natural biological significance. For the remaining nonmetallic elements boron, silicon, arsenic, selenium, fluorine, bromine and iodine, the details concerning their biological role are only partially known. However, it should be noted that, historically, the elements iodine and phosphorus were first obtained from biogenic material.

16.2 Boron

Very little is known about the biochemical function of boron, although a boron-containing natural product, 'boromycin', was isolated in 1967, and although borates seem to be essential for the normal growth and infection resistance of (domestic) plants, in particular of tobacco, beet and cabbage species [2]. An important function in the hormonal regulation of calcium and thus in cell division is assumed. Binding to glycolipid-containing cell membranes can be visualized via the known affinity of boron to form five-membered ring chelates with *cis*-hydroxy functions [3]. Due to the relatively high concentration in sea water boron has been found accumulated in some algae and sponges.

16.3 Silicon

Amorphous silica, $SiO_2 \cdot n\, H_2O$, was introduced in Section 15.3.3 as a solid constituent of marine organisms, land-living plants and animal skeletons. The pathogenic effect of silica-containing dust and fibers was also mentioned. As is the case with many other trace elements, severe deficiency symptoms become evident particularly during growth periods [3,4]. So far, no unambiguous information is available on the details of binding of silicates, e.g. during bone formation, mainly because of analytical problems; a regulatory function with regard to cross-linking and the amount of crystallization in connection with phosphoproteins (compare Figure 15.3) can be envisaged. There are antagonistic relationships between silicon and boron with respect to infection resistance of plants [2] and between silicon and aluminum due to the formation of inactive aluminosilicates (see Section 17.6).

16.4 Arsenic and PH_3

In the soluble form, $As(OH)_3$, of arsenic(III) oxide, As_2O_3, arsenic is a potent poison and carcinogenic substance for human beings. There is a close relationship between arsenates(V) and phosphates as a result of their position in the periodic system; however, due to the 'thiophilicity', i.e. the affinity especially of As^{III} for negatively charged S^{-II} ligands, the coordinatively unsaturated arsenic compounds are able to affect enzymes by blocking the redox-active sulfhydryl groups which are then no longer able to form conformation-determining disulfide bridges. In the case of acute arsenic poisoning as manifested by gastrointestinal cramps, a therapy with thiolate-containing chelate agents such as dimercaprol (2.1) is therefore indicated.

In organic, especially (bio)methylated form (compare Sections 3.2.4 and 17.3), arsenic compounds are less toxic and actually quite common, particularly in marine organisms such as algae, fish or crustaceans such as lobster [5,6]. S-Adenosyl methionine and methylcobalamin are able to methylate trivalent arsenic rapidly to give substances such as (16.1a); according to studies of certain moulds this reaction can proceed in an enantiospecific fashion [7].

$O=As(CH_3)_2R,$ R = OH, CH_3, CH_2CH_2OH, 5'-deoxyribosyl and derivatives

$(CH_3)_3As^+-R,$ R = 5'-deoxyribosyl and derivatives

$(CH_3)_3As^+-CH_2COO^-$ ('arsenobetaine')   (16.1a)

The low concentration of phosphate in sea water as caused by the precipitation with multivalent cations makes efficient uptake and accumulation mechanisms necessary which, however, may not be able to distinguish between phosphates and the related, only marginally less abundant arsenates (or vanadates, compare Figure 2.2). Based on the redox potentials and reactivity, reduction and biomethylation are possible for arsenic($+V$, $+III$) compounds but not for phosphates($+V$). This possible differentiation may be the reason for the existence of peralkylated arsenous oxides, arsonium salts and arsenobetaines (16.1a) as stable, excretable natural products. An alternative strategy for detoxification (see Section 17.1) consists in the energy-requiring transport out of the cell through anion-specific ATPases within the membrane. In arsenic-resistant microorganisms such an 'oxy anion pump' was identified which effects the excretion of arsenite, antimonite and arsenate, the latter being formed by a molybdenum-containing oxidase (see Table 11.1) [8].

In the 1970s it was found that a complete lack of arsenic can clearly lead to disorders in reproduction and growth of land-living animals [4], in particular via an impaired metabolism of the limiting amino acid methionine [3].

It was recently reported that phosphate which, like arsenate, is carried specifically as monohydrogen dianion HEO_4^{2-} by a transport protein (E = P, As [9]) can be converted to mutagenic phosphane, PH_3, according to (16.1b) under the special reducing conditions of methane formation from CO_2/HCO_3^- (16.1c; see also Figure 1.1 and Section 9.5).

$$HPO_4^{2-} + 10\ H^+ + 8\ e^- \rightarrow PH_3 + 4\ H_2O \tag{16.1b}$$

$$HCO_3^- + 9\ H^+ + 8\ e^- \rightarrow CH_4 + 3\ H_2O \tag{16.1c}$$

Implications of this result include the potential self-ignition of marsh gas (via traces of pyrophoric P_2H_4), the role of volatile biogenic PH_3 in the global phosphorus cycle, and the possible connection between a phosphate-rich diet and the incidence of intestinal cancer [10].

16.5 Bromine

Bromides have already been used in the nineteenth century as sedatives in the treatment of nervous disorders. Obviously, a modification of the transmembrane ionic nonequilibrium state is possible (see Chapter 13) when introducing this heavier homologue of chloride. However, molecular details of the effect of Br^- on ion channels or pumps are not yet known [1]. Due to the relative abundance of soluble bromide (and also iodide) in sea water, algae and other marine organisms often contain rather large amounts of organic bromine and iodine compounds which are synthesized with the help of heme- or vanadium-containing haloperoxidases (Section 11.4).

16.6 Fluorine

The cariostatic effect of trace amounts of fluorides has now been established for a long time; however, a definitive mechanism of action of this trace component can not yet be determined unambiguously despite numerous studies. A more effective remineralization, a 'hardening' of the tooth enamel surface, e.g. through formation of a particularly compact, acid-resistant crystalline layer under participation of fluoroapatite (compare Table 15.1), or the inhibition of caries-promoting enzymes such as enolases by fluoride which dissolves from solid-state depots are among the most commonly accepted mechanisms [1,11]. In this context, the usefulness of fluoridation of drinking water (1 ppm) remains widely controversial [12]. In contrast to the practice in many European countries this caries-preventive measure is common in parts of the United States, the United Kingdom, Canada and Australia, despite considerations of possible toxicity (fluorosis). The ability of fluorides to efficiently bind to many (heavy) metal ions in unphysiological high oxidation states and thus render them bioavailable is an additional argument against fluoridation of drinking water.

For the prevention of dental caries, fluoride is used in trace amounts in tooth pastes and other medical preparations. Preferred are fluoride-containing compounds such as monofluorophosphate PO_3F^{2-}, SnF_2, AlF_3 or organoammonium fluorides which promise efficient binding by the hydroxyapatite surface of teeth. For children, fluoride-containing tablets and gels ensure an effective incorporation into the growing tooth enamel. However, fluoride exhibits a rather small biooptimal concentration range (Figure 2.3) of only about one order of magnitude and the switch from beneficial activity to toxic effect is therefore a matter of concern (see above). For instance, organisms that are rich in fluoride such as the antarctic krill (a planktonic crustacean) are thus not suitable for human consumption. Some plants, bacteria and caterpillars accumulate toxic fluoroorganic substances like FH_2C-COO^- and use it as a deterrent [13].

Due to the importance of Ca^{2+} for so many biochemical processes (Section 14.2) the marked insolubility of calcium fluoride, CaF_2 ($pK_{sp} = 10$), renders acute ($\rightarrow$tissue necrosis) and chronic fluoride poisoning very serious. Fluorosis manifests itself by discoloration of teeth, deformation of the skeleton, renal failure and muscle weakness; calcium gluconate is indicated as antagonist in case of an acute poisoning. In spite of the only moderately acidic character of hydrofluoric acid ($pK_a \approx 3.2$) the action of acidic fluoride solutions on tissue, in particular on skin, leads to poorly healing wounds which have to be treated immediately. Here again the reason is the deactivation of Ca^{2+} which is essential for several steps of the coagulation mechanism. Numerous metalloenzymes are inhibited by binding of F^- to the metal center.

16.7 Iodine

The heaviest (and rarest) stable halogen which was first isolated from the ashes of marine algae by B. Courtois around 1812 had already been recognized as an essential element for higher organisms by the middle of the nineteenth century. This observation was facilitated by the pronounced accumulation of this very characteristic element in the thyroid gland, where it is present in the form of *polyiodinated* small organic compounds, the thyroid hormones thyroxine (tetraiodothyronine) and the even more active triiodothyronine (16.2).

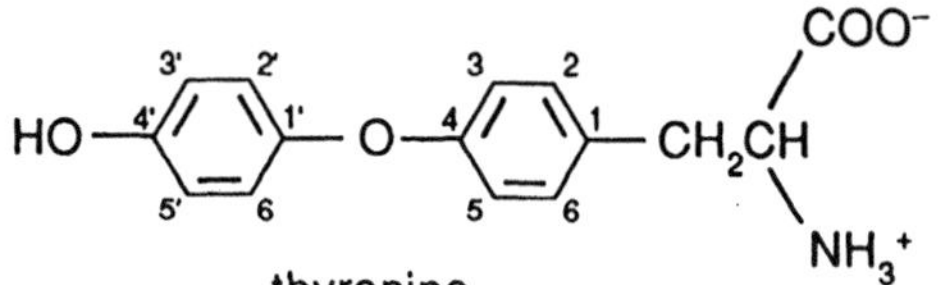

thyronine

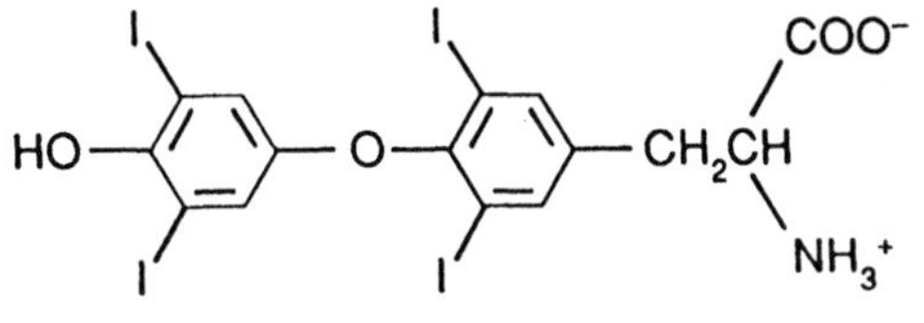

thyroxine
(3,5,3',5'-tetraiodothyronine, T$_4$)

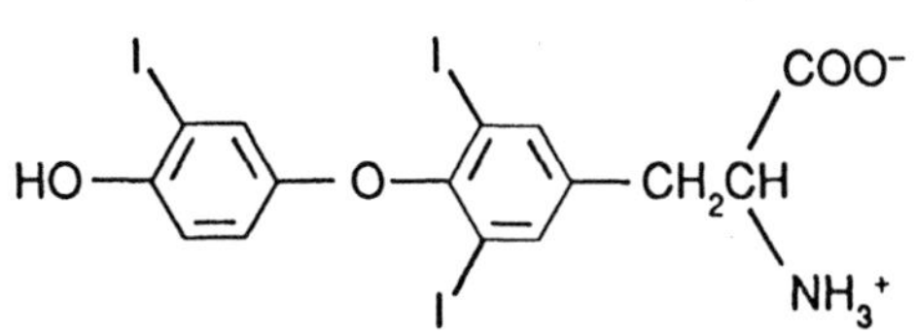

3,5,3'-triiodothyronine, T$_3$ (16.2)

The extreme twofold, i.e. physiological *and* intramolecular enrichment of iodine is quite remarkable, especially considering the central role of thyroid hormones in the control of energy metabolism and associated processes, from the biosynthesis of ATPases to the molting of birds. Familiar physiological disorders occur with regard to reduced thyroid activity (feeling 'cold', tiredness, even cretinism as caused by thyroid deficiency during the early growth period) and with respect to hyperactivity (feeling 'hot', restlessness, nervousness). Low thyroid activity due to iodine deficiency may be compensated by excessive growth of the organ (goiter, struma) with an increased tendency for tumor formation; this kind of disorder is not uncommon in regions far removed from the sea but can be counteracted by supplementing iodide preparations. In some countries, the drinking water is therefore iodinated (Italy) or the common salt is made available only in iodinated form (Switzerland, former East Germany); the trace element cobalt inhibits the uptake of iodine by thyronine. Due to the extreme localization of iodine in the human body the tumors of the thyroid can be successfully diagnosed and treated using the radioactive isotopes ^{131}I and [123]I (see Section 18.3.1).

What is the particular function of this rare element in the hormones? In its carbon-bound form, iodine is not redox-active at physiological potentials. Furthermore, metal ions are not significantly influenced by the compounds (16.2). Any explanation of the specific role of iodine has to take into account that bound iodide as the heaviest stable halide is an unusually large, spherical substituent; the ionic radius of I^- of about 220 ppm is unmatched by other monovalent elemental ions. In fact, when iodide in the 3,5,3′ positions of thyronine is substituted by approximately spherical methyl groups or, especially in the 3′ position, by the even larger isopropyl group $CH(CH_3)_2$, a hormonal activity similar to that of the polyiodinated species is found. Therefore, a receptor structure as shown in Figure 16.1 is assumed for the thyroid hormones in which especially *one* large spherical 'lock' cavity is preformed for the substituent in the 3′ position of the hormone 'key' [14,15].

A correspondence according to Figure 16.1 was possibly developed when organisms were still able to utilize the relatively abundant iodide from sea water (compare Figure 2.2). The iodination of thyronine which is derived from the amino acid tyrosine

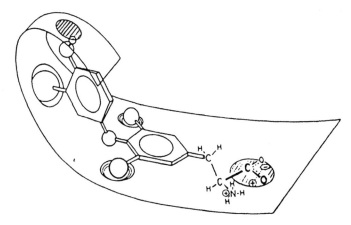

Figure 16.1
Assumed fit of the 3,5,3′-triiodothyronine (T_3) hormone into the matching receptor (from [15])

is an electrophilic substitution at the phenolic, i.e. electron-rich aromatic nucleus which, in contrast to alkylations, is obviously still possible under physiological conditions. Heme-containing thyreoperoxidases are mainly responsible for this reaction (see Section 6.3). The deiodination of T_4 to give the primarily active T_3 proceeds via a selenium-containing deiodinase [16].

16.8   Selenium

Without doubt, selenium as the heavier homologue of sulfur has recently become the most discussed essential nonmetallic element in biology and medicine [17–19]. Popular scientific and advertising publications make it appear as a miracle cure against cancer and the effects of old age. Although the chemistry of selenium qualitatively resembles that of sulfur, the selenium analogues of thiols, the selenols, feature a lower redox potential and can thus be more easily oxidized. On the other hand, the tetravalent state of selenous acid is thermodynamically stable in contrast to the metastable sulfites (see the stability diagrams in Figure 16.2). Finally, the reactivity of selenium compounds is generally higher than that of corresponding sulfur analogues due to longer bonds from Se to bound atoms.

Like fluorine, the trace element selenium features a rather small therapeutic width (compare Figure 2.3); deficiency symptoms (daily dosis $< 50 \, \mu g$) on one side and effects from poisoning on the other side ($> 500 \, \mu g$/day), e.g. due to overdoses of selenium drugs, lie close together. The high toxicity of selenium compounds has been known for a long time, characteristic symptoms being the loss of hair and the excretion of evil-smelling dimethylselenium, $(CH_3)_2Se$, through breath or skin. Dimethylselenium is formed via biomethylation and has a garlic-like odor which is noticeable even in the smallest concentrations. Disorders of the central nervous system as well as characteristic degenerations of keratinous, i.e. disulfide bridge-containing tissue of hairs and hooves, have also been observed for grazing animals which feed on particularly selenium-rich soil in Central Asia or in parts of the Western United States. In contrast, selenium deficiency symptoms can occur on extremely selenium-poor soil, especially for livestock with their specialized diet; muscular degeneration ('white muscle disease') in young animals as well as reproductive disorders have been observed [17]. Very similar symptoms have been found in the form of the 'Keshan' disease of adolescent humans, a fatal weakness of the heart muscle that occurred in Chinese regions with very selenium-deficient soil. Since selenium tends to bind soft heavy metal ions to an even larger extent than sulfur, there is an antagonism between the selenium and the heavy metal (e.g. copper) content of soil. A heavy metal detoxification function has not yet been established for selenoproteins, presumably because of the very small amounts involved; other, sulfur-rich proteins, the metallothioneins (Section 17.3), are available for this function. The Keshan disease can be successfully treated by administering trace quantities of Na_2SeO_3 or Na_2SeO_4 [17].

Selenium deficiency in mammals can also lead to liver necroses and an increased susceptibility for liver cancer; the absence of peroxide-destroying selenium enzymes (see below) in the eye lens may be connected with the occurrence of oxidatively induced glaucoma. In view of the antioxidative function of selenium-containing enzymes some epidemiological correlations for humans between the availability of

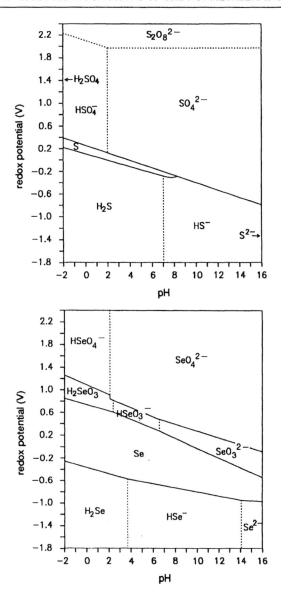

Figure 16.2
Stability (Pourbaix) diagrams for sulfur (top) and selenium (bottom) (according to [20])

selenium in drinking water and the abundance of breast or colon cancers have been published [18]. The antimutagenic effect of selenium compounds was also established in bacterial tests such as the Ames test, even though this element—like many others (see Chapter 19)—had previously been described as carcinogenic. In a normal diet, selenium deficiencies should not occur; despite the rarity of this element and an increased heavy metal load, the yearly requirement is only about 100 mg. Selenium-rich food includes certain mushrooms, garlic, asparagus, fish, and liver or kidney of animals. The availability of selenium varies considerably with geography and season, the pH of surface water playing an important role in addition to the soil content. Heavy metals like Cd, Cu, Cr, Pb, Hg and also the thiophilic zinc are selenium antagonists. A potential selenium source is sulfur-containing shampoos; the element was detected on the surface of hairs after several washings [21]. Due to the small therapeutic width of selenium its supplementation in the absence of medical supervision is not recommended.

The most abundant selenium-containing compound in organisms is selenocysteine [19] (see Table 2.5). The coding of this amino acid is quite unusual since the codon UGA used in the transcription was once only known as 'stop' or 'nonsense' command; in some organisms it is responsible for the synthesis of tryptophane. Furthermore, the integration of selenocysteine into proteins is different from that of other amino acids; it is assumed that the UGA codon is in fact multifunctional and can thus induce different responses according to the physiological requirements (flexible genetic code [22]). With regard to the mentioned concentration problem this flexibility could be part of a regulatory mechanism in view of the varying supply and demand of selenium.

In the known selenium-containing proteins selenocysteine appears only once in each polypeptide chain. Except for some substrate-specific bacterial formate, xanthine or nicotinic acid dehydrogenases, Ni/Fe/Se hydrogenase (Section 9.3) and glycine reductase (16.3 [23]), selenium occurs in mammals in iodothyronine deiodinase [16] and in the relatively well characterized glutathione peroxidase which catalyzes reaction (16.4).

$$\overset{\text{glycine}}{\underset{}{\text{reductase}}}$$

$${}^+NH_3-CH_2-COO^- + H_2PO_4^- + 2\,H^+ + 2\,e^- \longrightarrow {}^+NH_4 + CH_3-COOPO_3^{2-} + H_2O$$

glycine   acetyl phosphate

$$(16.3)$$

$$\overset{\text{glutathione}}{\underset{}{\text{peroxidase}}}$$

$$ROOH + 2\,G-SH \longrightarrow G-S-S-G + H_2O + ROH$$

G–SH: glutathione, γ-Glu-Cys-Gly (16.4)

Glutathione is a tripeptide which can dimerize via oxidative disulfide formation at the central cysteine part. The potentially membrane-damaging lipid hydroperoxide intermediates, ROOH, which may be formed during incomplete O_2 conversion are consumed in reaction (16.4) by a very rapid oxidation of glutathione, G—SH, to the disulfide G—S—S—G. Together with the tripeptide glutathione this Se-containing peroxidase enzyme thus functions as an antioxidant, similar to other peroxidases, superoxide dismutases (16.5), the vitamins C and E (16.6), or other heteroatom-rich compounds. A list of some major antioxidants [24,25] arranged according to their specific functions is shown in Table 16.1 (compare 4.6, 5.2 and 16.5).

The glutathione peroxidases are tetrameric proteins with subunits of 21 kDa each. These selenoproteins were first isolated from erythrocytes, i.e. from cells with high 'oxidative stress'. Despite various precautions (Table 16.1) there is a considerable

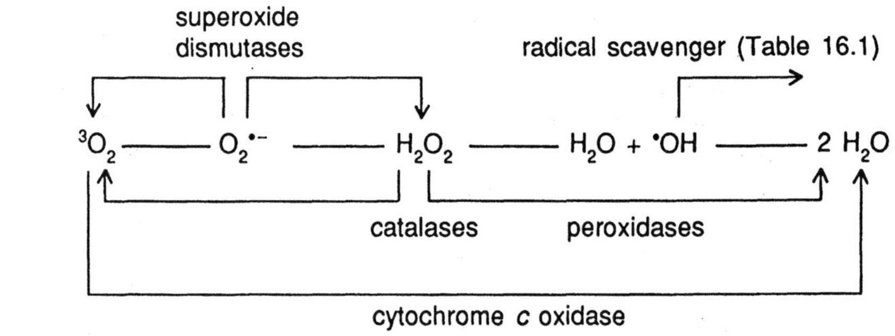

$$
\begin{array}{c}
\text{superoxide} \qquad\qquad\qquad\qquad \text{radical scavenger (Table 16.1)} \\
\text{dismutases} \\[4pt]
{}^{3}O_2 \longrightarrow O_2{}^{\bullet-} \longrightarrow H_2O_2 \longrightarrow H_2O + {}^{\bullet}OH \longrightarrow 2\,H_2O \\[4pt]
\text{catalases} \qquad \text{peroxidases} \\[6pt]
\text{cytochrome } c \text{ oxidase}
\end{array}
\qquad (16.5)
$$

$$
\text{vitamin E} \quad \xrightarrow[\;-\,ROOH\;]{\;+\,ROO^{\bullet}\;} \quad \text{(stable, inactive)}
$$

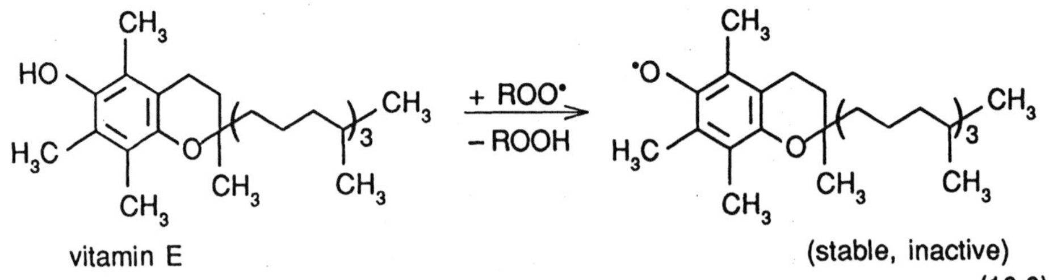

$$(16.6)$$

Table 16.1 Biological antioxidants

antioxidatively active compounds	section or (formula)	targets
peroxidases, catalases (Fe, Mn, V, Se + glutathione)	6.3 11.4 16.8	ROOH, HOOH
superoxide dismutases (Cu, Zn, Fe, Mn)	10.5	$O_2{}^{\bullet-}$, $HO_2{}^{\bullet}$
vitamin C (ascorbate) ceruloplasmin (in plasma)	(3.12) 10	${}^{\bullet}OH$
vitamin E (α-tocopherol) β-carotene (in membranes)	(16.6)	$ROO^{\bullet}$
transferrin	8.4.1	${}^{\bullet}OH$
S-rich compounds, e.g. metallothionein	17.3	${}^{\bullet}OH$
uric acid	(11.13)	${}^{\bullet}OH$
gold-containing compounds (therapeutical use)	19.4	${}^{1}O_2$ (hypothetical)

extent of uncontrolled, e.g. iron(II)-catalyzed oxidation of the long alkyl chains of fatty acids through $^3O_2$ and its transformation products (4.6, 5.2). According to the simplified scheme (16.7), a particularly effective autoxidation of lipid molecules can proceed via a radical chain reaction with branching made possible through peroxides and transition metals [26]. Lipid autoxidation significantly impairs membrane stability and function, e.g. through cross-linking. Since the functioning of compart-mentalizing membranes is absolutely essential for all organisms, the corresponding disorders have severe consequences. The inhibition of autoxidative chain reactions has thus become an important requirement for the existence of organisms, particularly in an oxygen-containing atmosphere. According to recent hypotheses, the processes of aging and some forms of uncontrolled tumor growth are caused by the inadequate inhibition of oxidative tissue degradation via free radicals ('free radical theory of aging' [27,28]), as apparently evident from the correlation between average lifetime and radical-producing O_2 turnover per units of body weight and time.

$$\text{initiation:} \quad R{-}H + {}^3O_2 \xrightarrow{\text{(Fe)}} R^\bullet + HO_2^\bullet$$

$$\text{chain propagation:} \quad R^\bullet + O_2 \longrightarrow ROO^\bullet$$

$$ROO^\bullet + R{-}H \longrightarrow R^\bullet + ROOH \quad \text{(alkyl hydroperoxide)}$$

$$\text{branching step:} \quad 2\,ROOH \xrightarrow{\text{Fe,Cu}} ROO^\bullet + RO^\bullet + H_2O$$

$$\text{termination:} \quad R^\bullet + R^\bullet \longrightarrow R{-}R \quad \text{(cross-linking)} \tag{16.7}$$

The molecular mechanism of lipid hydroperoxide reaction with selenium-contain-ing glutathione peroxidase has not yet been completely elucidated. First of all, it is assumed that mainly the ionized, i.e. the selenolate, form, $R{-}Se^-$, of seleno-cysteine reacts at pH 7. The degradation of ROOH proceeds very rapidly, nearly diffusion-controlled, thus justifying the biological use of the otherwise problematic trace element selenium. A primary oxygen abstraction probably leads to the state of a selenenic acid, RSeOH. Scheme (16.8) shows a catalysis mechanism which does not include further oxidation to tetravalent selenium even though that state can be produced upon action of excess peroxide on glutathione peroxidase *in vitro*. There is no clear evidence from X-ray absorption or photoelectron spectroscopy for a change in the oxidation state of selenium under physiological conditions.

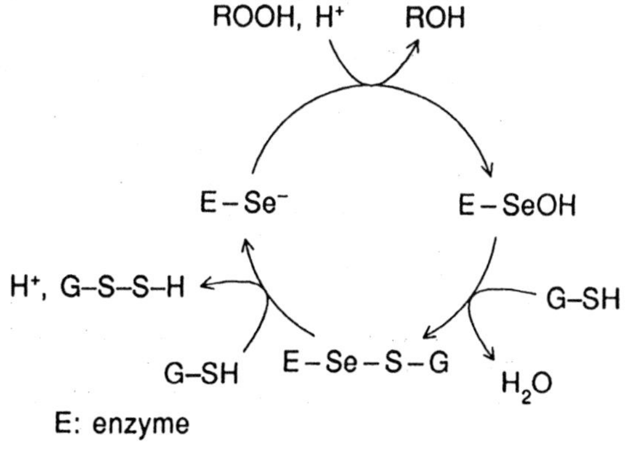

$$\tag{16.8}$$

A diselenide formation can be excluded because of the distance in the structurally sufficiently characterized tetrameric protein. As an intermediate of mechanism (16.8) a selenide/sulfide bridge between the enzyme and glutathione substrate has to be considered, model compounds with Se—S bonds could indeed be synthesized. Glutathione and the protein are fixed relative to each other by charged amino acid residues so that a high specificity of the enzyme for this particular thiol results; a specificity with respect to the peroxide substrate does not exist and would not be biologically useful. Meanwhile, first attempts are being made to obtain systems with a reactivity similar to that of glutathione peroxidase by introducing selenocysteine into other proteins [29].

References

1. K. L. Kirk, *Biochemistry of the Elemental Halogens and Inorganic Halides*, Plenum Press, New York, 1991.
2. E. Bengsch, F. Korte, J. Polster, M. Schwenk and V. Zinkernagel, Reduction in symptom expression of belladonna mottle virus infection of tobacco plants by boron supply and the antagonistic action of silicon, *Z. Naturforsch.*, **44c**, 777 (1989).
3. F. H. Nielsen, Nutritional requirements for boron, silicon, vanadium, nickel, and arsenic: current knowledge and speculation, *FASEB J.*, **5**, 2661 (1991).
4. N. T. Davies, An appraisal of the newer trace elements, *Phil. Trans. Roy. Soc. London*, **B294**, 171 (1981).
5. K. J. Irgolic, Arsenic in the environment, in *Frontiers in Bioinorganic Chemistry* (ed. A. V. Xavier), VCH Verlagsgesellschaft mbH, Weinheim, 1986, p. 399.
6. O. M. Ni Dhubhghaill and P. J. Sadler, The structure and reactivity of arsenic compounds: biological activity and drug design, *Struct. Bonding (Berlin)*, **78**, 129 (1991).
7. P. Gugger, A. C. Willis and S. B. Wild, Enantioselective biotransformation of ethyl-n-propylarsinic acid by the mould *Scopulariopsis brevicaulis*: asymmetry synthesis of (R)-ethylmethyl-n-propylarsine, *J. Chem. Soc., Chem. Commun.*, 1169 (1990).
8. B. P. Rosen, C.-M. Hsu, C. E. Karkaria, P. Kaur, J. B. Owolabi and L. S. Tisa, A plasmid-encoded anion-translocation ATPase, *Biochim. Biophys. Acta*, **1018**, 203 (1990).
9. H. Luecke and F. A. Quiocho, High specificity of a phosphate transport protein determined by hydrogen bonds, *Nature (London)*, **347**, 402 (1990).
10. G. Gassmann and D. Glindemann, Phosphine (PH_3) in the biosphere, *Angew. Chem. Int. Ed. Engl.*, **105**, 761 (1993).
11. C. Dawes and J. M. ten Cate (eds.), International symposium on fluorides: mechanism of action and recommendations for use, *J. Dent. Res. (Special Issue)*, **69**, 505–831 (1990).
12. M. Diesendorf, The mystery of declining tooth decay, *Nature (London)*, **320**, 125 (1986).
13. M. Meyer and D. O'Hagan, Rare fluorinated natural products, *Chem. Br.*, **28**, 785 (1992).
14. E. C. Jorgensen, Thyroid hormones and analogs, *Horm. Proteins Pept.*, **6**, 57 and 107 (1978).
15. N. M. Alexander, Iodine, in *Biochemistry of the Essential Ultratrace Elements* (ed. E. Frieden), Plenum Press, New York, 1984, p. 33.
16. M. J. Berry, L. Banu and P. R. Larsen, Type I iodothyronine deiodinase is a selenocysteine-containing enzyme, *Nature (London)*, **349**, 438 (1991).
17. A. Wendel (ed.), *Selenium in Biology and Medicine*, Springer-Verlag, Berlin, 1989.
18. G. N. Schrauzer, (ed.), *Selenium*, Wiley, Chichester, 1990.
19. T. C. Stadtman, Biosynthesis and function of selenocysteine-containing enzymes, *J. Biol. Chem.*, **266**, 16257 (1991).
20. B. Douglas, D. H. McDaniel and J. J. Alexander, *Concepts and Models of Inorganic Chemistry*, 2nd edn, Wiley, New York, 1983, pp. 579, 580.
21. G. Tölg and R. P. H. Garten, Large anxiety for small quantities—significance of analytical chemistry in the modern industrialized society as exemplified by trace determination of elements, *Angew. Chem. Int. Ed. Engl.*, **24**, 485 (1985).

22. H. Engelberg-Kulka and R. Schoulaker-Schwarz, A flexible genetic code, or why does selenocysteine have no unique codon?, *Trends Biochem. Sci.*, **13**, 419 (1988).
23. R. A. Arkowitz and R. H. Abeles, Isolation and characterization of a covalent seleno-cysteine intermediate in the glycine reductase system, *J. Am. Chem. Soc.*, **112**, 870 (1990).
24. E. D. Harris, Regulation of antioxidant enzymes, *FASEB J.*, **6**, 2675 (1992).
25. H. Sies, Strategies of antioxidant defense, *Eur. J. Biochem.*, **215**, 213 (1993).
26. J. M. C. Gutteridge and B. Halliwell, The measurement and mechanism of lipid peroxidation in biological systems, *Trends Biochem. Sci.*, **15**, 129 (1990).
27. W. A. Pryor, Free radical involvement in chronic diseases and aging. The toxicity of lipid hydroperoxides and their decomposition, *ACS Symposium Series* 277, 1985, p. 77.
28. R. L. Rusting, Why do we age?, *Sci. Am.*, **267**, 86 (1992).
29. Z.-P. Wu and D. Hilvert, Selenosubtilisin as a glutathione peroxidase mimic, *J. Am. Chem. Soc.*, **112**, 5647 (1990).

17 The Bioinorganic Chemistry of the Quintessentially Toxic Metals

17.1 Overview

The previous chapters have demonstrated that many 'inorganic' elements (in the form of their chemical compounds) are essential for life. However, even such essential substances will be poisonous if only the dosage is high enough (the 'Paracelsus principle', compare Figure 2.3). With regard to toxicity [1–7], two other, non-bioessential groups of inorganic elements can be distinguished: those that have not (yet) been recognized as relevant for life due to low abundance or bioavailability (e.g. insolubility at pH 7), and those elements for which *exclusively* negative effects have been found so far (Figure 17.1). Among the latter group of elements are the 'soft', thiophilic heavy metals mercury, thallium, cadmium and lead.

Many potentially toxic heavy metals such as Sb, Sn, Bi, Zr, the lanthanoides, Th or Ag are quite insoluble in their normal oxidized chemical forms under physiological conditions which includes an oxidizing atmosphere, pH 7 and a rather high chloride concentration. This is also true for the light metals titanium and aluminum; however, the anthropogenically caused pH lowering in the soil has locally increased the bioavailability of Al^{3+}. Since Al^{3+} is a strongly polarizing 'hard' cation with a high charge-to-radius ratio it is able to tightly coordinate and thus deactivate proteins and nucleic acids (Section 17.6). Enzymes can also be blocked by the coordination of other harmful metals, e.g. to active sites or to essential sulfhydryl groups; furthermore, there is the possibility of substitution of 'natural' metal centers in

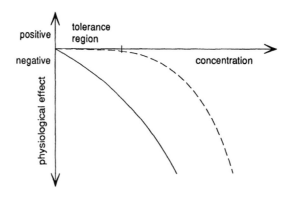

Figure 17.1
Representative dose-response diagrams for nonessential elements (compare Figure 2.3)

metalloenzymes by 'foreign' metals with similar but not identical chemical characteristics: $Zn \leftrightarrow Cd$; $Ca \leftrightarrow Pb$, Cd or ^{90}Sr; $K \leftrightarrow Tl$ or $^{134/137}Cs$; $Mg \leftrightarrow Be$ or Al; $Fe \leftrightarrow {}^{239}Pu$ (see 18.3).

Even before the additional man-made introduction of toxic substances into the environment, the organisms had to cope with such 'poisonous' elements as cadmium or mercury, either in the form of a continuous stress situation (cf. the composition of sea water, Figure 2.2) or during suddenly occurring catastrophic events such as a volcanic eruption. Thus, for *each* element there is a global cycle which is usually influenced also by the biosphere; the ecological system as a whole exists in an energetic and material flow equilibrium—in the same way as single individuals do (Figure 2.1). For some elements such as lead, mercury, arsenic or chromium, the type of compound, e.g. the ligation or oxidation state, plays a crucial role for the toxicity. For instance, in many but not all cases (As, see Section 16.4), the (bio)alkylated cations R_nM^+ are more toxic than the metals proper or their hydrated cations, M^{m+}. Organisms have developed various, generally energy-consuming detoxification strategies in order to remove unwanted inorganic substances [3,8]:

(a) Enzymatic transformations may occur from toxic to less toxic states ($Hg^{2+} \rightarrow Hg^0$; $As(OH)_3 \rightarrow HAsO_4{}^{2-}$ or $^+AsR_4$, Section 16.4) or to volatile compounds which can then be released into the environment ($SeO_3{}^{2-} \rightarrow Me_2Se$).

(b) Special membranes can hinder the passing of highly charged ions into particularly endangered regions such as brain, fetus or cell nucleus; the toxic species may be bound at the surface.

(c) Ion pumps may remove undesired substances such as $AsO_4{}^{3-}$ (Section 16.4) from sensitive cellular regions or render them less harmful through complexation or precipitation with a suitable partner (e.g. $Cd^{2+} + S^{2-} \rightarrow CdS\downarrow$ [9]).

(d) High molecular weight compounds such as the metallothionein proteins may bind toxic ions up to a certain storage capacity and thus remove them from circulation.

There are various legal assessments of the toxicity, e.g. of heavy metal species; as an example, Table 17.1 contains maximum permissible levels in solid sewage waste from the German *Klärschlammverordnung*.

However, these and similar standards have to be judged as values derived in the course of a *political assessment*; they can sometimes be exceeded even in natural soil and often do not take into consideration the chemical form of the element (speciation). According to Figure 2.4, the reaction of a population to varying concentrations of a substance is not uniform but shows a typical 'S-shape' for the idealized case of only quantitative differences [6,7]; statistical relations of this kind form the basis for the LD_{50} values (lethal dose for 50% of the population) used in toxicological evaluations. Within a population of more complex organisms and certainly in comparisons between different species there may also be qualitative differences in the reaction toward 'toxic elements'.

Therapeutic detoxification, e.g. through complexation of toxic metal ions, is indicated particularly in acute cases of poisoning; there are different chelating agents (2.1) for each of the metal ions according to their characteristics. The preferences of Zn^{2+}, Cd^{2+} and especially Cu^{2+} for N,S ligands, of arsenic and mercury for exclusive S-coordination, or of Pb^{2+} and Cd^{2+} for mainly S-containing polychelating ligands (compare 17.1) are quite typical. As the complexes have to be

Table 17.1 Typical legal standards[a] for the metal content of solid sewage waste

metal (ionic form)	maximally permissible level (mg/kg of air-dried waste)
zinc (Zn^{2+})	200
copper (Cu^{2+})	60
chromium (Cr^{3+})	100
lead (Pb^{2+})	100
nickel (Ni^{2+})	50
arsenic (As^{3+})	10
cadmium (Cd^{2+})	1.5
mercury (Hg^{2+})	1

[a]From the German 'Klärschlammverordnung'

stable in the physiological pH range and should be excretable with the urine, most practical chelating ligands for detoxification contain additional hydrophilic groups. However, many of these ligands show only limited selectivity and thus undesired side effects; the application of chelate drugs can thus be considered only as an emergency measure [10].

In the following, the bioinorganic basis of toxic effects of the four 'soft' metals lead, cadmium, thallium and mercury and of the 'hard' metals aluminum, beryllium and chromium will be discussed in detail; radioactive isotopes will be described in Chapter 18.

17.2 Lead

Historically, lead is the 'oldest' recognized toxic metal and it is also the one that has been most extensively spread into the environment by mankind [11]. In contrast to mercury and cadmium it is not particularly rare in the earth's crust (Figure 2.2), its relatively easy mining and processing, its apparent resistance against corrosion and the not easily recognized toxicity had made it a highly valuable metal in ancient civilizations. A logarithmic plot of the world lead production (Figure 17.2) illustrates the correspondence between the mining of the classical precious metals, silver and gold, and the inevitable co-production of lead; in the process of 'cupellation', heated raw metal is treated with air to remove the 'less' noble lead as liquid (m.p. 884 °C) and rather volatile lead(II) oxide, PbO [13].

After exhaustion of the known lead reserves in the final centuries of the Roman Empire, the 'discovery' of the 'New World' with its huge deposits of noble metals caused a significant increase in the world lead production which was furthered by

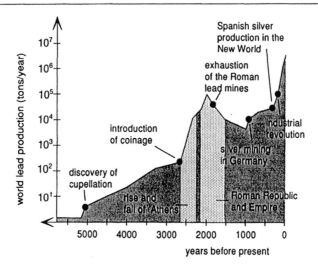

Figure 17.2
Logarithmic representation of the world lead production since the beginning of civilization (according to [12])

an increasing demand for the element in printing and weapons technology (lead ammunition). Additional demand arose after the industrial revolution when lead was beginning to be used in accumulators ('batteries'), bearing alloys, solder, optical glasses, pigments ('white lead' $PbCO_3 \cdot Pb(OH)_2$, 'red lead' $Pb_3O_4 = Pb_2[PbO_4]$), radiation protection material and fuel additives. Only in the 1980s has the world lead production levelled off; however, it still ranks sixth behind iron, aluminum, copper, manganese and zinc with regard to worldwide metal production.

Core samples taken from the inland ice of Greenland show that the global air pollution with lead compounds has increased more than a hundredfold during the last 3000 years. From an estimated concentration of 0.4 ng Pb/m^3 in prehistoric times the lead content has risen to as much as 500–10000 ng/m^3 in population centers [11]. In comparison of global cycles of the elements, lead has thus experienced the largest anthropogenic increase; its global elemental flux today has risen by a factor of about 20 in relation to the 'natural', i.e. prehistoric state.

Organometallic species such as the fuel additive tetraethyl lead, $Pb(C_2H_5)_4$, and also some inorganic lead compounds are distinguished by a rather high volatility and a tendency for rapid global distribution through the atmosphere; the same holds for cadmium and mercury compounds. However, the major part of lead pollution occurs in the form of oxides which are bound to very small particles [11] and may thus be removed, e.g. from food, by careful washing.

In aqueous solution, the behavior of lead is quite peculiar. With a redox potential of $E(Pb^{2+}/Pb) = -0.13$ V vs. normal hydrogen electrode (NHE), lead is not a noble metal; however, at pH 7 the potential $2H^+/H_2$ is -0.42 V so that lead is not attacked by oxygen-free water. Although oxygen from air would be able to oxidize lead even in neutral solution, there are insoluble basic lead carbonates and lead sulfates which form a protective layer at the metal surface, hydrogen carbonate and sulfate stemming from natural waters and carbon dioxide from air. In the days of the Roman Empire,

Figure 17.3
Medieval lead smelting requires the consumption of milk products such as butter (in the forefront) as a remedy against lead poisoning (from G. Agricola: *De re metallica*, 1556)

a popular method of wine 'improvement' was to take low-grade wine containing fruit acids and evaporate it to obtain a sweetening grape syrup, *sapa*. In the resulting acidic solutions, a considerable amount of the lead lining of the vessels used was apparently dissolved (up to 15–30 mg Pb/l); corresponding effects can also be observed for crystal glass containing Pb^{2+}. According to contemporary studies, the above-mentioned method of sweetening may have led to chronic lead poisoning in excess of the present-day limit of 500 μg/day.

Historically, lead poisoning ('saturnism') is not only connected with ancient Rome but also with the intense medieval mining and smelting activity in Central Europe (Figure 17.3). As a counter-measure, the consumption of butter was practised at that time; more recent attempts of detoxification include the combination of 2,3-dimercapto-1-propanol (BAL) with Ca(EDTA) (Table 2.4) and the development of specific, thiohydroxamate-containing ligands (17.1) [14]. The list of lead poisoning through painters' pigments (F. Goya), in the printing business and in highly polluted mining areas, e.g. of Eastern Europe, is long. In the 1840s, a large expedition to find the Northwest Passage was doomed not least because of slow lead poisoning of the crew due to faulty soldering of the newly developed food cans [15,16].

thiohydroxamate:

$$\overset{\displaystyle{}^{-}\text{O}}{}\diagdown\overset{}{\underset{\displaystyle R}{\overset{+}{N}}}=\overset{}{\underset{\displaystyle R'}{C}}\diagup\overset{\displaystyle \text{S}^{-}}{}$$

(17.1)

As with many other elements, the physiological retention time of lead and its compounds depends strongly on the location in the body. In blood and soft tissues (liver, kidney) retention times of about one month are observed; the lead compounds are excreted with urine, sweat or as components of (sulfide-containing) hairs and nails [11]. The strong bonding between heavy metals and sulfide-rich keratin in the very persistent hairs and nails allows good forensic proof of heavy metal poisoning; in some cases, the slow growth of these keratinous tissues may even allow the history of the poisoning to be traced back. The major part of incorporated lead is stored in the bone tissue due to similar (in)solubility properties of Pb^{2+} and Ca^{2+} compounds [11,15]; the residence time may then reach 30 years or more with possible effects on the development of degenerative processes such as osteoporosis.

The triethyl lead cation, $(C_2H_5)_3Pb^+$, is formed from tetraethyl lead by the dissociation of a carbanion. As the related organomercury cations, RHg^+ (Section 17.5), and triorganotin cations, R_3Sn^+, these organometallic compounds may cause severe disorders of the central and peripheral nervous system (cramps, paralysis, loss of coordination). The toxicity of organometallic cations results from the permeability of membranes, including the very discriminating blood–brain barrier [17], for such lipophilic species. Poisoning with inorganic lead compounds, on the other hand, primarily causes hematological and gastrointestinal symptoms such as colic. Even relatively low concentrations of lead inhibit the zinc-dependent 5-aminolevulinic acid dehydratase (ALAD; 12.18) which catalyzes an essential step in the porphyrin and thus also heme biosynthesis, viz. the pyrrole ring formation to porphobilinogen (Section 12.4). The occurrence of unreacted 5-aminolevulinic acid (ALA) in the urine is thus a very characteristic indication of lead poisoning [11]. Since the incorporation of iron into the porphyrin ligand with the help of the —SH group containing heme synthetase (ferrochelatase; [18]) is similarly inhibited by Pb^{2+}, there are also typical porphyrin precursors, the 'protoporphyrins', which can be detected in the urine after chronic lead poisoning. Accordingly, a form of anemia results before the long-term neurotoxic symptoms, especially a mental retardation in children, become evident [11,14,15,19]. Further toxic effects of lead poisoning include reproductive disorders like sterility and miscarriages. The replacement of organic lead compounds as anti-knock additives in fuel has meanwhile led to a significant decrease in lead pollution, especially in North America [11,20]. Lead exhibits a varying isotopic composition, depending on the source (lead as the end product of *two* radioactive decay series, see 18.1), and thus allows for detailed analyses of origin and distribution [15,20].

17.3 Cadmium

In its ionic form, Cd^{2+} (ionic radius 95 pm) shows great chemical similarity with two biologically very important metal ions, viz. the lighter homologue Zn^{2+} (74 pm), and Ca^{2+} (100 pm) which comes very close in size. Accordingly, cadmium as the 'softer' (Figure 2.6) and more thiophilic metal may displace cysteinate-coordinated zinc from its enzymes and even replace it in special cases [21], while it can also substitute for calcium, e.g. in bone tissue. Cadmium is generally regarded to be far more toxic than lead (compare Table 17.1). Chronic cadmium poisoning can cause embrittlement of bones and extremely painful deformations of the skeleton; such symptoms have been observed on a large scale as 'Itai-Itai disease' in Japan where particularly older women with a disposition for osteoporosis were affected after cadmium-containing waters had been used to irrigate rice fields during the 1950s. Although skeletal damage is primarily caused indirectly via impairment of renal functions after cadmium poisoning, the biomineral part of skeletons from Itai-Itai patients eventually contained up to 1% cadmium, a calcium-deficient diet having aggravated the toxic demineralization effect. Once cadmium is stored in the skeleton, its biological retention time is on the order of decades.

Cadmium is being used as a component of the Ni/Cd 'batteries', in color pigments (CdS or CdSe), in stabilizers for plastics and for metal surface treatment; furthermore, cadmium is a side product of zinc smelting. Cadmium can be incorporated through food with the liver and kidney of slaughter animals and with wild mushrooms which are particularly rich in this element. For as yet unknown reasons, cadmium absorption is especially effective via tobacco smoke, the cadmium content in the blood of smokers being much higher than that of nonsmokers. High concentrations of cadmium compounds have proven to be carcinogenic in animal experiments.

In the human body cadmium is also concentrated in the liver and kidney where, as in many other organisms, small (6 kDa) and with up to 30% unusually cysteine-rich proteins preferentially bind the soft heavy metal ion Cd^{2+} in addition to Zn^{2+} and Cu^+. These 'metallothionein' proteins feature a highly conserved sequence homology, in particular with regard to the cysteine residues (Figure 17.4); it can thus be assumed that they had been optimized early in the evolution and that they possess an essential function [22,23]. In addition to the liver and kidneys, the small intestines, the pancreas and the testicles of mammals contain larger amounts of these proteins. Structural studies of the mammalian form have been carried out using X-ray diffraction [24], EXAFS, UV, CD and ^{113}Cd NMR spectroscopy (^{113}Cd: $I = 1/2$, 12.3% natural abundance). It was found that a total of seven metal centers, each of them four-coordinate, can be bound in two clusters by nine and eleven cysteinate residues, respectively (Figure 17.4).

Peptide synthesis of the cysteine-rich domains of metallothioneins has demonstrated that such small synthetic oligopeptides can similarly bind metals like Cd^{2+} and can thus indicate approaches toward modified metallothioneins. Crustaceans and microorganisms such as yeast contain metallothioneins with slightly different cluster compositions and metal selectivities. In some plants and microorganisms there are cysteine-rich peptides, the phytochelatins H—$[\gamma$-Glu-Cys$]_n$-Gly—OH ($n = 2$–11), which also participate in heavy metal, i.e. Cu, Zn or Cd homeostasis [25], without forming metallothionein-type clusters [26].

The metallothioneins are presumably multifunctional proteins, the principal function depending on the organism and on the protein variant:

(a) First, the not universally accepted detoxification function involves mainly cadmium(II) but also other thiolate-preferring heavy metal centers such as copper(I), silver(I) and mercury(II). In comparison to cadmium(II) these ions favor even lower coordination numbers of three or two, so that twelve or even eighteen metal centers can be bound per metallothionein protein [22].

(b) Second, a storage or 'buffering' function for the essential elements zinc and copper may be assumed for metallothionein in unpoisoned organisms where such metals are bound instead of cadmium. Thus, the ubiquitous metallothioneins can serve

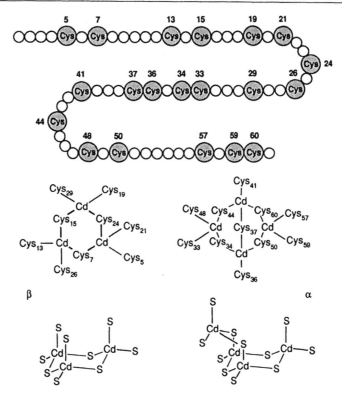

Figure 17.4
Typical cysteine arrangement in the amino acid sequence of mammalian metallothionein (top), binding of twelve terminal and eight bridging cysteinate residues to a total of seven Cd centers (middle) and idealized structures of the two resulting clusters (bottom): [3Cd–3S] chair (β cluster, M_3S_9), and the [4Cd–5S] adamantane fragment (α cluster, M_4S_{11})

in the homeostasis and transport of metal ions with thiolate affinity and with a preference for the coordination number four.

(c) A third function of very thiolate-rich proteins can be to trap and thus deactivate oxidizing free radicals such as ·OH (compare Table 16.1). The Cu^I-containing or metal-free metallothioneins are extremely oxidation sensitive but cannot easily form disulfide bridges due to their specific protein structure.

It is still open as to whether and how organisms can react to heavy metal poisoning with an increased production of specific metal-binding proteins such as the metallothioneins or phytochelatins [25]. For instance, the expression of preferentially Cu-binding yeast metallothionein seems to be controlled by a copper cluster-containing metalloregulatory protein [27,28] (compare also Section 17.5).

Yeast cells have developed another strategy for Cd detoxification. In the presence of cadmium, very small peptide-stabilized CdS particles are excreted which, due to their size of about 200 nm, can be regarded either as very large metal clusters or semiconducting solid-state particles [9].

In contrast to many other heavy metals and toxic elements, cadmium *does not* easily pass into the central nervous system or the fetus because, in its ionized form and under physiological conditions, it cannot be bioalkylated to form potentially membrane-penetrating organometallic compounds such as R_2Cd or RCd^+ [29–31]. The instability of such species in aqueous media is related to the electropositive character of Cd as reflected by the rather low oxidation potential $E = -0.40$ V of the element; in contrast, compounds of the elements Hg $(+0.85$ V), Se $(+0.20$ V), Te $(-0.02$ V), Pb $(-0.13$ V), Sn $(-0.15$ V), As $(-0.22$ V) and possibly Tl $(-0.34$ V) can be bioalkylated (potentials are given for the oxidation to the lowest stable oxidation state of the respective elements at pH 7). As discussed in Section 3.2.4, the cobalamin-catalyzed biomethylation of the less electropositive elements proceeds by a carbanion mechanism, that of the more oxidizable elements presumably via radicals.

17.4 Thallium

Like cadmium, thallium is a thiophilic heavy metal; it forms a stable, monovalent cation, Tl^+, under physiological conditions. As a 'substitute' for the similar K^+ ion, Tl^+ may penetrate membranes via potassium ion channels and pumps to reach sensitive areas and cause disorders. There are three reasons for this particular behavior: Tl^{3+} (the other stable oxidation state) is easily reduced to Tl^+ $(E_0 = +1.28)$, the ionic radii of Tl^+ (159 pm) and K^+ (151 pm, each for coordination number eight) are quite similar and Tl^+ shows a high affinity for inorganic S^{-II} ligands, similar to Ag^+ but with a slightly better solubility of the chloride. Long-term symptoms of severe thallium poisoning are paralysis and impaired sensory perception ($\rightarrow$neurotoxicity); the first, typical signs, however, are gastroenteritis and the loss of hair.

Due to the similarity of K^+ and Tl^+ the latter can reach nearly all regions of organisms where it can then be trapped and slowly released as a 'depot poison'. Many K^+-associated enzymes such as amino acid and enzyme synthetases, the Na^+/K^+-ATPase (Section 13.4), and most of the pyruvate metabolism which supplies energy to the nerve cells (Section 14.1, Figure 14.2) are inhibited by thallium [32]. Suitable countermeasures after thallium poisoning include dialyses, high supplementation with K^+ in combination with administering large quantities of mixed-valent iron cyanide complexes such as Prussian Blue, $Fe_4[Fe(CN)_6]_3$, in colloidal form; a chelate therapy is explicitly contraindicated [33]. Due to their 'open' structure [34], the Prussian Blue-type complexes are able to function as nontoxic cation exchangers, i.e. they do not release significant amounts of cyanide but can bind large monocations such as K^+ or Tl^+ and thus remove these from the organism.

17.5 Mercury

As with lead, mercury is a very ancient environmental contaminant. Pliny the Elder described the high mortality rate of workers in mercury mines. Although the metal was later used as a cure in the elemental or oxidic form for the treatment of syphilis, in disinfecting preparations or as a component of fungicides, the extreme toxicity of many mercury compounds only became evident through worldwide publicized incidents of mass poisoning after the release of Hg-containing catalyst wastes into Minamata Bay in Japan (between 1948 and 1960) and after grain seeds still contaminated with organomercury fungicides had been used for flour production in Iraq (1972). Accordingly, the use of this metal, e.g. in chloralkali electrolysis cells for the industrial production of chlorine and sodium hydroxide solution, is largely being phased out. During the last decades, mercury has become a very thoroughly studied example of a toxic heavy metal. Many problems of heavy metal pollution such as:

(a) the strong dependence of toxicity on the chemical form of the element,
(b) the influence of human activity on the global cycle,
(c) the kinetics of metabolic transformations and the distribution in the human body or
(d) the genetics of the latent 'resistance' of microorganisms against heavy metal poisoning

have been examined here in detail.

Regarding the elemental form, mercury is a relatively 'noble' heavy metal, i.e. it does not corrode in normal atmosphere, it is liquid at room temperature and comparatively volatile. The saturation vapor pressure is about 0.1 Pa, corresponding to 18 mg Hg/m^3, which is significantly higher than the typical permissible limit of 0.1 mg/m^3. Acute poisoning with metallic mercury is very rare; however, chronic inhalation of Hg vapor as furthered by the extensive use of this metal in physics and chemistry have been described in detail, often by the affected scientists themselves. In particular, the typical neurological symptoms of mercury poisoning gradually appear when working in insufficiently ventilated rooms where spilled metallic mercury had been in contact with air for a longer period of time. The symptoms have been described vividly by Alfred Stock (1876–1946) who developed the high-vacuum technique with Hg metal valves [35] for the preparation of very sensitive gases such as the smaller boranes. In addition to externally visible signs like a dark lining of the teeth (which also occurs after lead poisoning), the diminished blood circulation in the extremities is followed by impaired concentration and coordination, tremors (the 'mad hatter' syndrome of the hatters using $Hg(NO_3)_2$ for felt treatment; compare Lewis Carroll's *Alice's Adventures in Wonderland*), memory loss and, at a very high level of exposure, loss of hearing, blindness and death. After recognizing the reasons for his severe illness, Stock installed a ventilation system in his laboratories and, after a period of recuperation, was even able to continue some of his scientific work [35]. Despite its 'noble' character, metallic mercury dissolves to a small extent in oxygen-containing water or blood. In amalgamated (alloyed) form, particularly as a component of Ag-, Sn-, Zn- and sometimes Cu-containing tooth fillings, mercury is far less volatile or water soluble; the potential long-term health hazards through the additional Hg incorporation via amalgam fillings continues to be controversially discussed.

In its usual ionic form as Hg^{2+} ion, mercury is immediately toxic since this species is easily soluble at pH 7 and does not form insoluble compounds with those anions that are abundant in bodily fluids. An especially toxic form is represented by the organometallic cations RHg^+ and in particular by the methylmercury cation which can be formed in the organism from Hg^{2+} through carbanionic biomethylation (17.2 and Section 3.2.4) [29,31]. This specific toxicity is connected to the ambivalent lipophilic/hydrophilic character of such water-soluble organometallic cations which allows them to penetrate the very tightly constructed membrane partitions between the nervous system or the growing fetus and the rest of the organism. The placental membrane as well as the blood–brain barrier [17] restrict the access of many substances unless they are, just like the social drugs ethanol and nicotine or the R_nM^+ ions, relatively small, feature a hydrophilic *and* lipophilic molecular region and can thus move freely in the aqueous plasma and through the nonpolar membrane barriers.

$$RS-Hg-SR$$

$$2 H_2O \longleftarrow$$

$$\longleftarrow 2 RS^-$$

$$H_2O \rightarrow Hg^{2+} \leftarrow OH_2 \qquad\qquad RS-Hg-CH_3$$

methylcobalamin (see 3.4)

$$Cl^- \longleftarrow$$

aquo-cobalamin H_2O $\longleftarrow RS^-$ (e.g. enzyme-coordinated cysteinate)

$$H_2O \rightarrow {}^+Hg-CH_3 \; \rightleftharpoons \; Cl-Hg-CH_3$$

methylcobalamin Cl^- (neutral, nondissociated, lipophilic)

$$\longrightarrow \text{aquocobalamin}$$

$$H_3C-Hg-CH_3$$

(boiling point 96 °C)

(17.2)

As a noble metal and similar to silver or gold, mercury forms relatively covalent bonds with chloride. Following an incorporation of organomercury compounds, e.g. with food, the little dissociated molecules RHgCl can thus be formed in the stomach with its high content of hydrochloric acid; due to their lipophilicity, these molecules can then be efficiently resorbed. Mercury compounds are distinguished by a preference for very low coordination numbers of the metal center, the coordination number two with a linear arrangement being highly favored. As a consequence, mercury is not very susceptible to simple chelate coordination; even the detoxification function of metallothionein is thus not as efficient for this very thiophilic heavy metal Hg^{2+} as it is, for example, for Cd^{2+} with its strongly preferred four-coordination.

As the equilibria in (17.2) illustrate, the toxicity of mercury compounds is based on the strong affinity of this metal for the deprotonated forms of thiol ligands such as cysteine [36]; therefore, thiols, RSH, with sulfhydryl groups, —SH, are also called mercaptans (*mercurium captans*). Even in the absence of chelate effects, the resulting complex formation constants between 10^{16} and 10^{22} are very high; on the other hand, oxygen donor ligands are only weakly bound by Hg^{II}. Mercury compounds affect all protein structures and especially enzymes in which cysteines significantly influence the activity as metal coordinating, redox-active or conformation-determining groups (via disulfide bridges). It is characteristic, though, that the RHg^+-thiolate bond is kinetically labile, despite an extremely favorable equilibrium situation, so that exchange, recomplexation and eventually excretion are possible when better suited detoxification ligands such as dimercaprol (2.1) or dimercaptosuccinic acid, HOOC—CH(SH)—CH(SH)—COOH, are being offered. The kinetic lability of two-coordinate mercury complexes is a consequence of easily attainable transition states for an associative substitution (compare 14.7) with three- or four-coordinate metal. However, this kinetic lability also indicates that toxic mercury compounds are rapidly distributed in the body to those parts that feature the highest affinity toward this heavy metal; in fact, mercury is mainly found in the liver, kidney and the nervous system, especially the brain.

The binding of CH_3Hg^+ to nucleobases is also ambivalent and strongly dependent upon reaction conditions; scheme (17.3) shows the structurally documented example of an 8-aza-modified adenine [37]. The ability of RHg^+ to substitute the protons of primary amines in neutral and basic solution is quite remarkable. Furthermore, alkylmercury species can form relatively inert bonds to carbon centers of nucleobases which may account for the mutagenic effect of organomercurials.

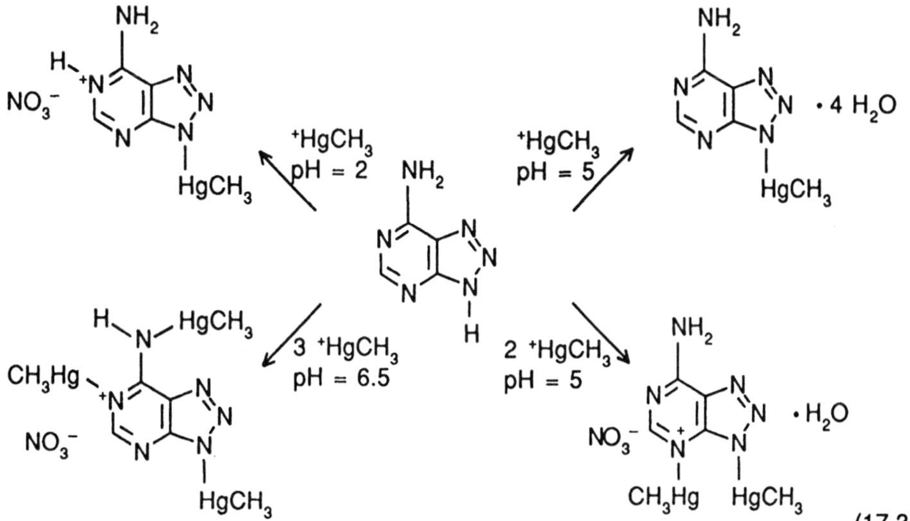

(17.3)

The relatively low immediate toxicity of elemental mercury and its volatility have enabled bacteria to develop a resistance mechanism toward soluble Hg compounds [38]. This mechanism has now been intensely studied in view of the increasing heavy metal pollution in the environment; corresponding organisms have thus been isolated from Hg-polluted soil or waters, e.g. in Boston harbor (USA). The detoxification system is effective against 'heavy metal stress' and is now quite well understood with respect to genetic control; the synthesis of the required proteins (MerA, MerB, MerP, MerT) is triggered and controlled by a metal-selective and gene-regulating sensor protein 'MerR'.

MerR is a metalloregulatory protein [28] which is sensitive even to nanomolar concentrations of Hg^{2+} and which is remarkably selective for Hg^{2+} as compared to the neighboring ions in the periodic system such as Cd^{2+} or Au^+. MerR binds mercury presumably in a three-coordinate form using cysteinate ligands [36]. It does not only control its own synthesis but, after metal binding, the activation of an RNA polymerase for the synthesis of further proteins. Of these, MerP is used in the periplasmic binding of dissolved mercury (Hg^{2+}, RHg^+) and MerT effects the transmembrane transport into the cell. Two 'processing' enzymes are especially interesting: a specific Hg^{II} reductase (MerA) and organomercury lyase (MerB).

Organomercury lyase is a relatively small (22 kDa) monomeric protein which accelerates the cleavage of the otherwise kinetically inert Hg—C bonds by a factor of 10^6–10^7 (17.4). This enzyme also cleaves some tetraorganotin compounds, albeit with much lower efficiency; it contains four conserved cysteine residues, presumably for mercury binding. A mechanistic hypothesis (17.5 [38]) shows the possible function of several cysteine ligands in a concerted substitution of R^- by Cys^- which requires an increased metal coordination number in transition state (3).

$$RHgX + H^+ + X^- \rightleftharpoons R–H + HgX_2 \qquad X = R'S, Hal \qquad (17.4)$$

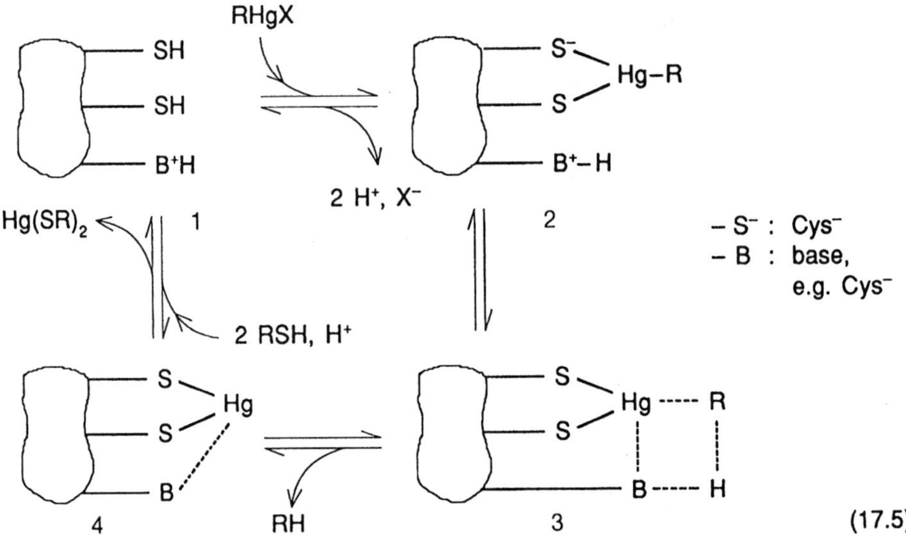

$$(17.5)$$

Reduction of coordinated Hg^{II} to the volatile and less toxic elemental mercury is effected by a special Hg^{II} reductase, a dimeric flavin- and NADPH-containing protein with a molecular mass of 2×60 kDa. The equilibrium of the catalyzed reaction (17.6) depends on the entatic stress which is forced upon the normally very stable bis(thiolato)mercury(II) complexes by the enzyme.

$$Hg(SR)_2 + NADPH + H^+ \rightleftharpoons Hg + NADP^+ + 2 RSH \qquad (17.6)$$

Flavine adenine dinucleotide (FAD) and four cysteine residues are present in the active center at the interface between two protein subunits, α_1 and α_2. An increase in the metal coordination number as well as an activating Hg coordination to *both* subunits are being discussed (17.7) [38], preceding the reducing action of the dihydro-flavin ($FADH^-$). The mechanism of this final rapid two-electron transfer from $FADH^-$ to specially coordinated Hg^{II} is still unclear.

The global cycle of mercury is largely determined by the volatility of the element and its compounds, particular of dimethylmercury (boiling point 96 °C) which is formed after complete biomethylation; today, the natural and (locally concentrated) anthropogenic sources contribute to about equal extent. The accumulation of organomercury compounds in marine animals which sometimes amounts to many orders of magnitude is quite astonishing and involves particularly predatory fishes

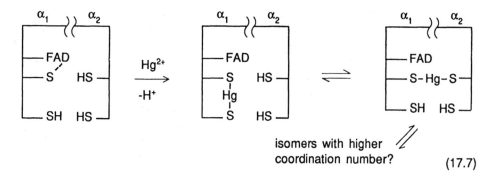

$$(17.7)$$

at the end of the food chain; therefore, these animals can represent a major source of Hg in the human diet. The biological half-life of $MeHg^+$ in such fishes is estimated to be several years. Less well known is the role of bacteria in the transformation of mercury compounds into volatile forms either through methylation in anoxic environments or through reduction to the metal (17.6), which is the dominating form in the atmosphere (half-life about one year). Figure 17.5 shows a simplified scheme of the Hg cycle with emphasis on the marine ecosystem.

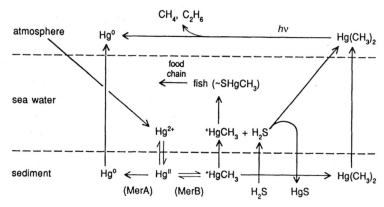

Figure 17.5
Marine (organo)mercury cycle (simplified)

17.6 Aluminum

Only since about 1975 has aluminum received scientific and public interest as a 'harmful' and 'toxic' metal. One reason for this concern is recent forest damage which can partly be attributed to the acidification of soil by 'acid rain' and the resulting release of Al^{3+}. Another motive to investigate the role of aluminum in organisms is the still controversial X-ray microanalytical discovery of aluminosilicate accumulation in certain brain-tissue areas of patients suffering from the Alzheimer syndrome. In both cases, satisfying relationships between cause and effect have not yet been conclusively established; however, the above-mentioned hypotheses have revived a long-neglected research area [39–41].

In molar proportions, aluminum is the most abundant 'true' metallic element in the earth's crust (Figure 2.2) and, after iron, the second most produced metal. However, according to all accounts it is an element with almost no 'natural' biochemical function. The main reason for this may be the very low solubility of Al^{3+} especially at pH 7 when it is almost completely present as insoluble hydroxide, $Al(OH)_3$, or its condensation products, $AlO(OH)$ and Al_2O_3, respectively (compare 8.20). In contrast to iron, aluminum is exclusively trivalent in its physiologically relevant compounds. Figure 17.6 illustrates that the very small amounts of remaining soluble species between pH 5 and 7 are cationic and anionic hydroxo complexes. Below pH 5, the dominating species is the hydrated Al^{3+} ion which is a strongly ligand-polarizing Lewis acid due to its high charge-to-radius ratio $(3+)/(50\,pm)$. In the absence of complexing agents, e.g. in soil containing little organic material, this ion can be released through sufficiently acidic precipitation and thus be made bioavailable.

Depending on the organic and mineral composition, soils may be buffered to different extents against added acid: neutral to slightly basic pH values are prevalent in calcareous soil, according to the hydrogen carbonate buffer system; silicates such as the aluminum-containing feldspar buffer at about pH 5–6.5. During 'weathering' processes, alkali and alkaline earth metal ions are released from the complex silicates and less soluble oxides result. The aluminum-containing clay minerals are attacked under release of $Al^{3+}(aq)$ in exchange for H^+ only below pH 5; even lower pH values are necessary to dissolve the more compact aluminosilicates and, finally, the aluminum and iron oxides proper.

The hydrated Al^{3+} ion, which can replace Mg^{2+} with great efficiency due to its higher charge and smaller ionic radius, and also hydroxyaluminosilicates are widely believed to be among the soluble tree-damaging agents. The detrimental reaction is not the replacement of the magnesium ion in chlorophyll, as is often assumed; the far higher Lewis acidity, stronger ligand coordination and much slower ligand substitution rate of Al^{3+} relative to Mg^{2+} causes alterations, in particular inhibitions of Mg^{2+}-regulated biochemical processes. Obviously, the vast number of Mg^{2+}/enzyme-induced phosphate transfer reactions (compare Section 14.1) can be blocked by Al^{3+}. The very high affinity of Al^{3+} toward polyanionic phosphate groups precludes

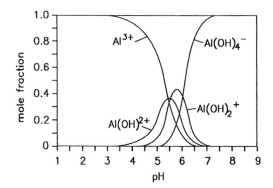

Figure 17.6
Molar contributions of various *soluble* complexes (each hydrated) in Al^{III}-containing aqueous solutions at different pH values (total solubility minimum at pH 6)

an efficient catalysis, i.e. a process with rapid restoration of the substrate-free catalyst system. For instance, the ATP tetraanion binds Al^{3+} stronger than Mg^{2+} by a factor of 10^5. Even the exchange of water molecules in the first coordination sphere of the hydrated complex proceeds more than five orders of magnitude slower for the aluminum system than in the Mg^{2+} aqua complex. The effective thermodynamic *and* kinetic inhibition by Al^{3+} of Mg^{2+}-dependent enzymatic processes involving kinases, cyclases, esterases, ATPases or phosphatases is therefore not unexpected [42].

The reduced growth of tree roots due to increased ratios of aluminum to calcium and aluminum to magnesium is one of the consequences of soil acidification. In addition, both H^+ and atmospheric nitrogen in the form of NH_4^+ are antagonists with regard to Mg^{2+} (compare Figure 2.5), further reducing the bioavailability of this important ion.

With respect to its coordination behavior, Al^{3+} as a small, highly charged metal ion prefers complexation with negatively charged oxygen-containing ligands. In the blood plasma, i.e. at low phosphate concentrations, chelate complexation occurs mainly with the partly or completely deprotonated citrate anion, $^-OOC-CH_2-C(OH)(COO^-)-CH_2-COO^-$ [42,43]. With about 5 $\mu g/l$ plasma fluid, the normal concentration of Al^{III} in blood is quite low and hard to determine precisely due to the ubiquity of aluminum. Once the metal ion, e.g. in chelate-ligated form, has passed the first barriers in the gastrointestinal tract, Al^{3+} is stored and transported mainly by transferrin (Section 8.4.1), which is hardly surprising considering the chemical similarity of Al^{3+} and Fe^{3+}. Differences exist mainly with respect to the more rapid release of Al^{3+} versus Fe^{3+} through transferrin [44]. When an efficient excretion through the kidneys is no longer possible, aluminum may form small, less highly charged complexes which can apparently cross the membrane barriers that usually hinder a highly charged ion from reaching, for example, the nervous system. With regard to a possible role of Al^{3+} in neuropathological symptoms, the strong tendency of this ion for coordination with deprotonated 1,2-dihydroxy aromatic systems such as the catecholamine neurotransmitters epinephrine or dopa has been established (compare 8.6 and 10.8) [45].

In connection with Alzheimer's disease there have been conflicting analytical reports concerning the possible accumulation of aluminum compounds, e.g. as aluminosilicates, in Alzheimer-typical neurofibrillary tangles or amyloid protein-containing plaques in certain regions of the brain, e.g. the hippocampus [42,46]. A direct causative involvement of aluminum in the Alzheimer syndrome seems to be unlikely now although it can be shown that Al^{3+} can cross-link polynucleotides and that Alzheimer and Down syndrome patients exhibit reduced transferrin activity [47].

Encephalopathy, dementia and even mortality have been established for hemodialysis patients who were subjected to an increased level of aluminum. Some of these patients had received aluminum compounds in the dialysis fluid to prevent hyperphosphataemia; in other cases, dialysis patients had accumulated dissolved aluminum which remained after water treatment to remove phosphates (normal concentration: 10 μg Al/l, typical permissible level: 200 $\mu g/l$). In view of the large quantities of water required for dialysis and in the absence of functioning kidneys, the high total amounts of Al^{3+} were then only insufficiently excreted. Chelate drugs such as deferrioxamine (2.1 and 8.8) which originally target Fe^{3+} have also been used in cases of Al^{3+} poisoning. Other symptoms of Al^{3+} overload include several forms of anemia (Al^{3+}/Fe^{3+} antagonism) and disorders of the bone metabolism; an accumulation of

Al^{3+} with its very high affinity toward phosphates was observed in the bone-forming osteoblasts. The tendency of Al^{3+} to form strong bonds with fluoride and the Al/Si antagonism also have to be mentioned. There is no reason, however, for an exaggerated anxiety, for example, with respect to the use of aluminum-containing kitchenware, despite some minute dissolution, for instance, when heating solutions containing fruit acids. Aluminum is not only a major component of many natural soils but is also present in relatively large concentrations in beverages such as tea (tea leaves accumulate Al) or beer and as a component of processed food; it is normally rapidly excreted via functioning kidneys. Large amounts of aluminum compounds can be found in chewing gum, tooth pastes and some strong antacids which are used to neutralize excess gastric acid.

17.7 Beryllium

There is a diagonal relationship between aluminum and divalent beryllium in the periodic table. Accordingly, the chemistry of both elements is rather similar, although Be^{2+} is slightly more soluble at pH 6–7 than Al^{3+} due to its lower charge and despite its extremely small ionic radius of 27 pm for the preferred coordination number four. This very rare element has become environmentally relevant through technical combustion and the use of beryllium-containing compounds and alloys which exhibit superior mechanical and nuclear properties [48]. Exposure to beryllium, typically inhaled in the form of dust, may lead to pulmonary diseases ('berylliosis', 'chronic beryllium disease' [49]) and lung cancer.

Once incorporated, beryllium is excreted only very slowly from organisms due to its spatially very concentrated dipositive charge. A long-term deposit of beryllium is situated in the phosphate-containing bone tissue. As a lighter homologue of magnesium, divalent beryllium can reach the cell nucleus where it may exert proven mutagenic, i.e. DNA-altering, as well as carcinogenic, i.e. tumor-inducing, effects. Gene transcription and expression are strongly impaired. A beryllium-induced inhibition of RNA or DNA synthesis which is otherwise catalyzed by Mg^{2+} ions has also been demonstrated (compare Section 14.1). Be^{2+} can drastically change the activity of phosphatases and kinases [48], the contraproductively strong binding of this ion inhibiting the catalysis of (de)phosphorylation reactions (Section 14.1). Finally, beryllium interacts with the immune system [49], its compounds being allergenic contact toxins. Only rudimentary studies exist regarding a possible chelate therapy of beryllium poisoning [50,51]. The permissible levels which are typically a hundredfold lower than even those of cadmium (compare Table 17.1) illustrate the extreme potential health hazard of this most toxic of all nonradioactive elements [50].

17.8 Chromium as Chromate(VI)

Under physiological conditions, the persistent oxidation states of chromium are Cr^{III} and Cr^{VI}, the latter as chromate, CrO_4^{2-}, at pH 7. Similar to trivalent aluminum and iron, trivalent chromium exists as insoluble hydroxide at neutral pH and can thus be absorbed by organisms and utilized as an essential element only in specially ligated form (compare Section 11.5). On the other hand, the chromates(VI) which

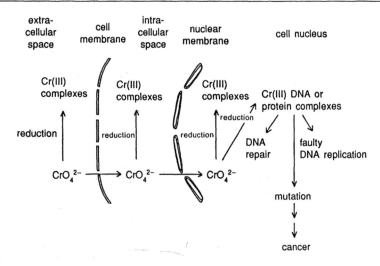

Figure 17.7
Flow diagram for uptake and reduction of chromate in different cellular compartments

have been recognized as skin irritants ever since the beginning of their industrial use were classified as potential carcinogenic substances. The uptake of CrO_4^{2-} by organisms seems to be surprising at first because the redox potential for reaction (17.8) is about +0.6 V at pH 7, i.e. chromate is only metastable in the presence of reducing organic compounds under physiological conditions. Nevertheless, the required number of three electrons (compare the metastability of permanganate, MnO_4^-, in water) leaves only a few special biochemical reduction systems as efficient reductants for chromate [52,53]. These include heme- and flavoproteins (NADH-dependent), thiols such as glutathione, GSH (compare 16.8), and ascorbate (3.12). According to EPR studies, highly reactive Cr^V states with one unpaired 3d electron may be formed in these reactions [54,55].

$$CrO_4^{2-} + 4\ H_2O + 3\ e^- \rightleftharpoons Cr(OH)_3 + 5\ OH^- \qquad (17.8)$$

Due to its structural similarity with the sulfate ion, SO_4^{2-}, chromate can overcome membrane barriers and reach the cell nucleus, unless rapidly reduced (Figure 17.7). In the cell nucleus, chromate can oxidatively damage genetically important components. The substitutionally more labile and stronger oxidizing Cr^V or Cr^{IV} intermediates formed during one-electron reduction steps and the simultaneously produced $RS^\bullet$ and $^\bullet OH$ radicals can directly attack at the DNA and effect bond cleavage, cross-linking and, as a consequence, faulty gene expression. Furthermore, the resulting substitutionally *inert* Cr^{III} can irreversibly bind to phosphate-containing DNA or free nucleotides and thus also affect genetic functions (compare Section 19.2). Due to the ligand field stabilization for a high-spin d^3 configuration in octahedral symmetry, the chromium(III) complexes are generally distinguished by an enormous kinetic stability; hydrated Cr^{3+} exchanges its water ligands more slowly, by a factor of 10^6, than even hydrated Al^{3+}.

References

1. O. Hutzinger (ed.), *The Handbook of Environmental Chemistry*, Springer-Verlag, Berlin, 1980.
2. I. Bodek *et al.* (eds.), *Environmental Inorganic Chemistry*, Pergamon Press, New York, 1988.
3. K. J. Irgolic and A. E. Martell (eds.), *Environmental Inorganic Chemistry*, VCH, Deerfield Beach, 1985.
4. R. B. Martin, Bioinorganic chemistry of metal ion toxicity, in *Metal Ions in Biological Systems* (ed. H. Sigel), Vol. 20, Marcel Dekker, New York, 1986, p. 21.
5. H. G. Seiler, H. Sigel and A. Sigel, *Handbook on Toxicity of Inorganic Compounds*, Marcel Dekker, New York, 1987.
6. J. E. Fergusson, *The Heavy Elements: Chemistry, Environmental Impact and Health Effects*, Pergamon Press, Oxford, 1990.
7. J. E. Fergusson, *Inorganic Chemistry and the Earth*, Pergamon Press, Oxford, 1982.
8. S. Silver, T. K. Misra and R. A. Laddaga, Bacterial resistance to toxic heavy metals, in *Metal Ions and Bacteria*, (eds. T. J. Beveridge and R. J. Doyle), Wiley, New York, 1989, p. 121.
9. C. T. Dameron, R. N. Reese, R. K. Mehra, A. R. Kortan, P. J. Carroll, M. L. Steigerwald, L. E. Brus and D. R. Winge, Biosynthesis of cadmium sulfide quantum semiconductor crystallites, *Nature (London)*, **338**, 596 (1989).
10. M. M. Jones, Newer chelating agents for *in vivo* toxic metal mobilization, *Comments Inorg. Chem.*, **13**, 91 (1992).
11. R. L. Boeckx, Lead poisoning in children, *Anal. Chem.*, **58**, 274A (1986).
12. D. M. Settle and C. C. Patterson, Lead in albacore: guide to lead pollution in Americans, *Science*, **207**, 1167 (1980).
13. J. O. Nriagu, Cupellation: the oldest quantitative chemical process, *J. Chem. Educ.*, **62**, 668 (1985).
14. K. Abu-Dari, F. E. Hahn and K. N. Raymond, Lead sequestering agents. 1. Synthesis, physical properties, and structures of lead thiohydroxamato complexes, *J. Am. Chem. Soc.*, **112**, 1519 (1990).
15. W. Kowal, O. B. Beattie, H. Baadsgaard and P. M. Krahn, Did solder kill Franklin's men?, *Nature (London)*, **343**, 319 (1990).
16. O. Beattie and J. Geiger, *Frozen in Time*, Bloomsbury Publishers Ltd, London, 1987.
17. G. W. Goldstein and A. L. Betz, The blood–brain barrier, *Sci. Am.*, **255**(3), 70 (1986).
18. H. A. Dailey, C. S. Jones and S. W. Karr, Interaction of free porphyrins and metallo-porphyrins with mouse ferrochelatase. A model for the active site of ferrochelatase, *Biochim. Biophys. Acta*, **999**, 7 (1989).
19. E. K. Silbergeld, Mechanism of lead neurotoxicity, or looking beyond the lamppost, *FASEB J.*, **6**, 3201 (1992).
20. K. J. R. Rosman, W. Chisholm, C. F. Boutron, J. P. Candelone and U. Görlach, Isotopic evidence for the source of lead in Greenland snows since the late 1960s, *Nature (London)*, **362**, 333 (1993).
21. N. M. Price and F. M. M. Morel, Cadmium and cobalt substitution for zinc in a marine diatom, *Nature (London)*, **344**, 658 (1990).
22. M. J. Stillman, F. C. Shaw and K. T. Suzuki (eds.), *Metallothioneins: Synthesis, Structure, and Properties of Metallothioneins, Phytochelatins, and Metal–Thiolate Complexes*, VCH Verlagsgesellschaft mbH, Weinheim, 1992.
23. J. H. R. Kägi and A. Schäffer, Biochemistry of metallothionein, *Biochemistry*, **27**, 8509 (1988).
24. W. F. Furey, A. H. Robbins, L. L. Clancy, D. R. Winge, B. C. Wang and C. D. Stout, Crystal structure of Cd,Zn metallothionein, *Science*, **231**, 704 (1986).
25. E. Grill, E. L. Winnacker and M. H. Zenk, Phytochelatins, a class of heavy-metal-binding peptides from plants, are functionally analogous to metallothioneins, *Proc. Natl. Acad. Sci. USA*, **84**, 439 (1987).
26. H. Strasdeit, A.-K. Duhme, R. Kneer, M. H. Zenk, C. Hermes and H.-F. Nolting, Evidence for discrete $Cd(SCys)_4$ units in cadmium phytochelatin complexes from EXAFS spectroscopy, *J. Chem. Soc., Chem. Commun.*, 1129 (1991).
27. K. H. Nakagawa, C. Inouye, B. Hedman, M. Karin, T. D. Tullius and K. O. Hodgson,

Evidence from EXAFS for a copper cluster in the metalloregulatory protein CUP2 from yeast, *J. Am. Chem. Soc.*, **113**, 3621 (1991).

28. T. V. O'Halloran, Transition metals in control of gene expression, *Science*, **261**, 715 (1993).

29. H. Sigel and A. Sigel (eds.), *Biological Properties of Metal Alkyl Derivatives*, in *Metal Ions in Biological Systems*, Vol. 29, Marcel Dekker, New York, 1993.

30. J. S. Thayer, Global bioalkylation of the heavy elements, in *Metal Ions in Biological Systems* (eds. H. Sigel and A. Sigel), Vol. 29, Marcel Dekker, New York, 1993, p. 1.

31. S. Krishnamurthy, Biomethylation and environment transport of metals, *J. Chem. Educ.*, **69**, 347 (1992).

32. K. T. Douglas, M. A. Bunni and S. R. Baindur, Thallium in biochemistry, *Int. J. Biochem.*, **22**, 429 (1990).

33. D. A. Labianca, A classic case of thallium poisoning and scientific serendipity, *J. Chem. Educ.*, **67**, 1019 (1990).

34. F. Herren, P. Fischer, A. Ludi and W. Hälg, Neutron diffraction study of prussian blue, $Fe_4[Fe(CN)_6]_3 \times H_2O$. Location of water molecules and long-range magnetic order, *Inorg. Chem.*, **19**, 956 (1980).

35. E. K. Mellon, Alfred E. Stock and the insidious 'Quecksilbervergiftung', *J. Chem. Educ.*, **54**, 211 (1977).

36. J. G. Wright, M. J. Natan, F. M. MacDonnell, D. M. Ralston and T. O'Halloran, Mercury(II)-thiolate chemistry and the mechanism of the heavy metal biosensor MerR, *Prog. Inorg. Chem.*, **38**, 323 (1990).

37. W. S. Sheldrick and P. Bell, Characterization of metal binding sites for 8-azaadenine. Formation and X-ray structural analysis of methylmercury(II) complexes, *Inorg. Chim. Acta*, **123**, 181 (1986).

38. M. J. Moore, M. D. Distefano, L. D. Zydowsky, R. T. Cummings and C. T. Walsh, Organomercurial lyase and mercuric ion reductase: nature's mercury detoxification catalysts, *Acc. Chem. Res.*, **23**, 301 (1990).

39. B. Corain, M. Nicolini and P. Zatta, Aspects of bioinorganic chemistry of aluminium(III) relevant to the metal toxicity, *Coord. Chem. Rev.*, **112**, 33 (1992).

40. R. C. Massey and D. Taylor (eds.), *Aluminium in Food and the Environment*, Royal Society of Chemistry, London, 1989.

41. R. B. Martin, Bioinorganic chemistry of aluminum, in *Metal Ions in Biological Systems* (ed. H. Sigel), Vol. 24, Marcel Dekker, New York, 1988, p. 1.

42. J. D. Cowburn, G. Farrar and J. A. Blair, Alzheimer's disease—some biochemical clues, *Chem. Br.*, **26**, 1169 (1990).

43. T. L. Feng, P. L. Gurian, M. D. Healy and A. R. Barron, Aluminum citrate: isolation and structural characterization of a stable trinuclear complex, *Inorg. Chem.*, **29**, 408 (1990).

44. H. M. Marques, Kinetics of the release of aluminum from human serum dialuminum transferrin to citrate, *J. Inorg. Biochem.*, **41**, 187 (1991).

45. R. Kiss, I. Sovago and R. B. Martin, Complexes of 3,4-dihydroxyphenyl derivatives. 9. Al^{3+} bonding to catecholamines and tiron, *J. Am. Chem. Soc.*, **111**, 3611 (1989).

46. J. P. Landsberg, B. McDonald and F. Watt, Absence of aluminium in neuritic plaque cores in Alzheimer's disease, *Nature (London)*, **360**, 65 (1992).

47. G. Farrar, P. Altmann, S. Welch, O. Wychrij, B. Ghose, L. Lejeune, J. Corbett, V. Prasher and J. Blair, Defective gallium-transferrin binding in Alzheimer disease and Down syndrome: possible mechanism for accumulation of aluminum in brain, *Lancet*, **335**, 747 (1990).

48. D. N. Skilleter, To be or not to be—the story of beryllium toxicity, *Chem. Br.*, **26**, 26 (1990).

49. L. S. Newman, To Be^{2+} or not to Be^{2+}: immunogenetics and occupational exposure, *Science*, **262**, 197 (1993).

50. D. F. Evans and C. Y. Wong, Nuclear magnetic resonance studies of beryllium complexes in aqueous solution, *J. Chem. Soc., Dalton Trans.*, 2009 (1992).

51. O. Kumberger, J. Riede and H. Schmidbaur, Beryllium coordination to bio-ligands: isolation from aqueous solution and crystal structure of a hexanuclear complex of Be^{2+} with glycolic acid, $Na_4[Be_6(OCH_2CO_2)_8]$, *Z. Naturforsch.*, **47b**, 1717 (1992).

52. P. H. Connett and K. E. Wetterhahn, *In vitro* reaction of the carcinogen chromate with cellular thiols and carboxylic acids, *J. Am. Chem. Soc.*, **107**, 4284 (1985).

53. R. N. Bose, S. Moghaddas and E. Gelerinter, Long-lived chromium(IV) and chromium(V)

metabolites in the chromium(VI) glutathione reaction: NMR, ESR, HPLC, and kinetic characterization, *Inorg. Chem.*, **31**, 1987 (1992).

54. P. O'Brien, J. Pratt, F. J. Swanson, P. Thronton and G. Wang, The isolation and characterization of a chromium(V) containing complex from the reaction of glutathione with chromate, *Inorg. Chim. Acta*, **169**, 265 (1990).

55. R. P. Farrell and P. A. Lay, New insights into the structures and reactions of chromium(V) complexes: implications for chromium(VI) and chromium(V) oxidations of organic substrates and the mechanisms of chromium-induced cancers, *Comments Inorg. Chem.*, **13**, 133 (1992).

18 Biochemical Behavior of Inorganic Radionuclides: Radiation Risks and Medical Benefits

18.1 Overview

Independently of the consequences of human activities, most organisms do not only have to coexist with 'toxic' elements and their compounds but also with naturally occurring radioactive isotopes and the resulting radiation. All forms of high-energy 'ionizing' radiation (α, β, γ, X-ray and neutron radiation) may cause the cleavage of chemical bonds, thus damaging enzymes or genetic material either directly or indirectly, for instance, through the hydroxyl radical $\cdot$OH which is typically formed from H_2O, the main component of organisms, under the influence of radiation [1]. The inherent radical scavenging and repair mechanisms of organisms which are active in counteracting 'oxidative stress' (Section 16.8) are also effective to a certain degree with respect to radiation damage. Copper complexes, including the variably applied superoxide dismutase (compare Section 10.5 [2]), or sulfur compounds such as cysteine or cysteamine (= 2-mercaptoethylamine, $H_2N-CH_2-CH_2-SH$) can contribute to reduce biological radiation damage when administered *prior* to the exposure, either through radical scavenging or rapid one-electron reduction of ionized species. Other therapeutic measures in the event of imminent incorporation of radioactive substances are the saturation of depots in the body with nonradioactive material, e.g. by taking 'iodide tablets', or the undelayed and selective complexation and excretion by use of (e.g. Sr, Pu) specific chelate ligands. In the early days of the nuclear sciences, radioactive material was handled less cautiously as evidenced by Marie Curie's death through radiation-induced cancer and similar incidents [3]. Today, the large-scale technical production, testing and dismantling of nuclear weapons and the minutely described global effects of the reactor accident in Chernobyl (Ukraine) in 1986 have very much heightened the sensitivity of the public toward radiation risks, with stringent consequences also for the handling of diagnostically and therapeutically useful radionuclides.

Radioactive isotopes are distinguished chemically from stable isotopes of the same element only through the generally small isotope effect on reaction rates which results from the different atomic or molecular masses. The extremely low detection limit for many radiating isotopes thus allows a fairly reliable temporal and spatial observation of physiologically important (^{32}P, ^{22}Na, ^{35}S) but also 'unphysiological' elements in organisms. On the other hand, even minute amounts of radioactive isotopes may cause severe genetic damages. The extent of these damages and the overall radiobiological effect depend on many factors such as the volatility of the isotope-containing compound, the possible binding to transporting particles (carriers), the effectivity of absorption by the organism, the kind and energy of the emitted radiation (α or β particle radiation vs. electromagnetic γ radiation), the localization of the isotope-containing metabolite in the organism, the radioactive ('physical') *and* the biological half-life, i.e. the average time of exposure. The latter is very much determined by the chemical speciation of the corresponding element, by the kind of exposure and by the distribution in the organism; the variations known from 'normal' toxicology (Figure 2.4) within and between populations also apply in this case. Before discussing the medical uses of radioactive isotopes as tracers in diagnosis or as possible tumor-destroying therapeutics, we shall briefly review the naturally occurring radionuclides and the main 'inorganic' isotopes resulting from nuclear fission reactions in their interactions with organisms [4,5].

18.2 Natural and Man-Made Radioisotopes Outside Medical Applications

In addition to the β-emitting isotopes ^{3}H (tritium, half-life 12.3 a) and ^{14}C (5730 a) which are continuously generated at high altitudes via cosmic radiation and neutron capture, another very bioessential inorganic element, potassium, is naturally present in form of its radioactive isotope ^{40}K (0.012% natural abundance). Due to its long half-life of 1.3 billion years, the amount of ^{40}K remaining after nuclear synthesis during the formation of the solar system is still large enough that an adult human contains about 15 mg of this isotope which slowly decays via β and γ radiation. Further radionuclides present in the natural environment such as ^{87}Rb (27.8% natural abundance, 4.9×10^{10} a), ^{115}In (95.7%, about 10^{15} a), ^{176}Lu (2.6%, 3.6×10^{10} a) or ^{187}Re (62.6%, 5×10^{10} a) do not contribute significantly to the natural radiation exposure due to the rarity of the elements, very long half-lives, weak radiation and the decay to stable isotopes.

The long-lived isotopes of thorium, ^{232}Th (100% natural abundance, 1.4×10^{10} a), and uranium, ^{235}U (0.72%, 7.0×10^8 a) and ^{238}U (99.3%, 4.46×10^9 a), have to be assessed differently because these heavy radioactive isotopes are mainly α emitters. This 'soft' radiation involving fairly heavy particles is easily shielded against; however, it is very hazardous once corresponding radionuclei are incorporated, the depth of penetration, e.g. in lung tissue, being several micrometers. On the other hand, the long-lived thorium and uranium isotopes produce decay series (18.1) with highly radioactive intermediates, e.g. isotopes of Ra, Rn, Po, Bi and Pb. Recent discussion has focused on the risk posed by the volatile, α-emitting noble gas element radon inside insufficiently vented rooms and buildings; the longest-lived isotope ^{222}Rn has a half-life of 3.8 days. According to recent estimations, approximately half of today's radiation exposure may be attributed to the volatile radon and its daughter isotopes [6]. The mentioned uranium and thorium isotopes which are quite abundant in natural minerals (see Figure 2.2) and thus also in the underground of buildings (depending on the geological formation) are responsible for the exposure to radon. The energy produced in the radioactive decay of the more abundant mineral components ^{40}K, ^{232}Th and ^{238}U is a main factor for the thermal 'equilibrium' of the earth and for geological dynamics; these isotopes, in particular the very soluble potassium ion, are also the main radioactive components in sea water.

$$^{238}U \xrightarrow{-\alpha} \, ^{234}Th \rightarrow \, ^{234}Pa \rightarrow \, ^{234}U \xrightarrow{-\alpha} \, ^{230}Th \xrightarrow{-\alpha} \, ^{226}Ra \xrightarrow{-\alpha} \, ^{222}Rn \xrightarrow{-\alpha} \, ^{218}Po \xrightarrow{-\alpha}$$

$$^{214}Pb \rightarrow \, ^{214}Bi \rightarrow \, ^{214}Po \xrightarrow{-\alpha} \, ^{210}Pb \rightarrow \, ^{210}Po \rightarrow \, ^{210}Bi \xrightarrow{-\alpha} \, ^{206}Pb$$

$$^{232}Th \xrightarrow{-\alpha} \, ^{228}Ra \rightarrow \, ^{228}Ac \rightarrow \, ^{228}Th \xrightarrow{-\alpha} \, ^{224}Ra \xrightarrow{-\alpha} \, ^{220}Rn \xrightarrow{-\alpha} \, ^{216}Po \xrightarrow{-\alpha}$$

$$^{212}Pb \rightarrow \, ^{212}Bi \rightarrow \, ^{212}Po \xrightarrow{-\alpha} \, ^{208}Tl \rightarrow \, ^{208}Pb \tag{18.1}$$

In connection with early extensive nuclear weapons testing, with the reactor accident in Chernobyl, and the discussion about technical reprocessing of spent nuclear fuel, some longer-lived isotopes resulting from the fission of ^{235}U and ^{239}Pu have become generally known. In the following, these are listed according to increasing nuclear masses. Characteristic for nuclear fission is that these isotopes are concentrated around mass numbers of about 90 and 135.

(a) ^{85}Kr (half-life 10.7 a, $\beta + \gamma$ radiation): as a chemically inert noble gas, krypton poses a great problem with respect to its retention in reprocessing plants. This non-polar and thus rather lipophilic substance is distributed in the body such that it is found especially in fatty tissue. As is the case with many other radionuclides, leucemia is a typical consequence of exposure. In contrast to ^{85}Kr, metastable ^{81m}Kr produced in a controlled nuclear process is a very short-lived (13 s) pure γ emitter and can thus be used in the scintigraphic depiction of the respiratory organs; its daughter isotope ^{81}Kr (2×10^5 a) also does not emit particles.

(b) ^{89}Sr (51 d); ^{90}Sr (29 a, only β radiation): divalent strontium which may be transported via particles or leaching is deposited mainly in the mineralized part of bone tissue due to its similarity with the lighter homologue calcium. Like Ca^{2+}, Sr^{2+} is very slowly exchanged in the biomineral, the biological half-life being in the order of one decade. Therefore, the long-term effect of ^{90}Sr on bone marrow function and blood formation is to cause leucemia, the bone marrow region being generally regarded as one of the most radiation-sensitive parts of the human body. To complete the negative profile of ^{90}Sr, it is only β emitting and, therefore, difficult to detect; furthermore, the similarly β emitting ^{90}Y (64 h) is formed as the daughter isotope. The excretion of incorporated ^{90}Sr may be promoted by administering Ca^{2+} (EDTA) complexes (compare 2.1) as long as there has been no binding in the solid material of the bone. Extraction procedures with macrocyclic ligands of the crown ether type (compare 13.4) have been considered for the separation of ^{90}Sr in reprocessing plants.

(c) ^{103}Ru (39 d); ^{106}Ru (1a, $\beta + \gamma$): ruthenium is the heavier homologue of iron and can partially replace it in biorelevant compounds. According to Table 5.1, radioactive ruthenium is mainly found in the hemoglobin of erythrocytes.

(d) ^{131}I (8 d, $\beta + \gamma$, 364 keV gamma energy): the extreme efficiency of the human body with regard to iodine accumulation (compare Section 16.7) is disadvantageous here because a continuously high exposure may eventually cause tumors of the thyroid gland. As iodide (I^-) or iodate (IO_3^-), the element is very soluble in water, allowing an effective access of radioactive iodine via food and drinking water. In the case of imminent exposure, however, the resorption may be inhibited by saturating the iodine demand with an excess of nonradioactive iodide. The biological half-life of iodine is about 140 days for humans, much longer than the physical half-life of 8 days. ^{131}I is an important radioisotope for medical applications (see below), either in ionic form or after covalent binding to organic carrier molecules.

(e) ^{132}Te (78 h, $\beta + \gamma$): as the heavier homologue of sulfur and selenium, tellurium features an affinity for heavy metals and is thus found mainly in the liver and kidneys.

(f) ^{133}Xe (5 d, $\beta + \gamma$): similar to ^{85}Kr. Other, exclusively γ-emitting xenon isotopes are used as radiodiagnostics, especially in the imaging of lung ventilation and of the circulatory system (see Section 18.3).

(g) ^{134}Cs (2 a); ^{137}Cs (30 a; $\beta + \gamma$, 660 keV gamma energy): cesium is a heavier, larger homologue of the bioessential potassium. Despite the different ionic radius of Cs^+ (174 pm) vs. K^+ (151 pm, each for coordination number eight), there is apparently no effective biological 'rejection' of this ion, presumably due to the more than millionfold lower abundance of Cs^+, e.g. in sea water, relative to K^+. In its ionic compounds, the cesium cation is water soluble and thus very bioavailable through food or drinking water, in a similar fashion as the iodide anion. Like other alkali metal cations (compare Section 13.2), Cs^+ can be bound and thus retained for some time by ionophores from lichen and mushrooms such as the anions of (nor)badion A (18.2 [7]).

norbadion A $n = 1$
badion A $n = 2$ (18.2)

In mammals, radioactive cesium is found to a large extent in the intracellular regions of the muscle; however, like the mimicked potassium but unlike iodide it is widely distributed over the whole body. Astonishingly, the biological half-life of Cs^+ in humans amounts to about four months and is thus longer than that of the related potassium ions. While the absorption of long-lived $^{137}Cs^+$ from radiation-polluted soil by plants may be repressed by applying K^+-rich fertilizer [8], the large-scale cesium separation from minutely contaminated whey was accomplished in Germany after the Chernobyl accident using inorganic ion exchangers of the Prussian Blue type (compare also the procedure for Tl^+ detoxification, Section 17.4).

(h) ^{140}Ba (13 d, $\beta + \gamma$): barium is the heavier homologue of strontium; its affinity to bone tissue can thus be anticipated. It should be remembered, however, that Ba^{2+} forms a very insoluble sulfate.

(i) ^{144}Ce (284 d, $\beta + \gamma$): cerium can occur in trivalent or tetravalent form. Since the ions of the lanthanoid elements are generally large, it may substitute for iron because of a similar charge-to-radius ratio.

(j) ^{147}Pm (2.6 a, $\beta + \gamma$): also a lanthanoid element. Lanthanoid(3+) ions can replace Ca^{2+} in enzymes (see Section 14.2).

Considering the problem of long-term storage of nuclear waste, the isotopes ^{137}Cs, ^{90}Sr and ^{147}Pm are the main radioactive decay components of completely exhausted nuclear fuel material after 20 years or more.

The most important longer-lived *heavy* isotopes from the natural uranium/thorium decay series (18.1) are characterized in the following:

(a) ^{210}Pb (22.3 a, β): compare Section 17.2.

(b) ^{210}Po (138 d, α): heavier homologue of tellurium (see above).

(c) ^{222}Rn (3.8 d, α): as a noble gas, radon is not bound to particulate matter; its distribution in the atmosphere, hydrosphere and in the body is thus essentially unhindered. In particular, the two polonium daughter isotopes of ^{222}Rn, viz. ^{218}Po and ^{214}Po (see 18.1) as very short-lived and thus intense α emitters may trigger carcinogenesis of the respiratory organs; after nuclear transformation of the gaseous 'intermediate' radon they can remain in the tissues, bound to solid particles (polonium is a chalcogen). As in the case of cadmium poisoning (compare Section 17.3), smoking is known to increase the health hazard disproportionally ($\rightarrow$synergism). Radon poisoning [6] is not only a problem for mine workers, particularly in the mining of uranium, relatively high concentrations have also been measured inside buildings which are situated above geologically old, uranium-containing (granite) underground, for instance in Central Switzerland, Sweden or parts of the Eastern United States. A practical corrective measure is the efficient ventilation below a hermetically sealed basement which, however, cannot protect against the traces of radon in drinking water, in natural gas or emanating from building materials. Inevitably, the air ventilation system required in those cases results in loss of heating energy. In fact, the largest part of radioactive background exposure in the industrial nations comes from 'natural' radon [6], a notion that has long escaped public attention, not in the least due to concentration on other matters such as the reactor accident in Chernobyl.

(d) ^{226}Ra (1600 a, $\alpha + \gamma$): radium is the heaviest element of the alkaline earth metal group. The very radiotoxic element [3] thus acts as a calcium analogue and, in the case of exposure, is enriched in the bone tissue (Ra^{2+} as 'bone-seeker') where it can affect the formation of blood cells and cause leucemia.

(e) ^{232}Th (1.4×10^{10} a); ^{235}U (7.0×10^8 a); ^{238}U (4.5×10^9 a): under physiological conditions α-emitting thorium and uranium occur in high oxidation states, e.g. as thorium(IV) compounds or uranyl (UO_2^{2+}) complexes such as $[UO_2(CO_3)_3]^{4-}$ which is found in sea water (Figure 2.2). The radiotoxicity of these very slowly decaying isotopes is not as high as that of the other elements mentioned in this section.

(f) ^{239}Pu (2.4×10^4 a, α): although plutonium occurs, to a very small extent, in 'natural' uranium ores, it is essentially a 'man-made' element which features relatively long-lived isotopes. The use of ^{239}Pu as a source of radiation and especially as nuclear fuel in civil and military areas has led to a large global production and thus to the potential for release of significant amounts of this isotope into the environment. Together with ^{226}Ra, ^{222}Rn and ^{90}Sr, ^{239}Pu belongs to the category of the most radiotoxic isotopes. In addition to the long radioactive half-life of ^{239}Pu (there are other, even longer-lived plutonium isotopes) which still guarantees sufficiently harmful numbers of decay processes per second, the physiological half-life of about 70 years is so high that self-detoxification through excretion is ineffective. After release via nuclear explosions or a reactor accident, plutonium mainly occurs in oxidic form bound to particulate matter and, after contact with water, as colloidal hydroxide; a special role in solubilizing highly charged Pu ions is due to coordinating carbonate. Mammals can absorb plutonium-contaminated particles through the lungs and tumor formation may already occur at this stage. After possible dissolution, this metal can use the iron transport system and thus reach liver and bone tissue from where it is very hard to remove.

The special problem of decontamination by chelate therapy is related to the similarity between plutonium in its typical oxidation states $+III/+IV$ and the biochemically so important $Fe^{+II/+III}$ system; in contrast to uranium, plutonium exhibits a rather stable trivalent state. The plutonium redox pair features a higher positive charge than the iron redox system. However, this is compensated by the larger ionic radius of this actinoid element; therefore, a good correspondence results with respect to the charge-to-radius ratios. This correspondence is even more relevant because the redox potentials of both pairs are comparable (18.3).

ratios charge/radius (pm)
(each for coordination number six)

$$\text{Pu}^{3+}: 3/100 = 0.030 \qquad\qquad \text{l.s. } \text{Fe}^{2+}: 2/61 = 0.033$$
$$\text{h.s. } \text{Fe}^{2+}: 2/78 = 0.026$$

$$\Updownarrow \quad E_0 = 0.98 \text{ V} \qquad\qquad\qquad \Updownarrow \quad E_0 = 0.77 \text{ V}$$

$$\text{Pu}^{4+}: 4/86 = 0.047 \qquad\qquad \text{l.s. } \text{Fe}^{3+}: 3/55 = 0.055 \qquad (18.3)$$

It is thus a challenge for coordination chemistry to develop and provide plutonium-selective chelate ligands which, at the same time, affect the iron metabolism as little as possible. Fairly successful catecholate systems such as '3,4,3-LICAMC' (2.1) or corresponding macrocyclic ligands have been designed to bind large, highly charged metal ions which prefer high coordination numbers [9].

18.3   Bioinorganic Chemistry of Radiopharmaceuticals

18.3.1   Overview

Radiopharmaceuticals are mainly used for diagnostic purposes, i.e. to obtain information on pathological states of organs which should thus be selective targets of these compounds. In order to accomplish this function without posing an unnecessary radiation risk, the nuclides used must not be α or β particle emitters and their gamma energies should preferably lie between 100 and 250 keV, the region that is best accessible to scintillation counters and thus very sensitive to external detection (radioscintigraphy; single photon emission computer tomography, SPECT). In contrast to other imaging techniques such as magnetic resonance, ultrasound or computer tomography which are better suited to represent structural details, radioscintigraphy has the advantage of depicting *time-resolved physiological processes*. The physical half-life of the isotope should therefore be long enough to allow a controlled production, administering and sufficient distribution in the organism (competition between physical and biological half-lives). On the other hand, the radioactive decay time should be short enough (< 8 d) so that the radiation is sufficiently intense for detection over a short period of time, levelling off rapidly after the actual diagnosis and thus allowing a repetition of the procedure during medical treatment. The administered amount and the additional radiation dose can then be kept as small as possible. Considering the limited number of radionuclides and the wide variation of physical half-lives, there is only a rather small number of really useful isotopes available. A look at the periodic system immediately shows that metallic elements with their particular coordination chemistry should provide most of the physically suitable radioisotopes for medicine [10] (see Table 18.1). Furthermore, an uncomplicated handling of radiopharmaceuticals is desirable, e.g. very short-lived isotopes may cause problems for the transport from the reactor source to the clinical institution.

To fulfill its purpose, it is necessary that the radioisotope in its administered form is selectively taken up by single organs or by tumors. It is remarkable that tumor tissue with its altered and generally increased metabolism can accumulate certain inorganic compounds. Increased membrane permeability, local changes in pH, or variations of electrolyte ion or bioligand concentrations are being discussed as possible reasons. Up to now, the following three isotopes have mainly been used in radiodiagnostics:

(a) ^{131}I (8 d), mainly as iodide with its nearly exclusive selectivity for the thyroid gland;

(b) ^{67}Ga (78 h), which has mostly been used in its trivalent form as slowly hydrolyzing citrate complex (compare citrate binding of the lighter homologue aluminum in blood, Section 17.6) and which is employed in localizing inflammatory processes [11] due to its being transported by transferrins (Section 8.4.1), and

(c) ^{99m}Tc (6 h), which, in the form of several different complexes, is being used on a large scale for the imaging of various regions of the body (see below).

In attempts to improve the selectivity and thus reduce radiation dosage, the interest has also focused on the possibility of tagging radionuclides to monoclonal antibodies and thus to combine a very *sensitive detection methodology* with a *specific transport medium* (radioimmunoassay, immunoscintigraphy). While the nonmetallic radioactive iodine isotopes have to be covalently bound to antibody molecules, a labeling with metal isotopes requires antibody-anchored chelate ligands. Since the metal ions involved are typically large (Table 18.1), polydentate chelate ligands such as the potentially octadentate EDTA analogue diethylenetriamine pentaacetate (DTPA, 18.4) or functionalized macrocycles are usually employed [12,13]. Due to their convenient physical half-lives, the γ-emitting isotopes ^{67}Ga, ^{97}Ru, ^{99m}Tc and ^{111}In are under consideration for such advanced diagnostic techniques.

diethylenetriamine pentaacetate
(X⁻ = COO⁻)   (18.4)

For the attractive prospect of a local radiation *therapy* with longer-lived high-energy (α or β) *particle*-emitting isotopes such as ^{67}Cu, ^{90}Y, ^{186}Re, ^{188}Re or ^{212}Bi [12–14], the selectivity ascertained by monoclonal antibodies is particularly desirable; however, the problem of insufficient excretion due to accumulation in the liver and

Table 18.1 Important imorganic radionuclides in diagnostics and therapy

radionuclide	physical half-life	main component of γ-energy (keV)	typical use
^{43}K (β,γ)	22 h	373	heart diagnosis
^{57}Co (γ)	271 d	122	B$_{12}$ diagnosis
^{67}Cu (β,γ)	62 h	93, 185	radio(immuno)therapy
^{67}Ga (γ)	78 h	93, 185, 300	tumor diagnosis
^{81}Rb (β^+,γ)	4.6 h	190, 446	heart diagnosis
^{90}Y (β,γ)	64 h	556	radio(immuno)therapy
^{97}Ru (γ)	69 h	216, 325, 461	tumor and liver diagnosis
^{99m}Tc (γ)	6 h	140	(see Chap. 18.3.2)
^{111}In (γ)	67 h	171, 245	radioimmunology
^{123}I (γ)	13 h	159	thyroid diagnosis
^{129}Cs (γ)	32 h	372, 411	heart diagnosis
^{131}I (β,γ)	8 d	364	thyroid (diagnosis, therapy)
^{153}Sm (β,γ)	46 h	103	radio(immuno)therapy
^{169}Yb (γ)	32 d	198	diagnostics
^{186}Re (β,γ)	89 h	137	radio(immuno)therapy
^{188}Re (β,γ)	17 h	155	radio(immuno)therapy
^{192}Hg (γ)	5 h	157, 275, 307	diagnostics
^{201}Tl (γ)	73 h	68 - 80	heart diagnosis
^{203}Pb (γ)	2 d	279	diagnostics
^{212}Pb (β,γ)	11 h	239	radio(immuno)therapy
^{212}Bi (α,β,γ)	1 h	727	radio(immuno)therapy

kidneys remains to be solved. Coordination chemistry ought to provide possible solutions through the development of specifically tailored complex ligands.

The pharmaceutically useful radionuclides may be isotopes of essential, non-essential, toxic or even exclusively radioactive elements such as technetium. Considering the extremely minute quantities (ng region) for the detection of γ radiation, the danger from the chemical toxicity of *diagnostic* agents has to be regarded as rather low, even, for example, with ^{201}Tl. The situation is different for *radiotherapeutic* agents, of which only the iodine isotopes have proven useful due to their extraordinary selectivity. The chemistry of the less familiar elements from Table 18.1 under physiological conditions is not always well known; frequently, complex hydrolysis reactions of the administered compounds in the organism play an important role for transport, metabolism and radiotherapeutic usefulness.

Stable inorganic isotopes such as ^{10}B (19.8% natural abundance) or ^{97}Ru (15.7%) can have an indirect radiotherapeutic effect. They feature a very high cross-section for the capture of slow, 'thermal' neutrons from a neutron source and can thus be selectively converted to α-emitting nuclear excited isotopes like $^{11}B*$ in the (tumor) tissue (boron neutron capture therapy, BNCT [15]). Boron-modified biomolecules and polyborane clusters have been tested for this purpose, the selective transport to the tumor remaining a major problem [16,17].

A very sensitive radioimaging method with exceptional spatial and short-time resolution is positron emission tomography (PET) which requires β^+-emitting and thus short-lived isotopes like ^{11}C (20 min), ^{18}F (110 min), ^{82}Rb (76 s), ^{62}Cu (9.7 min) or ^{68}Ga (68 min).

18.3.2 Technetium—a 'man-made bioinorganic element'

In quantitative terms, technetium compounds are by far the most important radio-pharmaceuticals used today (market share >80% [10,18–20]). The longest-lived isotope of this element is ^{98}Tc with a half-life of about 4 million years. Since technetium does not occur as a product of very slow decay series such as (18.1), the amount produced in nuclear synthetic reactions during the formation of the solar system has totally disappeared. Surprisingly, however, the supply of technetium is higher today than that of the stable heavier homologue rhenium; its price is also lower than that of gold or of the platinum metals. The reason is that the relatively long-lived isotope ^{99}Tc (2.1×10^5 a, only β emitter) is found to about 6% among the products of uranium fission.

For radiomedical purposes, the isotope ^{99}Tc is important not in its slowly β-decaying ground state but in a metastable, nuclear excited state, i.e. as exclusively γ-emitting ^{99m}Tc with a diagnostically useful half-life of six hours. At 140 keV the γ emission lies in a physiologically and technically very suitable region; also, the daughter isotope ^{99}Tc is a pure β emitter which does not disturb the γ detection. However, the major reason for the popularity of this radioisotope in radiodiagnostics is the availability of an easily operable technetium 'reactor' or 'generator' which allows the convenient preparation of applicable solutions in a normal clinical environment [18]. The starting isotope is radioactive ^{99}Mo (as molybdate MoO_4^{2-}; compare Section 11.1) which can be generated from neutron addition to nonradio-active ^{98}Mo. The half-life of 66 h for ^{99}Mo allows a controlled planning of its clinical use; the resulting monoanionic pertechnetate, $[^{99m}TcO_4]^-$, is separated from the dianionic molybdate, $[^{99}MoO_4]^{2-}$, through elution with physiological saline at an ion exchange column (18.5). The monoanion formed after the radioactive decay is easily eluted in nanomolar to micromolar concentrations whereas the dianion is retained on the column.

$$^{98}Mo \xrightarrow{+n} {}^{99}Mo \xrightarrow{-\beta^-} {}^{99m}Tc \xrightarrow{-\gamma} {}^{99}Tc \xrightarrow{-\beta^-} {}^{99}Ru$$

half-lives: **66 h**   **6 h**   210 000 a

$$(18.5)$$

The chemical basis of the pertechnetate reactor which has been used since about 1965 is that technetium differs from its lighter homologue manganese by being relatively stable and only weakly oxidizing in the heptavalent form. In contrast to manganese, the thermodynamically stable oxidation states of technetium in water and in the absence of special ligands include only the very soluble pertechnetate (+VII), the insoluble hydrous oxide $TcO_2 \cdot n\ H_2O$, and the metallic element Tc (Figure 18.1). Pertechnetate itself can be used for the imaging of the thyroid gland as it is a large monoanion like iodine; however, most applications require a reduction of this state in the presence of various special ligands. Tin(II) compounds have been established as efficient reductants even though there is a possible association of the technetium complexes with the resulting tin(IV) species. Reduced technetium in the oxidation states (+I) to (+V) forms relatively inert complexes with chelate ligands containing O, N, phosphate and phosphane donor centers; in addition, there is a

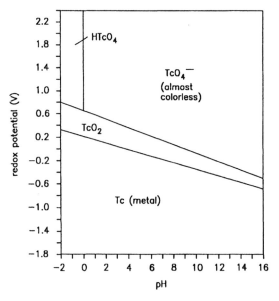

Figure 18.1
Stability diagram of technetium (according to [21])

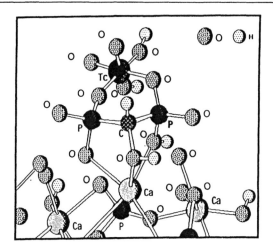

Figure 18.2
Proposed linkage between a six-coordinate technetium complex and the 001 face of hydroxy-apatite via a hydroxymethylene diphosphonate ligand, $^-2O_3P—CH(OH)—PO_3^{2-}$ (from [19])

pronounced tendency for cluster formation (compare the neighboring element molybdenum, (11.1)). In addition to complex formation, the labeling of proteins or colloidal particles and the incorporation into erythrocytes are common clinical practice [10,19].

Inert polymeric complexes of technetium with diphosphonate ligands, $^-2O_3P—CR_2—PO_3^{2-}$ ($R = H$, CH_3, OH), have been particularly useful for the imaging of bone tissue. Diphosphonates are the analogues of polyphosphates but are more difficult to hydrolyze; they can be excreted rather rapidly due to their slow enzymatical degradation. It is assumed that the bifunctional phosphonate ligands allow a linking between the Tc tracer and the growing hydroxyapatite crystals (Figure 18.2), similar to the connection between collagen and hydroxyapatite (see Figure 15.3).

Many 'normal' complexes of technetium in the oxidation states $(+V)$ to $(+I)$ with N,O chelate ligands such as EDTA or DTPA (2.1 and 18.4) are retained in the organs of eventual metabolism and excretion like the kidneys, liver or spleen. Similar to radioactive iodide I^-, monoanionic $[^{99m}TcO_4]^-$ can be used in the examination of the thyroid gland; on the other hand, monocationic technetium complexes with lower-valent metal $(+I, +III)$ and π acceptor ligands such as phosphanes, arsanes, dioximes or isonitriles are suitable to examine cardiovascular disorders [10]. In addition to larger *atomic* ions such as radioactive $^{81}Rb^+$ or $^{201}Tl^+$, molecular monocations such as $[Tc(CNR)_6]^+$ are accepted to some extent by heart muscle cells instead of K^+ [22].

In view of the optimal physical and practical properties of ^{99m}Tc there is now a continuing search for technetium radiotracers with better selectivity, e.g. for tumor tissue or for perfusion of the brain; complexes containing small ligands with nonpolar substituents should be suited to efficiently cross the blood–brain barrier. The great challenge for coordination chemistry is therefore to provide intelligent *chemical* (i.e. *ligand*) *modifications* of the *physically suited* isotope in order to exactly target specific *physiological processes* in the organism.

References

1. D. Schulte-Frohlinde, Early events in the radiation biology of *E. coli* cells, *Radiat. Phys. Chem.*, **34**, 173 (1989).
2. J. R. J. Sorenson, Copper complexes as 'radiation recovery' agents, *Chem. Br.*, **25**, 169 (1989).
3. R. M. Macklis, The great radium scandal, *Sci. Am.*, **269**(2), 78 (1993).
4. E. Pochin, *Nuclear Radiation: Risks and Benefits*, Clarendon Press, Oxford, 1985.
5. I. Bodek, W. J. Lyman, W. F. Reehl and D. H. Rosenblatt (eds.), *Environmental Inorganic Chemistry*, Pergamon Press, New York, 1988.
6. A. F. Gardner, R. S. Gillett and P. S. Phillips, The menace under the floorboards?, *Chem. Br.*, **28**, 344 (1992).
7. D. C. Aumann, G. Clooth, B. Steffan and W. Steglich, Complexation of cesium-137 with the top pigment of maronenroehrlings (*Xerocomus badius*), *Angew. Chem. Int. Ed. Engl.*, **28**, 453 (1989).
8. E. Marshall, Fallout from Pacific tests reaches congress, *Science*, **245**, 123 (1989).
9. J. Xu, T. D. P. Stack and K. N. Raymond, An eight-coordinate cage: synthesis and structure of the first macrotricyclic tetraterephthalamide ligand, *Inorg. Chem.*, **31**, 4903 (1992).
10. S. Jurisson, D. Berning, W. Jia and D. Ma, Coordination compounds in nuclear medicine, *Chem. Rev.*, **93**, 1137 (1993).
11. R. L. Hayes and K. F. Hübner, Basis for the clinical use of gallium and indium radionuclides, in *Metal Ions in Biological Systems* (ed. H. Sigel), Vol. 16, Marcel Dekker, New York, 1983, p. 279.
12. D. Parker, Tumour targeting with radiolabelled macrocycle-antibody conjugates, *Chem. Soc. Rev.*, **19**, 271 (1990).
13. L. Yuanfang and W. Chuanchu, Radiolabelling of monoclonal antibodies with metal chelates, *Pure Appl. Chem.*, **63**, 427 (1991).
14. P. Bläuenstein, Rhenium in nuclear medicine: general aspects and future goals, *New J. Chem.*, **14**, 405 (1990).
15. J. H. Morris, Boron neutron capture therapy, *Chem. Br.*, **27**, 331 (1991).
16. R. F. Barth, A. H. Soloway, R. G. Fairchild, Boron neutron capture therapy for cancer, *Sci. Am.*, **263**(4), 68 (1990).
17. M. F. Hawthorne, The role of chemistry in the development of cancer therapy by the boron-neutron capture reaction, *Angew. Chem. Int. Ed. Engl.*, **32**, 950 (1993).
18. M. Molter, Technetium-99m–Die Basis der modernen nuklearmedizinischen *in vivo*-Diagnostik, *Chem. Ztg.*, **103**, 41 (1979).
19. M. J. Clarke and L. Podbielski, Medical diagnostic imaging with complexes of ^{99m}Tc, *Coord Chem. Rev.*, **78**, 253 (1987).
20. T. C. Pinkerton, C. P. Desilets, D. J. Hoch, M. Mikelsons and G. M. Wilson, Bioinorganic activity of technetium radiopharmaceuticals, *J. Chem. Educ.*, **62**, 985 (1985).
21. B. Douglas, D. H. McDaniel and J. J. Alexander, *Concepts and Models of Inorganic Chemistry*, 2nd edn, Wiley, New York, 1983, p. 502.
22. E. Deutsch, W. Bushong, K. A. Glavan, R. C. Elder, V. J. Sodd, K. L. Scholz, D. L. Fortman and S. J. Lukes, Heart imaging with cationic complexes of technetium, *Science*, **214**, 85 (1981).

19 Chemotherapy with Compounds of Some Nonessential Elements

19.1 Overview

According to the principles outlined in Section 2.1 it is self-evident that compounds of the essential elements can be therapeutically useful. As with pharmaceuticals containing only 'organic' molecules as active ingredients, it may be that a special chemical compound of such an element is particularly active, either the administered substance itself or a metabolic product (compare Table 6.2). In addition to particle-emitting, i.e. primarily physically effective radiotherapeutic agents with inorganic isotopes, there is now an increasing number of chemotherapeutically active compounds of those inorganic elements which are nonessential according to present knowledge. Among these are arsenic compounds [1] such as the salvarsan derivatives $As_n Ar_n$ (Ar: aryl), the discovery of which was a landmark in chemotherapy (P. Ehrlich, 1908 Nobel prize for medicine), some long-known antiseptic preparations containing mercury, silver or boron, or certain bismuth complexes which have proven to be specifically effective against infectious forms of gastritis [2,3]. The clinically used 'inorganic' drugs based on nonessential elements [3–6], which are described in the following, show a more complex mechanism of action than the early bactericidal preparations. The focus will be mainly on platinum complexes as used in cancer treatment, on gold compounds in the therapy of rheumatoid arthritis, and on lithium which is being used in the treatment of psychiatric disorders such as schizophrenia.

19.2 Platinum Complexes in Cancer Therapy

19.2.1 Discovery, Application and Structure–Effect Relationships

The cytostatic effect of *cis*-diamminedichloroplatinum(II), 'cisplatin' (19.1), a square planar complex due to the d^8 configuration of the metal (compare Section 3.2.1 and Figure 2.10), was serendipitously discovered by B. Rosenberg in the 1960s. Studying the influence of weak alternating currents on the growth of *E. coli* bacteria he used ostensibly inert platinum electrodes. The result of these experiments was an inhibition of cell reproduction without simultaneous inhibition of bacterial growth which eventually led to the formation of long, filamentous cells [7]. In the course of the following more detailed studies it was found that not the electric current itself but trace amounts of *cis*-configurated chloro complexes such as (19.1) or (19.2) resulting from oxidation of the platinum electrode were responsible for this biological effect. Like gold, oxidized platinum forms very stable complexes with halides or pseudohalides, and the otherwise very noble metals are thus more easily oxidized in the presence of such ligands. Familiar examples include the dissolution of these metals in concentrated nitric *and* (chloride-containing) hydrochloric acid, 'aqua regia', or the cyanide process in the oxidative leaching of gold ores. In the presence of chloride and ammonium/ammonia as components of typical buffered culture media, there was apparently sufficient material oxidatively dissolved from the platinum electrode; the potential for the formation of the primarily resulting complexes $[PtCl_{4,6}]^{2-}$ from Pt is about 0.7 V. The observed filamentous growth of bacteria indicates the potential anti-tumor activity of the corresponding substances [8] via an inhibition of cell division (cytostatic effect). Out of a large number of platinum complexes [9–12] and other metal compounds [5,6,13,14] tested *in vivo* and in partial clinical screening, cisplatin (19.1) had for a long time shown by far the best results.

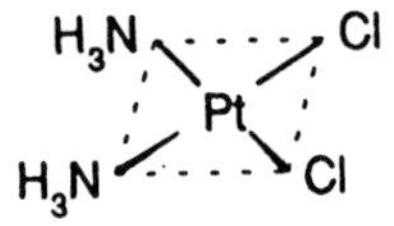

$$H_3N \cdots\ Cl$$
$$Pt$$
$$H_3N \cdots\ Cl \tag{19.1}$$

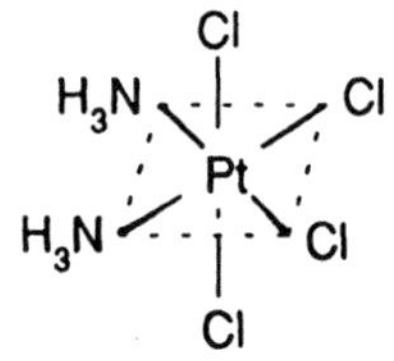

$$Cl$$
$$H_3N \cdots\ Cl$$
$$Pt$$
$$H_3N \cdots\ Cl$$
$$Cl \tag{19.2}$$

Complex (19.1) which has been approved as a drug since about 1978 is still being used, alone or in combination with other cytostatic agents such as bleomycin (19.12), vinblastin, adriamycin, cyclophosphamide or doxorubicin (19.6), against testicular or ovarian cancers and increasingly against bladder, cervical and lung tumors as well as tumors in the head/neck area. Over the years, the prospects for a complete cure of testicular and bladder cancer have vastly improved (>90%), mainly due to the use of cisplatin and other 'second-generation' platinum drugs. Since 1983, cisplatin has been the cytostatic drug with the highest turnover in the United States; annual revenues are in excess of US$ 100 million, and about 30 000 patients per year have regularly been treated successfully. For a long time, this compound has topped the list of the most successful patent applications granted to American universities (here: Michigan State University).

The most common side effects of a cisplatin therapy include kidney and gastro-intestinal problems, including nausea, which may be attributed to the inhibition of enzymes through coordination of the heavy metal platinum to sulfhydryl groups in proteins. Accordingly, a treatment with sulfur compounds such as sodium diethyl-dithiocarbamate (19.3) or thiourea and subsequent diuresis may counteract these symptoms. In contrast to many other cytostatic agents, however, cisplatin causes only minor, reversible damages to the spinal cord region. Second-generation

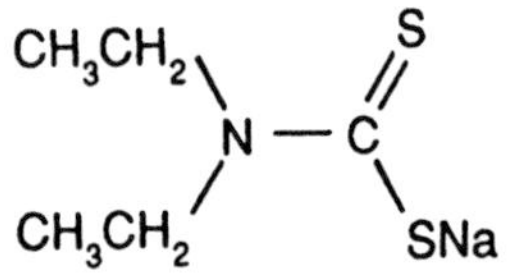

$$CH_3CH_2 \diagdown\ \ \ \ S$$
$$N-C \diagup$$
$$CH_3CH_2 \diagup\ \ \ \ SNa \tag{19.3}$$

analogues of cisplatin with at least the same effectivity but lower therapeutical dosage and also reduced side effects have been clinically tested. They include compounds such as 'carboplatin', 'spiroplatin' and 'iproplatin' (19.4 [9–11]). Carboplatin turned out to be the most useful analogue of cisplatin; it has been available in the United Kingdom since about 1990. Compared to cisplatin it shows similar activity against ovarian and lung tumors but reduced side effects with respect to the peripheral nervous system and the kidneys (differential toxicity). The non-coplanar structure of carboplatin due to its tetrahedrally configured spiro carbon atom is presumably responsible for the delayed degradation to potentially damaging derivatives. At 37 °C the retention half-life in blood plasma is 30 hours for carboplatin compared with only 1.5–3.6 hours for cisplatin [11].

carboplatin: *cis*-diammine(1,1-cyclo-butanedicarboxylato)platinum(II)

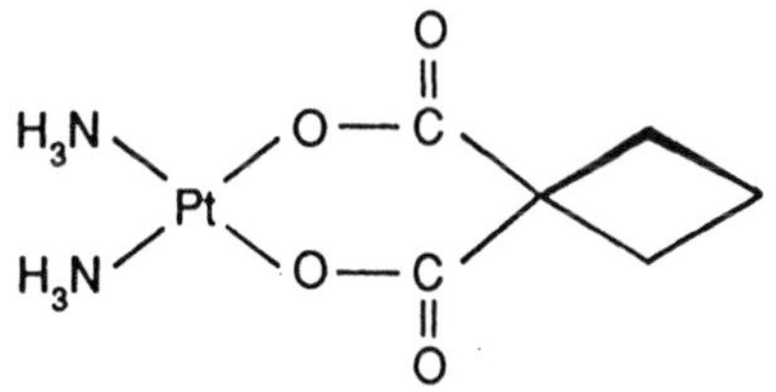

spiroplatin: aqua-1,1-bis(amino-methyl)cyclohexanesulfatoplatinum(II)

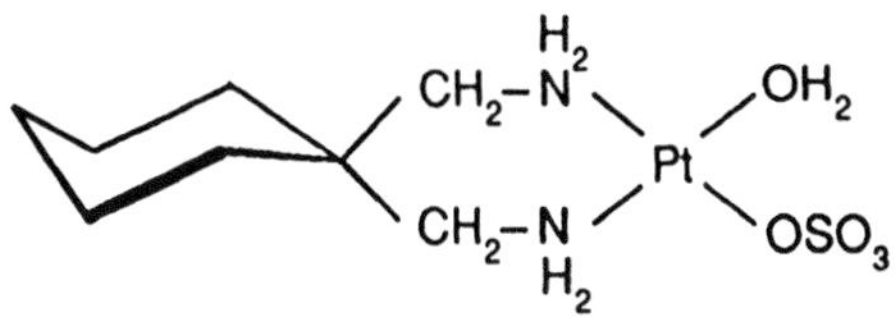

iproplatin, CHIP: *cis*-dichlorobis(isopropyl-amine)-*trans*-dihydroxoplatinum(IV)

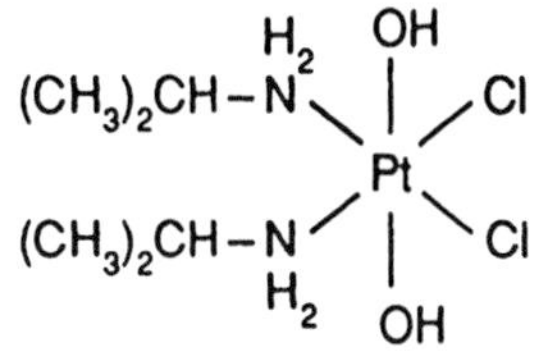

(19.4)

The large number of rather easily synthesized platinum complexes which were tested for anticancer activity has allowed the following structure–effect relationships to be formulated:

(a) Both square-planar platinum(II) as well as octahedrally configured platinum(IV) complexes show cytostatic activity; however, that of the platinum(IV) compounds is usually lower. It has been assumed that the 'active' Pt^{IV} complexes are reduced to Pt^{II} derivatives *in vivo*, possibly by cysteine.

(b) In general, continuous cytostatic activity has been found only for compounds with *cis* configuration; most but not all [15] *trans* isomers seem to be ineffective.

(c) Active complexes contain two nonleaving (NL) groups in *cis* positions and two monodentate or one bidentate ligand.

(d) Amine ligands are the preferred nonleaving groups; they must contain at least one N—H bond and thus a possibility for hydrogen bond formation. The N—H bonds in coordinated primary or secondary amines can have several functions [12]; they can facilitate the approach of the molecule to DNA and contribute to the base-specific formation and stabilization of the resulting adducts (see Figure 19.3).

(e) The ligands X corresponding to the general formula cis-$Pt^{II}X_2(NL)_2$ (compare 19.5) for divalent or cis-$Pt^{IV}X_2Y_2(NL)_2$ for tetravalent platinum are typically anions that exhibit an intermediate bond stability with platinum and are thus exchangeable on a therapeutical/physiological time-scale. Examples for X are halides, carboxylates (often as chelating ligands), sulfates, aqua or hydroxo ligands (compare 19.4). $Trans$-positioned OH^- groups are often used for platinum(IV) compounds (19.4) to increase the water solubility; with a maximum of 0.25 g per 100 ml H_2O cisplatin itself is not particularly soluble. Complexes with very labile ligands X are toxic while very inert Pt—X bonds render the corresponding substances inactive.

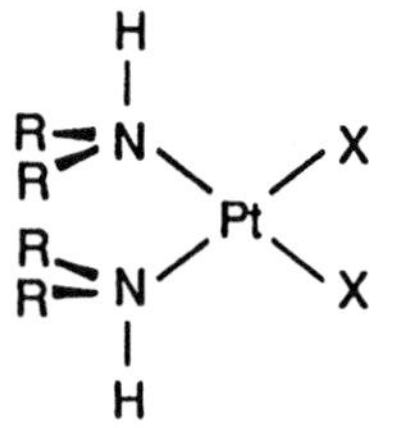

$$
\begin{array}{c}
H \\
| \\
\begin{matrix} R \\ R \end{matrix} \!\!\Rightarrow\! N \diagdown \qquad \diagup X \\
\qquad\qquad Pt \\
\begin{matrix} R \\ R \end{matrix} \!\!\Rightarrow\! N \diagup \qquad \diagdown X \\
| \\
H
\end{array}
$$

(19.5)

(f) As a rule, active complexes are neutral and may thus initially penetrate cell membranes more easily than charged compounds.

In 'third-generation' platinum-containing cytostatic agents, the actually effective Pt complex is coupled to a functional carrier molecule [10]. The carrier should improve the selectivity for tumor tissue by providing required material for the rapidly growing tumor cells. Inherent cytotoxic activity of the ligand, e.g. of the amine doxorubicin (19.6), may contribute in a synergistic manner.

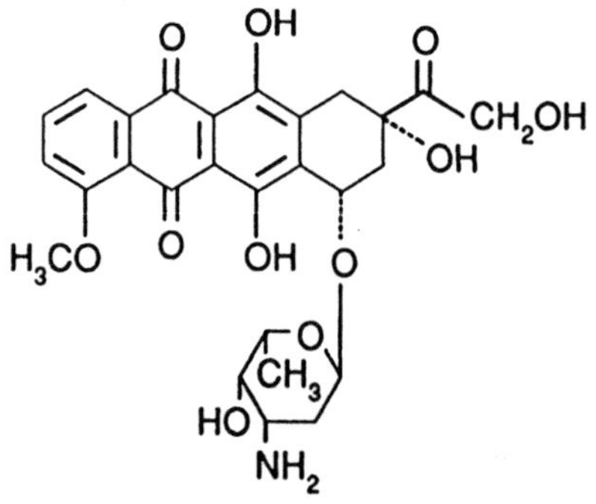

doxorubicin   (19.6)

Cytostatic activity has been established for many platinum complexes via screening methods; however, extensive (and expensive) clinical tests often remain to be done. In addition to reducing side effects, current efforts are being devoted to improved solubility, retarded excretion, diminished therapeutic resistance, and more selective transport through cell membranes. The latter is relevant for a different activity spectrum, especially in the treatment of lung cancer or with regard to antiviral effects. Another target is cytostatic platinum drugs for oral application since cisplatin and similar compounds (19.4) are rapidly hydrolyzed in gastric fluid and have to be administered via various kinds of infusions.

19.2.2 Cisplatin: Mode of Action

Tumor cells are distinguished from normal body cells by the loss of genetic control of their life span. Likewise, the feedback mechanisms with regard to the existence of neighboring cells are impaired which leads to the uncontrolled growth of tumor tissue. In normal cells, these processes are restrained and regulated by proto-oncogenes, and cancer may thus result from changes of these genes or their expression. According to this basic concept of carcinogenesis, cisplatin is believed to exert its cytostatic effect primarily through coordination with DNA in the cell nucleus while reactions in other regions, e.g. with serum proteins, cause undesired side effects. From experiments with second-generation compounds it is known that the mode of action may actually be quite complex; depending on the type of platinum compound there is more or less inhibition of DNA, RNA or protein synthesis.

The pathways of cisplatin in the human body are schematically outlined in Figure 19.1. After injection (oral administering is not possible due to hydrolysis in highly acidic gastric juice), cisplatin may be bound to plasma proteins and then renally excreted (30–70%); the remaining fraction is transported by the blood in an unaltered form. After passive transport of neutral cisplatin through cell membranes of different organs or tumor cells, it is rapidly hydrolyzed due to the markedly lower chloride concentration in intracellular regions (compare Figure 13.3). Within cells, about 40% of the platinum is present as cis-$Pt(NH_3)_2Cl(H_2O)^+$. This hydrolysis product (see 19.7) of cisplatin is kinetically labile since H_2O is a much better leaving group with respect to Pt^{II} than Cl^-; it is thus assumed that cis-$Pt(NH_3)_2Cl(H_2O)^+$ is a particularly active form of the cytostatic agent. The positive charge of this substituted complex supports such an assumption because it would be more likely to approach and coordinate to the negatively charged DNA.

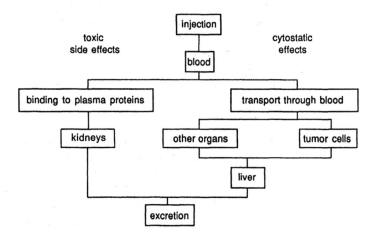

Figure 19.1
Simplified metabolic pathways of cisplatin in the human body

$$cis\text{-}Pt(NH_3)_2Cl_2$$

$$+Cl^- \Big\Uparrow -Cl^- \quad \begin{array}{l} k_1 = \\ 6.3\times10^{-5}\,s^{-1} \end{array}$$

$$cis\text{-}Pt(NH_3)_2Cl(OH) \overset{+\,H^+}{\underset{-\,H^+}{\rightleftharpoons}} cis\text{-}Pt(NH_3)_2Cl(H_2O)^+$$

$$(pK_a = 6.4)$$

$$+Cl^- \Big\Uparrow -Cl^- \quad \begin{array}{l} k_1 = \\ 2.5\times10^{-5}\,s^{-1} \end{array}$$

$$cis\text{-}Pt(NH_3)_2(OH)_2 \overset{+\,H^+}{\underset{-\,H^+}{\rightleftharpoons}} cis\text{-}Pt(NH_3)_2(H_2O)(OH)^+ \overset{+\,H^+}{\underset{-\,H^+}{\rightleftharpoons}} cis\text{-}Pt(NH_3)_2(H_2O)_2{}^{2+}$$

$$pK_a = 7.2 \qquad\qquad \Big\Updownarrow \qquad\qquad pK_a = 5.4$$

$$\text{oligomers} \tag{19.7}$$

The retention times of platinum are different for the individual organs and decrease in the order kidney > liver > genitals > spleen > bladder > heart > skin > stomach > brain [11]. After interaction with the DNA in the cells of these organs, the degradation products are excreted via the liver and kidneys.

For a binding of metal ions and their complexes to DNA or, more generally, to nucleotides, there are several different coordination sites available. The metal centers can bind to the negatively charged oxygen atoms of (poly)phosphate groups (compare Section 14.1) or to the nitrogen and oxygen atoms of purine and pyrimidine bases (Section 2.3.3). Planar complexes [16] or those containing large π systems as ligands should also be able to intercalate between two base pairs (compare Figure 2.12), possibly even in a sequence specific fashion. Finally, coordinated ligands with, for example, amine or hydroxyl functions may form hydrogen bonds with proton acceptor components of the polynucleotides.

Interactions of metal ions or metal complexes with nucleic acids [17] generally play an important role:

(a) in sustaining the conformation of the tertiary structure of polyelectrolytes such as DNA or RNA through electrostatic effects,
(b) in the nucleic acid metabolism, particularly in phosphoryl transfer (nuclease and polymerase activity, see Section 14.1),
(c) in the regulation (Section 17.5), replication and transcription of genetic information (see also Section 12.6),
(d) for efforts directed at specific DNA cleavage with synthetic probes (restriction enzyme analogues, compare Section 2.3.3) and
(e) for metal-induced mutagenesis [18].

Such mutations can be due to geometric distortions of the DNA [19] through unphysiological cross-linking or to the stabilization of a wrong nucleobase tautomer after metal coordination (see 2.13 [18]). Metal complex/nucleic acid interactions and their physiological consequences can thus be quite varied; even in the case of the extensively studied platinum compounds they are far from being fully understood [11,12,20,21]. Replication and genetic transcription may be impaired with regard to their accuracy ('fidelity') or the ability to recognize and repair defective sites [22]. A significant aspect for the mutagenicity of compounds, even of the essential metals, is their behavior towards discriminating mechanisms such as the membrane barrier protecting the chromosomal area (compare Figure 17.7). As potent cytostatic agents, the above-mentioned platinum compounds may also cause mutations if present in higher concentrations. Based on such observations, the low but detectable platinum emissions from exhaust catalysts in vehicles are being controversially discussed with regard to their potential health hazard.

After loss of Cl$^-$, the cationic species resulting from *cis*-Pt(NH$_3$)$_2$Cl$_2$ were shown to form coordinative bonds to nitrogen atoms of nucleobases; *in vitro* these include bonds to N7 of guanine, N1 and N7 of adenine and N3 of cytosine (2.11). N1 of adenine and N3 of cytosine are engaged in hydrogen bonding within the DNA framework; for various nucleotide oligomers serving as DNA models, the highest binding affinity was found between N7 of guanine and platinum (see Figure 19.2).

A coordinatively unsaturated metal complex fragment with two open sites in the *cis* position may bind in different ways to double-stranded DNA (Figure 19.2). Since complexes with only one labile ligand such as diethylenetriamineplatinum(II) chloride, Pt(dien)Cl$^+$ (19.8), are therapeutically inactive, the monofunctional platinum species presumably serve only as intermediates. The possible alternatives of coordinative interaction between DNA and (NL)$_2$Pt^{2+} include chelate complex formation, e.g. via coordination to nitrogen and oxygen atoms of *one* guanine base, *intrastrand* cross-linking of two nucleobases of a *single* DNA strand, *interstrand* cross-linking of two different strands of one DNA molecule, or the metal-induced attachment of a protein to DNA. Experiments have shown that the chelate complex formation with O6 and N7 (Figure 19.2) of a free guanine base is principally possible; however, it is not favored within the DNA double helix. Interstrand cross-linking and protein–DNA interactions also make only minor contributions to the overall platinum/DNA adduct formation in the case of Pt(NH$_3$)$_2$$^{2+}$, the situation (and the cytostatic spectrum) being different with more recently developed dinuclear complexes such as *cis*-Cl$_2$(H$_3$N)Pt[H$_2$N(CH$_2$)$_4$NH$_2$]-*cis*-Pt(NH$_3$)Cl$_2$ [23]. Most of the

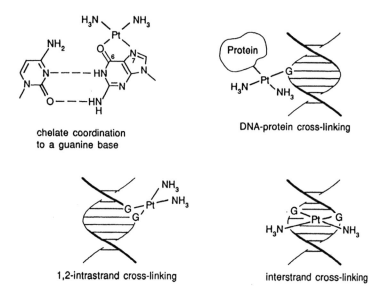

chelate coordination
to a guanine base

DNA-protein cross-linking

1,2-intrastrand cross-linking

interstrand cross-linking

Figure 19.2
Possible kinds of bonding between *cis*-Pt(NH$_3$)$_2$$^{2+}$ and guanosine (G) in double-stranded DNA

retained diammineplatinum(II) forms bonds to two neighboring N7-coordinated guanosine (G) nucleotides on the same DNA strand (1,2-intrastrand d(GpG) cross-linking; d: deoxy form of ribose; p: phosphate); 1,2-intrastrand d(ApG) cross-linking (each via N7) has also been observed [24].

$$\left[\begin{array}{c} H \\ \diagup N \diagdown NH_2 \\ Pt \\ \diagdown N \diagup Cl \\ H_2 \end{array}\right]^+$$

(19.8)

The binding of the platinum complex fragment to DNA leads to changes in structure and overall characteristics. Some sequence-specific DNA-cleaving enzymes can no longer attack the DNA at the platinum-coordinated oligoguanosine sequences; DNA synthesis from a single-strand template is inhibited by the coordination of cis- and trans-Pt(NH$_3$)$_2$$^{2+}$. Contrary to cis-Pt(NH$_3$)$_2$$^{2+}$, which coordinates to neighboring guanosine nucleotides d(GpG), the trans isomer prefers guanosine coordination in d(GpNpG) sequences, where N denotes any other nucleotide. The monofunctional cationic complex Pt(dien)Cl$^+$ (19.8), on the other hand, does not inhibit DNA synthesis. Substitution reactions at platinum are usually slow so that cis/trans isomerization is not relevant on the physiological time-scale.

The structural changes of DNA after coordination of Pt(NH$_3$)$_2$$^{2+}$ fragments may be quantitatively assessed via physical measurements, relating, for example, to the thermal stability. Binding of the cis complex destabilizes the DNA double helix and thus results in a lowered melting point; binding of trans-Pt(NH$_3$)$_2$$^{2+}$ or Pt(dien)$^{2+}$, on the other hand, leads to an increase of the melting point, suggesting a stabilization of the double helix by interstrand cross-links or hydrogen bonds. Electron microscopy shows that platinum-containing DNA is shorter by an average of 170 pm per coordinated cis-Pt(NH$_3$)$_2$$^{2+}$ and by about 100 pm per coordinated trans isomer. These relatively minor structural distortions are interpreted as kinks in the biopolymer (19.9) [19].

$$Pt \left< \begin{array}{cc} 5' & 3' \\ & \\ & \\ 3' & 5' \end{array} \right.$$

(19.9)

Direct structural comparisons of intact and platinum-coordinated DNA fragments have been made using NMR spectroscopy in solution and X-ray structure analyses of crystallized material. The structure of cis-Pt(NH$_3$)$_2$[d(pGpG)] in Figure 19.3 shows a square-planar platinum center surrounded by two cis-positioned NH$_3$ ligands and two N7 nitrogen atoms of the two guanine bases. While the nucleobases are situated nearly parallel to each other (stacking) in an intact DNA, they form a dihedral angle of about 80 ° in this model (Figure 19.3) for platinum-coordinated DNA, suggesting the possibility of a significantly disturbed double helix structure. The intramolecular

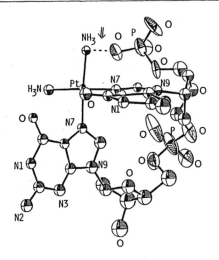

Figure 19.3
Molecular structure of *cis*-Pt(NH$_3$)$_2$[d(pGpG)] with an indicated ammine-phosphate hydrogen bond (one of four crystallographically independent molecules in the unit cell; from [24])

hydrogen bond between a coordinated ammine ligand and the terminal phosphate is remarkable with respect to the requirement for such ligands in active platinum cytostatica (Figure 19.3).

A coordination to neighboring, nearly orthogonally oriented guanine bases (each via N7) is also evident in the complex of *cis*-Pt(NH$_3$)$_2$$^{2+}$ with d(CpGpG) where hydrogen bonds are also formed [25]. It should be realized, however, that weak intra- and intermolecular interactions including solvent effects may favor different conformations in solution and in the solid state [12]. A kinking is not observed if only one guanosine is present in the oligonucleotide chain. For the complex between *trans*-Pt(NH$_3$)$_2$$^{2+}$ and the DNA fragment d(A*pGpG*pCpCpT) coordination was observed between the metal and the N7 nitrogen atoms of the adenine and the second, i.e. *the next but one* guanine base; the possibility of a cross-linking of two different nucleobases was shown for *cis*-Pt(NH$_3$)$_2$$^{2+}$ in combination with anionic 9-methylguanine and 1-methyluracil [26].

An explanation of the mode of action of platinum cytostatic drugs has to take into account the dynamics of physiological processes and the striking difference between *cis* and *trans* isomers. Using ^{195}Pt NMR spectroscopy ($I = 1/2$, 33.8% natural abundance), the process of incorporation of *cis*- and *trans*-diamminedichloroplatinum(II) into a short double-stranded DNA with 30–50 base pairs and a molecular mass of about 25 kDa was studied. The rate determining step is the substitution of Cl$^-$ by H$_2$O (compare 19.7), followed by the formation of a monofunctional Pt/DNA adduct in a rapid reaction. The rate of the actual cross-linking depends on the substitution of the second chloride ligand [16,27].

Since *cis*- as well as *trans*-Pt(NH$_3$)$_2$$^{2+}$ alter the double helical structure of DNA and its replication, the question remains why it is only the *cis* complex that shows cytostatic activity. The *trans* isomer is resorbed more rapidly than cisplatin; however, the concentration of the DNA-coordinated *trans* complex begins to decrease after six hours whereas the *cis* isomer is then still accumulated in the cell nucleus. Only very little *trans* compound is still coordinated to DNA after 24 hours. These results indicate that the changes in the DNA structure caused by the *trans* isomer are sensed differently by the endogeneous repair mechanisms from those caused by coordination of cisplatin [22]. In cells where these repair mechanisms were partly blocked through mutations, the cytotoxicity of both *cis*- and *trans*-Pt(NH$_3$)$_2$Cl$_2$ was higher. A monofunctionally coordinated *trans*-configurated metal complex fragment —Pt(NH$_3$)X(NH$_3$), X = Cl$^-$ or H$_2$O, is likely to be attacked quite easily; external chloride or sulfur ligands with their pronounced *trans*-labilizing effect could particularly contribute to a rapid dissociation and excretion of the *trans*-configurated metal complex fragment [28]. The specific binding of a chromosomal 'high-mobility group' (HMG-1) protein to *cis*-Pt(NH$_3$)$_2$$^{2+}$-containing DNA [29] suggests a flawed genetic information transfer, either via altered transcription or through faulty recognition and thus shielding from DNA repair processes.

19.3   Cytotoxic Compounds of Other Metals

Following the therapeutical success of cisplatin and similar platinum-containing anti-tumor drugs, a large number of complexes with other metals have been studied and, in several cases, subjected to clinical tests [5,6]. Many of these complexes are neutral, have two moderately labile leaving groups in *cis* position and feature groups that are susceptible to intercalation. Since bonds to platinum(II) are quite inert, the nonleaving groups at other, substitutionally more labile metal centers have to be coordinated in a kinetically stable fashion, e.g. either through polyhapto-binding of organic ligands or through chelate ligation.

Examples of such complexes are the metallocenes and metallocene dichlorides (19.10 [13,14]) which have shown cytostatic activity against various tumors in cell experiments. Redox-stable titanocene dichloride (19.10, M = Ti) in particular proved to be a potent agent against breast, lung and intestinal (colon) cancer tissue. In contrast to cisplatin, titanocene dichloride exhibits side effects with regard to the liver when used in therapeutically necessary amounts. The detailed mode of action is unknown; a binding to N7 or, via chelate formation, to N7 and O6 of purine nucleotides was detected *in vitro*. While substitution of the chloride ligands has little effect on the anti-tumor activity, changes at the cyclopentadienide rings lead to inactive compounds. In addition to similarly constructed diorganotin(IV) compounds such as R$_2$SnX$_2$, the compound bis(1-phenyl-1,3-butanedionato)diethoxytitanium-(IV), budotitane (19.11) [30], has turned out to be a very promising and clinically tested cytostatic agent against colon cancer; it also exhibits the above-mentioned characteristics.

Anti-tumor-active ruthenium, rhodium and gold compounds with amine, phosphane and carboxylate ligands are distinguished by substitutionally rather inert metal

M = Ti, V, Nb, Mo   (19.10)

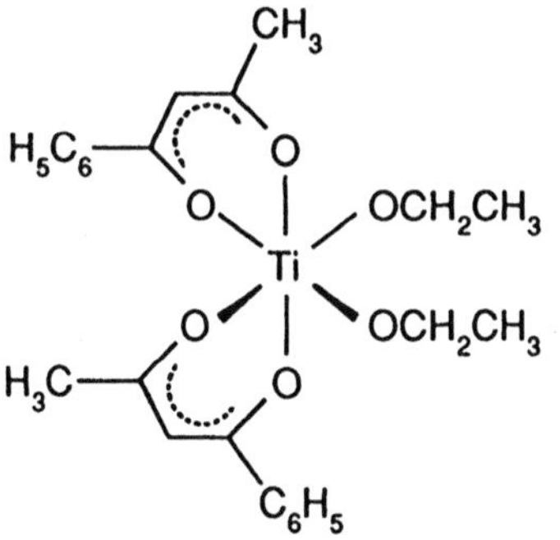

budotitane (19.11)

centers. Metal–metal bonded dinuclear complexes such as $Rh_2(O_2CCH_3)_4$ are also being considered for that purpose [5,6].

Iron, ruthenium and copper complexes with potentially DNA-intercalating organic ligands are known especially in the form of complexes with the antibiotic bleomycin (19.12, Fe) or with ligands such as 1,10-phenanthroline (2.14 and Figure 2.12, Ru or Cu). The presence of redox-active metal ions is essential for the oxidative or photoinduced DNA cleavage caused by these reagents; for example, a metal-based activation of radical-forming O_2 involving oxoferryl intermediates (see Sections 6.2 to 6.4) [31] is assumed for antineoplastic iron(II)/bleomycin complexes (19.12).

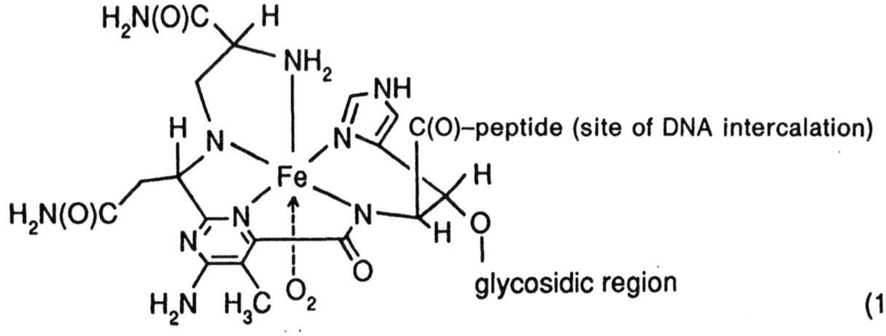

(19.12)

Bis(1,10-phenanthroline)copper(I) is the best-known example of a metal-containing chemical nuclease; its peroxide-induced ability of phosphodiester cleavage in DNA or RNA may be applied in sequence specific fashion by using carrier systems [32,33]. In similar complexes of ruthenium (2.14) which may be activated by irradiation, the partially selective interaction with parts of the DNA probably occurs via electrostatic effects and shape adaption ('molecular recognition') rather than via direct intercalation (Figure 2.12). Such chemical analogues of restriction enzymes have a potential for various fields requiring DNA analysis and modification.

19.4 Gold-Containing Drugs Used in the Therapy of Rheumatoid Arthritis

19.4.1 Historical Development

Gold was already used for therapeutical purposes in ancient civilizations; for instance, a Chinese prescription dating from the sixth century AD describes in detail the dissolution of metallic gold for the use in elixirs aimed at achieving immortality. As can be reconstructed today [34], the oxidative dissolution of this noble metal involved potassium nitrate, KNO_3, containing iodate, IO_3^-, as impurity, which can be reduced to iodide by reductants such as $FeSO_4$ or organic material. In the presence of I^-, the potential for oxidation of gold is lowered by approximately 1 V and $[AuI_2]^-$ is formed in this process. Gold was propagated as a universal remedy and its protective function against leprosy has frequently been mentioned. In 1890 Robert Koch found that gold(I) cyanide, AuCN, inhibits the growth of bacteria that cause tuberculosis; however, a systematic use of this compound was impossible due to its high toxicity. In 1924 Mollgaard applied a thiosulfato complex of monovalent gold in an attempt to cure tuberculosis and a few years later gold(I) thioglucose (19.13) was beginning to be used in the therapy of rheumatic fever. However, it was only much later that similar compounds were subjected to systematic clinical tests and received proper attention [4,35,36].

$$Na_3[O_3S_2-Au-S_2O_3]$$

trisodiumgold(I)bis(thiosulfate)
('sanocrysin')

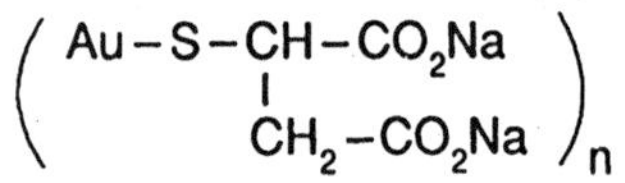

$$\left(\begin{array}{l} Au-S-CH-CO_2Na \\ \quad\quad\quad\; | \\ \quad\quad\; CH_2-CO_2Na \end{array}\right)_n$$

disodiumgold(I)thiomalate ('myochrisin')

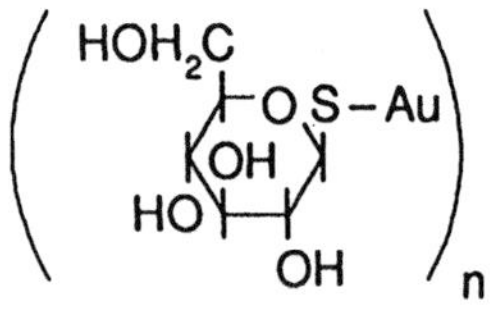

gold(I)thioglucose ('solganol')

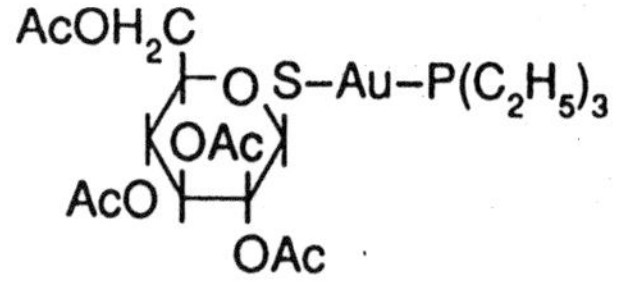

(2,3,4,6-tetrakis-*O*-acetyl-1-thio-*β*-D-
glucopyranosido)gold(I)triethyl-
phosphine ('auranofin', 'ridaura®')   (19.13)

19.4.2 Gold Compounds as Anti-Rheumatic Agents

Only the monovalent form of gold has any therapeutical importance. The aqua complex of gold(I) is unstable and disproportionates according to $3\,Au^I \rightarrow 2\,Au^0 + Au^{III}$; however, the monovalent state can be stabilized through 'soft', polarizable ligands such as CN^-, PR_3 or thiolates, RS^-, under formation of preferentially linear d^{10} metal complexes. Most of the gold(I) compounds with the desired biological activity contain thiolate ligands; gold(III) would act as a strong oxidant under these conditions.

The most important representatives of gold-containing anti-rheumatic drugs are listed in (19.13).

The linear arrangement of the sulfur ligands around the two-coordinate metal center occurs in all cases, including solganol and myochrisin; in these latter species, there are oligomeric structures with six or eight units and bridging sulfur centers. A ring arrangement of the oligomers has been discussed; however, XAS studies rather point to open chain structures [37]. Solganol and myochrisin are water-soluble but insoluble in hydrophobic environments and thus have to be administered intramuscularly to prevent hydrolysis in acidic gastric fluid. The lipophilic auranofin, on the other hand, can be administered orally with about 25% resorption. After long-term therapy with myochrisin and auranofin, constant levels of 30–50 μg/ml blood are attained; in the blood, myochrisin is mainly bound to albumin in the serum while auranofin is equally distributed between serum and erythrocytes.

During the increasingly practised therapy with gold compounds to retard active rheumatic processes, there are not uncommon side effects resembling allergic reactions on skin and mucous membranes as well as gastrointestinal and renal problems. These effects restrict a gold therapy to only about two-thirds of the patients; it is assumed that some side effects are due to the formation of Au^{III} compounds. Since gold(I) forms thermodynamically stable complexes with sulfur ligands, the addition of sulfur-containing chelating agents such as penicillamine or dimercaprol (2.1) and of antihistamines or adrenocorticosteroids may reduce the toxicity of the gold drugs.

19.4.3 Hypotheses on the Mode of Action of Gold-Containing Anti-Rheumatic Drugs

Arthritis is an inflammation of the tissue which surrounds the joints. It is assumed that the damage is caused by the action of hydrolytic enzymes from lysosomes (rheumatoid arthritis as an autoimmune reaction). Examinations of the tissue show that gold is preferentially accumulated in the joints and is stored in the lysosomes of macrophages, forming 'aurosomes'. Inhibition of the lysosomal enzyme activity can be rationalized assuming a coordination of gold to the thiolate groups, RS^-, which are present in the enzymes. In vitro, disodiumgold(I) thiomalate readily reacts with other thiolates, RS^-, under release of thiomalate. This process is analogous to the rapid exchange reaction of the similar, linearly coordinated and thus kinetically labile Hg^{II} thiolate complexes (associative substitution mechanism, see Section 17.5).

According to a different hypothesis, gold(I) compounds are able to inhibit the formation of undesired antibodies in the collagen region.

Yet another hypothesis suggests that irreversible damage of the joints may result from lipid oxidation with subsequent degradation of proteins by the free radicals formed (compare Section 16.8). Superoxide ions, $O_2^{\cdot-}$, which can be produced by activated phagocytes [38] (compare Section 10.5), play an important role in this process. It was shown that several oxidants can relatively easily convert $O_2^{\cdot-}$ to reactive, not-spin-inhibited singlet dioxygen, $^1O_2$ (5.3). In this context it is important to note that gold(I) compounds should be able to deactivate this excited singlet state of dioxygen due to the particularly high spin-orbit coupling constant of this heavy element (intersystem crossing; [39]).

19.5   Lithium in Psychopharmacological Drugs

For several decades manic-depressive (bipolar) psychoses have been treated with lithium salts, often in form of exactly measured lithium carbonate. The Li^+ ion is therapeutically valuable because it counteracts *both* phases in the typically cyclic course of this disorder [40,41].

Difficulties in the therapy with lithium compounds result from the relatively high toxicity of the metal and the resulting very small therapeutic width. While a concentration of about 1 mmol Li^+ per liter of blood is necessary for a successful treatment, a 2 mM concentration may already cause toxic side effects, in particular in the renal and nervous systems (tremor). Concentrations of 3 mM and higher may eventually be lethal for the patient. For this reason, lithium carbonate or other salts of Li^+ are administered orally in several carefully controlled doses per day. On acute poisoning, the blood has to be purified using Na^+-containing dialysis fluids.

Li^+ is relatively rare in the earth's crust and in sea water. It is also the lighter and, with an ionic radius of 60 pm for the coordination number four, significantly smaller and thus more polarizing homologue of Na^+, a fact that could directly explain the neurological (side) effects (compare Chapter 13). On the other hand, Li^+ shares a diagonal relationship and the same physiologically important affinity towards phosphate ligands with the slightly larger Mg^{2+}. There are several hypotheses with regard to the specific antipsychotic mode of action for lithium in which the effect on the cellular information system (via binding to phosphates) is a focal point [40,42]; inhibitions of the inositol/phosphate metabolism, of an adenyl cyclase, or of a guanine nucleotide-binding protein, a 'G protein', are being discussed.

References

1. O. M. N. Dhubhghaill and P. J. Sadler, The structure and reactivity of arsenic compounds: biological activity and drug design, *Struct. Bonding (Berlin)*, **78**, 130 (1991).
2. W. A. Herrmann, E. Herdtweck and L. Pajdla, 'Colloidal bismuth subcitrate' (CBS): isolation and structural characterization of the active substance against *Helicobacter pylori*, a causal factor of gastric diseases, *Inorg. Chem.*, **30**, 2581 (1991).
3. M. J. Abrams and B. A. Murrer, Metal compounds in therapy and diagnosis, *Science*, **261**, 725 (1993).
4. P. J. Sadler, Inorganic chemistry and drug design, *Adv. Inorg. Chem.*, **36**, 1 (1991).
5. B. K. Keppler, Metal complexes as anticancer agents. The future role of inorganic chemistry in cancer therapy, *New J. Chem.*, **14**, 389 (1990).
6. B. K. Keppler (ed.), *Metal Complexes in Cancer Chemotherapy*, VCH Verlagsgesellschaft mbH, Weinheim, 1993.
7. B. Rosenberg, L. van Camp and T. Krigas, Inhibition of cell division in *E. coli* by electrolysis products from a platinum electrode, *Nature (London)*, **205**, 698 (1965).
8. B. Rosenberg, L. van Camp, J. E. Trosko and V. H. Mansour, Platinum compounds: a new class of potent antitumor agents, *Nature (London)*, **222**, 385 (1969).
9. B. Lippert and W. Beck, Platin-Komplexe in der Krebstherapie, *Chem. Unserer Zeit*, **17**, 190 (1983).
10. A. Pasini and F. Zunino, New cisplatin analogs: on the way to better carcinostatics, *Angew. Chem. Int. Ed. Engl.*, **26**, 615 (1987).
11. P. Umapathy, The chemical and biological consequences of the binding of the antitumor drug cisplatin and other platinum group metal complexes to DNA, *Coord. Chem. Rev.*, **95**, 129 (1989).
12. J. Reedijk, The relevance of hydrogen bonding in the mechanism of action of platinum antitumor compounds, *Inorg. Chim. Acta*, **198–200**, 873 (1992).

13. I. Haiduc and C. Silvestru, Metal compounds in cancer chemotherapy, *Coord. Chem. Rev.*, **99**, 253 (1990).
14. P. Köpf-Maier and H. Köpf, Non-platinum-group metal antitumor agents—history, current status, and perspectives, *Chem. Rev.*, **87**, 1137 (1987).
15. M. Coluccia *et al.*, A trans-platinum complex showing higher antitumor activity than the cis congeners, *J. Med. Chem.*, **36**, 510 (1993).
16. W. I. Sundquist and S. J. Lippard, The coordination chemistry of platinum anticancer drugs and related compounds with DNA, *Coord. Chem. Rev.*, **100**, 293 (1990).
17. T. D. Tullius (ed.): *Metal–DNA Chemistry*, ACS Symposium Series 402, 1989.
18. B. Lippert, H. Schöllhorn, U. Thewalt, Metal-stabilized rare tautomers of nucleobases. 4. On the question of adenine tautomerization by a coordinated platinum(II), *Inorg. Chim. Acta*, **198–200**, 723 (1992).
19. T. P. Kline, L. G. Marzilli, D. Live and G. Zon, Investigations of platinum amine induced distortions in single- and double-stranded oligodeoxyribonucleotides, *J. Am. Chem. Soc.*, **111**, 7057 (1989).
20. B. Lippert, Platinum nucleobase chemistry, *Prog. Inorg. Chem.*, **37**, 1 (1989).
21. S. J. Lippard, Chemistry and molecular biology of platinum anticancer drugs, *Pure Appl. Chem.*, **59**, 731 (1987).
22. S. L. Bruhn, J. H. Toney and S. J. Lippard, Biological processing of DNA modified by platinum compounds, *Prog. Inorg. Chem.*, **38**, 477 (1990).
23. B. Van Houten, S. Illenye, Y. Qu and N. Farrell, Homodinuclear (Pt,Pt) and Heterodinuclear (Ru,Pt) metal compounds as DNA-protein cross-linking agents; potential suicide DNA lesions, *Biochemistry*, **32**, 11794 (1993).
24. S. E. Sherman, D. Gibson, A. H. J. Wang and S. J. Lippard, Crystal and molecular structure of cis-[Pt(NH$_3$)$_2${d(pGpG)}], the principal adduct formed by cis-diamminedichloroplatinum(II) with DNA, *J. Am. Chem. Soc.*, **110**, 7368 (1989).
25. G. Admiraal, J. L. van der Veer, R. A. G. de Graaff, J. H. J. den Hartog and J. Reedijk, Intrastrand bis(guanine) chelation of d(CpGpG) to cis-platinum: an X-ray single crystal structure analysis, *J. Am. Chem. Soc.*, **109**, 592 (1987).
26. G. Frommer, H. Schöllhorn, U. Thewalt and B. Lippert, Platinum(II) binding to N7 and N1 of guanine and a model for a purine-N^1, pyrimidine-N^3 cross-link of cisplatin in the interior of a DNA duplex, *Inorg. Chem.*, **29**, 1417 (1990).
27. S. E. Sherman and S. J. Lippard, Structural aspects of platinum anticancer drug interactions with DNA, *Chem. Rev.*, **87**, 1153 (1987).
28. O. Krizanovic, F. J. Pesch and B. Lippert, Nucleobase displacement from trans-diamineplatinum(II) complexes. A rationale for the inactivity of trans-DDP as an antitumor agent?, *Inorg. Chim. Acta*, **165**, 145 (1989).
29. P. M. Pil and S. J. Lippard, Specific binding of chromosomal protein HMG1 to DNA damaged by the anticancer drug cisplatin, *Science*, **256**, 234 (1992).
30. B. K. Keppler, C. Friesen, H. G. Moritz, H. Vongerichte and E. Vogel, Tumor-inhibiting bis(β-diketonato) metal complexes: Budotitane, *Struct. Bonding (Berlin)*, **78**, 97 (1991).
31. L. Que, Jr, Oxygen activation at nonheme iron centers, in *Bioinorganic Catalysis* (ed. J. Reedijk), Marcel Dekker, New York, 1993, p. 347.
32. D. S. Sigman, T. W. Bruice, A. Mazumder and C. L. Sutton, Targeted chemical nucleases, *Acc. Chem. Res.*, **26**, 98 (1993).
33. D. S. Sigman, C. B. Chen and M. B. Gorin, Sequence-specific scission of DNA by RNAs linked to a chemical nuclease, *Nature (London)*, **363**, 474 (1993).
34. C. Glidewell, Ancient and medieval chinese protochemistry, *J. Chem. Educ.*, **66**, 631 (1989).
35. S. J. Berners-Price and P. J. Sadler, Gold drugs, in *Frontiers in Bioinorganic Chemistry*, VCH Verlagsgesellschaft mbH, Weinheim, 1986, p. 376.
36. K. C. Dash and H. Schmidbaur, Gold complexes as metallodrugs, in *Metal Ions in Biological Systems* (ed. H. Sigel), Vol. 14, Marcel Dekker, New York, 1982, p. 179.
37. R. C. Elder and M. K. Eidsness, Synchrotron X-ray studies of metal-based drugs and metabolites, *Chem. Rev.*, **87**, 1027 (1987).
38. A. R. Cross and O. T. G. Jones, Enzymic mechanisms of superoxide production, *Biochim. Biophys. Acta*, **1057**, 281 (1991).
39. E. J. Corey, M. M. Mehrotra and A. U. Khan, Antiarthritic gold compounds effectively quench electronically excited singlet oxygen, *Science*, **236**, 68 (1987).

40. N. J. Birch and J. D. Phillips, Lithium and medicine: inorganic pharmacology, in *Adv. Inorg. Chem.*, **36**, 49 (1991).
41. G. N. Schrauzer and K.-F. Klippel (eds.), *Lithium in Biology and Medicine*, VCH Verlagsgesellschaft mbH, Weinheim, 1990.
42. S. Avissar, G. Schreiber, A. Danon and R. H. Belmaker, Lithium inhibits adrenergic and cholinergic increases in GTP binding in rat cortex, *Nature (London)*, **331**, 440 (1988).

Bibliography

A. Following is a list of texts on bioinorganic chemistry which treat at least parts of the topics described in this volume. Furthermore, most recent textbooks on inorganic chemistry contain a brief chapter devoted to this subject.

A. M. Fiabane and D. R. Williams, *The Principles of Bio-inorganic Chemistry*, The Chemical Society, London, 1977.

E. I. Ochiai, *Bioinorganic Chemistry: An Introduction*, Allyn and Bacon, Boston, 1977.

M. N. Hughes, *The Inorganic Chemistry of Biological Processes*, 2nd edn, Wiley, New York, 1981.

R. W. Hay, *Bio-inorganic Chemistry*, Ellis Horwood, Chichester, 1984.

E. I. Ochiai, *General Principles of Biochemistry of the Elements*, Plenum Press, New York, 1987.

M. N. Hughes and R. K. Poole, *Metals and Micro-organisms*, Chapman and Hall, London, 1989.

J. J. R. Frausto da Silva and R. J. P. Williams, *The Biological Chemistry of the Elements*, Clarendon Press, Oxford, 1991.

M. J. Kendrick, M. T. May, M. J. Plishka and K. D. Robinson, *Metals in Biological Systems*, Ellis Horwood, Chichester, 1992.

B. Many journals, including those with a broader spectrum within the sciences, contain progress reports on bioinorganic topics. The following journals and series are particularly devoted to this subject:

(a) *Journal of Inorganic Biochemistry* (Elsevier, Amsterdam).

(b) *BioMetals* (ed. G. Winkelmann) (Rapid Communications of Oxford).

(c) *Advances in Inorganic Biochemistry* (eds. G. L. Eichhorn and L. G. Marzilli) (Elsevier, Amsterdam).

(d) *Metal Ions in Biological Systems* (ed. H. Sigel) (Marcel Dekker, New York).

(e) *Metal Ions in Biology* (ed. T. G. Spiro) (Wiley, New York).

(f) *Perspectives on Bioinorganic Chemistry* (eds. R. W. Hay, J. R. Dillworth and K. B. Nolan) (Jay Press, Hampton Hill).

(g) *Biochemistry of the Elements* (ed. E. Frieden) (Plenum Press, New York).

C. The following books and journal issues appeared, for example, after corresponding international conferences and contain reviews of different aspects of bioinorganic chemistry by several authors:

(a) *J. Chem. Educ.*, **62**, 916–1001 (1985) (*Bioinorganic Chemistry—State of the Art*).

(b) *Recl. Trav. Chim. Pays-Bas*, **106**, 165–439 (1987).

(c) *Chem. Scr.*, **28A**, 1–131 (1988) (*Biophysical Chemistry of Dioxygen Reactions in Respiration and Photosynthesis*).

(d) *J. Inorg. Biochem.*, **36**, 151–372 (1989).

(e) G. L. Eichhorn (ed.), *Inorganic Biochemistry*, Elsevier, Amsterdam, 1973.

(f) A. W. Addison, W. R. Cullen, D. Dolphin and B. R. James (eds.), *Biological Aspects of Inorganic Chemistry*, Wiley, New York, 1977.

(g) R. J. P. Williams and J. R. R. Frausto da Silva (eds.), *New Trends in Bio-inorganic Chemistry*, Academic Press, London, 1978.

(h) P. Harrison (ed.), *Metalloproteins*, Parts 1 and 2, Verlag Chemie, Weinheim, 1985.

(i) A. V. Xavier (ed.), *Frontiers in Bioinorganic Chemistry*, VCH Verlagsgesellschaft mbH, Weinheim, 1986.

(j) S. Otsuka and T. Yamanaka (eds.), *Metalloproteins; Chemical Properties and Biological Effects*, Elsevier, Amsterdam, 1988.

(k) R. Dessy, J. Dillard and L. Taylor (eds.), *Bioinorganic Chemistry*, ACS Symposium Series 100, 1971.

(l) K. N. Raymond (ed.), *Bioinorganic Chemistry*, Vol. II, ACS Symposium Series 162, 1977.

(m) A. E. Martell (ed.), *Inorganic Chemistry in Biology and Medicine*, ACS Symposium Series 140, 1980.

(n) L. Que, Jr (ed.), *Metal Clusters in Proteins*, ACS Symposium Series 372, 1988.

(o) K. Burger (ed.), *Biocoordination Chemistry*, Ellis Horwood, Chichester, 1990.

(p) T. J. Beveridge and R. J. Doyle (eds.), *Metal Ions and Bacteria*, Wiley, New York, 1989.

(q) S. L. Lippard (ed.), *Prog. Inorg. Chem.*, **38**, 1–516 (1990).

(r) A. G. Sykes (ed.), *Adv. Inorg. Chem.*, **36**, 1–486 (1991).

(s) J. Reedijk (ed.), *Bioinorganic Catalysis*, Marcel Dekker, New York, 1993.

(t) *Chem. Rev.*, **94**, 567–856 (1994) (*Metal-Dioxygen Complexes*).

D. The references in the individual chapters were selected according to the following criteria:

(a) accessibility in libraries, i.e. articles from established periodicals were preferred over less available texts or conference reports,

(b) topicality, i.e. most articles originate from the years 1980–1993,

(c) availability of useful further references to detailed research articles or earlier reviews.

Review articles or books bearing on several aspects are usually cited at the beginning of the chapters; the sequence of citations corresponds to the individual references in the main text.

Glossary

(Short definitions of some terms that are not further explained in the text; see also index and list of inserts.)

allosteric effect: the modulation of active sites in a protein through interactions, e.g. the binding of molecules, at distant locations of the protein

ambidentate: denotes the ability of a ligand to bind via different coordination sites

anisotropic: directionally dependent

antibonding orbital: occupation by electrons leads to a weakening of the involved chemical bonds

apoprotein, apoenzyme: the major though inactive portion of a protein which is responsible for the selectivity of the holoenzyme (3.2) but requires prosthetic groups, a metal ion or a coenzyme, for activity

assimilatory: refers to the energy-consuming uptake of matter which is then used in metabolic processes

autolysis: self-degradation, e.g. of a hydrolyzing enzyme (autoproteolysis)

autotrophic: independent of 'organic' material with respect to the production of matter and energy

catecholates: the dianions of 1,2-dihydroxybenzene (catechol) and derivatives

cGMP (cyclic guanosine monophosphate), cAMP, etc.: cyclic nucleotides formed through ring closure between the phosphate and the 3'-OH group

charge transfer: transfer of charge between different parts of a molecule, especially after excitation with light

chelate complex: coordination compound in which a metal center and a polydentate ligand form at least *one* ring system

chiral: refers to a three-dimensional object which can occur in two nonsuperimposable mirror-image forms (enantiomers)

chromophor: the specific part of a larger molecule which is mainly responsible for long-wavelength light absorption

circular dichroism (CD): different absorption by optically active substances of the two (right and left) components of linearly polarized light

cluster: aggregation of several metal centers in close proximity

coenzyme: relatively small part of the holoenzyme which often may be reversibly separated from the apoenzyme but is essential for the type of catalyzed reaction

condensation reaction: linkage of two or more chemical compounds under release of small molecules like H_2O

dehydrogenases: oxidoreductase enzymes with coupled electron/proton ('hydrogen') transfer (see insert 'Oxidation')

diamagnetism: magnetic behavior which results exclusively from the polarization of electrons (even paired ones)

dielectric constant: parameter which describes the tendency of a medium to reduce the forces acting between charged particles

diffusion controlled reaction: when each collision between reactants leads to a successful reaction the frequency of collisions between freely diffusing particles remains the only factor limiting the reaction rate

dissimilatory: refers to the energy-producing breakdown of matter by organisms

doublet state ($S = 1/2$): the configuration with *one* unpaired electron in a molecule containing an odd total number of electrons

'early' transition metals: metals of the groups 3 to 6 in the periodic system (groups 8 to 11 are 'late' transition metals)

electrophile: particle with a high reactivity toward electron-rich reaction partners

enantiomer: *one* of the mirror-image isomers of a chiral compound. Differentiation according to the absolute configuration (R or S); however, for amino acids and carbohydrates the L/D designation is still being used

endergonic: energy-consuming (positive ΔG)

end-on: coordination of a ligand through a terminal donor center (compare 5.6)

entatic state: high-energy 'strained' state of an enzymatic catalyst which is complementary to the transition state geometry of the substrate

entropy: thermodynamic quantity describing the degree of randomness or disorder in a system

enzymes: 'biocatalysts' whose categorization depends on the type of chemical reaction catalyzed. The most common and best characterized enzymes are *oxidoreductases*, catalyzing redox reactions, group-transferring *transferases* ($\rightarrow$substitution reactions) and *hydrolases*.

epitaxial: constructed in defined layers which are oriented according to a crystalline template

extinction coefficient: quantifies the extent of absorbed electromagnetic radiation

ferredoxins: group of electron-transferring enzymes with Fe/S centers (Sections 7.1 to 7.4)

'forbidden' process: reaction or physical process (transition between states) with a low probability due to not-fulfilled quantum mechanical selection rules, non-compatible symmetry or an indispensable spin conversion (opposite: 'allowed' process)

η^n: denotes the 'hapticity' of a ligand, i.e. the number n of atoms directly coordinating to a metal center. If several metal centers are bound, one such prefix is used for each metal center ($\eta^n : \eta^m \ldots$)

'hard': little polarizable (see insert 'Hard' and 'soft' coordination centers)

Hund's rule: two or more orbitals of the same energy ('degenerate' orbitals) are occupied by electrons in such a way that the state with the lowest energy features the maximum total spin quantum number

hydrogen bond: binding of the two-coordinate proton to two small donor atoms (O, N, F, anionic S) with the effect of inter- or intramolecular linking and structuring

hydroxylases: see oxygenases

I: total spin quantum number for an atomic nucleus (isotope)

inner-sphere reaction: process in which both reaction partners partly 'penetrate' each other in the transition state, leading to a transient increase of coordination numbers, e.g. through common bridging atoms (opposite: outer-sphere reaction)

intercalation: deposition of molecules into the free space of a layered structure such as the base stacks of DNA

ISC (inter-system crossing): change between states of different multiplicity, e.g. from singlet to triplet

isotropic: directionally independent (spherical symmetry)

kinetic: refers to the *rate* of a reaction (time dependence; opposite: thermodynamic)

kink: structural irregularity in an otherwise regularly arranged polymer

lability: low stability with regard to dissociation or substitution due to low activation energy

Lewis acid: electron-pair acceptor

Lewis base: electron-pair donor

lysosomes: membrane-bound compartments in the cytoplasm which serve in controlled degradation and recycling processes

metastability: 'stability' due to low reactivity (high activation energy) but not due to low total energy

met form: the form of an oxygen-utilizing protein in which the metal is oxidized but not oxygenated

molecular orbital: quantum mechanical one-electron wave function which describes the distribution of a certain electron within the molecule

morphological: refers to the macroscopic shape (structure, form)

μ_n: denotes a ligand bridging n metal centers

nucleophile: particle with a high reactivity toward electron-poor reaction partners

nucleoside: combination of nucleobase and pentose carbohydrate (2.11)

nucleotide: combination of nucleobase, carbohydrate and (oligo)phosphate (2.11)

oxidases: oxidoreductase enzymes which primarily catalyze the electron transfer from the substrate to an electron acceptor such as dioxygen (see insert 'Oxidation')

oxy form: O_2-carrying form of an oxygen-transporting protein (opposite: deoxy form)

oxygenases (hydroxylases): oxidoreductase enzymes in which electron transfer is coupled with oxygen transfer (see insert 'Oxidation')

paramagnetism: magnetic property which results from the orientation of electrons in an external field if the total (spin and orbital) angular momentum is not zero

primary structure of a protein: 'linear' sequence of amino acids in the polypeptide chain

pseudorotation: intramolecular structural rearrangement between isomeric trigonal–bipyramidal conformations via a square-pyramidal transition state involving some nondissociative ligand exchange (axial/equatorial)

quarternary structure: three-dimensional arrangement of different peptide chains (protein subunits) in an oligomeric protein

Raman effect: observation of molecular vibrational frequencies as energy differences in scattered electromagnetic radiation (see insert 'Resonance Raman spectroscopy')

S: total spin quantum number for electrons in a multielectron system

second messenger: information mediator, e.g. between a hormonal signal (primary activation) and the cellular reaction

secondary structure of a protein: local conformation of the peptide chain, described e.g. as α-helical arrangement

semiquinone: one-electron reduction product (radical) of a quinonoid compound

side-on: coordination of a ligand via one or several bonds (see 5.6)

single crystal: crystal with macroscopically uniform orientation of the unit cells (periodicity, long-range order; see insert Structure determination by X-ray diffraction)

singlet state ($S = 0$): the state with no unpaired electrons (complete spin pairing) in compounds with an even total number of electrons

'soft': easily polarizable (see insert 'Hard' and 'soft' coordination centers)

spin crossover: change of the total spin quantum number, e.g. the transition from a high-spin to a low-spin configuration, as induced by external factors

spin–orbit coupling: interaction of spin and orbital angular momentum of the electron. The corresponding interaction constant increases strongly with increasing atomic number of the elements

SQUID (superconducting quantum interference device): very sensitive equipment for measuring magnetic susceptibilities

superexchange: spin–spin interaction between not directly linked centers through a connecting atomic or molecular bridge (see insert Spin–spin coupling)

susceptibility: material-specific proportionality constant correlating magnetization and the strength of the external magnetic field (see insert Spin–spin coupling)

tautomers: rapidly equilibrating isomers which are distinguished, for example, by different positions of hydrogen atoms

template effect: defined organization of an assembly of initially separated ligand components through coordination and specific interligand bond formation at a template metal center

tertiary structure of a protein: three-dimensional shape of a naturally folded single polypeptide chain (monomeric protein)

tetrahedral splitting: at comparable ligand field 'strength' the splitting of the d orbitals in tetrahedral symmetry (two stabilized, three destabilized orbitals) is only $4/9 = 44\%$ of the octahedral splitting (three stabilized, two destabilized orbitals; see also 12.4)

thermodynamic: refers to the equilibrium aspect of a reaction (time-independent view; opposite: kinetic)

three-point adhesion: requirement for unambiguous orientation in space and for stereoselectivity

thylakoid: lamellar 'disk' formed by pigment-containing internal membranes in chloroplasts

transcription: high-fidelity formation of a DNA-complementary messenger-RNA

triplet state ($S = 1$): state exhibiting two unpaired electrons with parallel spin orientation (even total number of electrons)

vesicle: small globular object, filled with fluid and separated by a membrane hull

List of Inserts

'Hard' and 'Soft' Coordination Centers   14
The 'Entatic State' in Enzymatic Catalysis   22
Electron Spin States in Transition Metal Ions   29
Electron Paramagnetic Resonance I   44
Organic Redox Coenzymes   50
Structure Determination by X-ray Diffraction   65
X-ray Absorption Spectroscopy   72
Spin–Spin Coupling   75
Resonance Raman Spectroscopy 101
Mössbauer Spectroscopy 102
Optical Isomerism in Octahedral Complexes 155
Electron Paramagnetic Resonance II 192
'Oxidation' 218
Heteroatom Nuclear Magnetic Resonance (NMR) 271

Index

A

α-lactalbumin 295
α-oxidation 119
α radiation 351–353, 355, 359
a₃ complex 207
acetaldehyde 114, 258, 259
acetic acid 178
acetyl phosphate 325
acetyl-CoA 52
acetyl-CoA synthase 172, 178
acetyl-coenzyme A 178
acetylene 227, 233
acid rain 343
aconitase 128, 136, 138, 293
acrylamide 146
acrylonitrile 146
actin 300
activated acetic acid 179
activation energy 22, 23
adenine 33, 34, 369
ADP (adenosine diphosphate) 107, 287, 300
adrenalin 199
adriamycin 364
aequorin 294
aerobactin 152, 153, 157
aging 212, 327
agrobactin 157, 159
alcohol 218, 242, 257, 259
alcohol dehydrogenase (ADH) 243, 245, 257, 258
aldehydes 259
aldehyde oxidase 129, 216, 218, 259
aldol condensation 257
aldolase 257
algae 235, 236
alkaline phosphatase 243, 288
alkyl hydroperoxide 327
allosteric effect 381
allosteric regulation 96, 142
aluminosilicates 319, 343–345
Alzheimer syndrome 255, 343, 345
amalgam fillings 339
amavadin 236–237
ambidentate 33, 381
amine oxidases 187, 191, 202, 204
amine oxides 218

amino acids 17–20, 113, 144
amino acid homology 175
amino acid sequences 136, 141, 164, 260, 337
5-aminolevulinic acid (ALA) 335
5-aminolevulinic acid dehydratase (ALAD) 243, 257, 335
aminopeptidases 254, 255
ammine 371
ammonia 173, 205, 224, 225, 229, 231, 313
ammonium (NH_4^+) 230, 231, 345
amorphous silica 304, 306, 309, 313, 314, 318
amphoteric system 248
amylotrophic lateral sclerosis (ALS) 190, 212
anemia 12, 40, 151, 203, 335, 345
angina pectoris 123
angiotensin converting enzyme (ACE) 251
anhydride 253
anisotropic 381
anisotropic EPR spectra 192
annexins 297
antabuse 259
antacids 346
antagonism 10, 11, 190, 219, 286
antenna pigments 28, 61
anti 19
anti-Parkinson drug 198
antibiotics 32, 151, 275, 277, 373
antibodies 172
antibonding 42
antibonding orbital 381
antiferromagnetic coupling 75, 76, 93, 104, 117, 133, 189, 196
antioxidants 82, 187, 190, 203, 211, 323, 325, 326
antiport 281, 282
antirheumatic drugs 375
antiviral activity 366
apatite 307, 310
apoenzyme 39, 40, 245
apoferritin 168
apoprotein 42, 381
aqua regia 364
aquacobalamin 41
aragonite 303, 304, 306, 313

archaebacteria   2, 24, 180, 216
arginine   123, 210, 252, 255
arsenate   218, 319
arsenic   3, 319
arsenite   218
arsenite oxidase   216
arsenobetaine   319
arthritis   375
ascorbate   17, 161, 190, 326
ascorbate oxidase   190, 202–204
ascorbic acid   50
aspartate   19, 142, 164, 166, 168, 209, 254,
        291, 297
aspartate transcarbamoylase   258
aspartic acid   18
assimilation   178, 224, 381
associative process   290
astacin   254
atmosphere   82
Atmungsferment   206
ATP (adenosin triphosphate)   57, 107,
        226, 228, 229, 234, 267, 271, 281–284,
        287–290, 300, 345
ATP synthesis   66
ATPase enzymes   188, 281, 283, 290, 294,
        319, 322, 345
auranofin   374, 375
autoimmune disease   212
autoimmune reaction   375
autolysis   381
autoproteolysis   253–256
autoprotolysis   249
autotrophic   381
autoxidation   327
autoxygenation   118
azide   211
azurin   190, 192, 194

B

β-carotene   326
β-lactams   144
β-methylaspartic acid   46
β radiation   351–355, 358–360
b/f-complex 69, 133
back electron transfer   67
back-donation   88
back-scattering   72
bacterial leaching   131
bacteriochlorophyll   58–60, 65, 64–66
bacterioferritin   161, 166
bacteriopheophytin   64, 66
bactopterins   220
badion A   354
BAL   13, 334
bananatrode   198
banded iron formations   82

barbiturates   113
baryte   304, 315, 316
base-off   43
bc_1 complex   108, 133
benzene   114
benzo[a]pyrene   114, 115
berylliosis   346
bicarbonate   98, 249–251, 313
bifunctional catalysis   248
bioalkylation   40
bioavailability   6, 8, 10, 83, 188, 215, 217,
        235, 293, 295, 330, 344, 345
bioceramics   311
biocides   237
biocompatible   311
biocomposites   307
biomethylation   52, 319, 323, 338, 339, 342
biomimetic   303
biomineral   267, 303–309, 312–314, 353
biomineralization   15, 303, 307, 314, 315
bioorganic chemistry   1
biopolymer   370
biosensor   198
biosynthesis   26, 52, 53, 152, 160, 162,
        226, 257, 335
biotechnological process   180
biotin   218
biotin-5-oxide   218
bioweathering   131
bis(salicylaldehyde)ethylenediimine   53
bleomycin   364, 373
blood   375
blood clotting   256, 293
blood coagulation   300
blood plasma   268, 270, 345
blood serum   151
blood transfusion   154
blood-brain barrier   52, 335, 340, 361
'blue' copper centers   187, 189, 192–194,
        196
'blue' oxidases   190, 202
Bohr effect   98
bone deformations   312
bone formation   318
bone marrow   161, 165, 353
bone tissue   335, 353–356, 361
bone-seeker   355
bones   303, 304, 308, 309, 311
borates   318
boromycin   318
boron neutron capture therapy
        (BNCT)   359
botulinum   256
bovine spleen phosphatase   144
breast cancer   324
bromides   320
bromoperoxidase   237

budotitane 372, 373
buffering 246

C

Ca^{2+}-ATPases 297, 312
Ca^{2+} pump 283, 294
Ca^{2+} channel 294
cadmium poisoning 335, 355
cage reaction 121
calciferol 113
calcimycin 294, 295
calcite 304, 306, 307, 309, 313
calcium antagonists 294
calcium phosphates 309–311
calmodulin 123, 297, 300
calsequestrin 296
Calvin cycle 66, 68
cAMP 297
camphor 115
cancer 323, 347, 351, 363–367, 372
carbamate 173, 286
carbanions 42, 43, 46, 52
carboanhydrase (CA) 174, 243, 246–248, 283, 312
carbocations 42, 43
carbohydrates 17
carbon dioxide (CO$_2$) 2, 56, 57, 66, 68, 82, 107, 173, 175, 178–180, 216, 246, 248–251, 303, 307, 313, 320
carbon monoxide (CO) 95, 123, 178, 179, 226, 227
carbonate 164, 249, 294, 303, 312
carbonic acid 98, 248
carbonylation 52, 183
carboplatin 364, 365
carboxamide 253, 256
carboxylic ester 253
carboxypeptidase A (CPA) 174, 243, 251–253, 255
carcinogenesis 355, 367
carcinogenic 33, 319, 336, 347
cardiomyopathy 12
caries 12, 307, 311, 312, 320
cariostatic effect 320
carotenes 59, 63
carrier mechanism 277, 278
catalases 56, 75, 89, 90, 119, 120, 326
catalysis 16, 20, 22, 23, 56, 74, 107, 115, 123, 124, 128, 175, 196, 242, 246, 248, 254, 256, 286, 290, 292
catechol 198
catechol dioxygenase 144, 187, 202
catechol oxidase 200
catecholates 145, 152, 153, 157, 200, 205, 356, 381
cation exchangers 338
celestite 304, 306, 315

central nervous system 164
ceruloplasmin 188, 190, 202–204, 326
cGMP (cyclic guanosine monophosphate) 381
channel pore 278
charge recombination 67
charge separation 63, 64, 67, 68, 71, 73
charge transfer 101, 381
chelate 23, 26, 27, 31–33, 40, 53, 138, 139, 152, 153, 188, 200, 220, 236, 268, 270, 273, 291, 296, 319, 340, 345, 351, 357, 360, 369, 372, 381
chelate drugs 332
chelate therapy 12, 338, 356
chemical nucleases 373
chemical shift 271
chemical speciation 8
chemiosmotic effect 107, 267
chemoautotrophic metabolism 130
chemolithotrophic metabolism 130, 216
chemotherapy 34, 363
Chernobyl 351, 353–355
chirality 155, 381
chitin 303
chiton molluscs 308, 309, 315
chloride 71, 73, 79, 119, 271, 283, 320, 330, 340, 367, 371, 372
chlorin 24
chlorinated hydrocarbons 114
chlorophyll 24, 57, 59–61, 63, 286
chloroplasts 57, 133
chlorpromazine 113
cholera 151
chromatography 59
chromophor 61, 101, 381
chromosome 188
circular dichroism (CD) 155, 381
cisplatin (cis-diamminedichloroplatinum(II)) 2, 4, 34, 363–367, 371, 372
citrate 137, 165, 309, 312, 345, 357
citric acid 157
clay minerals 344
cluster 131, 134, 135, 217, 336, 337, 381
CO dehydrogenase 172, 178, 179, 183
CO oxidoreductase 178
CO$_2$ disposal 283
CO$_2$ fixation 286
cobalamin 24, 338
cobaloxime 53, 54
cobester 53
coboglobin 93
codein 113
coenzymes 17, 39, 40, 189, 205, 381
coenzyme A 47
coenzyme B12 39, 40, 141
coenzyme M 53, 180, 181

collagen 188, 205, 303, 309, 310, 361, 375
collagenases 219, 242, 254
colon cancer 324
complementary color 59
composite materials 303, 305, 307
concentration gradient 16, 268, 269, 281, 284, 294
condensation 242, 244, 256, 381
conformation 66, 245
connective tissue 205, 256
cooperative effect 91, 96, 97, 104, 105, 198
coordination geometry 20
coordination polymerization 98
coprogen 157, 158
corals reefs 313
coronary diseases 151
correlation diagram 118
corrin 24, 40, 41
corrinoid 52, 178
cosmic radiation 352
Costa complex 53, 54
creatine phosphate 287
cross-linking 191, 205, 327, 347, 368–371
cross-links 370
crown ether 32, 274, 353
crude oil 26, 27, 235
cryptands 274
cryptates 32, 274
crystal growth 66, 308
Cu_A 206, 207
Cu_B 207
cupellation 332
Curie law 76
cyanide 208, 211, 364
cyanobacteria 68, 176
cyanocobalamin 41
cyclases 236, 297, 345
cyclic guanosine monophosphate (cGMP) 123, 280, 297, 299
cyclophosphamide 364
cysteamine 351
cysteinate 115, 133–136, 192–195, 205, 246, 255, 258, 262, 336, 341
cysteine 18, 19, 109, 146, 175, 177, 204, 236, 261, 264, 325, 336, 337, 340, 342, 351, 365
cystic fibrosis 283
cytochrome 67, 90, 107, 110, 112, 122, 187, 217, 219, 220, 221
cytochrome a 207
cytochrome a_3 207
cytochrome c 89, 108–111, 119
cytochrome c oxidase 89, 90, 108, 191, 192, 205–208, 326
cytochrome c peroxidase (CCP) 112, 119, 120
cytosine 33, 34, 369

cytostatic drugs 371
cytostatic effect 364

D

d orbitals 29, 30
dark current 280
dark reaction 66
decalcification 246
decarbonylation 53
decay series 335, 352, 359
decontamination 131, 356
deferrioxamine 13, 151–154, 345
deficiency symptoms 11, 12
dehydratases 48, 293
dehydration 291
dehydrogenases 218, 258, 325, 381
dehydrolase 128
deiodinase 323, 325
delocalization 133–136, 189, 195, 196, 205, 207, 229
demineralization 306, 311, 312
denitrification 123, 205, 224, 225, 226
dental caries 307, 311, 312, 320
deoxyhemerythrin 103
deoxyhemocyanin 197
2'-deoxyribose 141
dephosphorylation 282, 288, 300
depolarization 299
depot poison 338
desaminase 48
desferal 13
desulfurization 183
detergent molecules 65
deterrent 314
detoxification 4, 82, 98, 112, 114, 119, 122, 211, 263, 306, 319, 331, 332, 336, 337, 340, 354
diabetes 12, 238
diagnostics 237, 359
diagonal relationship 79, 232, 235, 346, 376
dialkylglycine decarboxylase 269
diamagnetism 381
diatomaceous earth 314
diatoms 308
diazene 230, 231
dibenzo-1,4-dioxines 114
dicalcium phosphate dihydrate 311
dielectric constant 382
diethylenetriamine pentaacetate 357, 358
diethylenetriamineplatinum(II) chloride 369
differential toxicity 365
diffusion control 210, 246, 268, 269, 327, 382
digitoxigenin 282

dihydrogen 57, 174, 175, 178, 227, 228
dihydropterin 221
dimercaprol 13, 319, 340, 375
2,3-dimercapto-1-propanol 334
dimercaptosuccinic acid 340
dimethylmercury 342
dimethylselenium 323
dimethylsulfoxide (DMSO) 218, 223
dinitrogen 175, 225, 226, 229, 230
dinitrogen fixation 175, 230
dinitrogenase 226, 229, 234
dinitrogenase reductase 228, 229, 234
dinosaurs 172
diol dehydratase 46
dioxines 114, 122
dioxygen 16, 68–73, 76, 78, 82–86, 93,
    98–101, 104, 105, 116, 119, 146, 151,
    187, 191, 196, 197, 206, 207
dioxygenases 144, 187, 191
diphosphonate 361
dismutation 209
disorders 188, 190
disproportionation 119, 191, 209
dissimilation 382
dissipative system 6
distal histidine 92, 95
disulfide 141, 181, 183
disulfide bridges 337, 340
disulfiram 259
dithiol 141, 143
dithiolene 220, 221
diuresis 364
DNA 15, 33, 34, 257, 260, 261, 264, 267,
    286, 347, 365–371
  biosynthesis 141
  cleavage 373
  polymerase 257
  repair 128, 139, 262, 263, 347, 372
doming 26
dopa 198, 199, 345
dopamine 191, 199
dopamine β-monooxygenase 191, 198,
    199, 202
dopaquinone 200
dose-response diagrams 9, 10, 330
double helix structure 370
doublet state 382
Down syndrome 345
doxorubicin 364, 366
drug design 251
DTPA 357, 361

E

early transition metals 215, 382
EDTA 13, 334, 357, 361
EF-hand 297–299

effective concentration 22
egg shells 308
electrolytes 267, 270, 271
Electron Nuclear Double Resonance
    (ENDOR) 45
Electron Paramagnetic Resonance
    (EPR) 44
electron reservoir 78, 175
Electron Spin Resonance (ESR) 44
electron transfer 46, 67, 105, 107,
    109–112, 128, 133–135, 144, 169, 183,
    187–190, 196, 205, 207, 218, 219, 223
electrophile 382
elemental composition 7, 8
elimination reaction 291
emission 67
enamel prisms 312
enantiomeric 155, 382
enantiomorphous complexes 34
encephalopathy 345
endergonic 382
end-on 85–88, 101, 105, 226, 229, 382
endocytosis 165
endonuclease III 128, 136, 139
endopeptidase 254, 255
ENDOR (Electron Nuclear DOuble
    Resonance) 45, 66, 136, 146, 177,
    189, 194, 195, 229
endosome 164
ene-1,2-dithiolate 220
energy profiles 22
energy transfer 59, 61
enolases 291, 292, 320
entatic state 20, 22, 23, 36, 111, 135, 196,
    245, 268, 280, 293, 300, 382
enterobactin 152, 157–160
entropy 6, 382
enzymatic browning 198
enzymes 16, 20, 56, 382
enzyme/substrate complex 23
ephedrine 113
epinephrine 199, 345
epitaxial 382
epoxidation 114
EPR (Electron Paramagnetic
    Resonance) 42, 44, 45, 66, 71–74, 77,
    78, 86, 93, 133–136, 172, 176–179, 189,
    192, 195, 196, 205, 207, 219, 222, 223,
    235, 272, 291, 347
equilibrium 6, 85
erythrocytes 91, 95, 98, 161, 162, 191,
    209, 210, 246, 268, 283, 325, 353, 361
escape 121
essential 7, 9, 12, 242, 359
esterases 252, 345
ethane 233
ethanol 114, 257–259, 340

ethanolamine ammonia lyase   46
ethylene   198, 227, 233
eukaryotes   180
evolution   128, 130, 188, 233
EXAFS (Extended X-ray Absorption Fine
    Structure)   72–74, 146, 166, 173, 177,
    179, 217, 219, 220, 222, 228, 231, 234,
    237, 257, 336
excited state   59
exciton   59, 61
exhaust catalysts   369
extinction coefficient 382
extraction   129

F

F430   24, 53, 173, 182, 183
fac   155, 156
[2Fe-2S]   133, 140
[3Fe-4S]   136–138, 140, 177
[4Fe-4S]   134–136, 138–139, 177, 229
Faraday balance   76
fatty acids   113, 119, 144, 146, 327
Fe/S centers   128, 131, 132, 139, 217
Fe/S clusters   124, 134–135, 172, 175
Fe/S proteins   108, 140
feedback   293, 296–297, 300, 367
feldspar   344
FeMo-cofactor   217, 228
FeMo-protein   228
Fenton's reagent   49
fermentation   175
ferredoxins   63, 67, 128, 130, 133–136, 382
ferrichromes 152–154
ferrihydrite   166, 304, 306, 309, 315
ferrioxamines   151, 153–156
ferritin   89, 109, 161–162, 165–169, 303
ferrochelatase   161, 335
ferromagnetic coupling   75
ferroxidase   168, 188, 203
fertilizer   225
fidelity   368
first messenger   293
flavin adenine dinucleotide (FAD)   51
flavin mononucleotide (FMN)   51
flavins   17, 50, 112, 128, 217, 342
flavoenzymes   107
flavoproteins   124
flow equilibrium   6, 281, 331
fluorescence   294
fluorescence indicators   272
fluoridation   320
fluoroapatite   304, 312, 320
fluorosis   320, 321
folic acid   50
food cans   334
'forbidden' process   382
forest damage   71

formate   218, 250
formate dehydrogenase   216
formylmethanofuran dehydrogenase   218
Förster mechanism   61
fossils   303
fuel additives   333
fumarate reductase   131
fungicides   338, 339
FUR   152

G

γ radiation   102, 351–360
g factor   44, 193
G-protein   376
GAL4   243, 262
galactose oxidase   191, 202, 204
gastric fluid   244, 270, 367, 375
gastritis   363
gastroenteritis   338
gates   280
gelatinase   254
geobiotechnology   130
geochemical biomineralization   131
glaucoma   323
global cycles   331, 333, 339, 342
glucose   57, 235
glucose tolerance factor   238
glutamate   19, 142, 143, 166, 168, 246,
    251–253, 291, 297
glutamate mutase   46
glutamic acid   18, 46, 47
glutamine   194
glutaric acid   226
glutathione   238, 260, 325, 328, 347
glutathione peroxidase   325–328
glycerol dehydratase   46
glycerol(1,2)-diol dehydrogenase   258
glycine   325
glycine reductase   325
glycogen   297, 300
glycoproteins   163, 282, 309
glyoxalase   243, 260
goiter   322
gold   131, 326, 373
gout   219
gramicidin   275–279
gravity sensors   304, 309, 313
greenhouse effect   180, 246, 303
greigite   315
growth hormone   242
guanine   33, 34, 369–371
guanine nucleotide-binding protein
    376
guanosine   369, 370
guanosine monophosphate   123
guanylate cyclase   123
gypsum   304

H

H cluster 139
H$^+$ diffusion 311
H$^+$/K$^+$-ATPase 283
H cluster 139, 175
H/D exchange 177
H$_2$ases 174
H$_2$O$_2$ 119, 120
hair 335
haloperoxidases 236, 237, 320
hangover syndrome 259
'hard' 14, 17, 382
HCO$_3^-$/Cl$^-$ antiport system 283
heart diagnosis 358
heavy metals 330
heavy metal
  poisoning 312, 337, 339
  pollution 341
  stress 341
heme 24, 64, 88, 90, 104, 107 -110, 112,
    128, 151, 187
heme d 107
heme d_1 124
heme dioxygenase 118
heme oxygenase 98, 123
heme peroxidase 119, 120
hemerythrin 88, 100, 103, 141, 187
hemochromatosis 151
hemocyanin 88, 100, 104, 187, 191, 196,
    197, 198
hemodialysis 345
hemoglobin 88–98, 109, 122, 162, 187
hemolymph 191
hemoproteins 105, 107, 109, 122, 123
hemosiderin 89, 161, 165, 169
herbicides 71
heteroaromaticity 26
heterocycles 218
Hg(II) reductase 341, 342
high potential iron sulfur protein
    (HiPIP) 130, 135, 136
high-mobility group (HMG-1) protein 372
high-spin 30, 31
histamine 113
histidine 18, 66, 73, 96, 104, 109, 120, 134,
    142–146, 164, 166, 187, 194, 204–210,
    228, 244–251, 255, 258, 261, 263
holoenzyme 40
homeostasis 8, 150, 336, 337
homocitrate 228
homocysteine 52
homolysis 42, 43, 48
hormones 17, 113, 198, 251, 262
hormone/receptor complexes 242, 262
horseradish peroxidase (HRP) 109,
    119–122
human growth hormone (hGH) 262

human immunodeficiency virus
    (HIV) 256, 262
Hund's rule 30, 75, 84, 382
hydrazine 228–231
hydride 258–260
hydrocorphin 182, 183
hydrogen bond 31, 65, 95, 136, 141, 159,
    164, 210, 247, 249, 251, 253, 365,
    368–371, 382
hydrogen carbonate (HCO$_3^-$) 249, 250
hydrogen ion conductivity 312
hydrogen peroxide (H$_2$O$_2$) 69, 83, 119,
    120, 209
hydrogen sulfide 57
hydrogen sulfite 250
hydrogenases 128, 129, 139, 141, 172,
    174–178, 181, 183, 227, 228, 325
hydrogenation 227
hydrolases 23, 251, 256
hydrolysis 173, 242–249, 252–254,
    286–288, 367
hydroperoxide 83, 204
hydroperoxo ligand 104
hydroporphyrins 24
hydroquinones 66
hydrothermal vents 216
hydroxamates 152, 153, 156, 157
hydroxocobalamin 41
hydroxyapatite 303–306, 309–312, 320,
    361
6-hydroxydopa quinone 204
hydroxyl radical ($\cdot$OH) 69, 83, 326, 337,
    347, 351
hydroxylamine 123, 205
hydroxylases 112, 143, 198, 217, 218, 382
hyperfine coupling 74, 172, 189, 192, 207
hyperphosphataemia 345
hyperpolarization 280
hyperthermophilic 130, 216
hypobromite (—OBr) 237
hypochlorite (—OCl) 119

I

imaging 354, 357, 359
imidazolate 209
imidazole 18, 95, 187
immobilization 99
immune system 163, 212, 242, 346
immunoscintigraphy 357
immunotoxic reaction 114
in-plane 26
induced fit 253
industrial waste 131
inertia sensors 316
infectious diseases 151, 163
information transfer 16, 267, 268
infusion 367

inner-sphere electron transfer   66, 382
inositol-1,4,5-triphosphate   299
insulin   196, 235, 238, 242, 243, 262, 263
intercalation 383
interdependence   10
intermediate-spin   31
interstrand   369
intrastrand   369
inter-system crossing (ISC)   63, 375, 383
invertebrates   88, 100
inverted region   67
ion channels   16, 269, 277, 279
ion exchange   311, 354, 360
ion pumps   16, 281, 307, 312, 331
ionizing radiation   212, 351
ionophores   31, 273–276, 294, 354
iproplatin   364, 365
iron deficiency   151
iron-sensor   128, 138
iron store   165
iron sulfides   130
iron-sulfur proteins   89, 112, 130
isobutyryl-CoA mutase   47
isocitrate   137
isocyanide   227
isomer shift   102
isomerase   128, 291
isomerization   50
isomorphous substitution   36, 66, 245
isopenicillin-N synthase   146
isotropic 383
isozymes   246
Itai–Itai disease   336

J

Jahn–Teller effect   189, 192, 193

K

keratin   188, 335
Keshan disease   323
ketone   259
kinases   236, 288–291, 300, 345, 346
kinetic 383
kink   370, 383
Klärschlammverordnung   332
Knallgas   175
krill   321

L

lability 383
laccase   187, 190, 202, 203
lactoferrin   162, 163, 164
late transition metals   215
lead poisoning   334, 335, 339
lead smelting   334
leprosy   151

leucemia   353, 355
leucine   166
leukotrienes   144, 146
Lewis acidity   244, 286, 344, 383
Lewis basicity   383
lichen   235
life   6
ligand field
    splitting   117
    stabilization   262
    theory   29, 31
    transition   101, 189, 192
ligand-to-metal charge transfer
    (LMCT)   101, 133, 144–146, 164, 189,
    195, 197, 219
ligases   257
light absorption   26
light-harvesting   59–62, 68, 69
lignin peroxidase   119, 122
limestone   303, 313
lipids   17
lipid hydroperoxides   325
lipid oxidation   375
lipoic acid   51
lipoxygenases   146
liver alcohol dehydrogenase (LADH)   258,
    259
low-spin   30, 31
luminescence   67
lung cancer   346
lyases   48
lysine   19
D-α-lysine mutase   47
L-β-lysine mutase   47
lysosomes   169, 375, 383

M

M cluster   228, 229
macro nutrients   6, 267
macrochelates   289
macrocycle   23, 31, 268, 270, 273–276, 357
mad hatter syndrome   339
magnesium therapy   286
magnetic coupling   78
magnetic susceptibility   76, 77, 174
magnetite   304, 306, 309, 315
magnetosomes   315
magnetotactic bacteria   306, 309, 315
malaria   98
manganese nodules   75
manic-depressive psychoses   376
many-electron process   124
marsh gas   320
mass elements   6, 267
material sciences   303
matrix   307, 309, 313
matrix metalloproteinases (MMPs)   254, 255

melanins 198, 200
membranes 15, 32, 65–69, 71, 134,
    176, 206, 207, 269, 271, 275–278, 286,
    327, 335, 347, 366
membrane barrier 368
membrane proteins 277, 278, 281
membrane receptors 151, 159
menaquinone 51
Menke's 'kinky hair' syndrome 188
mer 155, 156
mercaptans 340
2-mercaptoethylamine 351
7-mercaptoheptanoyl-O-
    phosphothreonine 180
mercury cycle 343
mercury poisoning 339
mescaline 113
met-form 103
metallocenes 372
metalloenzymes 22, 23, 321
metalloregulatory proteins 337, 341
metallothioneins 188, 192, 243, 263, 326,
    331, 336, 337, 340
metamphetamine 113
metastability 383
met form 383
methane (CH_4) 2, 143, 175, 178–180
methane monooxygenase (MMO) 143, 144,
    187
methanofuran 180, 181
methanogenic bacteria 2, 52, 172, 180
methanol 259
methanotrophic 143
methionine 18, 52, 109, 115, 166, 194, 319
methionine synthetase 52
methoxatin 51
methyl-coenzyme M 180
methyl-coenzyme M reductase 172,
    180–183
methylcobalamin 39–42, 319, 340
2-methyleneglutarate mutase 46
9-methylguanine 371
methylmalonyl-CoA mutase 47, 48
methylmercury cation 52
2-methylthioethanesulfonate 180
1-methyluracil 371
microbial degradation 180, 202
microsomes 112
milk 162, 219
Minamata Bay 338
mineralization 308, 312, 313
mispairing 33
mitochondria 56, 133, 206, 209
mitochondrial membrane 107, 108
mixed-valent 74, 133, 134, 136, 138, 144,
    168, 205, 207
Mo-dependent nitrogenase 234

model compound 3, 35, 36, 54, 183, 197
molecular modeling 251
molecular orbital 383
molecular recognition 32, 273, 373
molybdate(VI) 215
molybdoenzymes 218
molybdopterin 179, 217, 220
monensin A 276, 277
monoclonal antibodies 357, 359
monofluorophosphate 320
monooxygenases 112, 120, 144, 191, 196,
    198, 200
monooxygenation 115, 144, 202
morphine 113
morphological 383
Mössbauer spectroscopy 93, 102, 133, 134,
    136, 167, 179, 229, 272, 314
Z,Z-muconic acid 145
mucosa cells 313
mugineic acid 152, 160, 161
multiplicity 84
muscle cells 284, 294
muscle contraction 293, 299
muscle fibers 300
muscle relaxation 123
mushrooms 235, 236
mutagenesis 368
mutases 46
mutation 210
mycobactin 157, 158
myeloperoxidase 119
myochrisin 374, 375
myoglobin 88–92, 95, 97, 109, 120, 122
myosin 300
myosin ATPase 300

N

Na^+/Ca^{2+} exchange 282
Na^+/H^+ antiport 283
Na^+/K^+-ATPase 235, 281, 282, 291, 292,
    338
NAD kinase 297
NAD(P)H 66, 68, 107, 342
NAD^+/NADH system 50, 108, 112, 175,
    258, 259, 260
NADP 297
NADP oxidoreductase 129
nails 335
nanoparticles 167
natural gas 180
negative catalysis 23, 50
Nernst equation 153
nettle plants 314
neurotoxicity 259, 338
neurotransmitters 123, 144, 198, 280,
    345
neutrophils 212

nickel allergy   172
nicotianamine   160
nicotine   340
nicotinic acid   238
nif genes   232
nitrate   123, 216, 224, 225, 250
nitrate reductase   216, 218, 222, 225
nitrification   224, 225
nitrile hydratase   146
nitrite (NO_2^-)   122, 123, 205, 216, 225
nitrite reductases   123, 124, 187, 191, 205, 255
nitrogen cycle   191, 205, 224, 225
nitrogen fixation   224–226, 230, 232
nitrogenase   128, 129, 136, 139, 175, 217, 226–228, 231, 232
nitroprusside   123
nitrosamines   114
nitrosyl complexes   123, 205
nitrovasodilators   123
NMR (nuclear magnetic resonance)   66, 136, 235, 271, 272, 291, 294, 336
NO (nitric oxide, nitrogen monoxide, nitrosyl radical)   122, 123, 205, 225–227
NO synthase   123, 297
NO_2 225
Nobel prizes   26, 32, 36, 40, 57, 65, 92, 273, 279, 290, 363
non-blue copper   189
'non-blue' oxidases   191, 202
non-innocent   86
non-leaving groups   365
nonactin   275, 276
nonheme iron enzymes   144
nonheme iron proteins   128
norbadion A   354
norephedrine   191
norepinephrine   113, 198, 199
norepinephrine *N*-methyltransferase   199
nuclear fission   352, 353
nuclear fuel   355
nucleases   256, 286, 296, 368
nucleation   307, 308, 312
nucleic acids   33, 368
nucleobases   33, 341, 369
nucleophile   383
nucleophilic substitution   290
nucleoside   383
nucleoside triphosphates   288
nucleotides   33, 141, 280, 296, 299, 368, 383
nutrients   1, 10, 11

O

octacalcium phosphate (OCP)   311
octopus   198
oesophageal cancer   314

opsin   280
optical isomerism   155
orbital degeneracy   192
ores   131
organomercurials   341
organomercury cycle   343
organomercury lyase   341, 342
organometallics   16, 40, 41, 52
organometallic cations   335, 339
ornithine mutase   47
ortho-hydroxylation   198
osteoblasts   311, 312, 346
osteoclasts   312
osteocytes   312
osteoid   312
osteoporosis   307, 336
Ostwald rule   311
otoconia   313
otoliths   313
ouabain   282
out-of-plane   26, 29, 54, 96, 116, 120
ovotransferrin   163
oxene   118
oxidases   187, 217, 383
oxidation   218
oxidation states   93, 94
oxidative addition   42, 182
oxidative leaching   364
oxidative stress   325, 351
oxides   117, 118, 131
oxidoreductases   23, 218
oxoferryl(IV)   117, 120, 207
oxometalates   215
oxotransferase activity   217
oxy anion pump   319
oxy form   383
oxygen transfer   112, 215, 216, 217, 218, 222
oxygen-evolving complex (OEC)   67, 69–71, 76, 78
oxygenases   187, 218, 383
oxygenation   218
oxyhemerythrin   103
oxyhemocyanin   197
ozone   84, 225

P

π back-bonding   16, 111, 226
π back-donation   86, 123
π conjugation   26
π/π interaction   204, 289
P cluster   139, 228–229
P-450   89, 90, 109–123, 143, 144, 187, 198, 218
parabactin   157, 159
paramagnetism   383
paramagnetic shift   272
parathyroid hormone   312

parvalbumins 296, 297
'patch-clamp' method 280
pearls 303
penicillamine 13, 188, 375
peptidases 252–255
peptide bonds 296
peptides 183–184, 253
periodic table 4
peroxidases 56, 75, 89, 90, 117–122, 144,
   187, 326
peroxides 83–86, 101, 104, 119, 151, 197,
   200, 212, 218, 327
phagocytes 212, 375
pharmaceuticals 113, 363
pharmacology 3, 113
phase transformations 309
phenacetin 113
1,10-phenanthroline 373
phenobarbital 113
phenolate 19
phenols 119–122, 191, 198
phenothiazine 113
phenylalanine 144, 198, 199, 251
phenylalanine 4-monooxygenase 199, 202
pheophytin 63
phosphane (PH_3) 319, 361
phosphatases 144, 236, 256, 288, 290, 311,
   345, 346
phosphate 17, 144, 161, 166, 167, 289,
   294, 300, 309–310, 312, 319, 344–346,
   361, 368, 371, 376
   esters 144
   transfer 286, 288, 290, 344
phosphoenolpyruvate 291, 292
2-phosphoglycerate 292
phospholipid double layer 275
phospholipid membranes 57
phosphorus cycle 320
phosphoryl group 141
phosphoryl transfer 256, 368
phosphorylase kinase 297
phosphorylation 71, 107, 108, 175, 206,
   207, 282, 288, 346
photosynthesis 56, 64, 67, 68, 69, 82, 178,
   187, 190, 246, 281, 307, 313
photosynthetic membrane 60
photosystems 63, 69
photosystem I 63, 68–70
photosystem II 63, 68–73
phycobilin 59
phycocyanin 59
phycoerythrin 59
phytochelatins 336, 337
phytoferritin 161
phytolithes 314
phytoplankton 57
phytosiderophores 160

picket fence porphyrins 99
pigments 58, 333–336
pioneer plants 226
placer gold 306
plankton 315, 321
plant hormone 198
plastocyanin 63, 69, 190, 192, 194, 204
plastoquinone 51, 63, 69, 71
^{31}P NMR spectroscopy 287, 309
polarized light 155
pollution 339
polychlorinated biphenyls (PCBs) 114
polycondensation 168
polyelectrolytes 368
polymerases 98, 368
polyphenols 190
polyphosphates 288
Popeye 133
porphin 24
porphobilinogen 257
porphyrins 24, 66, 117
positron emission tomography (PET) 359
primary structure of a protein 383
prostaglandines 113
proteases 251, 254–256
protein crystallography 65
protein folding 65
proteinase 254, 255
proteinase K 296
proton gradient 107
proton-shuttle 96
protoporphyrins 335
protoporphyrin IX 88, 90
proximal histidine 92, 95, 98
Prussian Blue 338, 354
pseudohalides 93, 211
pseudorotation 291, 383
$Pt(dien)^{2+}$ 370
$Pt(dien)Cl^+$ 369
pterin 199, 221
pump storage model 268, 269
purines 368
purine metabolism 219
purple acid phosphatase 144, 288
purple bacteria 57, 63, 67–69
push-pull effect 248
push-pull mechanism 174, 226
pyrimidines 368
pyrite 130
pyrroloquinoline quinone (PQQ) 51, 204,
   205
pyruvate 179, 338
pyruvate kinase 269, 291, 292

Q

quadrupolar splitting 102
quasi-macrocycles 31, 270, 276

quaternary structure   98, 383
quercetinase   187, 191
Quin 2AM   294
quinine   98
quinone pool   66
o-quinones   191, 200

R

R state   96–98
rack mechanism   23
radiation damage   351
radical   16, 19, 26, 34, 40, 42–49, 52, 54,
      82, 85, 114, 120, 141–143, 151, 212,
      263, 326, 327, 337, 347, 351, 373, 375
radical anion   66, 83, 117, 118, 120, 209
radical cage   49
radical cation   66, 69, 119, 120, 122
radical chain reaction   327
radical scavenger   45
radioactive background exposure   355
radioactive isotopes   271, 322, 351
radiodiagnostics   159, 354, 357, 359
radioimmunoassay   357
radioimmunotherapy   358
radioisotopes   159, 352, 354, 357
radionuclides   351–353, 357–359
radiopharmaceuticals   356–359
radioscintigraphy   357
radiotherapeutic agents   359
radiotoxicity   355, 356
radula   315
Raman effect   101, 383
Raney nickel   183
reaction center   57, 60–68, 112
rearrangement   46, 49
'rebound'   117
receptor   276
recognition   34, 261, 262, 297, 372
Recommended Dietary Allowances
      (RDA)   12, 13
recycling   4, 131
redox coenzymes   50
redox pairs   16
redox potential   42–43, 85, 107–110, 128,
      130, 133, 140, 153, 184, 188, 196, 207,
      217, 319, 323, 333, 347, 356
reductive elimination   42
regulatory protein   152
reorganization energy   111
repair mechanism   351, 372
replication   368
reprocessing   353
resistance   339, 341
resonance Raman spectroscopy   101, 117,
      197
resonance transfer   61
resorption   150, 161, 162

respiration   56, 83, 119, 187, 212, 246, 281
respiratory burst oxidase   212
respiratory chain   107–109, 191, 206
restriction enzyme analogues   368
restriction enzymes   373
retention time   335, 336, 368
retina   280
rheumatic fever   374
rheumatoid arthritis   212, 363, 373
rhodopsin   280
rhodotorulic acid   153, 157
ribonucleases   236
ribonucleotide reductase   47–50, 56, 89,
      141, 142
ribose   141
ribozymes   286
ribulose-1,5-bisphosphate carboxylase
      (rubisco)   286
rice   160, 180
rickets   307
ridaura   374
Rieske center   63, 130, 133, 134
RNA   33, 257, 286, 368
RNA polymerase   257, 341
rod cells   280
ruberythrin   133
rubredoxin   130–133, 140, 195
ruffling   26
ruminants   180, 190, 219
rusticyanin   196

S

S states   73
S-adenosyl methionine   52, 319
S100-proteins   297
saddle conformation   40
salen   53, 54, 85
salvarsan   363
sanocrysin   374
sarcoplasmic reticulum   284, 294, 299
saturation curve   91, 98
saturnism   334
schizophrenia   363
scintillation counters   357
sea water   8, 313, 319, 322, 331, 352, 355
sea-urchin   313
seaweeds   236
second messenger   293, 296, 383
secondary structure of a protein   383
sedatives   320
sediments   180
selenenic acid   327
selenocysteine   18, 19, 178, 325, 328
self assembly   142
semiquinone   66, 145, 205, 383
sensors   192, 227, 275, 303–306
sensor protein   341

sensory cell   280
serine   19
serum albumin   188
serum proteins   367
serum transferrin   164
sewage waste   331
shells   303, 304, 306, 312
shunt pathway   116
sickle cell anemia   98
side effects   364, 367, 375, 376
side-on   85–88, 105, 383
siderophores   151–153, 156–159
silica   309
silicic acid   313, 314
single crystals   65, 384
single electron transfer   85, 93
single photon emission computer
    tomography (SPECT)   357
singlet dioxygen   84, 375
singlet state 384
siroheme   107, 124
site-directed mutagenesis   3, 36, 112
size selectivity   32, 274
skeletons   303, 307
snakes   256
social drugs   259, 340
sodium diethyldithiocarbamate   364
'soft'   14, 17, 384
soil acidification   345
solar radiation   84
solder   333, 334
solganol   374, 375
solid state NMR   314
soy bean lipoxygenase   146
spacer   111
special pair   28, 62, 66, 67, 112
speciation   8, 10, 331, 352
specific rotation   155
spin crossover   96, 112, 182, 384
spin-flipping   76
spin–orbit coupling   63, 68, 194, 375, 384
spin pairing   30
spin quantum number   76
spin–spin coupling   75, 100, 136, 138, 143,
    144, 167, 196
spin state   29
spin-verbot   85
spinach   130, 133
spiroplatin   364, 365
spleen   161, 162, 165, 169
spring-tension model   96
squid nerve   268
SQUID susceptometry   76, 172, 384
stability diagrams   85, 233, 324, 360
stacking   204, 370
statoconia   313
statoliths   313

stellacyanin   194, 196
steroids   112, 113, 272
steroid hormones   262
stromelysin   254
structure determination   65
structure-effect relationships   365
struma   322
sulfate   215, 216, 219, 347
sulfide   17, 124, 128, 131, 133, 134, 136, 218
sulfide reductase   136
sulfite   122, 216, 219
sulfite oxidase   216, 218, 219, 222
sulfite reductase   124, 129
sulfonamides   250
sulfoxides   218
sulfur cycle   218
sulfuranyl radical   183
summer smog   225
sunlight   58, 59
superexchange   75, 104, 133, 167, 384
supernucleophile   42, 45
superoxide   69, 83–86, 93, 209, 212, 218,
    375
superoxide dismutases (SODs)   56, 75,
    187, 190, 191, 209–212, 243, 245, 326,
    351
superparamagnetism   167
supersaturated   308, 313
supramolecular chemistry   273
susceptibility   384
symport process   281
syn   19
synchrotron radiation   72
synergism   10
synergistic anion   164
synthetases   257

T
T state   96–98
taste receptors   280
tautomers   368, 384
tautomeric forms   33
teeth   303, 304, 306, 308, 312
template   115, 274, 307, 384
tertiary structure   368, 384
tetanus   256
tetraethyl lead   333, 335
tetragonal distortion   118
tetrahedral splitting   384
tetrahydrofolic acid   (THFA)   51, 180
tetrahydromethanopterin   180, 181
tetrahydropterins   144, 221
tetraiodothyronine   321
tetraorganotin compounds   342
tetrapyrroles   23, 26, 59
TF IIIA   243, 260, 262
therapeutic resistance   366

therapeutic width   10, 323, 325, 376
thermodynamic 384
thermolysin   21, 243, 254, 255, 296
thiobacilli 131
thiocyanate 211
thiohydroxamate 334
thiophilicity 319, 330
thioredoxin 141
thiosulfate sulfur transferase   136
thiourea 364
three state hypothesis 135
three-point fixing   22, 61, 384
threonine 19
thylakoid membrane   57, 69, 384
thymine 33, 34
thyreocalcitonin 312
thyreoperoxidases   119, 323
thyroid gland   353, 357, 358, 360, 361
thyroid hormones   119, 321, 322
thyronine   321, 322
thyroxine 321
tissue necrosis 321
toadstool   236, 237
tooth enamel   311, 312, 320, 321
tooth paste   320, 346
topaquinone (TPQ)   204
topoisomerases 291
toxicity   12, 82, 164, 330–332, 335,
        339–341, 359, 374, 376
toxicology   52, 113
toxins   119, 256
trace elements   2, 6–8, 12, 40, 238
transcription   260, 262, 368, 372, 384
transcription-regulating factors   242
transferrins   89, 151, 161–165, 345, 357
transgenic organisms   163
transition state   22, 196, 245, 283, 288, 290
triiodothyronine   321, 322
triplet   63, 76, 84, 85, 117, 196, 197, 384
tripod ligand   180
troponin C   298–300
troponins 297
tryptophane   19, 111, 120, 204, 289, 325
tryptophyl radical   143
tuberculosis   151, 374
tumor   322, 327, 346, 353, 357–359,
        364–367
    diagnosis 358
    metabolism 254
tunicates   24, 173, 235–236
tunichlorin   24, 173
tunichrome 236
tunnel effect 111
two-electron reactivity 204
type 1 copper   centers   189, 192, 196, 204,
        205
type 2 copper centers   189, 196, 204

type 3 copper centers   189, 196, 204
tyrosinase   187, 191, 198–201
tyrosinate   144, 145, 164, 204
tyrosine   18, 19, 63, 70, 73, 74, 119, 120,
        144, 166, 199, 200, 252, 322
tyrosine 3-monooxygenase   199
tyrosyl radical   142, 143, 204

U

ubiquinone   51, 66, 107, 108
ultratrace elements   172
Umpolung 244
uniport process   281
uphill-catalysis 57
uracil 33
uranium fission   359
urea 173
urea amidohydrolase   173
urease   172–174
uric acid   216, 222, 326
urothione 220
uteroferrin 144

V

V-dependent nitrogenase   233
valinomycin   32, 275, 276
vaterite   304, 309
vesicle 384
vinblastin 364
vitamin B_{12} 39, 41
vitamin C   161, 326
vitamin D   113, 293
vitamin D3   114
vitamin E   326
vitamin H   218
vitamin K   51
vitamin K_1 63
voltage-controlled channels   280

W

water   8, 17, 57, 68, 69, 76–79, 83,
        244–249, 251, 254, 256
water hardness equilibrium   307
water oxidase   79
water oxidation   69, 74, 78, 119
weathering 344
weddelite 304
western diamond rattlesnake   256
whewellite 304
Wilson's disease   188

X

X-ray absorption near edge structure
        (XANES)   72, 177
X-ray absorption spectroscopy (XAS)   71,
        72, 77, 178, 375

X-ray crystallography 39
X-ray diffraction 65
xanthine 216, 222
xanthine oxidase 129, 216, 218–222
xenobiotic 112, 119

Y

yeast 258, 262

yeast alcohol dehydrogenase (YADH) 258

Z

Z scheme 70
Z-aconitate 137, 138
zinc finger (*zif*) 260–262
zucchini 202, 204
zymogen 255

LaVergne, TN USA
06 July 2010
188279LV00009B/3/P